T0226126

Microcrystalline and Nanocrystalline Semiconductors—1998

MATERIALS RESEARCH SOCIETY
SYMPOSIUM PROCEEDINGS VOLUME 536

Microcrystalline and Nanocrystalline Semiconductors—1998

Symposium held November 30–December 3, 1998, Boston, Massachusetts, U.S.A.

EDITORS:

Leigh T. Canham
Defence Research Agency
Malvern, United Kingdom

Michael J. Sailor
University of California, San Diego
La Jolla, California, U.S.A.

Kazunobu Tanaka
Joint Research Center for Atom Technology
Ibaraki, Japan

Chuang-Chuang Tsai
Applied Komatsu Technology
Santa Clara, California, U.S.A.

Materials Research Society
Warrendale, Pennsylvania

CAMBRIDGE UNIVERSITY PRESS
Cambridge, New York, Melbourne, Madrid, Cape Town,
Singapore, São Paulo, Delhi, Mexico City

Cambridge University Press
32 Avenue of the Americas, New York NY 10013-2473, USA

Published in the United States of America by Cambridge University Press, New York

www.cambridge.org
Information on this title: www.cambridge.org/9781107413740

Materials Research Society
506 Keystone Drive, Warrendale, PA 15086
http://www.mrs.org

© Materials Research Society 1999

This publication is in copyright. Subject to statutory exception
and to the provisions of relevant collective licensing agreements,
no reproduction of any part may take place without the written
permission of Cambridge University Press.

This publication has been registered with Copyright Clearance Center, Inc.
For further information please contact the Copyright Clearance Center,
Salem, Massachusetts.

First published 1999
First paperback edition 2013

Single article reprints from this publication are available through
University Microfilms Inc., 300 North Zeeb Road, Ann Arbor, MI 48106

CODEN: MRSPDH

ISBN 978-1-107-41374-0 Paperback

Cambridge University Press has no responsibility for the persistence or
accuracy of URLs for external or third-party internet websites referred to in
this publication, and does not guarantee that any content on such websites is,
or will remain, accurate or appropriate.

CONTENTS

Preface xiii

Materials Research Society Symposium Proceedings xv

PART I: LIGHT EMISSION FROM NANOCRYSTALLINE SILICON

Light Emitting Micropatterns of Porous Semiconductors 3
D.J. Lockwood, P. Schmuki, and L.E. Erickson

Strongly Superlinear Light Emission and Large Induced
Absorption in Oxidized Porous Silicon Films 9
H. Koyama and P.M. Fauchet

Enhancing the External Quantum Efficiency of Porous
Silicon LEDs Beyond 1% by a Post-Anodization
Electrochemical Oxidation 15
B. Gelloz, T. Nakagawa, and N. Koshida

Integration of Electrically Isolated Porous Silicon LEDs
for Applications on CMOS Technology 21
K.D. Hirschman, L. Tsybeskov, C.C. Striemer, S. Chan, and
P.M. Fauchet

Photoluminescence From Single Porous Silicon Chromophores 27
M.D. Mason, G.M. Credo, K.D. Weston, and S.K. Buratto

Auger Effect Seen in the Porous Silicon Fast Luminescent Band 33
R. M'ghaïeth, I. Mihalcescu, H. Maâref, and J.C. Vial

Defects and Phonon Assisted Optical Transitions in Si
Nanocrystals 39
G. Allan, C. Delerue, and M. Lannoo

Luminescence Study of Self-Assembled, Silicon Quantum Dots 45
S. Miyazaki, K. Shiba, N. Miyoshi, K. Etoh, A. Kohno, and M. Hirose

Formation Process of Si Nanoparticles Formed by Laser
Ablation Method 51
T. Makimura, T. Mizuta, T. Ueda, and K. Murakami

Correlation Between Defect Structures and Light Emission
in Si-Nanocrystal Doped SiO_2 Films 57
K. Sato, Y. Sugiyama, T. Izumi, M. Iwase, Y. Show, S. Nozaki,
and H. Morisaki

Light Emitting Nanostructures in Implanted Silicon Layers 63
R. Plugaru, J. Piqueras, B. Mendez, G. Craciun, and N. Nastase

Formation and Luminescent Properties of Oxidized Porous
Silicon Doped With Erbium by Electrochemical Procedure 69
  V. Bondarenko, N. Vorozov, L. Dolgyi, V. Yakovtseva, V. Petrovich,
  S. Volchek, N. Kazuchits, G. Grom, H.A. Lopez, L. Tsybeskov,
  and P.M. Fauchet

Long Luminescence Lifetime of 1.54 µm Er^{3+} Luminescence
From Erbium-Doped Silicon Rich Silicon Oxide and Its Origin 75
  S-Y. Seo, J.H. Shin, and C. Lee

Local Structure and Er^{3+} Emission From Pseudo-Amorphous
GaN:Er Thin Films 81
  S.B. Aldabergenova, M. Albrecht, A.A. Andreev, C. Inglefield,
  J. Viner, P.C. Taylor, and H.P. Strunk

PART II: PROPERTIES OF NANOCRYSTALLINE
SEMICONDUCTORS AND PERIODIC STRUCTURES

*Perspectives of Porous Silicon Multilayer Technology 89
  M. Thönissen

Porosity-Induced Optical Phonon Engineering in III-V
Compounds 99
  I.M. Tiginyanu, G. Irmer, J. Monecke, H.L. Hartnagel, A. Vogt,
  C. Schwab, and J-J. Grob

Ultrasound Emission From Porous Silicon: Efficient
Thermo-Acoustic Function as a Depleted Nanocrystalline
System 105
  N. Koshida, T. Nakajima, M. Yoshiyama, K. Ueno, T. Nakagawa,
  and H. Shinoda

Production of Silicon Nanocrystals by Thermal Annealing of
Silicon-Oxygen and Silicon-Oxygen-Carbon Alloys: Model
Systems for Chemical and Structural Relaxation at $Si-SiO_2$
and $SiC-SiO_2$ Interfaces 111
  G. Lucovsky, D. Wolfe, and B. Hinds

Porous Silicon Multilayer Mirrors and Microcavity Resonators
for Optoelectronic Applications 117
  S. Chan, L. Tsybeskov, and P.M. Fauchet

Silicon-Based UV Detector Prototypes Using Luminescent
Porous Silicon Films 123
  L. Peraza, M. Cruz, A. Estrada, C. Navarro, J. Avalos, L.F. Fonseca,
  O. Resto, and S.Z. Weisz

Effects of Various Hydrogen Dilution Ratios on the Performance
of Thin Film Nanocrystalline/Crystalline Silicon Solar Cells 129
  Y.J. Song and W.A. Anderson

*Invited Paper

Integration of Multilayers in Er-Doped Porous Silicon Structures
and Advances in 1.5 µm Optoelectronic Devices 135
    H.A. Lopez, S. Chan, L. Tsybeskov, H. Koyama, V.P. Bondarenko,
    and P.M. Fauchet

Structural Characterization of nc-Si/a-SiO$_2$ Superlattices
Subjected to Thermal Treatment 141
    G.F. Grom, L. Tsybeskov, K.D. Hirschman, P.M. Fauchet,
    J.P. McCaffrey, H.J. Labbé, and D.J. Lockwood

PART III: BIOLOGICAL APPLICATIONS AND SURFACE
CHEMISTRY OF NANOCRYSTALLINE SEMICONDUCTORS

In-Vivo Assessment of Tissue Compatibility and Calcification
of Bulk and Porous Silicon 149
    A.P. Bowditch, K. Waters, H. Gale, P. Rice, E.A.M. Scott,
    L.T. Canham, C.L. Reeves, A. Loni, and T.I. Cox

*Surface Chemistry of Porous Silicon 155
    J-N. Chazalviel and F. Ozanam

Functionalization of Porous Silicon Surfaces by Solution-Phase
Reactions with Alcohols and Grignard Reagents 167
    N.Y. Kim and P.E. Laibinis

Functionalization of Porous Silicon Surfaces Through
Hydrosilylation Reactions 173
    J.M. Buriak, M.P. Stewart, and M.J. Allen

Enhancement in Efficiency and Stability of Oxide-Free Blue
Emission From Porous Silicon by Surface Passivation 179
    H. Mizuno and N. Koshida

Modification of Visible Light Emission From Silicon Nanocrystals
as a Function of Size, Electronic Structure, and Surface Passivation 185
    M.V. Wolkin, J. Jorne, P.M. Fauchet, G. Allan, and C. Delerue

Nanostructure of Porous Silicon Using Transmission Microscopy 191
    M.H. Nayfeh, Z. Yamani, O. Gurdal, and A. Alaql

Electro-Polymerization in Porous Silicon Films 197
    R.K. Soni, L.F Fonseca, O. Resto, A. Guadalupe, and S.Z. Weisz

PART IV: SYNTHESIS AND SPECTROSCOPY
OF NANOCRYSTALLINE SEMICONDUCTORS

Microsized Structures Fabricated With Nanoparticles as
Building Blocks 205
    Y. Tian, A.D. Dinsmore, S.B. Qadri, and B.R. Ratna

*Invited Paper

Electron and Hole Relaxation Pathways in II-VI Semiconductor
Nanocrystals 211
V.I. Klimov, Ch. Schwarz, X. Yang, D.W. McBranch, C.A. Leatherdale,
and M.G. Bawendi

Strong Photoluminescence in the Near-Infrared From
Colloidally-Prepared HgTe Nanocrystals 217
M.T. Harrison, S.V. Kershaw, M.G. Burt, A.L. Rogach, A. Kornowski,
A. Eychmüller, and H. Weller

Efficient Luminescence From GaAs Nanocrystals in SiO$_2$ Matrices 223
Y. Kanemitsu, H. Tanaka, S. Mimura, S. Okamoto, T. Kushida,
K.S. Min, and H.A. Atwater

Plasmons on Luminescent Porous Silicon Prepared With Ethanol
and Critical Point Drying 229
O. Resto, L.F. Fonseca, S.Z. Weisz, A. Many, and Y. Goldstein

PART V: SYNTHESIS AND PROPERTIES OF MICROCRYSTALLINE
AND NANOCRYSTALLINE SEMICONDUCTORS

Nanoparticle Precursors for Electronic Materials 237
D.S. Ginley, C.J. Curtis, R. Ribelin, J.L. Alleman, A. Mason,
K.M. Jones, R.J. Matson, O. Khaselev, and D.L. Schulz

Low-Temperature ECR-Plasma Assisted MOCVD Microcrystalline
and Amorphous GaN Deposition and Characterization for
Electronic Devices 245
Z. Hassan, M.E. Kordesch, W.M. Jadwisienzak, H.J. Lozykowski,
W. Halverson, and P.C. Colter

Nanostructured Arrays Formed by Finely Focused Ion Beams 251
R.A. Zuhr, J.D. Budai, P.G. Datskos, A. Meldrum, K.A. Thomas,
R.J. Warmack, C.W. White, L.C. Feldman, M. Strobel, and K-H. Heinig

Optically Detected Magnetic Resonance and Double Beam
Photoluminescence of CdS/HgS/CdS Nanoparticles 257
H. Porteanu, A. Glozman, E. Lifshitz, A. Eychmüller, and H. Weller

Optical and Electrical Properties of CdTe Nanocrystal Quantum
Dots Passivated in Amorphous TiO2 Thin Film Matrix 263
A.C. Rastogi, S.N. Sharma, and S. Kohli

Optical Properties of Self-Organized InGaAs/GaAs Quantum
Dots in Field-Effect Structures 269
A. Babinski, T. Tomaszewicz, A. Wysmolek, J.M. Baranowski,
C. Lobo, R. Leon, and D. Jagadish

Giant Anisotropy of Conductivity in Hydrogenated Nanocrystalline
Silicon Thin Films 275
A.B. Pevtsov, N.A. Feoktistov, and V.G. Golubev

The Inverted Meyer-Neldel Rule in the Conductance of
Nanostructured Silicon Field-Effect Devices 281
  R.E.I. Schropp and H. Meiling

In Situ Studies of the Vibrational and Electronic Properties
of Si Nanoparticles 287
  J.R. Fox, I.A. Akimov, X.X. Xi, and A.A. Sirenko

X-ray Reflectivity Study of Porous Silicon Formation 293
  V. Chamard, G. Dolino, and J. Eymery

Light Induced ESR Measurements on Microcrystalline
Silicon With Different Crystalline Volume Fractions 299
  J. Müller, F. Finger, P. Hapke, and H. Wagner

VLS Growth of Si Nanowhiskers on an H-Terminated
Si{111} Surface 305
  N. Ozaki, Y. Ohno, S. Takeda, and M. Hirata

Thermodynamics and Kinetics of Melting and Growth of
Crystalline Silicon Clusters 311
  P. Keblinski

Microstructure and Size Distribution of Compound Semiconductor
Nanocrystals Synthesized by Ion Implantation 317
  A. Meldrum, S.P. Withrow, R.A. Zuhr, C.W. White, L.A. Boatner,
  J.D. Budai, I.M. Anderson, D.O. Henderson, M. Wu, A. Ueda,
  and R. Mu

Synthesis of Boron Carbide Nanowires and Nanocrystal Arrays
By Plasma Enhanced Chemical Vapor Deposition 323
  D. Zhang, B.G. Kempton, D.N. McIlroy, Y. Geng, and
  M.G. Norton

Germanium Nanostructures Fabricated by PLD 329
  K.M. Hassan, A.K. Sharma, J. Narayan, J.F. Muth, and C.W. Teng

PART VI: OXIDE AND CHALCOGENIDE
SEMICONDUCTORS

In Situ X-ray Diffraction Study of Lithium Intercalation in
Nanostructured Anatase Titanium Dioxide 337
  R. Van de Krol, E.A. Meulenkamp, A. Goossens, and J. Schoonman

Novel Electronic Conductance CO_2 Sensors Based on
Nanocrystalline Semiconductors 341
  M-I. Baraton, L. Merhari, P. Keller, K. Zweiacker, and J-U. Meyer

Growth Kinetics of Quantum Size ZnO Particles 347
  E.M. Wong, J.E. Bonevich, and P.C. Searson

Synthesis and Characterization of Mn Doped ZnS Quantum
Dots From a Single Source Precursor 353
  M.A. Malik, P. O'Brien, and N. Revaprasadu

In Situ Diagnostics of Nanomaterial Synthesis by Laser
Ablation: Time-Resolved Photoluminescence Spectra and
Imaging of Gas-Suspended Nanoparticles Deposited
for Thin Films 359
    D.B. Geohegan, A.A. Puretzky, A. Meldrum, G. Duscher, and
    S.J. Pennycook

Thiol-Capped CdSe and CdTe Nanoclusters: Synthesis by a
Wet Chemical Route, Structural and Optical Properties 365
    A.L. Rogach, A. Eychmüller, J. Rockenberger, A. Kornowski,
    H. Weller, L. Tröger, M.Y. Gao, M.T. Harrison, S.V. Kershaw,
    and M.G. Burt

A Novel Route for the Preparation of CuSe and $CuInSe_2$
Nanoparticles 371
    M.A. Malik, P. O'Brien, N. Revaprasadu, and G. Wakefield

The Preparation and Characterization of Nanocrystalline
Indium Tin Oxide Films 377
    J. Aikens, H.W. Sarkas, and R.W. Brotzman, Jr.

Enhanced Photoluminescence for ZnS Nanocrystals Doped
With Mn^{2+} Close to Carboxyl Groups and/or Sn^{2-} Vacancies 383
    T. Isobe, T. Igarashi, M. Konishi, and M. Senna

Microstructure and Sensing Properties of Cryosol Derived
Nanocrystalline Tin Dioxide 389
    S.M. Kudryavtseva, A.A. Vertegel, S.V. Kalinin, L.I. Kheifets,
    J. Van Landuyt, L.L. Meshkov, S.N. Nesterenko, E.S. Rembeza,
    and A.M. Gaskov

Chemical Solution Deposited Copper-Doped CdSe
Quantum Dot Films 395
    N. Chandrasekharan, S. Gorer, and G. Hodes

Characterization of ZnSe Nanoparticles Prepared Using
Ultrasonic Radiation Method 401
    J. Xu, W. Ji, S-H. Tang, and W. Huang

Nanoparticle-Based Contacts to CdTe 407
    D.L. Schulz, R. Ribelin, C.J. Curtis, D.E. King, and D.S. Ginley

Surface Stoichiometry of CdSe Nanocrystals 413
    J. Taylor, T. Kippeny, J.C. Bennett, M. Huang, L.C. Feldman,
    and S.J. Rosenthal

Femtosecond Interfacial Electron Transfer Dynamics of
CdSe Semiconductor Nanoparticles 419
    C. Burda, T.C. Green, S. Link, and M.A. El-Sayed

PART VII: MICROCRYSTALLINE AND POLYCRYSTALLINE SEMICONDUCTORS

*Structural and Electronic Properties of Laser Crystallized
Silicon Films 427
T. Sameshima

Nucleation Processes in Si CVD on Ultrathin SiO_2 Layers 439
T. Yasuda, D.S. Hwang, K. Ikuta, S. Yamasaki, and K. Tanaka

GaAs Micro Crystal Growth on an As-Terminated Si (001)
Surface by Low Energy Focused Ion Beam 445
T. Chikyow and N. Koguchi

Amorphous/Microcrystalline Phase Control in Silicon Film
Deposition for Improved Solar Cell Performance 451
J. Koh, H. Fujiwara, Y. Lee, C.R. Wronski, and R.W. Collins

Preparation and Characterization of Microcrystalline and
Epitactially Grown Emitter Layers for Silicon Solar Cells 457
K. Lips, J. Platen, S. Brehme, S. Gall, I. Sieber, L. Elstner,
and W. Fuhs

Probing the Elementary Surface Reactions of Hydrogenated
Silicon PECVD by In Situ ESR 463
S. Yamasaki, C. Malten, T. Umeda, J-I. Isoya, and K. Tanaka

Evaporated Polycrystalline Germanium for Near Infrared
Photodetection 469
L. Colace, G. Masini, F. Galluzzi, and G. Assanto

"Mirrorless" UV Lasers in ZnO Polycrystalline Films and Powder 477
H. Cao, Y.G. Zhao, H.C. Ong, J.Y. Dai, X. Liu, E.W. Seelig,
and R.P.H. Chang

Phase Segregation in SIPOS: Formation of Si Nanocrystals 481
A. Vilà, J.R. Morante, B. Caussat, P. Barathieu, and E. Scheid

Low-Temperature Deposition of Polycrystalline Silicon
Thin Films Prepared by Hot Wire Cell Method 487
M. Ichikawa, J. Takeshita, A. Yamada, and M. Konagai

Enhancement of the Amorphous to Microcrystalline Phase
Transition in Silicon Films Deposited by SiF_4-H_2-He Plasmas 493
G. Cicala, M. Losurdo, P. Capezzuto, G. Bruno, T. Ligonzo,
L. Schiavulli, C. Minarini, and M.C. Rossi

Dechanneling Study of Nanocrystalline Si:H Layers Produced
by High Dose Hydrogen Irradiation of Silicon Crystals 499
V.P. Popov, A.K. Gutakovsky, I.V. Antonova, K.S. Khuravlev,
E.V. Spesivtsev, I.I. Morosov, and G.P. Pokhil

*Invited Paper

Optical Analysis of Plasma Enhanced Crystallization of
Amorphous Silicon Films 505
  L. Montès, L. Tsybeskov, P.M. Fauchet, K. Pangal, J.C. Sturm,
  and S. Wagner

LEDS Based on Oxidized Porous Polysilicon on a Transparent
Substrate 511
  C.C. Striemer, S. Chan, H.A. Lopez, K.D. Hirschman, H. Koyama,
  Q. Zhu, L. Tsybeskov, P.M. Fauchet, N.M. Kalkhoran, and L. Depaulis

Effect of Hydrogen Plasma Treatments at Very High Frequency
on p-type Amorphous and Microcrystalline Silicon Films 517
  E. Centurioni, A. Desalvo, R. Pinghini, R. Rizzoli, C. Summonte,
  and F. Zignani

Robust Exciton Polariton in a Quantum-Well Waveguide 523
  M. Shirai, K. Hosomi, T. Mishima, and T. Katsuyama

Characterization of Laser Ablated Germanium Nanoclusters 527
  S. Vijayalakshmi, F. Shen, Y. Zhang, M.A. George, and H. Grebel

Theoretical Investigation of Effective Quantum Dots Induced
by Strain in Semiconductor Wires 533
  K. Shiraishi, M. Nagase, S. Horiguchi, and H. Kageshima

Electron Beam Excited Plasma CVD for Silicon Growth 539
  K. Okitsu, M. Imaizumi, T. Ito, K. Yamaguchi, M. Yamaguchi, T. Hara,
  M. Ban, M. Tokai, and K. Kawamura

Pressure Induced Structural Transformations in Nanocluster
Assembled Gallium Arsenide 545
  S. Kodiyalam, A. Chatterjee, I. Ebbsjö, R.K. Kalia, H. Kikuchi,
  A. Nakano, J.P. Rino, and P. Vashishta

Study of Electrical Properties of Ge-Nanocrystalline Films
Deposited by Cluster-Beam Evaporation Technique 551
  S. Banerjee, H. Ono, S. Nozaki, and H. Morisaki

Development of a Porous Silicon Based Biosensor 557
  K-P.S. Dancil, D.P. Greiner, and M.J. Sailor

Author Index 563

Subject Index 567

PREFACE

This book contains the papers presented at Symposium F, "Microcrystalline and Nanocrystalline Semiconductors," at the 1998 MRS Fall Meeting in Boston, Massachusetts. It was the fifth MRS symposium held on this topic since 1989, and this proceedings contains 87 of the 166 papers presented. In addition to the now traditional themes of synthesis, structure, and optoelectronic properties of nano- and microcrystalline semiconductors, properties leading to new optical and biological applications were also reported.

In the area of synthesis, techniques have emerged for the production of nanoparticles with narrow size distributions and more well-behaved electronic properties. Wet chemical techniques to build core-shell structures (a nanoparticle of one semiconductor coated with a thin layer of a material having a larger bandgap) have yielded more chemically robust materials with higher photoemission quantum yields. Solution and gas phase methods to synthesize Si and Ge nanocrystallites have also been perfected. Chemical techniques have also been brought to bear on the modification of semiconductor interfaces, and many contributions described chemical reactions to functionalize or stabilize nanocrystalline materials. The interest in producing stable materials with narrow size distributions stems from the desire to design longer-lived electroluminescent devices and to impart specific functionality to the material for sensor, micromachine, or photonic applications.

Earlier symposia have focused on electroluminescent devices made of nanocrystalline and porous silicon. High-efficiency electroluminescent devices that operate at visible wavelengths are desired for display applications. Steady progress has been made in this area, and a porous Si device with an electroluminescence efficiency above 1.1% was reported in the symposium. Although the efficiencies are improving, long-term stability of these devices is still a problem. Communications applications require devices that emit at wavelengths in the near infrared, and a room-temperature device incorporating an emitting Er species in a nanocrystalline Si matrix was reported. Along with these more familiar optoelectronics themes, several new properties and applications of nanocrystalline materials were represented in the symposium.

One emerging theme involves the fabrication of passive optical elements made from nanostructured semiconductors. Multilayer structures can be used to artificially narrow the broad emission band from Si nanocrystallites or other emissive species to very narrow linewidths, providing a convenient means of fabricating devices with tunable optical properties. It has been found that multilayered structures, such as Bragg reflectors, can be easily synthesized by pulsed electrochemical corrosion of silicon. The porous silicon layers produced display a periodic porosity which provides the required optical characteristics. Such structures are currently of interest for applications in optical filtering and switching, and for silicon-based lasers.

Application of nanocrystalline materials to biology and medicine is another emerging theme that was explored in this symposium. Properly prepared luminescent quantum particles were reported to have characteristics for biological tagging and counting experiments superior to conventional fluorescent molecules. Highly sensitive biological sensors based on a chemically modified porous Si etalon structure were also reported. Finally, in-vivo studies of bulk and porous Si showed that their soft tissue biocompatibility is comparable to Ti, a well-established biomaterial.

Many contributions on microcrystalline semiconductors were concerned with the development of efficient solar cells. The recent fabrication of efficient (>10%) solar cells from microcrystalline TiO_2 films (the Graetzel cell) were discussed in several presentations, and crucial elements of microstructure and surface chemistry were explored. Although the initial TiO_2 cells were liquid junction devices, progress towards the development of an efficient, all solid-state device was presented. New syntheses and studies of ternary phase semiconductors of interest for solar cell applications were also presented. Symposium A at this Meeting focused entirely on polycrystalline Si, and so the representation of this field in our symposium was lower than in previous years.

A significant number of contributions reported observations of new properties of microcrystalline and nanocrystalline semiconductors. The coming of age of high-sensitivity near-field and far-field spectroscopic microscopy as an analytical tool has yielded some fascinating results on nanoparticles. Intermittent emission (blinking) phenomena from nanocrystallites seems to be a general trend; such observations were reported on group IV and II-VI materials. The intermittency is thought to be related to carrier trapping events at the nanoparticle interfaces. Unusual nonlinear optical properties and magnetic effects were reported for several systems. One of the more surprising observations was that nanocrystalline porous Si layers can emit a strong ultrasonic signal under an applied bias. The effects of size on the structural properties (lattice dimensions as well as microstructure) of nanocrystalline materials were also covered.

There was an unusually large number of authors associated with Symposium F who won MRS awards at this Meeting. The MRS best poster award for the Tuesday night poster session held at the Marriott Hotel went to Herman A. Lopez for his paper titled "Integration of Multilayers in Er-Doped Porous Silicon Structures for Optoelectronic Devices." The MRS best poster award for the Wednesday night poster session held at the Marriott Hotel also went to a Symposium F participant, Alkiviathes Meldrum, for his paper titled "Microstructure and Size Distribution of Compound Semiconductor Nanocrystals Synthesized by Ion Implantation." Bryan D. Huey was a Graduate Student Gold Award winner for his talk "Electronic Properties of Individual Grain Boundaries in ZnO and $SrTiO_3$ Bicrystals and Polycrystals." There were two Graduate Student Silver Award winners from Symposium F: Moonsub Shim for his presentation "Dielectric Dispersion Studies of Semiconductor Nanocrystal Colloids: Evidence for a Ground State Dipole Moment," and Eva M. Wong, who made two contributions to Symposium F, the poster "Photoluminescence of Zinc Oxide Quantum Particle Thin Films," and the talk "The Growth Kinetics of Nanocrystalline Zinc Oxide Particles From Colloidal Suspensions."

In conclusion, the organizers would like to thank the authors, speakers, and session chairs of Symposium F, the MRS staff for their excellent back-up facilities available during the Meeting, and the symposium aides who provided tireless assistance to the organizers and session chairs. Our special thanks go to Ms. Christie A. Canaria for her help in preparing and indexing this volume. Finally, the organizers wish to thank the sponsors listed below for financial assistance:

British Telecom Labs
NEC Corporation
Hitachi Central Research Laboratory

Leigh T. Canham
Michael J. Sailor
Kazunobu Tanaka
Chuang-Chuang Tsai

January 1999

MATERIALS RESEARCH SOCIETY SYMPOSIUM PROCEEDINGS

Volume 507— Amorphous and Microcrystalline Silicon Technology—1998, R. Schropp, H.M. Branz, M. Hack, I. Shimizu, S. Wagner, 1999, ISBN: 1-55899-413-0

Volume 508— Flat-Panel Display Materials—1998, G. Parsons, C-C. Tsai, T.S. Fahlen, C. Seager, 1998, ISBN: 1-55899-414-9

Volume 509— Materials Issues in Vacuum Microelectronics, W. Zhu, L.S. Pan, T.E. Felter, C. Holland, 1998, ISBN: 1-55899-415-7

Volume 510— Defect and Impurity Engineered Semiconductors and Devices II, S. Ashok, J. Chevallier, K. Sumino, B.L. Sopori, W. Götz, 1998, ISBN: 1-55899-416-5

Volume 511— Low-Dielectric Constant Materials IV, C. Chiang, P.S. Ho, T-M. Lu, J.T. Wetzel, 1998, ISBN: 1-55899-417-3

Volume 512— Wide-Bandgap Semiconductors for High Power, High Frequency and High Temperature, S. DenBaars, J. Palmour, M.S. Shur, M. Spencer, 1998, ISBN: 1-55899-418-1

Volume 513— Hydrogen in Semiconductors and Metals, N.H. Nickel, W.B. Jackson, R.C. Bowman, R.G. Leisure, 1998, ISBN: 1-55899-419-X

Volume 514— Advanced Interconnects and Contact Materials and Processes for Future Integrated Circuits, S.P. Murarka, M. Eizenberg, D.B. Fraser, R. Madar, R. Tung, 1998, ISBN: 1-55899-420-3

Volume 515— Electronic Packaging Materials Science X, D.J. Belton, M. Gaynes, E.G. Jacobs, R. Pearson, T. Wu, 1998, ISBN: 1-55899-421-1

Volume 516— Materials Reliability in Microelectronics VIII, J.C. Bravman, T.N. Marieb, J.R. Lloyd, M.A. Korhonen, 1998, ISBN: 1-55899-422-X

Volume 517— High-Density Magnetic Recording and Integrated Magneto-Optics: Materials and Devices, J. Bain, M. Levy, J. Lorenzo, T. Nolan, Y. Okamura, K. Rubin, B. Stadler, R. Wolfe, 1998, ISBN: 1-55899-423-8

Volume 518— Microelectromechanical Structures for Materials Research, S. Brown, J. Gilbert, H. Guckel, R. Howe, G. Johnston, P. Krulevitch, C. Muhlstein, 1998, ISBN: 1-55899-424-6

Volume 519— Organic/Inorganic Hybrid Materials, R.M. Laine, C. Sanchez, C.J. Brinker, E. Giannelis, 1998, ISBN: 1-55899-425-4

Volume 520— Nanostructured Powders and Their Industrial Application, G. Beaucage, J.E. Mark, G.T. Burns, D-W. Hua, 1998, ISBN: 1-55899-426-2

Volume 521— Porous and Cellular Materials for Structural Applications, D.S. Schwartz, D.S. Shih, A.G. Evans, H.N.G. Wadley, 1998, ISBN: 1-55899-427-0

Volume 522— Fundamentals of Nanoindentation and Nanotribology, N.R. Moody, W.W. Gerberich, N. Burnham, S.P. Baker, 1998, ISBN: 1-55899-428-9

Volume 523— Electron Microscopy of Semiconducting Materials and ULSI Devices, C. Hayzelden, C. Hetherington, F. Ross, 1998, ISBN: 1-55899-429-7

Volume 524— Applications of Synchrotron Radiation Techniques to Materials Science IV, S.M. Mini, S.R. Stock, D.L. Perry, L.J. Terminello, 1998, ISBN: 1-55899-430-0

Volume 525— Rapid Thermal and Integrated Processing VII, M.C. Öztürk, F. Roozeboom, P.J. Timans, S.H. Pas, 1998, ISBN: 1-55899-431-9

Volume 526— Advances in Laser Ablation of Materials, R.K. Singh, D.H. Lowndes, D.B. Chrisey, E. Fogarassy, J. Narayan, 1998, ISBN: 1-55899-432-7

Volume 527— Diffusion Mechanisms in Crystalline Materials, Y. Mishin, G. Vogl, N. Cowern, R. Catlow, D. Farkas, 1998, ISBN: 1-55899-433-5

Volume 528— Mechanisms and Principles of Epitaxial Growth in Metallic Systems, L.T. Wille, C.P. Burmester, K. Terakura, G. Comsa, E.D. Williams, 1998, ISBN: 1-55899-434-3

Volume 529— Computational and Mathematical Models of Microstructural Evolution, J.W. Bullard, L-Q. Chen, R.K. Kalia, A.M. Stoneham, 1998, ISBN: 1-55899-435-1

Volume 530— Biomaterials Regulating Cell Function and Tissue Development, R.C. Thomson, D.J. Mooney, K.E. Healy, Y. Ikada, A.G. Mikos, 1998, ISBN: 1-55899-436-X

Volume 531— Reliability of Photonics Materials and Structures, E. Suhir, M. Fukuda, C.R. Kurkjian, 1998, ISBN: 1-55899-437-8

MATERIALS RESEARCH SOCIETY SYMPOSIUM PROCEEDINGS

Volume 532— Silicon Front-End Technology—Materials Processing and Modelling,
N.E.B. Cowern, D.C. Jacobson, P.B. Griffin, P.A. Packan, R.P. Webb,
1998, ISBN: 1-55899-438-6

Volume 533— Epitaxy and Applications of Si-Based Heterostructures, E.A. Fitzgerald,
D.C. Houghton, P.M. Mooney, 1998, ISBN: 1-55899-439-4

Volume 535— III-V and IV-IV Materials and Processing Challenges for Highly Integrated
Microelectonics and Optoelectronics, S.A. Ringel, E.A. Fitzgerald, I. Adesida,
D. Houghton, 1999, ISBN: 1-55899-441-6

Volume 536— Microcrystalline and Nanocrystalline Semiconductors—1998, L.T. Canham,
M.J. Sailor, K. Tanaka, C-C. Tsai, 1999, ISBN: 1-55899-442-4

Volume 537— GaN and Related Alloys, S.J. Pearton, C. Kuo, T. Uenoyama, A.F. Wright, 1999,
ISBN: 1-55899-443-2

Volume 538— Multiscale Modelling of Materials, V.V. Bulatov, T. Diaz de la Rubia, R. Phillips,
E. Kaxiras, N. Ghoniem, 1999, ISBN: 1-55899-444-0

Volume 539— Fracture and Ductile vs. Brittle Behavior—Theory, Modelling and
Experiment, G.E. Beltz, R.L. Blumberg Selinger, K-S. Kim, M.P. Marder, 1999,
ISBN: 1-55899-445-9

Volume 540— Microstructural Processes in Irradiated Materials, S.J. Zinkle, G. Lucas,
R. Ewing, J. Williams, 1999, ISBN: 1-55899-446-7

Volume 541— Ferroelectric Thin Films VII, R.E. Jones, R.W. Schwartz, S. Summerfelt, I.K. Yoo,
1999, ISBN: 1-55899-447-5

Volume 542— Solid Freeform and Additive Fabrication, D. Dimos, S.C. Danforth, M.J. Cima,
1999, ISBN: 1-55899-448-3

Volume 543— Dynamics in Small Confining Systems IV, J.M. Drake, G.S. Grest, J. Klafter,
R. Kopelman, 1999, ISBN: 1-55899-449-1

Volume 544— Plasma Deposition and Treatment of Polymers, W.W. Lee, R. d'Agostino,
M.R. Wertheimer, B.D. Ratner, 1999, ISBN: 1-55899-450-5

Volume 545— Thermoelectric Materials 1998—The Next Generation Materials for Small-Scale
Refrigeration and Power Generation Applications, T.M. Tritt, M.G. Kanatzidis,
G.D. Mahan, H.B. Lyon, Jr., 1999, ISBN: 1-55899-451-3

Volume 546— Materials Science of Microelectromechanical Systems (MEMS) Devices,
A.H. Heuer, S.J. Jacobs, 1999, ISBN: 1-55899-452-1

Volume 547— Solid-State Chemistry of Inorganic Materials II, S.M. Kauzlarich,
E.M. McCarron III, A.W. Sleight, H-C. zur Loye, 1999, ISBN: 1-55899-453-X

Volume 548— Solid-State Ionics V, G-A. Nazri, C. Julien, A. Rougier, 1999,
ISBN: 1-55899-454-8

Volume 549— Advanced Catalytic Materials—1998, P.W. Lednor, D.A. Nagaki, L.T. Thompson,
1999, ISBN: 1-55899-455-6

Volume 550— Biomedical Materials—Drug Delivery, Implants and Tissue Engineering,
T. Neenan, M. Marcolongo, R.F. Valentini, 1999, ISBN: 1-55899-456-4

Volume 551— Materials in Space—Science, Technology and Exploration, A.F. Hepp,
J.M. Prahl, T.G. Keith, S.G. Bailey, J.R. Fowler, 1999, ISBN: 1-55899-457-2

Volume 552— High-Temperature Ordered Intermetallic Alloys VIII, E.P. George, M. Yamaguchi,
M.J. Mills, 1999, ISBN: 1-55899-458-0

Volume 553— Quasicrystals, J-M. Dubois, P.A. Thiel, A-P. Tsai, K. Urban, 1999,
ISBN: 1-55899-459-9

Volume 554— Bulk Metallic Glasses, W.L. Johnson, C.T. Liu, A. Inoue, 1999,
ISBN: 1-55899-460-2

Volume 555— Properties and Processing of Vapor-Deposited Coatings, M. Pickering,
B.W. Sheldon, W.Y. Lee, R.N. Johnson, 1999, ISBN: 1-55899-461-0

Volume 556— Scientific Basis for Nuclear Waste Management XXII, D.J. Wronkiewicz,
J.H. Lee, 1999, ISBN: 1-55899-462-9

Prior Materials Research Society Symposium Proceedings available by contacting Materials Research Society

Part I

Light Emission From Nanocrystalline Silicon

LIGHT EMITTING MICROPATTERNS OF POROUS SEMICONDUCTORS

D.J. LOCKWOOD*, P. SCHMUKI**, L.E. ERICKSON*
*Institute for Microstructural Sciences, National Research Council of Canada,
Ottawa, ON, Canada K1A 0R6
**Swiss Federal Institute of Technology, ETH/EPFL, Dept. of Materials Science, LC-DMX,
CH-1015 Lausanne, Switzerland

ABSTRACT

We report a principle that allows writing visible light emitting semiconductor patterns of arbitrary shape down to the sub-micrometer scale. We demonstrate that porous semiconductor growth can be electrochemically initiated preferentially at surface defects created in an n-type substrate by Si^{++} focused ion beam bombardment. For n-type material in the dark, the electrochemical pore formation potential (Schottky barrier breakdown voltage) is significantly lower at the implanted locations than for an unimplanted surface. This difference in the threshold voltages is exploited to achieve the selectivity of the pore formation process. Visible light emitting patterns of porous Si and GaAs have been created in this way. At present, the size of the structures is limited only by the diameter of the writing ion beam, and pattern diameters in the 50–200-nm range are possible.

INTRODUCTION

Silicon is technologically the most important semiconductor material. However, applications in semiconductor photonics appeared unlikely due to its indirect electronic band gap.[1] Therefore, the discovery of electrochemically formed visible light emitting porous Si[2] has recently stimulated intense research activity.[3] The main reason for this tremendous interest is the prospect of light emitting devices (LEDs) made of porous Si.[1,4]

Here we report a principle[5,6] that allows writing visible light emitting semiconductor patterns of arbitrary shape down to the sub-micrometer scale. Porous semiconductor growth can be electrochemically initiated preferentially at surface defects created in an n-type substrate by Si^{++} focused ion beam (FIB) bombardment. For n-type material in the dark, the electrochemical pore formation potential (Schottky barrier breakdown voltage) is significantly lower at the implanted locations than for an unimplanted surface. This difference in the threshold voltages is exploited to achieve the selectivity of the pore formation process.

Prior approaches to patterning porous Si include: Ga ion implantation followed by an anisotropic wet etch process;[7] a lithographic process using a silicon nitride mask and electrochemical polarization;[8] light induced carrier generation in n-type material together with electrochemical polarization;[9] an ion implantation process (both masked and maskless) to dope and amorphize silicon, which suppresses anodization (negative pattern);[10] and by an ion milling process to create surface defects, which enhance the etch (or anodization) rate in the milled areas.[8] In contrast to these processes, we implant Si^{++} ions to avoid any doping effect in Si wafers, implant the ions at a sufficiently high energy to avoid surface sputtering effects, and use a maskless selective electrochemistry technique that produces porous Si only in the implanted regions.

Mat. Res. Soc. Symp. Proc. Vol. 536 © 1999 Materials Research Society

EXPERIMENT

The ions were implanted into n-type Si and GaAs (100) wafers (doped with 5×10^{15} cm^{-3} As and 2.5×10^{17} cm^{-3} Si, respectively) at room temperature using a 100 kV JEOL 104UHV FIB system. Si^{++} ions at 200 keV were selected from the AuSi ion beam using an $\mathbf{E} \times \mathbf{B}$ mass filter. The nominal beam width was 100 nm. Patterns comprised of squares 50×50 μm^2 and alphabetic characters were implanted with ion doses from 3×10^{13} to 3×10^{16} ions/cm^2. The patterns could be observed in an optical microscope or a scanning electron microscope (SEM) after implantation, but before polarization, for doses exceeding 1×10^{15} ions/cm^2. This may be attributed to amorphization of the substrate and/or to defect-induced surface bulging.

The implanted samples were electrochemically polarized in 20% HF for Si and 1 M HCl for GaAs using a staircase potential sweep, in steps of 10 mV every 5 s (for Si) or every 2 s (for GaAs) in the anodic direction (further details can be found in Refs. 5, 6, and 11). The potentials were measured with respect to a saturated calomel electrode (SCE). This process was done in the dark to avoid light-induced carrier generation in the sample. Under these conditions (similar to the diode behaviour of a p–n junction under reverse bias), the holes (h$^+$) necessary in the case of Si for Si^0 oxidation and dissolution as (SiF$_6$)$^{2-}$, only become sufficiently available at the Si surface at the potential where Schottky barrier breakdown occurs.[5] A similar process occurs in GaAs.[12] After the formation of the porous structures, the samples were intensively rinsed with deionized water, and then blown dry in an argon stream.

The room temperature photoluminescence (PL) of the samples was excited with 15 mW of 457.9-nm argon laser light, dispersed with a Spex 14018 double spectrometer, and detected with a cooled RCA 31034A GaAs photomultiplier. Scanning electron micrographs were acquired using a JEOL 840A SEM equipped with a Digiscan image-acquisition archiving system.

RESULTS

Pore Formation

Figure 1 shows a polarization curve of an n-type GaAs (100) sample in 1 M HCl. The potential region with a blocking characteristic (low current flow), typical of n-type material/electrolyte junction in the dark, is obtained in the anodic direction up to a potential of approximately 3.1 V. At higher potentials a steep current increase is observed and previous investigations showed that at this potential the first pores are formed in the material.[13] Therefore, this potential value is called the pore formation potential (PFP).

Also included in Fig. 1 is a polarization curve of a sample that was scratched with a diamond scribe. The scratch is only several millimetres long, but a significant increase in the current—of more than a decade—is obtained at approximately 1.2 V; *i.e.*, this sample shows an initial PFP significantly lower than that of the intact sample. The inset in Fig. 1 shows an SEM image of the surface of a scratched sample polarized to 2.5 V (*i.e.*, below the PFP of an intact sample). The image reveals that pore formation took place only in the region of the scratch. The rest of the surface remained intact. Hence, dissolution processes occurred only locally at the scratch and the additional current flow, compared with an intact sample, must be attributed to this surface location. These very high local current densities indicate that Schottky barrier breakdown is facilitated at surface defects.

More controlled local defect generation can be achieved by using FIB implantation. This is shown, for example, in Fig. 2, which gives a typical current/voltage characteristic for a reference (nonimplanted) n-type Si sample in 20% HF. Starting from the open circuit potential, with an in-

4

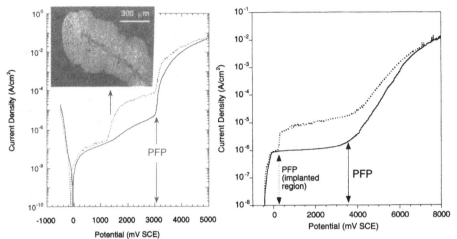

Figure 1. Current–voltage (polarization) curve of n-type GaAs (100) in 1 M HCl (solid line) and a polarization curve acquired with a sample that was intentionally scratched (dashed line). The curves were acquired in the dark. The inset shows the morphology of a scratched sample after polarization from -0.5 to 2.5 V.

Figure 2. Current-voltage (polarization) curve of n-type Si (100) in 20% HF (solid line) and the polarization curve acquired with a ϕ=300 μm capillary electrode on a 50-μm square implanted with 3×10^{14} cm^{-2} Si^{++} (dashed line). The curves were acquired in the dark.

creasing potential the current increases up to a plateau region. In this region, the current is controlled by electrons that overcome the charge carrier depletion region (Schottky barrier) at the semiconductor/electrolyte interface by thermal activation. At a potential of approximately 3.8 V, the current steeply increases. In this region, local dissolution processes occur on the Si surface which, after extended polarization, lead to a porous surface. There are several factors that influence the PFP such as the concentration of the anion in the electrolyte or the temperature. The predominant factors are, however, the conduction type and doping concentration of the substrate, as expected from the Schottky approach.

The defects created by the FIB have a drastic effect on the PFP. This is clear from the second curve included in Fig. 2, which shows a polarization curve acquired with a microelectrode pipette (ϕ=300 μm^2) that was placed on a 50-μm implanted square (dose of 3×10^{14} cm^{-2}). In this case, a first significant current increase appears at +0.25 V. Although the implanted area is only about 3% of the total area exposed to the electrolyte, the current density in the plateau region is more than a decade higher than for the reference sample. Hence, the effective current density in the implanted region is about 300 times higher than on the intact Si surface; *i.e.*, it becomes comparable to current densities observed with the reference sample above its PFP. Figure 3 shows an optical micrograph of a polarized sample where letters were written with the FIB at a dose of 3×10^{14} cm^{-2}. The letters show the typical porous Si interference colours ranging from red to green. The letter R was obtained with a single FIB scan, which resulted in a linewidth of approximately 200 nm. Figure 4(a) shows an SEM micrograph of a 50-μm square implanted with the same dose (3×10^{14} cm^{-2}) and identically treated. From Fig. 4(b), it is apparent that porous Si has been formed within the square. The surrounding area is completely unattacked.

5

Figure 3. Optical micrograph of porous Si letters produced by 3×10^{14} cm^{-2} Si^{++} FIB implantation in n-type Si (100) and subsequent electrochemical polarization in 20% HF from -0.5 to 3.5 V. The R in NRC was outlined with a single (100-nm-wide) FIB line; the rest of the letters were uniformly implanted. The letters show the green and red interference colours typical of porous Si.

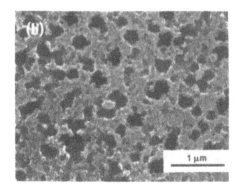

Figure 4. SEM image of a 50-μm square of Si implanted with 3×10^{14} cm^{-2} Si^{++} after polarization in 20% HF from -0.5 to 3.5 V. (b) Higher magnification of (a) within the square.

Photoluminescence

The PL spectrum shown in Fig. 5 was measured with an argon laser beam focused in the centre of the 50-μm square of Fig. 4. The PL spectrum peaks at 655 nm in the orange–red region of the spectrum and, in width and wavelength position, is typical of the PL response of porous Si.[3] The band shape shows irregularities that can be attributed to nanoscopic nonuniformities in the material porosity. Electrochemically treated areas next to the implanted patterns were investigated as a reference and, in every case, the unimplanted areas showed the spectral behaviour of a clean Si surface (no light emission in the visible range) as did implanted areas not yet electrochemically treated. This clearly indicates that the electrochemical formation of porous Si is responsible for the PL observed and not lattice defects or amorphization created by the implantation.

In Si, the highest PL intensity was observed from the sample implanted with a dose of 3×10^{14} cm^{-2} and polarized to 3.5 V (Fig. 5). Other conditions led to significantly lower PL intensities. The PL intensity and morphology of the samples were correlated, in that the sample with the highest amount of porosity with feature sizes in the nanoscopic range gave the highest PL intensity. This is consistent with a quantum confinement explanation for the red PL of porous Si.[1,3]

In GaAs, similar results were obtained. Figure 6 shows a PL spectrum taken within the attacked square of a sample implanted with 1×10^{15} ions cm^{-2} and polarized to 3 V, and a reference spectrum of the GaAs (100) surface next to the square. For the spectrum taken on the unimplanted

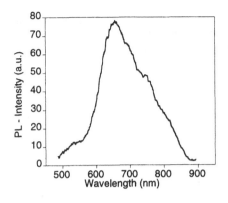

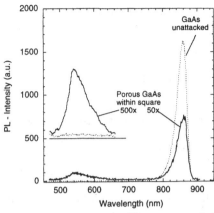

Figure 5. Room-temperature PL spectrum of porous Si acquired in the centre of the square shown in Fig. 4.

Figure 6. Room-temperature PL spectra acquired on a GaAs (100) sample implanted with 1×10^{15} cm^{-2} Si^{++} and polarized from -0.5 to 3.0 V. The solid line shows the response within the square and the dashed line from a surface location next to the square. The inset gives a magnification of the short wavelength region and is offset by 500 a.u.

part of the sample, an intense PL peak is observed at approximately 860 nm as expected for bulk GaAs (the sharp cutoff to higher wavelength is due to the photomultiplier response). The spectrum taken within the square shows an additional broad green–yellow band, peaked at approximately 540 nm. This spectral behaviour is consistent with previous work on porous GaAs.[14] For samples polarized to 1 and 2 V, no green PL peak was observed for any implantation dose. For a sample polarized to 3 V, the squares with dosages between 3×10^{14} ions cm^{-2} and 3×10^{15} ions cm^{-2} showed a significant green PL peak. The fact that for higher doses a nonuniform large-shape etching morphology occurred, which did not result in green PL, is consistent with an explanation for the visible PL in terms of a quantum confinement model, as postulated in earlier work.[14,15]

At present, although visible to the eye, the PL intensity observed in both Si and GaAs is relatively weak when compared to the "bulk" porous material and, therefore, further optimization of the formation parameters is needed. However, it is noteworthy that this process has the potential to write nanostructures in the 50–200-nm range, depending on the diameter of the writing ion beam.

CONCLUSIONS

In conclusion, we have shown how to produce laterally confined light-emitting Si or GaAs by a direct FIB writing process. We clearly demonstrate that a creation of surface defects followed by an electrochemical "development" treatment, tailored to trigger dissolution at defects, can be used to form visible light-emitting porous Si or GaAs selectively. At present, the size of the structures appears to be limited only by the diameter of the writing ion beam. Thus patterns in the 50–200-nm range seem quite possible.

Recently, electroluminescent devices based on large-scale porous Si structures have been reported.[4] The process described here could facilitate a drastic shrinkage in the dimensions of such devices and hence could be a basis for, or part of, a process leading to extremely high-resolution optoelectronic applications.

Additionally, the above findings show that surface lattice defects represent centres of enhanced dissolution, and hence represent the initiation site for pore formation, when conditions are established where Schottky barrier breakdown is the rate-determining step for the surface dissolution reaction.

ACKNOWLEDGEMENTS

We thank H.G. Champion, J.W. Fraser and H.J. Labbé for their meticulous help with the experiments and the Swiss National Science Foundation for financial support.

REFERENCES

1. D.J. Lockwood, Ed., *Light Emission in Silicon* (Academic Press, Boston, 1997).

2. L.T. Canham, Appl. Phys. Lett. **57**, 1046 (1990).

3. See, for example, A.G. Cullis, L.T. Canham, P.D.J. Calcott, J. Appl. Phys. **82**, 909 (1997).

4. K.D. Hirschmann, L. Tsybeskov, S.P. Duttagupta, P.M. Fauchet, Nature **384**, 338 (1996).

5. P. Schmuki, L.E. Erickson, D.J. Lockwood, Phys. Rev. Lett. **80**, 4060 (1998).

6. P. Schmuki, L.E. Erickson, D.J. Lockwood, J.W. Fraser, G. Champion, H.J. Labbé, Appl. Phys. Lett. **72**, 1039 (1998).

7. J. Xu and A.J. Steckl, Appl. Phys. Lett. **65**, 2081 (1994).

8. S.P. Duttagupta, C. Peng, P.M. Fauchet, S.K. Kurinec, T.N. Blanton, J. Vac. Sci. Technol. B **13**, 1230 (1995).

9. V.V. Doan and M.J. Sailor, Science **256**, 1791 (1992).

10. J.C. Barbour, D. Dimos, T.R. Guilinger, M.J. Kelly, S.S. Tsao, Appl. Phys. Lett. **59**, 2088 (1991).

11. P. Schmuki, L.E. Erickson, D.J. Lockwood, J.W. Fraser, B.F. Mason, G. Champion, H.J. Labbé, J. Electrochem. Soc. **146** (to be published in 1999).

12. P. Schmuki, D.J. Lockwood, H.J. Labbé, J.W. Fraser, M.J. Graham, in *Pits and Pores: Formation, Properties and Significance for Advanced Luminescent Materials,* edited by P. Schmuki, D.J. Lockwood, H.S. Isaacs, and A. Bsiesy (Electrochemical Soc., Pennington, NJ, 1997), p. 112.

13. P. Schmuki, J. Fraser, C.M. Vitus, M.J. Graham, H. Isaacs, J. Electrochem. Soc. **143**, 3316 (1996).

14. P. Schmuki, D.J. Lockwood, H.J. Labbé, J.W. Fraser, Appl. Phys. Lett. **69**, 1620 (1996).

15. D.J. Lockwood, P. Schmuki, H.J. Labbé, J.W. Fraser, in *Pits and Pores: Formation, Properties and Significance for Advanced Luminescent Materials,* edited by P. Schmuki, D.J. Lockwood, H.S. Isaacs, and A. Bsiesy (Electrochemical Soc., Pennington, NJ, 1997), p. 447.

STRONGLY SUPERLINEAR LIGHT EMISSION AND LARGE INDUCED ABSORPTION IN OXIDIZED POROUS SILICON FILMS

H. KOYAMA, P.M. FAUCHET
Department of Electrical and Computer Engineering, University of Rochester, Rochester, NY 14627

ABSTRACT

The optical properties of oxidized free-standing porous silicon films excited by a cw laser have been investigated. It is found that samples oxidized at 800-950 °C show a strongly superlinear light emission at an excitation intensity of ~10 W/cm^2. This emission has a peak at 900-1100 nm and shows a blueshift as the oxidation temperature is increased. These samples also show a very large induced absorption, where the transmittance is found to decrease reversibly by ≤ 99.7 %. The induced absorption increases linearly with increasing pump laser intensity. Both the superlinear emission and the large induced absorption are quenched when the samples are attached to materials with a higher thermal conductivity, suggesting that laser-induced thermal effects are responsible for these phenomena.

INTRODUCTION

The optical response of porous silicon (PSi) under high-intensity excitation conditions has been studied before [1-3]. It is reported that the intensity of the visible photoluminescence (PL) saturates, i.e., the PL efficiency decreases as the excitation intensity is increased. This behavior is attributed to Auger recombination [2,3] or the saturation of radiative recombination centers [1]. Regarding the optical absorption, several research groups have found a large absorption increase with increasing the excitation laser intensity [4,5]. Transmittance changes of 50-60 % are reported. The absorption by photoexcited carriers has been proposed as the mechanism responsible for this induced absorption.

In this study, we have examined the optical properties of oxidized PSi films under excitation by a moderately high power cw laser. Oxidized PSi exhibits a superior luminescence stability under either optical or electrical excitation [6-8], and therefore it is suitable for the study of high-excitation effects. We report here a strongly superlinear light emission [9] and a very large absorption increase [10] observed in samples oxidized at 800-950 °C. Both of these phenomena are attributed to thermal effects induced by the laser irradiation.

EXPERIMENT

PSi samples were prepared by the anodization of (100) p$^+$-Si wafers (0.008-0.012 Ωcm) in a solution of 50% HF: ethanol = 1:1 at 100 mA/cm^2 for 2.5 min. At the end of the anodization, the current density was abruptly increased to ~700 mA/cm^2 to lift off the PSi layers. The thickness of the PSi films is about 12 μm. These free-standing PSi films were then oxidized in dry O$_2$ at 800-950 °C for 10 min. The emission spectra were measured using a lock-in technique and a cooled Ge detector. The spectra were then corrected for the apparatus response using a

9

calibration lamp. Absorption measurements were carried out using a He-Ne laser (632.8 nm) as a probe beam. In both emission and absorption measurements, a multiline (457.9 - 514.5 nm) Ar⁺ laser was employed as the pump (excitation) beam. The spot diameter of the pump laser beam was ~ 1mm, while that of the probe beam (in transmission measurements) was slightly less than the spot size of the pump beam.

RESULTS

Superlinear Emission

Figure 1 shows the emission spectra of an oxidized PSi sample measured at various excitation intensities. At low excitation levels ($\leq$ 5 W/cm^2), this sample only exhibits a weak PL peaking at ~ 850 nm. With increasing excitation intensity, however, this PL is quenched, and then a very strong emission peaking at ~ 1050 nm appears. As shown in Fig. 2, this strong emission has a very sharp excitation-intensity dependence that can be expressed by a power law with n = 9.6. Figure 3 compares the spectra of three samples oxidized at different temperatures. As the oxidation temperature is increased, the emission becomes more pronounced and shifts to the blue.

This strongly nonlinear emission has only been observed in free-standing samples [9]. It was not observed in PSi samples anodized and oxidized under the same conditions but left on the Si substrate, for excitation intensities up to 40 W/cm^2. Furthermore, the strong emission from free-standing PSi samples was found to be quenched completely when some materials (metal, Si wafer, or even glass) were physically attached to the samples. Similar quenching was also observed by simply blowing air onto the sample surface. These observations suggest that this emission is due to a large temperature increase under moderately high power cw laser excitation.

Costa et al. [11] reported a similar sharp increase of emission intensity in Si nanoparticles prepared by plasma-enhanced chemical vapor deposition. Since this increase was only observed in vacuum,

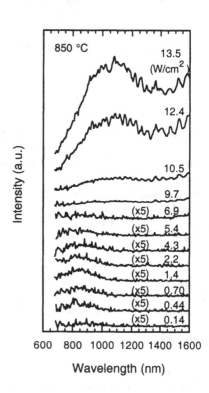

FIG. 1. Emission spectra of a PSi sample oxidized at 850 °C at various excitation intensities.

10

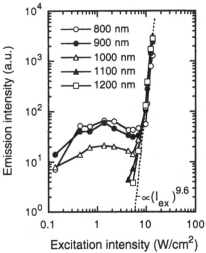

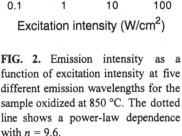

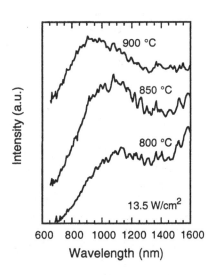

FIG. 2. Emission intensity as a function of excitation intensity at five different emission wavelengths for the sample oxidized at 850 °C. The dotted line shows a power-law dependence with $n = 9.6$.

FIG. 3. Emission spectra at 13.5 W/cm² for three samples oxidized at different temperatures.

they attributed it to thermal radiation. The thermal radiation from blackbodies is well formulated by Planck's law. To describe the thermal radiation from non-blackbody materials, we need to invoke Kirchhoff's law [12]. In this case, the thermal radiation intensity $I_{TR}(\lambda)$ emitted by a material with absorptivity $a(\lambda)$ (≤ 1) is given by

$$I_{TR}(\lambda) \propto a(\lambda) \exp(-hc/\lambda kT) / \lambda^5 \tag{1}$$

under the assumption that $\exp(hc/\lambda kT) >> 1$. Here h is Planck's constant, c is the speed of light, k is Boltzmann's constant, and T is the temperature of the material.

Assuming that T is proportional to the laser excitation intensity, we find that the temperature dependence of the measured emission intensity can be expressed as $I(T) \propto \exp(-\Delta E/kT)$ [9]. A fit to our data gives a value of $\Delta E \sim 1\text{eV}$, which is close to the observed emission peak energy. The existence of a peak at ~ 1050 nm in the emission spectrum can be explained in terms of a sharp increase in $a(\lambda)$ in this wavelength regime due to the presence of the fundamental absorption edge. The blue shift of the emission peak as a function of oxidation temperature (Fig. 3) may be consistent with the blue shift of the absorption edge that is expected where the particle size enters the quantum confinement regime. These results suggest that thermal radiation is a possible mechanism responsible for this emission.

The superlinear emission can also be explained in terms of thermally-assisted PL. According to Daub and Würfel [13], if the free-carrier absorption can be neglected, the total radiation intensity $I(\lambda)$ including both thermal radiation and PL (resulting from either direct or phonon-assisted indirect transitions) should follow the relationship

$$I(\lambda) \propto a(\lambda) \exp[-(hc/\lambda-\mu)/kT]/\lambda^5 \qquad (2)$$

provided that $\exp[(hc/\lambda-\mu)kT] >> 1$. Here μ is the difference between the electron and hole quasi-Fermi energies. When $\mu = 0$, the relationship (2) reduces to the thermal radiation formula (1). As easily seen, μ does not affect the wavelength dependence of the emission intensity. This means that the PL has exactly the same spectral features as thermal radiation in the wavelength range where band-to-band transitions dominate the absorptivity $a(\lambda)$. This makes it very difficult to distinguish between the contributions from thermal radiation and those from PL in the emission spectrum. The only difference between these two emissions exists in their intensities through the factor $\exp(\mu/kT)$. It should be noted that if $\mu \geq kT$, PL overcomes the thermal radiation. A sharp increase in PL efficiency with increasing temperature can be explained by, for example, assuming the thermal reexcitation of trapped carriers [9].

In our results shown in Fig. 2, no emission-wavelength dependence is observed in the slope of the emission intensity as a function of the excitation intensity. This appears to be inconsistent with the thermal radiation model and therefore support the PL model. Indeed, according to Eq. (1), the temperature dependence of the thermal radiation should be a function of the emission wavelength (in other words, the thermal radiation should show a blue shift as the excitation intensity is increased). We should note, however, that the presence of a large induced absorption results in a red shift of $a(\lambda)$, which can counteract the blue shift of the thermal radiation.

Induced Absorption

Figure 4 shows the measured transmittance of five PSi samples (one as-anodized sample and four oxidized samples) as a function of the pump laser intensity. It is clearly seen that in any of the samples the transmittance decreases strongly with increasing pump intensity. These changes are completely reversible, as long as the pump intensity does not exceed the damage threshold (~ 20 W/cm² for the oxidized

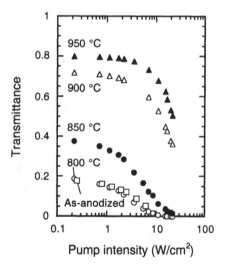

FIG. 4. Transmittance of five different PSi samples at 632.8 nm as a function of the pump Ar⁺ laser intensity.

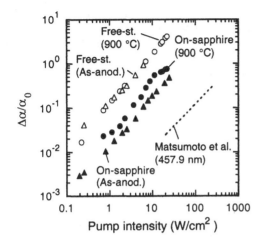

FIG. 5. Normalized induced absorption in free-standing samples (open symbols) and samples attached to a sapphire substrate (closed symbols) as a function of pump intensity. The dotted line shows the data reported by Matsumoto *et al.* (Ref. 5).

samples and ~ 7 W/cm² for the as-anodized sample). In the sample oxidized at 800 °C, we obtain a maximum reversible transmittance change of 99.7%.

In Fig. 5 we show the increase of the absorption coefficient $\Delta\alpha$, normalized to the linear absorption coefficient α_0. $\Delta\alpha/\alpha_0$ increases linearly with increasing pump intensity. This is in agreement with the results of Matsumoto *et al.* (dotted line) [5]. They measured the absorption in an as-anodized free-standing PSi sample made from a p⁻-Si substrate using the 457.9 nm line of an Ar⁺ laser as both the pump beam and the probe beam. Our normalized induced absorption is nearly two-orders-of-magnitude higher than theirs at the same pump intensity.

Since the absorption coefficient increases linearly with pump intensity, Matsumoto *et al.* [5] proposed that the absorption by the carriers generated by the pump beam is responsible for the induced absorption. However, the absorption increase due to the thermally-induced bandgap shrinking is proportional to the pump laser intensity [10]. In addition, it is observed that $\Delta\alpha$ decreases strongly when the samples are physically attached to a sapphire substrate, as shown by the closed symbols in Fig. 5. These results suggest that the large induced absorption in cw-laser-excited PSi samples originates from the bandgap shrinking induced by the high temperature achieved under laser irradiation. The large discrepancy between our results and those of Matsumoto *et al.* can be explained by the difference in the pump laser spot size (~ 1mm and 50 µm, respectively), since the temperature rise due to laser heating is proportional to both the laser intensity and the spot diameter [14].

CONCLUSIONS

We have reported two unique optical properties of oxidized free-standing PSi films observed under excitation by a moderately high power cw laser beam: a superlinear light emission and a large induced absorption. The superlinear emission exhibits a very sharp intensity increase as a function of the excitation intensity, which can be expressed by a power law with $n \sim 10$. Its spectrum has a peak at 900-1100 nm, and blueshifts with increasing sample oxidation temperature. These properties can be explained in terms of thermal radiation and/or thermally assisted PL. The induced absorption is very large, and a maximum reversible transmittance change of 99.7 % is obtained. The increase of the absorption coefficient $\Delta\alpha$ is proportional to the pump intensity. We suggest that the thermally induced bandgap shrinking resulting from cw-laser excitation is responsible for this absorption increase.

ACKNOWLEDGMENTS

This work was supported by the Army Research Office.

REFERENCES

1. Y. Kanemitsu, Phys. Rev. B **49**, 16845 (1994).

2. C. Delerue, M. Lannoo, G. Allan, E. Martin, I. Mihalcescu, J. C. Vial, R. Romestain, F. Muller, and A. Bsiesy, Phys. Rev. Lett. **75**, 2228 (1995).

3. D. Kovalev, B. Averboukh, M. Ben-Chorin, F. Koch, Al. L. Efros, and M. Rosen, Phys. Rev. Lett. **77**, 2089 (1996).

4. V. Klimov, D. McBranch, and V. Karavanskii, Phys. Rev. B **52**, R16989 (1995).

5. T. Matsumoto, M. Daimon, H. Mimura, Y. Kanemitsu, and N. Koshida, J. Electrochem. Soc. **142**, 3528 (1995).

6. V. Petrova-Koch, T. Muschik, A. Kux, B. K. Meyer, F. Koch, and V. Lehmann, Appl. Phys. Lett. **61**, 943 (1992).

7. J. L. Batstone, M. A. Tischler, and R. T. Collins, Appl. Phys. Lett. **62**, 2667 (1993).

8. L. Tsybeskov, S. P. Duttagupta, and P. M. Fauchet, Solid State Commun. **95**, 429 (1995).

9. H. Koyama, L. Tsybeskov, and P. M. Fauchet, J. Lumin. (in press).

10. H. Koyama and P. M. Fauchet, Appl. Phys. Lett. **73**, 3259 (1998).

11. J. Costa, P. Roura, G. Sardin, J. R. Morante, and E. Bertran, Appl. Phys. Lett. **64**, 463 (1994); J. Costa, P. Roura, J. R. Morante, and E. Bertran, J. Appl. Phys. **83**, 7879 (1998).

12. F. Grum and R. J. Becherer, *Optical Radiation Measurements Vol. 1*, Academic Press, New York, 1979, Chap. 4.

13. E. Daub and P. Würfel, Phys. Rev. Lett. **74**, 1020 (1995); J. Appl. Phys. **80**, 5325 (1996).

14. M. Lax, J. Appl. Phys. **48**, 3919 (1977).

ENHANCING THE EXTERNAL QUANTUM EFFICIENCY
OF POROUS SILICON LEDS BEYOND 1%
BY A POST-ANODIZATION ELECTROCHEMICAL OXIDATION

B. GELLOZ, T. NAKAGAWA AND N. KOSHIDA
Division of Electronic and Information Engineering, Faculty of Technology, Tokyo University
of Agriculture and Technology, Koganei, Tokyo 184 8588, Japan

ABSTRACT

External quantum efficiencies (EQEs) of electroluminescent devices based on porous
silicon (PS) reported to date are still below the minimum requirements for practical applications
such as display devices (1 %) and optical interconnection (10 %). Post-anodization anodic
oxidation of PS to enhance the EQE of electroluminescence from devices based on a thin
transparent indium tin oxide contact mounted on either porosified n^+-type silicon or p^+n^+-type
silicon has been investigated. Enhancement of EQE by more than 2 orders of magnitude has
been achieved on our devices. CW EQE of 0.51% has been obtained by using a single
anodically oxidized n^+-type porous layer. The device based on the p^+n^+ substrate yielded a CW
EQE of 1.1 %, with a power efficiency of 0.08%. It is the first time that EQE greater than 1% is
obtained. Furthermore, anodically oxidized devices show better stability than non-oxidized
devices. The anodic oxidation proceeds in such a way that it mainly decreases the size of non-
confined silicon in PS. The dramatic enhancement in EQE can therefore be explained by
preferential reduction of leakage carrier flow through non-confined silicon.

INTRODUCTION

The demonstration [1] that efficient visible photoluminescence (PL) could be obtained at
room temperature from porous silicon (PS) seemed to open up the possibility of silicon optoelectronics. A large amount of work has been undertaken in order to realize all solid state
electroluminescent devices. However, the external quantum efficiencies (EQE) reported to date
are still below the minimum requirements for practical applications such as display devices (1%)
and optical interconnection (10 %) [2]. Indeed, the first reported EQE was of the order of 10^{-5}
%, using a device based on thin semitransparent gold contact on PS [3]. The maximum CW
EQE obtained by using a single n^+-type porous layer and a porosified p^+n junction is 0.21% [4]
and 0.18% [5], respectively. Under pulsed-operation, EQE of 0.8% [6] has been reported, in
which a porosified p^+n junction structure has been used.

One of the reasons for the low EQE is mainly because most carriers flow via non-
confined silicon present in the porous silicon structure. These carriers do not contribute to EL.
Thus, this large amount of leakage component limits the EQE. In order to enhance the EQE, we
propose here the use of anodic oxidation of PS as a post-treatment. When anodic oxidation is
performed under constant current, it has been shown that oxidation occurs preferably at the
surface of non-confined silicon during the first stages of the process when using p^+-type PS [7].
Such a partial oxidation is expected to mainly decrease the size of non-confined silicon in PS.
Then, the number of carriers flowing through non-confined PS may be reduced, resulting in
higher EQE. The effect of anodic oxidation used as a post-treatment of PS on the electrical and
EL properties has been investigated for a device including a single n^+-type PS layer and a device
including a porosified p^+n^+ junction where the top contact is made on the p^+ side. In both cases,

15

the top contact is a thin transparent indium tin oxide (ITO) layer.

EXPERIMENT

p^+n^+ substrates are obtained from n^+ (111) (0.018Ω.cm) substrates. First, a 100nm thick silicon oxide layer is built by dry oxidation of the substrate. Then, front face implantation with boron at 35keV to a dose of $5.10^{15}cm^{-2}$ is performed. The projected range, R_p, is about 120nm. This implant is then activated by thermal annealing at 900C during 10min. Finally, the oxide layer is removed by dipping the substrates into an HF electrolyte. SEM pictures reveal a sharp change in the PS structure at a depth of 200nm.

N^+-type PS and p^+n^+ porous junctions are formed under 100mA/cm^2 during 5 min, under illumination from a 500W tungsten lamp mounted at a distance of 20 cm. Electrolyte is HF (55wt.%): ethanol=1:1. The total thickness of the porous layer is about 30 μm in both cases.

Having been rinsed with ethanol for 2 min, and without being dried, anodic oxidation of the porous layers is achieved by anodically polarizing (35 mA/cm^2) PS in an aqueous solution containing 1M sulfuric acid (H_2SO_4). This oxidation implies hole consumption and electron injection [8]. Therefore, EL can be observed during the treatment when oxidation takes place in luminescent crystallites after a while during which oxidation occurs only in non-confined silicon. The EL properties of our n^+ and p^+n^+ porous layers under anodic oxidation are found similar to that reported for p^+-type PS [7]. The best results for EQE are obtained when anodic oxidation is performed until the maximum of EL intensity during the treatment (290 s). Samples discussed in this paper has been oxidized in this condition. Then, the samples are rinsed with ethanol for 2 min, and dried. 600nm of ITO is then deposited by sputtering onto the porous layers for use as the top electrode. PL and EL measurements are conducted in N_2 atmosphere. PL spectra are measured using 325nm excitation beam from a He-Cd laser. Keithley source-measure unit (Model 238) is used as a voltage source and also as an ammeter of the diode current. The emitted light is detected via a monochromator (Nikon G250) and a photomultiplier tube (Hamamatsu R928) whose output current is measured using a picoammeter (Advantest TR8652). The polarity of the voltage bias is defined such that it is positive when the front ITO contact is positive with respect to the back Al contact, namely positive voltage corresponds to forward condition. All experiments are conducted under CW operation in N_2 atmosphere. EQE is calculated as the ratio of the number of photons emitted in the hemisphere by the number of charges flowing through the device. A photometer (collecting surface = 1cm^2) is used to count the emitted photons, at a distance of 4.7cm from the emitting device. The total number of photons emitted in the hemisphere is derived by assuming a uniform spatial emission. The device based on porosified n^+-type silicon is named device 1 and the device based on porosified p^+n^+-type silicon is named device 2.

RESULTS

First the spectral characteristics of the devices are considered. Figure I-(a) shows the PL spectra of device 1 with and without the oxidation treatment. As expected [9], anodic oxidation treatment induces a blue-shift of PL. The EL spectrum of device 1 oxidized, also represented on the same figure, fits very well the PL one. This indicates that the same carrier-recombination mechanisms is involved in the two phenomena. Therefore, excitons localized within silicon nanocrystallites in porous silicon are likely to be responsible also for the EL and oxygen atoms incorporated at the inner surface of PS may not introduce any new radiative recombination centers. The spectral characteristics of device 2 oxidized are represented in figure I-(b). The PL and EL spectra fits well on the low wavelength side whereas on the other side, the luminescence intensity is higher for the PL. This results in a PL peak (672nm) slightly red-shifted compared to the EL one (650nm). This slight difference is probably due to the superficial p⁺-type porous layer which is expected to emit weaker and red-shifted luminescence than the underlying n⁺-type PS. The PL should include both a significant component from the p⁺ superficial layer (where most of the laser beam is absorbed) and a component from the underlying n⁺-type porous layer. On the contrary, most of the EL should come from the underlying optically active n⁺-type porous layer since the low porosity p⁺ layer is poorly emissive. In this case, the PL spectrum should be slightly red-shifted compared to the EL one.

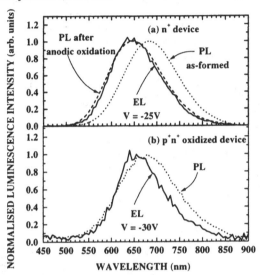

Figure I: Normalized EL and PL spectra for the oxidized n⁺-based device (a) and p⁺n⁺-based device (b). Also represented is the PL spectrum of the non-oxidized n⁺-based device.

Figure II shows the current and EL intensity as a function of applied voltage for several devices. Efficient EL can be obtained under either forward or reverse conditions. At first, let's consider the effect of anodic oxidation on the current density and the EL intensity in the case of the n⁺-type based device (filled-squares and filled-circles). The effect of oxidation is to significantly lower the current density. For instance, at -25V, its value is -210mA/cm² and -17.8µA/cm² when PS is non-oxidized and oxidized, respectively. Under reverse conditions, the difference in current density is of about 4 orders of magnitude whereas EL intensity does not seem to be significantly dependent on whether the sample is oxidized. Under forward conditions, the current density is 3 orders of magnitude lower and the EL intensity one order of magnitude greater when the PS is oxidized. This show that the EQE of the device is improved by about 4 orders of magnitude whether the device is operated under reverse or forward conditions. The measured value of the EQE of the oxidized sample is 0.51 %. This value is the highest EQE of a single PS layer light emitting device ever reported to date. The external power efficiency is 0.05 %.

The effect of anodic oxidation on the current density and the EL intensity in the case of the p⁺n⁺-

type based device can be seen from the hollow-diamond and hollow-triangle curves on figure II. Again the current density is greatly reduced on the treated device. The difference is about 2 orders of magnitude under reverse operation and about 3 orders of magnitude under forward operation. The EL intensity is about 1 order of magnitude greater under reverse operation and about 1 order of magnitude lower under forward operation. In this latter case however, if the curve slopes remain unchanged for greater voltages, the EL intensity of the oxidized device is expected to join that of the non-oxidized device. The EQE is increased about 3 orders of magnitude under reverse conditions. EQE of 1.1% (external power efficiency of 0.08%) has been measured for this device under reverse operation. This value is the highest reported to date. These results show that anodic oxidation used as post-treatment of porous silicon dramatically increases the EQE of our devices. Whether the porous layer is oxidized or not, the EQE is found higher under reverse operation than under forward operation. In the following, focus is put on results obtained under reverse conditions.

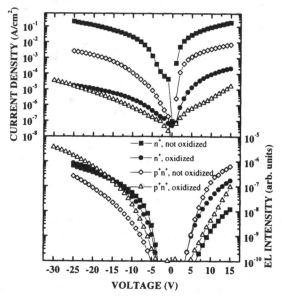

Figure II : Current density and EL intensity versus voltage for different devices whose porous layer has been either oxidized or not. Positive voltage refers to a forward biased diode.

Figure III shows the EL intensity versus the electrical current under reverse operation, for four different devices. Hollow circles and filled circles refer to device 2 oxidized and not oxidized, respectively. Hollow squares and filled squares refer to device 1 oxidized and not oxidized, respectively. The reduction of the current density due to the anodic oxidation treatment is very clear on this figure. The latter show that the EQE of device 2 is higher than the one of device 1 when the porous layers are oxidized and also when they are not. This result may be attributed to an improvement of the quality of the front contact due to the presence of the p^+-type superficial layer. Indeed, there should be a huge porosity gradient across the porous layers obtained from n^+ substrates. The porosity of PS near the top surface is very high, inducing an uneven surface. This could result in bad quality contact between the ITO and the porous layer. Moreover, high porosity layers can be easily damaged. The low porosity p⁺-type superficial layer should provide a smoother transition between the ITO and the optically active n⁺ PS, providing greater mechanical stability and a good electrical contact to the ITO.

The transport properties of our devices are not discussed in details here. However, our characteristics are very similar to those obtained previously using the same kind of n⁺-type porous layers, with gold contact and without oxidation [10]. Then, carrier generation mechanisms (impact excitation by energetic hot electrons and tunneling of valence-band electrons into the conduction-band of neighboring crystallites) described in reference 10 may be invoked to explain the electrical behavior of our devices.

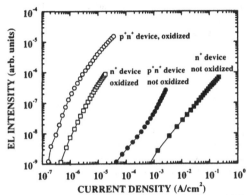

Figure III : EL intensity versus current density, for different devices which porous layer has been either oxidized or not oxidized. All characteristics correspond to reverse CW operation.

DISCUSSION

In order to understand the reasons for the enhancement of the EQE by anodic oxidation post-treatment, let's consider how this particular kind of oxidation occurs. The characteristics of the anodic oxidation depend on the doping level of the initial substrate from which PS is obtained. Our substrates are heavily doped n⁺-type silicon. Although the behavior of PS layers formed from this type of substrate under anodic oxidation had not been reported previously, we expected and found it to be similar to the previously reported one of p⁺-type PS [7], because of the similar structure of p⁺ and n⁺-type PS. The detailed mechanisms of anodic oxidation of p⁺-type luminescent porous silicon has been previously reported elsewhere [7]. Oxidation occurs at the internal surface, where holes supplied by the substrate are injected. As the experiment is performed under constant current, holes flows via the energetically-easiest paths. During the first stages of the process, hole injection, i.e. oxidation, occurs only in non-confined silicon. The formation of oxide at the surface makes charge-exchange between silicon and the electrolyte more difficult. The potential must correspondingly be increased in order to maintain the current constant. Hole injection in more energetic levels is then achieved. There is a period when injection occurs in confined crystallites. EL can be observed at this stage. EL intensity reaches a maximum when carrier injection in confined crystallites is optimal. Afterwards, oxidation becomes so high that electrical contact between non-confined PS and confined crystallites is broken. Carrier injection in confined silicon correspondingly decreases. That is why EL then decreases and disappears. Finally, oxidation takes place only in non-confined silicon.

When oxidation is performed up to the maximum of EL, non-confined silicon is much more oxidized than luminescent crystallites. This should significantly reduce the number of carriers flowing through non-confined silicon, optimizing carrier injection into luminescent crystallites. This results in enhanced EQE. It should be noted that thermal oxidation could not lead to such an enhancement since it occurs on the whole internal surface of PS (without selecting non-confined silicon from confined silicon). Moreover, contrary to thermal oxidation, anodic oxidation does not affect both the PS structure and passivation with hydrogen [11]. It also enhances localization of carriers in crystallites [9] and enhances the luminescence-homogeneity.

Anodic oxidation also enhances the stability of the devices during the first stages of operation when a constant identical voltage is applied to an oxidized and a non-oxidized device. The difference in stability may be the result of the difference in the current density. The rather high current density recorded with the non-oxidized devices may cause rapid thermal-induced degradation of EL compared to oxidized devices where current densities are lower.

CONCLUSIONS

The use of anodic oxidation of PS has improved the EL-EQE of our devices based on porosified n^+-type silicon and p^+n^+-type silicon by more than two orders of magnitude. CW EQE up to 0.51 % has been achieved by using a single anodically oxidized n^+-type porous layer. This value is the highest reported to date for a single PS layer light emitting device. This dramatic enhancement can be explained by the fact that the post-treatment based on anodic oxidation considerably reduces carrier flow via non-confined silicon which has been preferably oxidized. As a result, both carrier injection and localization in luminescent crystallites is enhanced. A CW EQE of 1.1% has been obtained from a device based on porosified and anodically oxidized p^+n^+-type silicon. This is the highest value of the EQE reported to date for an EL device based on PS. It is also the first time that an EQE greater than 1% is obtained. The role of the superficial p^+-type layer is probably to provide greater mechanical stability and better electrical contact to the ITO. The anodic oxidation treatment also improves the stability of the devices. Finally, one can notice that this treatment is available for all kind of PS based devices.

REFERENCES

1. L. T. Canham, Appl. Phys. Lett. **57**, 1046 (1990).
2. T. I. Cox, in *Properties of Porous Silicon*, EMIS Datareviews Series No.**18**, Ed. by L. T. Canham (INSPEC, The Institution of Electrical Engineers, UK, London,1997) pp. 290-310.
3. N. Koshida and H. Koyama, Appl. Phys. Lett. **60**, 347 (1992).
4. B. Gelloz, T. Nakagawa and N. Koshida, Appl. Phys. Lett. **73**, 14, 2021 (1998).
5. A. Loni, A. J. Simons, T. I. Cox, P.D.J. Calcott and L.T. Canham, Electronics Letters, **31**, 1288 (1995).
6. K. Nishimura, Y. Nago and N. Ikeda, Jpn. J. Appl. Phys. **37**, L303 (1998).
7. S. Billat, J. Electrochem. Soc. **143**, 3, 1055 (1996).
8. J. N. Chazalviel and F. Ozanam, Mater. Res. Soc. Symp. Proc. **283**, 359 (1992).
9. J. C. Vial, S. Billat, A. Bsiesy, G. Fishman, F. Gaspard, R. Herino, M. Ligeon, F. Madeore, I. Mihalcescu, F. Muller and R. Romestain, Physica B **185**, 593 (1993).
10. T. Oguro, H. Koyama, T. Ozaki and N. Koshida, J. Appl. Phys. **81**, 3, 1407 (1997).
11. M. A. Hory, R. Herino, M. Ligeon, F. Muller, F. Gaspard, I. Mihalcescu and J. C. Vial, Thin Solid Films **255**, 200 (1995).

INTEGRATION OF ELECTRICALLY ISOLATED POROUS SILICON LEDS FOR APPLICATIONS IN CMOS TECHNOLOGY

K.D. HIRSCHMAN*, L. TSYBESKOV, C.C. STRIEMER, S. CHAN AND P.M. FAUCHET†
Department of Electrical and Computer Engineering, University of Rochester, Rochester, NY 14627
* also Department of Microelectronic Engineering, Rochester Institute of Technology, Rochester, NY 14623
† also Laboratory for Laser Energetics, and The Institute of Optics, University of Rochester, Rochester NY 14627

ABSTRACT

Previously we reported on integrated porous silicon (PSi) based LEDs with local bipolar drivers that would have possible applications in an active-matrix configuration for display or optical signal transmission [1]. We now report further progress in device engineering and integration that has enabled the fabrication of improved light-emitting silicon devices based on oxide-passivated nanocrystalline silicon (OPNSi) that can be integrated with standard CMOS technology. Design of experiments methodology was used to direct device engineering experiments, providing a better understanding of how process parameters influence the resulting device performance. It was discovered that the formation of stacking fault defects in the LED region prior to anodization has a predominant effect on both carrier transport and electroluminescence (EL) capabilities of the devices. Process development work has resulted in the fabrication of OPNSi LEDs that offer many of the required attributes of a useful silicon light-emitter. The devices exhibit EL at a bias < 5V, diode-like rectifying I-V characteristics, good stability under DC bias, and uniform emission over the LED contact area. These LEDs have been combined with a new fabrication technique which enables the formation of electrically isolated integrated LEDs in the single-crystal substrate. Each diode structure is formed in its own individual well, enabling junction-isolated devices without a common substrate electrode. Such a process is required to realize integrated LEDs which can be individually addressed in an X-Y array configuration using full-rail voltage modulation. Details of the LED characteristics and device integration will be discussed.

INTRODUCTION

Advancements in device engineering and integration have enabled the fabrication of light-emitting silicon devices that have sufficient characteristics for certain applications and can be integrated with standard CMOS technology. The highest reported external power efficiency of PSi-based LEDs under DC operation exceeds 0.1% [2]. However, these devices are based on hydrogen passivated material, in which EL stability remains an issue. In addition, devices which can be integrated with standard microelectronic process technology must be able to withstand chemical and thermal processes which would be destructive to as-anodized PSi due to its extreme reactivity and inherently fragile structure. Partial oxidation of PSi produces silicon nanoclusters within an oxide matrix, or oxide-passivated nanocrystalline silicon (OPNSi). Properly prepared, this material exhibits appropriate light-emitting and carrier transport properties and is compatible with conventional processing techniques. Since our previous report [1], a significant effort has been directed toward advancing the integration of OPNSi LEDs. Process development and device engineering requirements identified two primary goals. The first goal was to develop a fabrication process which yields good working devices reproducibly. The second goal was to develop an integrated structure that is completely CMOS compatible and provides electrically isolated devices within the crystalline silicon substrate.

EXPERIMENTAL

Bulk-film OPNSi LED Fabrication

Bulk-film LEDs were fabricated by forming the OPNSi active layer over the entire wafer surface. These devices were used as an engineering tool for parameter optimization, producing device performance results in a fraction of the time required for the processing of complete integrated LEDs. Throughout the fabrication procedure there are several variables which have a direct influence on the performance of the

21

OPNSi LEDs. Several process improvements have been made in order to gain control over these factors and obtain reproducible results, enabling the use of designed experiments in order to determine the influence of factors under investigation. The primary factors studied include anodization conditions (current density and time), and post-anodization annealing (furnace/rapid thermal, temperature, time and ambient). Another factor which was not initially under investigation but found to be significant was the effect of surface stress prior to anodization. LED responses included electroluminescence (relative intensity and stability) and the electrical response (I-V characteristics and threshold conditions). Several small designed experiments (typically 3^2 factorial designs) were run, which translates to over 50 different treatment combinations for the preliminary investigation.

Through the use of designed experiments a nominal OPNSi LED process sequence has been developed. The bulk-film OPNSi LEDs are fabricated in three process stages; substrate preparation, active layer formation, and contact formation. The following description represents nominal process conditions used to fabricate the device shown in Fig. 1a. Substrate preparation involves forming a high concentration boron-doped backside layer on a 5-10Ωcm p-type wafer, then performing a 950°C steam oxidation for 3.5hours, growing approximately 5000Å SiO_2. The oxide is removed, followed by backside aluminum sputter deposition and sintering, thus assuring an ohmic backside contact. The wafer is then anodized in 49%HF/ethanol 1:1 volume solution using a constant current density of 3.5mA/cm² (resulting voltage of 0.38V +/- 0.02V) for 2min, forming a 70% porous PSi layer ~ 0.4μm thick (see Fig. 1b).

After backside aluminum removal, the wafer is annealed in a nitrogen ambient for 20min at 900°C. The oxygen available in the atmospheric furnace system (with 900°C slow push/pull) actually grows ~ 40Å SiO_2 on a bare silicon wafer. This results in oxide-passivation of the silicon nanocrystals, forming OPNSi. The active layer is then capped with 0.25μm polysilicon deposited via low pressure chemical vapor deposition (LPCVD) at 610°C, forming an excellent interface to the active light-emitting material (see Fig. 1b). The polysilicon is then selectively doped n^+ using lithography (cathode area ~ 0.3cm²) and a high-dose low-energy phosphorus ion-implant (10^{15}cm⁻² at 35keV), followed by an activation anneal at 900°C in nitrogen for 15min. The undoped backside polysilicon is removed using an SF_6 reactive-ion etch, and a dilute 50:1 HF etch is then used to remove any oxide formed on the frontside polysilicon during the activation anneal before contact formation. Aluminum contacts are then formed to the n+ poly cathode areas via sputter deposition through a shadow mask, and backside aluminum is deposited to reproduce the backside substrate anode contact.

Electrically isolated integrated OPNSi LEDs

The described LEDs have been combined with a new fabrication technique which enables the formation of electrically isolated integrated LEDs in the single-crystal substrate. This method forms each diode structure in its own individual well, enabling junction-isolated devices without a

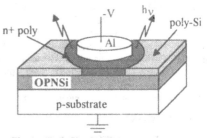

Fig. 1a. Bulk-film OPNSi LED cross-section

Fig. 1b. SEM x100K image of a bulk-film OPNSi LED after a 4sec buffered oxide etch to highlight active layer. Note the interface quality between the polysilicon and OPNSi layers.

22

Fig. 2a. p^+ and n^+ buried layer formation followed by growth of epitaxial silicon. The p^+ layer is used for an ohmic contact to the p-epi for anodization. The n^+ buried layer is used for device isolation.

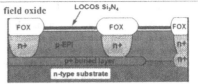

Fig. 2b. Integrated structure following LOCOS, which defines the active regions. The n^+ plug and n^+ buried layer merge to provide junction isolation between LEDs.

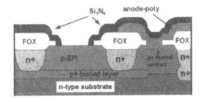

Fig. 2c. Integrated structure following formation of the anode grid, which extends to the wafer edge and provides a global contact for anodization. Nitride provides protection against the etchant during anodization, as well as electrical isolation between the cathode-biased electrolyte and the anode-poly.

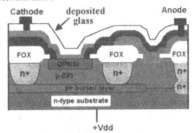

Fig. 2d. Final cross-section of the integrated LED structure. The cathode and anode are labeled for a forward-bias configuration, and can be reversed while maintaining electrical isolation between the substrate and neighboring LEDs.

Fig. 2. Cross-sections illustrating the various process stages during the fabrication of electrically isolated integrated OPNSi LEDs.

common substrate electrode. This structure can be integrated within a CMOS fabrication sequence, however some added process complexity is required due to the use of an epitaxial silicon layer. The process description is divided into several process subgroups (A through E), details of which are described along with corresponding cross-sections shown in Fig. 2.

A) Formation of n^+ & p^+ buried layers

The starting substrates were n-type, (100) orientation, 5-15Ωcm silicon wafers that were 100mm in diameter. Heavily doped buried layers (to be buried beneath lightly doped epitaxial silicon) were then formed using lithography and ion implant. Boron-doped p^+ buried layers were formed to serve as the anode for both anodization and electrical operation of the LED. Phosphorus doped n^+ buried layers were formed to provide junction isolation between neighboring LEDs using an up/down isolation scheme. After a high temperature anneal to activate and diffuse the dopants to the desired profile, the wafers were prepared for epitaxial silicon deposition. Epitaxial silicon with a thickness of 5μm and a resistivity of 7Ωcm was grown at atmospheric pressure using trichlorosilane at a temperature of 1150°C by Moore Technologies, San Jose, CA (see Fig. 2a).

B) LOCOS and up/down isolation

After epitaxy, a LOCalized Oxidation of Silicon (LOCOS) strategy was used to define the active device and field isolation regions. A thin 400Å pad oxide was grown, followed by an LPCVD silicon nitride film with a thickness of 1500Å. Lithography and reactive ion etching (RIE) was done to remove the nitride from the field regions. A high dose phosphorus implant was done to create n^+ surface regions, followed by a high-temperature drive-in / thick (0.8μm) field oxidation. This thermal step pushes the phosphorus n^+ plugs down from the top (under the field oxide), and also moves the n^+ buried layer up (thus the name up/down isolation), merging the two regions for device isolation (see Fig. 2b).

C) Anode formation and isolation for local anodization

A p^+ buried anode contact was formed by patterning and etching the 1000Å oxide, followed by a high-dose boron implant into the open contact regions. The anode-poly was deposited via LPCVD, and subsequently doped using a boron spin-on-glass dopant source at 1050°C. After oxide removal, lithography and RIE were done to pattern the anode-poly which remained as a global anode contact to each p^+ buried layer under the p-epi regions, forming an "anode grid" over the entire wafer surface. Another lithography step was used to pattern and etch the 1000Å oxide in order to form an isolation ring around the LED region. A 1500Å silicon nitride film was then deposited via LPCVD, making direct contact to the silicon in the isolation ring in order to protect the outside regions during anodization in the HF/ethanol mixture. The nitride also protected the anode-poly which needed to be isolated from the cathode-biased electrolyte. The nitride film was patterned and etched (via RIE) to open the LED areas for anodization. Nitride was also removed from the edge region of the wafer in order to expose the p^+ anode-poly. Aluminum was deposited and patterned to form a contact ring extending inward ~ 5mm from the edge of the wafer. A sinter step was then done at 450°C in forming gas (H_2/N_2 mixture) in order to insure an ohmic contact to the anode grid (see Fig. 2c).

D) OPNSi formation

The wafers were anodized in HF/ethanol using a specially designed electrochemical cell by biasing the anode grid using the aluminum wafer-edge contact (note that there is no backside contact). Porous silicon was locally formed in the open LED areas, each device having its own local p^+ buried layer anode contact. The aluminum wafer-edge contact was then removed using a method which protects the porous silicon from coming into contact with any chemicals. A passivation anneal (previously described) was then done, transforming the porous silicon into OPNSi.

E) Electrode formation & interconnects

The cathode-poly was deposited via LPCVD, followed by a phosphorus implant. The cathode-poly was then patterned and etched (via RIE). A lithography step was then done to pattern and protect the anode regions for each individual LED, followed by the removal of the nitride and anode-poly layers in between devices, thus separating the anode electrodes. A 3000Å spin-on oxide was then deposited and cured (also serving as the n^+ implant activation anneal), and contact cuts were patterned and etched to the cathode and anode polysilicon layers. Aluminum was then sputter deposited, patterned and sintered. See Fig. 2d for a complete cross-section of the electrically isolated integrated OPNSi LED structure.

RESULTS AND DISCUSSION

Bulk-film OPNSi LEDs

The analysis of the designed experiments showed some success in identifying significant factors, while others were not able to provide conclusive results due to the fact that several treatment combinations did not yield "working devices". A "working device" is classified as a device that has reasonable transport characteristics and exhibits EL. Earlier experiments had contradicting conclusions on the trends or impact of the factors under study on the electrical or EL responses. However more recent experiments confirmed that the nominal processing conditions (mentioned in the previous section) yielded the best working devices.

It was found that the integrity of the crystalline silicon substrate prior to anodization played an important role on the performance of the LEDs. The purpose of the thick steam oxidation step prior to anodization was to induce a significant amount of stress on the wafer surface, thus injecting silicon interstitials that agglomerate and form extrinsic stacking faults, known as *oxidation induced stacking faults* (OISFs) [3]. The anodization process can highlight dislocation lines and stacking fault defects that intersect the surface of the wafer. These appear as lines or small pits arranged orthogonal along <110> directions on (100) oriented wafers, and were easily observed under an optical microscope. The presence of these defects was determined to have a dominant effect on both the electrical and EL response. It is proposed that these defects result in a lower quality insulating layer, thus assisting transport to silicon

24

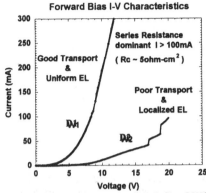

Fig. 3a. I-V Characteristics of bulk-film OPNSi LEDs processed similarly, except D1 had a process step which induced surface stress prior to anodization. The device area is 0.3cm².

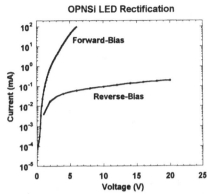

Fig. 3b. Rectifying characteristics of OPNSi LEDs fabricated like D1 shown in fig. 3a. OISFs prior to anodization do not seem to have a negative effect on current blocking in reverse bias.

crystallites within the oxide-passivated material. Although not exactly the same, this finding is quite similar to the results presented in ref. [2], in which the described LED process requires a significant amount of surface disorder prior to anodization. The presence of stacking fault defects has also been determined to cause increased oxide leakage and premature failure of thin gate oxides in MOS devices [4].

Figure 3a shows I-V characteristics of LEDs processed both with and without OISFs. Device D2 (without OISFs) demonstrates "poor transport" behavior, where current flow is inhibited until high bias, resulting in abrupt upshifts in current which crowds in very small regions, evidenced by localized pinpoint light emission. Device D1 (with OISFs) demonstrates "good transport" behavior, where the onset of current flow (linear scale) occurs at bias levels < 5V, and the current density is uniform, evidenced by uniform EL over the entire cathode area. EL threshold conditions are found at a forward bias of 3.5 - 4V, and current density < 100mA/cm² (now easily calculated since transport is uniform). The devices are very stable, withstanding current densities exceeding 2.5A/cm² without degradation. In addition, it is possible to fabricate OPNSi LEDs reproducibly, with excellent within-wafer uniformity of device characteristics.

Integrated Structures

The described integration strategy of electrically isolated OPNSi LEDs proved successful. The anodization process formed local PSi regions, with a 20% non-uniformity in thickness (range/mean) due to resistance of the anode grid (thus, thicker PSi on devices near the wafer edge). This would have not been possible if not for the ohmic anode contact made to each individual LED. The devices exhibited transport and EL much like that of the bulk-film devices. Although an intentional surface-stress step was not introduced in the integrated process, it is believed that the stress induced by the LOCOS isolation technique [5] was sufficient to yield working devices. Small diode array structures were designed and fabricated, with a circuit schematic shown in Fig. 4a and an actual LED array shown in Fig. 4b. Electrical isolation and device rectification (as shown in Fig. 3b) allows each device to be individually accessed by the addressing scheme shown in Fig. 4a, which shows that only one LED is in forward bias and the other LEDs are either in zero bias (same row or column) or reverse bias. Electrical isolation also allows the devices to be voltage-modulated using full-swing forward/reverse bias, as opposed to current-modulation which is required for common-substrate electrode devices. This improvement in the input waveform should significantly increase the modulation frequency capability of these devices [6], because during the turn-off cycle carriers are swept away by a reverse-bias electric field as opposed to lifetime-limited relaxation.

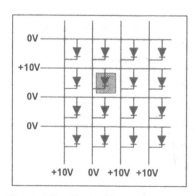

Fig. 4a. Circuit schematic of an LED array, where only one LED is put into forward bias. The other devices are either in zero bias or reverse bias.

Fig. 4b. Small 4x4 OPNSi LED array. The columns connect to the cathodes and the rows connect to the anodes of the devices. Each device has a light-emitting area of ~ 5000μm².

CONCLUSION

This work presents a significant advancement in the status of silicon-based light emitting devices. Experimental results have demonstrated that OPNSi LEDs can be fabricated reproducibly, and operated under reasonable current and voltage levels. It was determined that the fabrication procedure requires the formation of stacking fault defects caused by surface stress and the injection of silicon interstitials. For actual CMOS integration this would have to be performed in a controlled manner to avoid the formation of crystalline defects in the transistor regions. This issue requires further investigation, but could possibly be done using localized thermal stress or silicon implantation. The efficiency of the devices remains low (relative to other light-emitting technologies) and at this time they would not be competitive for display applications. However the ability to make stable light emitters integrated within standard silicon microelectronics without the need for exotic technologies is quite appealing. Additional efforts in process optimization and device integration with other optoelectronic components may eventually lead to several new applications of silicon-based optoelectronics.

ACKNOWLEDGMENTS

The authors wish to acknowledge David Sieloff, Kitty Corbett, Timothy Nguyen, Michael Tiner, Sri Samavedam and Michael Schippers for the materials characterization work done at the Mororola Advanced Products Research and Development Laboratory in Austin, Texas. Thanks to RIT undergraduate students Jason Benz and Martin Szwarc for assistance in device fabrication and testing. Special thanks to Dr. Lynn F. Fuller, Director of the RIT Center for Microelectronic Engineering, for use of the integrated circuit fabrication facilities, Diane Potter for her photographic work, and Jim Greanier for his machine-shop work. This research was supported in part by the Army Research Office.

REFERENCES

1. K.D. Hirschman, L. Tsybeskov, S.P. Duttagupta and P.M. Fauchet, *Nature* **384**, 338 (1996)
2. A. Loni, A.J. Simons, T.I. Cox, P.D.J. Calcott and L.T. Canham, *Electronics Lett.* **31**, 1288 (1995)
3. S. Wolf and R.N. Tauber, *Silicon Processing for the VLSI Era Vol. 1: Process Technology,* (Lattice Press, Sunset Beach, CA, 1986) pp. 45-46.
4. M. Liehr, G.B. Bronner and J.E. Lewis, *Appl. Phys. Lett.* **52**, 1892 (1988)
5. S. Wolf, *Silicon Processing for the VLSI Era Vol. 2: Process Integration*, (Lattice Press, Sunset Beach, CA, 1990) pp. 17-45.
6. A.J. Simons, T.I. Cox, A. Loni, P.D.J. Calcott, M.F. Uren and L.T. Canham, *Mat. Res. Soc. Symp. Proc.* **452**, 693, (1997)

PHOTOLUMINESCENCE FROM SINGLE POROUS SILICON CHROMOPHORES

M.D. Mason, G.M. Credo, K.D. Weston and S.K. Buratto
Department of Chemistry, University of California, Santa Barbara, CA 93106-9510
buratto@chem.ucsb.edu

ABSTRACT

We spatially isolate and detect the luminescence from individual porous Si nanoparticles at room temperature. Our experiments show a variety of phenomena not previously observed in the emission from porous Si including a distribution of emission wavelengths, resolved vibronic structure, random spectral wandering, luminescence intermittency and irreversible photobleaching. Our results indicate that the emission from porous Si nanoparticles originates from excitons in quantum confined Si, strongly influenced by the surface passivating layer of the Si nanocrystal.

INTRODUCTION

Anodic etching of Si wafers in aqueous HF has stimulated tremendous interest over the past several years due to the visible light emission from the porous Si formed.[1-4] Despite the wide variety of spectroscopic techniques (absorption, luminescence, Raman and infrared spectroscopies) applied to porous Si samples, a detailed understanding of the photoluminescence has yet to be reached. [3-7] A detailed description of the emission from porous Si is difficult from the current data because conventional spectroscopic techniques probe too large of a volume for the highly heterogeneous porous Si samples. [7-12] The poor spatial selectivity of the spectroscopy techniques results in data which is spatially-averaged, containing signals from a wide variety environments in the porous Si including different sizes of Si particles and different surface chemistry. [4] In the experiments described here, we spatially isolate and detect emission from single porous Si nanoparticles and simultaneously measure the particle size. [13] We observe a variety of phenomena not previously observed in the luminescence from porous Si including a distribution of luminescence wavelengths, resolved emission peaks, discrete jumps in intensity, random spectral wandering, and irreversible photobleaching. [14] We attribute the signals observed in our experiments to be emission from quantum-confined Si that is strongly influenced by the oxide species that passivate the surface.

EXPERIMENT AND RESULTS

Our experimental approach combines the techniques of single particle spectroscopy [15,16] and shear force microscopy, [17] an analog to attractive mode atomic force microscopy (AFM).

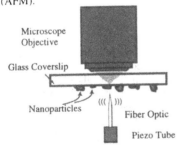

Figure 1. Experimental setup (not drawn to scale). The particle-covered glass slide is mounted upside down. Confocal images are obtained by collecting the luminescence resulting from scanning the sample with laser light focused by a microscope objective. Topography images are obtained by scanning the sample over the shear force fiber optic tip.

A diagram of our experimental configuration is shown in Figure 1. Samples of well-separated porous Si nanoparticles were prepared by spin casting a 5 μL aliquot of a dilute colloidal suspension of porous Si nanoparticles (~ 1 nM) onto a glass coverslip. Colloidal porous Si samples were prepared from bulk porous Si using the method of Heinrich, et al.[18] In our experiments, p-type Si was anodically etched in a 1:4 by volume solution of hydrofluoric acid (49%) and ethanol for 30 minutes at 20 mA/cm^2. The resulting porous silicon layer was mechanically removed from the surface and sonicated in hexanes to reduce particle size. The sonicated solution was filtered through a 200 nm syringe filter to yield a stock solution. We note that the porous Si nanoparticles do not dissolve well in hexanes and only the smallest particles are suspended in the stock solution. In all of our experiments we see very few particles with size greater than 20 nm.

Emission from single nanoparticles was imaged in the far field using a laser scanning confocal microscope with a high numerical aperture oil-immersion objective (1.3 NA) described in detail elsewhere.[19] For all experiments, either the 488 nm or 514 nm line of an Ar$^+$ laser was used as the excitation source. The excitation spot size, focused on the sample side opposite the immersion oil, was approximately 250 nm in diameter. Using a beam-splitter in the path of the collected fluorescence signal, we were able to acquire emission intensity images and emission spectra simultaneously. Figure 2A shows a photoluminescence image (15 x 15 μm^2) of one of our samples. Each 250 nm bright spot indicates an emitting nanoparticle. It is important to note that the size of bright spot represents the size of the illumination spot and not the size of the nanoparticle. The average density of a series of images similar to Fig. 2A was found to be 1.8 particles/100 μm^2.

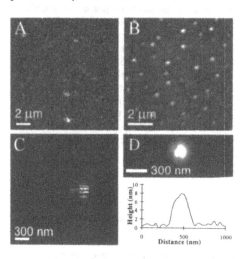

Figure 2. Representative images of porous Si nanoparticles deposited on silica. The fluorescence (LSCFM) in (A) shows three spatially isolated fluorophores on a linear gray scale with white being 15,000 cts/s. Part (B) is a shear-force topography image of the same sample as shown in a illustrating the actual number of nanoparticles dispersed on the sample. Part (C) is the fluorescence from a single nanoparticle exhibiting rapid on/off "blinking" behavior which is observed for ~20 % of nanoparticles investigated. Part (D) shows a topographic image and accompanying shear-force line trace of a single nanoparticle.

The size of each nanoparticle was determined using shear force microscopy, as depicted schematically in Fig. 1. In shear force microscopy,[20] a tapered optical fiber tip (diameter approximately 200 nm) was used to scan the surface of the nanoparticles on glass. This tip was attached to a small piezoelectric tube and dithered on resonance. The tip-sample distance was determined by monitoring the dither amplitude as the sample approached the tip. The dither amplitude was measured by scattered laser light synchronously with the dither frequency, and provided the input for the feedback loop of our scanning electronics which was set to maintain a constant height above the sample surface (approximately 10 nm). Figure 2B shows a shear force microscopy image (8 x 8 µm²) of the same sample. A bright spot in the topography image indicates a nanoparticle. We note that a blank glass sample is flat on this same height scale. As is apparent, there are many more nanoparticles present than are emitting. A series of images similar to Fig. 2B results in an average density of 65 particles/100 µm². Another important result of the data of Fig. 2 is that even though our samples are covered by a high density of particles (65 particles/100 µm²), only 2.8% of the porous Si nanoparticles are luminescent.

As is the case with all AFM techniques, the observed image is a convolution of the tip shape and the shape of the particle. If the size of the nanoparticle is much smaller than the tip itself then the nanoparticle images the tip rather than the reverse, and all features appear the same size in the lateral dimensions (as is the case in Fig. 2B). If we assume, however, that each nanoparticle is roughly spherical then the height of each feature in Fig. 2B is a much more accurate representation of the particle size (see Fig. 2D). [21] Using this measure, all of the nanoparticles in Fig. 2b are between 5 and 15 nm.

During imaging it was observed that the fluorescence emission of many of the Si nanoparticles appeared to blink "on" and "off" during the course of a scan (as illustrated in the image of Fig. 2C). The blinking behavior was examined more closely by positioning the excitation beam over a single particle and collecting the emission intensity vs. time (intensity time course). In addition to blinking "on" and "off," many of the nanoparticles also emitted at discrete intensity levels as illustrated in Figure 3. In Fig. 3A there are four distinct intensity levels (after background subtraction). We attribute this behavior to emission from a combination of three chromophores with the above intensity levels corresponding to emission from 3, 2, 1, or no chromophores.

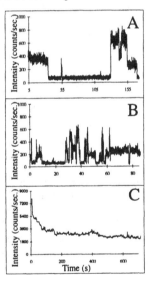

Figure 3. Emission intensity vs. time data for three porous Si particles. Parts (A) and (B) show discrete intensity jumps in the emission from two different ~10 nm nanoparticles. Part (C) is from a large particle (> 500 nm) which exhibits no intensity fluctuations.

The time course of Fig. 3B shows more rapid blinking. In general, we observe a decrease in the "on" times with increasing excitation intensity. This suggests that the observed blinking behavior is light induced. The time course of Fig. 3c is from a very large (> 500 nm) particle. No blinking is observed for large particles as expected for an ensemble of chromophores. The gradual decrease in fluorescence intensity observed in Fig. 3C is due to an irreversible photo-oxidation of the particle similar to that which occurs in bulk porous Si. [4]

The off periods in the time courses of Figs. 3A and 3B indicates a long-lived dark state, much longer than is expected for a surface trapped state or a triplet state both of which are expected to decay on the 1 – 100 ms time scales. The blinking behavior observed here is indicative of a charge-transfer state similar to that observed in the room temperature luminescence from single CdSe quantum dots. [15,16] This state is characterized by a charge existing in the surface passivating layer produced by Auger ionization and the remaining carrier existing in the quantum dot. Energy transfer to this "free carrier" provides an efficient non-radiative path for this state making it dark. An important consequence of this model is that it implies a strong coupling between excitons and the surface of the quantum dot. [22-24]

Further influence of the surface passivating layer is observed in the photoluminescence spectra from individual nanoparticles as shown in Figure 4.

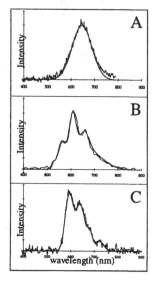

Figure 4. Room temperature emission spectra of three ~10 nm porous Si nanoparticles. Spectrum (A) shows no resolved structure, fits to a single gaussian lineshape and is representative of ~50% of the nanoparticles investigated. Spectra (B) and (C) each show clear structure and fit to the sum of 4 gaussians. The splittings between the gaussians are all around 150 meV and are attributed to vibronic coupling to Si-O-Si groups in the surface passivating layer of the quantum confined Si chromophore.

The broad lineshape observed in Fig. 4A (similar to the lineshape of bulk porous Si) is indicative of only about 50% of the Si nanoparticles studied. The remaining nanoparticles exhibit much more interesting emission spectra with narrow lines and resolved fine structure as illustrated by the spectra in Figs. 5b and 5c. The spectra of Figs. 5B and 5C each show four resolved peaks and fit well to the sum of four gaussians. Each gaussian has a FWHM of approximately 115 meV and the splitting between adjacent peaks is 160 meV (1300 cm^{-1}). In general λ_{max} is different for each nanoparticle and spans a range of over 100 nm for the nanoparticles observed. We attribute the difference in λ_{max} to difference in the size of the chromophore (or chromophores) in each nanoparticle which is representative of the quantum size effect. It is important to note that we do not obtain an accurate size of the emitting species, only the size of the entire particle.

The structure observed in the spectra in Figs. 4B and 4C is reminiscent of vibronic structure in molecular fluorescence. The size of the splitting, however, makes it unlikely that this fine structure is due to coupling to phonon modes in Si which are expected to be much smaller (around 55 meV).[4] The only candidate in the appropriate frequency range for vibronic coupling are Si-O-Si stretching modes which we observe in the 1100 - 1400 cm^{-1} range in the parent bulk porous Si. [24] Such modes exist only in the surface passivating layer of the Si chromophore.

We have also observed dynamics in the emission spectra of single porous Si nanoparticles. Figure 5 shows a series of spectra acquired sequentially over a period 10 min (60 s per spectrum). The peak wavekength varies over 5% with slight changes in the FWHM from spectrum to spectrum. A correlation between intensity fluctuations and the photoluminescence spectrum is also observed as seen in Figure 6. The intensities of the spectra in Fig. 6A, 6B and 6C are shown on the same scale while that of Fig. 6D is multiplied by a factor of 10 to help illustrate the large spectral shift.

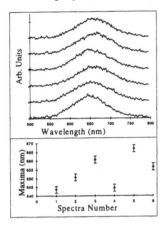

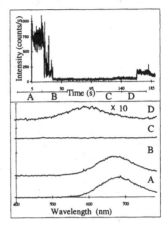

Figure 5. Sequential emission spectra and corresponding peak maxima. The emission intensity versus wavelength spectra in (A) were acquired sequentially over 60 second integration periods for a <20 nm porous silicon nanoparticle.

Figure 6. Emission intensity vs. time data and corresponding fluorescence spectra of a single porous Si particle (<50 nm). Spectra were collected sequentially using 30-second integration periods.

A probable explanation for the behavior demonstrated in Fig. 4 is that there are at least three chromophores contributing to the total emission of the nanoparticle; one weak emitter centered at $\lambda_{max} \approx 575$ nm (see Fig. 4D), and two stronger ones emitting near $\lambda_{max} \approx 650$ nm and $\lambda_{max} \approx 680$ nm (Figs. 4B and 4A), respectively. During period (A) all are emitting, during period (B) the spectrum is dominated by the lower energy chromophores, during period (C) none of the chromophores are emitting and during period (D) only the weakest (and highest energy) chromophore is emitting.

CONCLUSIONS

The observed blinking behavior and emission fine structure are indicative of emission from only a small number of emitting species. Thus, it is highly unlikely that emission from surface species such as siloxene (SiO_xH_y) could account for these phenomena due to the large number (>> 100) of such species present on the surface of a 10 nm particle. Thus the model for the luminescence of porous Si nanoparticles consistent with the results of our experiments is one in which the emission is strongly influenced by the surface of the Si quantum dot. Excitons in the Si quantum dot are further confined near the surface of the quantum dot and emission from these excitons is strongly coupled to vibrations in the surface passivating layer. This surface confinement also contributes to the blinking behavior by providing a means for transferring energy non-radiatively as discussed earlier in the text.

Our results also imply that controlled modification of the surface by species other than oxygen should produce dramatic differences in the emission yield, the blinking behavior and the observed vibronic structure. The characterization of these emission parameters from single nanoparticles with different surface terminating groups should provide important new insight into the role of the surface on the luminescence of porous Si and possibly lead to important breakthroughs in the application of porous Si in optical and opto-electronic devices.

ACKNOWLEDGEMENTS

This work is supported by the David and Lucile Packard Foundation (Packard Fellowship) and NSF (#CHE-9501773). G.M.C. acknowledges funding through a UCSB Graduate Opportunity Fellowship.

REFERENCES

1. L.T. Canham, *Appl. Phys. Lett.* **57**, 1046 (1990).
2. V. Lehman, U. Gösele, *Appl. Phys. Lett.* **58**, 865 (1991).
3. R.T. Collins, P.M. Fauchet, M.A. Tischler, *Physics Today* **50**, 24 (1997).
4. A thorough review of previous spectroscopy experiments and the current understanding of porous Si luminescence can be found in a very recent review article: A.G. Cullis, L.T. Canham, P.D.J. Calcott, *J. Appl. Phys.* **82**, 909 (1997).
5. S.M. Prokes, *J. Appl. Phys.* **73**, 407 (1993).
6. M.S. Hybertsen, *Phys. Rev. Lett.* **72**, 1514 (1994).
7. L. E. Brus, *J. Phys. Chem.* **98**, 3575 (1994).
8. W. L. Wilson, P. F. Szajowski, and L. E. Brus, *Science* **262**, 1242 (1993).
9. A. L. Efros, M. Rosen, B. Averboukh, D. Kovalev, M. Ben-Chorin, and F. Koch, *Phys. Rev. B* **56**, 3875 (1997).
10. F. Muller, et al., *J. Lumin.* **57**, 283 (1993).
11. J. C. Vial, et al., *IEEE Trans. Nuc. Sci.* **39**, 563 (1992).
12. P. Dumas, et al., *J. Vac. Sci. Technol. B* **12**, 2064 (1994).
13. G. M. Credo, M.D. Mason, S.K. Buratto, submitted.
14. M.D. Mason, G.M. Credo, K.D. Weston, *Phys. Rev. Lett.* **80**, 5405 (1998).
15. M. Nirmal et al., *Nature* **383**, 802 (1996).
16. S.A. Empedocles, D.J. Norris, M.G. Bawendi, *Phys. Rev. Lett.* **77**, 3873 (1996).
17. R.J. Cook, H.J. Kimble, *Phys. Rev. Lett.* **54**, 1023 (1985).
18. J.L. Heinrich, C.L. Curtis, G.M. Credo, K.L. Kavanagh, M.J. Sailor, *Science* **255**, 66 (1992).
19. K.D. Weston, S.K. Buratto, *J. Phys. Chem. A* **102**, 3635 (1998).
20. The shear force technique described in E. Betzig, P.L. Finn, J.S. Weiner, *Appl. Phys. Lett.* **60**, 2484 (1992) is commonly used as the distance regulation in near-field scanning optical microscopy (NSOM).
21. D. Sarid, *Scanning Force Microscopy : With Applications To Electric, Magnetic, And Atomic Forces* (Oxford University Press, New York 1991).
22. M.J. Sailor, E.J. Lee, *Adv. Mater.* **9**, 783 (1997).
23. L.E. Brus *Phys. Rev. B* **53**, 4649 (1996).
24. V.M. Dubin, F. Osanam, J.-N. Chazalviel, *Phys. Rev. B* **50**, 14867 (1994).

AUGER EFFECT SEEN IN THE POROUS SILICON FAST LUMINESCENT BAND

R. M'ghaïeth[1], I. Mihalcescu[2], H. Maâref[1], J.C. Vial[2]
[1]Laboratoire de Physique des Semiconducteurs, Faculté des Sciences de Monastir, Route de l'Environnement, 5000 Monastir, Tunisia.
[2]Laboratoire de Spectrométrie Physique, Université Joseph Fourier-Grenoble I, BP 87, 38402 St Martin d'Hères Cedex, France; Irina.Mihalcescu@ujf-grenoble.fr

ABSTRACT

Time resolved photoluminescence (PL) measurements are performed on oxidized and fresh porous silicon at room temperature. Comparing the evolution of the nanosecond time delayed PL in both cases, a new feature of the PL spectra is identified: the fast-red band, present as well in fresh or aged samples. The nonlinear excitation intensity dependence of this component is described by a simple model where, the Auger effect inside isolated silicon nanocrystallites plays the dominant role.

INTRODUCTION

Porous silicon has been intensively studied last years, for its capacity to be taken as a model system for nano-sized crystalline indirect semiconductor, interacting with an enormous internal surface. Its red (centered in general around 700nm) and slow (in microsecond time scale) band photoluminescence (PL) has been carefully characterized and various experimental results and models associate this luminescence to the internal crystalline core of the porous silicon structure. By studying the saturation[1] of this luminescence, the polarization memory[2], the electroluminescence tunability or the "holeburning" in the PL spectra[3], the weight, in silicon nanocrystallite (NC), of Auger recombination, was emphasized.

The aim of this paper is to present new experimental results that can be interpreted in the frame of Auger effect in silicon NC's, namely the nanosecond component of the PL decay. Time resolved photoluminescence are well suited for a direct proof of the Auger recombination: as the Auger effect in silicon nanocrystallites takes place only in minimum twice excited crystallites, we expect the Auger component amplitude to be a non-linear function of the excitation intensity. Theoretical calculations[4] predict an Auger recombination time in nanosecond range easy to distinguish from the usual decay time, which ranges in 10 μs scale. Unfortunately, a fast component (<10ns) was already found, in general (but not exclusively) for oxidized porous silicon. The origin of this blue-fast band, still debated, was attributed either to intrinsic direct recombination in small particles[5] or associated to the presence of oxide[6,7] and/or hydrocarbon pollution of this oxide[8].

To avoid a possible confusion, we previously investigated the difference between the fast luminescence of oxidized and fresh porous layer by comparing the nanosecond evolution of the delayed PL in both cases. In this way we found a new feature of the porous silicon luminescence: a nanosecond-red band, present in both fresh and oxidized (aged) samples. By studying then the dependence of the amplitude of this band on the excitation intensity we will identify it as the Auger component.

EXPERIMENTAL SETUP

Two type of porous silicon layers are studied, both anodized in the same conditions: starting from (100) p-type silicon wafer (4-6Ωcm), in 15% ethanoic HF solution with 10 mA/cm^2 current density in the darkness, for 1-2 minutes. One of the samples is studied just after the preparation and the other one was 1 year aged and anodically oxidized. In this way samples are

0.5-0.75 µm thick and have roughly 80 % of porosity.

The steady state PL was excited with the 363nm ray of a Hg lamp ($\cong$2mW/cm^2) and the time resolved PL, with the third harmonic of a Nd3+:YAG laser (355nm) with $\cong$ 5ns-pulse width and $\cong$ 5mJ-pulse energy. At the sample surface, the laser beam was focused on 0.1cm^2 surface, obtaining a maximal excitation energy density of about 15mJ/cm^2. As the excitation beam has somewhat gaussian profile in intensity, we selected only the central PL with a 0.02cm^2 pinhole in the front of the monochromator (Jobin-Yvon H20). Detection was done with a Hamamatsu R3809U-51 Microchannel Plate PM tube (response time 0.1ns) followed by Lecroy digital oscilloscope (400Mhz bandwidth). We estimate our time resolution (excluding laser jitter) at about 0.5ns. For the delayed PL, the oscilloscope was replaced with a sample&hold boxcar (5ns gate).

RED NANOSECOND BAND

Figure 1 compares, for fresh (a) and aged sample(c), the PL spectra changes (decay) in the first 50ns following the pulsed excitation. As can be seen, the fresh sample spectra are gaussian-like centered around 610nm. The decay seems to be roughly homogeneous, meaning that the PL intensity decreases with the delay time without wavelength dependence, the PL peak having only a small red-shift of about 10nm (fig. 1b). On the contrary, the (oxidized)aged sample

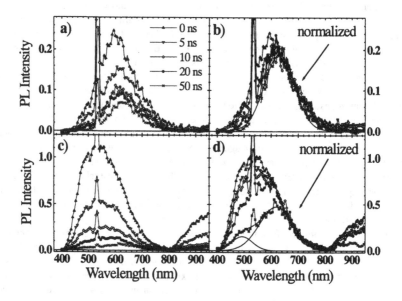

Figure 1: PL spectra for fresh (a) and oxidized sample (b) acquired at different delays after the excitation pulse : 0ns (▲), 5ns(▽), 10ns(◊), 20ns(●) and 50ns(∗). The same spectra are represented, respectively, in b) and d), normalized on the red side of the spectra. This type of representation emphasizes the existence of two components, one centered at $\cong$490nm, specific to oxidized sample, faster than the second one, centered at $\cong$610nm, common to both samples.

spectra are not gaussian, and the spectra shape changes with the delay time. In figure 1d we present the same spectra normalized on the red side of the spectra. In this way , one can easily identify , two spectral components, with different time evolution. The one centered at about 490nm, has a faster decay than the second one, centered at about 610nm, which has identical shape and delay time evolution to the fast component of the fresh sample. An example of gaussian fit of the PL spectrum for a fresh sample is presented in figure 1b. The corresponding decomposition of the PL shape for an aged sample in 2 gaussian lines is given in fig.1d.

The first fast-band, present only for the oxidized samples, centered around 490nm, has been already identified[5,7,9] and carefully characterized. It has been shown that this high-energy band has an independent evolution compared to the slow red-band, as a function of excitation wavelength or temperature. S.Komura and coworkers[10], have attributed this band to the radiative recombination via fixed energy levels as a result of oxygen-induced defects and Harris et al[11] have measured the lifetime of this component to about 0.9ns.

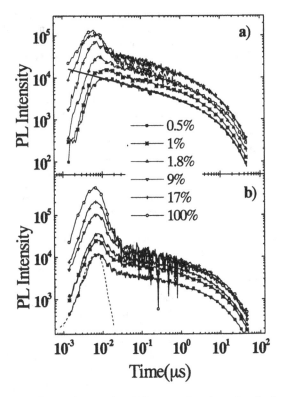

On the contrary, the second fast band, centered at about 600nm, that we identified in both samples (fresh or not) is present at only high excitation intensities and seems to be related to the slow red PL. In order to clarify the relation between both red bands, slow and fast, we studied their relative change, as a function of excitation intensity.

EXCITATION INTENSITY DEPENDENCE - AUGER SIGNATURE

Figure 2 show the modification of the PL decay, at 600nm, as a function of excitation intensity for the fresh (upper part of the figure) and oxidized (lower part) sample. We acquired the entire decay from nanosecond scale up to 100µs.

One can easily identify the two components of the PL decay :

1) the slow component, becomes predominant after 40ns following the excitation pulse. It is described by a stretched exponential function:

$$I_{PL}(t) = I_0 t^{(\beta-1)} \exp\left(-\frac{t}{\tau}\right)^{\beta}$$

Figure 2 : PL decay at $\lambda = 600$nm at a function of excitation intensity, for fresh sample (a) and oxidized sample (b). The excitation ($\lambda = 355$nm) is given in % of the maximal excitation intensity. The pulse shape of the excitation is given in b) by a dashed line. The solid line in a) is an example of a stretched exponential fit.

where I_0, β and τ are the fitting parameters (fig.2a shows an example of this type of fit). With the increase of the excitation intensity the decay keeps the same shape (β and τ stay constant), but the amplitude I_0 increases slightly, saturates and, for the highest excitation intensities, it starts to decrease. For both samples, the absolute values of the slow component amplitude are comparable, slightly lower for the oxidized sample.

2) the fast component is in the nanosecond region. It follows the excitation pulse rise and decays with a fall-time slower than that of the laser pulse, nearly 2.5ns. One may notice that aged sample have a fast component 3 times higher than the fresh one.

In order to have a more quantitative analysis of both component variation, we have roughly separated the two components: slow decay was fitted with a stretched exponential function, where parameters τ and β were kept constant, only the amplitude I_0 varied from a decay to another. Then, the fast component was obtained by subtraction of this fit from the entire PL.

Figure 3 represents the final result, the integrated area ($I = \int I_{PL}(t)dt$) of the slow component and of the fast component, for both samples. Note that the integrated area represents the total number of photons emitted by the fast or by the slow component.

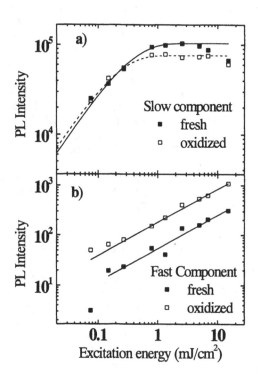

The slow component saturation (fig. 3a) with the excitation intensity has been already established[1,2,3] and attributed to the Auger recombination. Indeed, the Auger effect is a three-carrier interaction where an electron-hole pair (*e-h*) transfers its energy to a third nearby carrier. In silicon nanocrystallites the Auger recombination lifetime is fast, nanosecond[4] or less, and act as soon as two *e-h* pairs are excited inside the nanocrystallite. Then increasing the excitation intensity for more than one *e-h* pair per crystallite, all excess pairs will be lost in fast Auger recombination, only the last remaining pair will contribute to the slow component. The dependence of the slow component on the excitation intensity has been described by a saturation law, $I \propto (1 - \exp(-\sigma I_{ex}))$, integrated over the all layer thickness[12]. Here σ is the absorption cross section at the excitation wavelength (355nm) and I_{ex} the excitation energy (in photons/cm²). An example of this type of fit is

Figure3 : The integrated area, $I = \int I_{PL}(t)dt$, of the slow component (a) and fast component (b) for fresh (■) and oxidized sample (□), as a function of the excitation energy. The lines in figure a) represents the best fit with a saturation-like law (fit detail in text) for both samples, and in b) a power-like fit with an exponent 0.67.

given in Figure 3a for both samples, oxidized or not. The fitting parameters obtained are realistic, $\sigma \cong 0.6 \cdot 10^{-14}$ cm^2, $\alpha \cong 2.1 \cdot 10^4$ cm^{-1} for the fresh sample and $\alpha \cong 1.7 \cdot 10^4$ cm^{-1} for the oxidized one, in agreement with those obtained elsewhere[1].

The fast component is a power-like function of the excitation intensity. We will show that this dependence with an exponent of 2/3, is the signature of the Auger recombination. In confined semiconductors, Auger recombination has new behavior due to the discreteness of the excitation. When considering an optical excitation, a quantum dot can be either singly excited (and thus, no Auger effect has to be considered) or it can be doubly excited in which case an Auger effect opens a new route for the relaxation. If the excitation intensity remains below this limit, the intensity of the Auger component should be proportional to the square of excitation intensity. But for higher excitation intensities we start to have an analogue situation to the bulk semiconductors, for very high carrier concentration, where the average distance between carriers becomes less than the crystallite size. By doing an estimation of the total e-h pairs excited inside a silicon crystallite during the excitation pulse (see endnote 13), one can show that in our experiment we merely are in this last case.

The carrier density dynamics, in one crystallite, is thus roughly representative forof all porous layers, naturally, in the fast nanosecond regime. If N is the density of e-h pairs in one crystallite their relaxation will by given[14] by : $\dfrac{dN}{dt} = \sigma I_{ex}(t) - CN^3 - BN^2 - AN$, where the first term on the right defines the excitation, the second one, the Auger recombination, the third one, the nongeminate radiative recombination and the last one, the monomolecular recombination (radiative or not). As the fast component quantum yield is very low, one can safely neglect the bimolecular term. We also neglect the last term, which represents the slow component. The excitation pulse is larger than the fast component lifetime, meaning that in our experiment the fast component is roughly in a steady state regime. One obtains then a dependence of the average number of e-h pairs as a function of excitation intensity: $N = K(\sigma / C)^{1/3} * I_{ex}^{1/3}$, where K is a constant dependent only on the layer thickness and the absorption coefficient[15]. The number of emitted photons, in the fast component will be given by: $I_{fast} = BN^2 = BK(\sigma / C)^{2/3} * I_{ex}^{2/3}$. The power-like law dependence, obtained in this way, fits very well with what we have found experimentally (fig. 3b). This lead us to conclude that the red fast component dynamics is dominated by the Auger recombination.

The absolute value of the fast component intensity is then directly related to the ratio $B / C^{2/3}$, as an equivalent ratio of radiative processes (expressed by the bimolecular coefficient B) and the non-radiative ones (given by $C^{2/3}$, with C the Auger coefficient). A direct estimation starting from the experimental data is difficult to obtain, a very crude estimation gives $B / C^{2/3} \cong 2.5 \cdot 10^3$ cm^{-1}s$^{1/3}$ for the fresh layer and $8 \cdot 10^3$ cm^{-1}s$^{-1/3}$ for the oxidized one. Note that for bulk silicon one have something like $5 \cdot 10^5$ cm^{-1}s$^{-1/3}$. The last subject we must discuss is the absence of Auger rate dependence on the wavelength, expressed in our case by homogeneous-like fast PL spectra (fig.1). If, as is usually assumed, the photon energy represents directly the carrier confinement energy in nanocrystallites, the Auger rate does not have noticeable size dependence. We can only suggest that the absence of dependence can be due to the limited time resolution of our experimental setup.

CONCLUSION

For a long time, it has been unclear, if the slow and fast PL components are related or not to an emission coming from the quantum porous silicon. This paper identifies a nanosecond red band that together with the slow red emission are intrinsic. Here we show clearly that both components come from the same dots at different level of excitation, the Auger effect being the cause of the acceleration of the relaxation when the dots are highly excited.

REFERENCES

[1]. I. Mihalcescu, J.C. Vial, A. Bsiesy, F. Muller, R. Romestain, E. Martin, C. Delerue, M. Lannoo, and G. Allan, Phys.Rev.B **51**, 17 605 (1995)

[2]. Al. L. Efros, M. Rosen, B. Averboukh, D. Kovalev, M. Ben-Chorin, and F. Koch, Phys.Rev.B **56**, 3875, (1997)

[3]. D. Kovalev, H.Heckler, B. Averboukh, M. Ben-Chorin, M.Schwartzkopff and F. Koch, Phys.Rev.B **57**, 3741,(1998)

[4]. C. Delerue, M. Lannoo, G. Allan, E. Martin, I. Mihalcescu, J.C. Vial, R. Romestain, F. Muller, and A. Bsiesy, Phys.Rev.Lett. **75**, 2228 (1995)

[5]. Y. Kanemitsu, Phys.Rev.B **49**, 16845 (1994); P. Maly, F. Trojanek, and J. Kudrna, A.Hospodkova, S.Banas, V.Kohlova, J.Valenta, I.Pelant, Phys.Rev.B **54**, 7929 (1996).

[6]. S. M. Prokes, Appl.Phys.Lett. **62**, 3244 (1993); A.G. Cullis, L.T. Canham, G. M. Williams, P.W. Smith, and O.D. Dosser, J.Appl.Phys. **75**, 493 (1994); H. Tamura, M. Ruckschloss, T. Wirschem, and S. Verpeck, Appl.Phys.Lett. **65**, 1537 (1994);

[7]. D.I. Kovalev, I. D. Yaroshelskii, T. Muschik, V.Petrova-Koch, and F. Koch, Appl.Phys.Lett. **64**, 214 (1994)

[8]. L.T. Canham, A. Loni, P.D.J. Calcott, A.J. Simsons, C. Reeves, M.R. Houlton, J.P. Newey, K.J. Nash, T.I. Cox, Thin Solid Films, **276**, 112 (1996)

[9]. P. M. Fauchet, Phys.Stat.Sol. (b) **190**, 53 (1995)

[10]. S. Komuro, T. Kato, T. Morikawa, P. O'Keeffe, Y. Aoyagi, J.Appl.Phys. **80**, 1749 (1996)

[11]. C.I. Harris, M. Syvajarvi, J.P. Bergman, O. Kordina, A. Henry, B. Monemar and E. Janzen, Appl.Phys.Lett. **65**, 2451 (1994)

[12]. The depth dependence of the excitation intensity, due to the layer absorption at the excitation wavelength was considered: $I_{ex}(z) = I_{ex} \exp(-\alpha z)$ with α the absorption coefficient.

The intensity of luminescence becomes: $I \propto \int_{0}^{l} (1 - \exp(-\sigma I_{ex}(z))) dz$ with l, the layer thickness.

[13]. For a maximum pulse energy of about 10mJ/cm^2 one has $2 \cdot 10^{16}$ cm^{-2} photons which will be absorbed in the porous layer, in a proportion of $(1-\exp(-\alpha l)) \cong 0.8$ (e.g. for the fresh layer). Taking an average density of the silicon crystallites of about 10^{18} cm^{-3}, one obtains an average number of excited e^--h pair/crystallite of $\cong 200$. This corresponds to the highest excitation intensity, meaning that in our experiment the average number of e^--h pairs excited inside a crystallite during the laser pulse, is scanned from 1 to 200.

[14]. M. Ghanassi, M.C. Schanne-Klein, F.Hache, A.I. Ekimov, D.Ricard and C. Flytzanis, Appl.Phys.Lett. **62**, 78 (1993)

[15]. K = $1.5/\alpha^*(1-\exp(-2\alpha l/3)) \cong 0.3$cm for the fresh sample and K$\cong$0.25cm for the oxidized one.

DEFECTS AND PHONON ASSISTED OPTICAL TRANSITIONS IN Si NANOCRYSTALS

G. ALLAN, C. DELERUE and M. LANNOO
Institut d'Electronique et de Microélectronique du Nord, Département Institut Supérieur d'Electronique du Nord, BP 69, 59652 Villeneuve d'Ascq Cedex, France, delerue@isen.fr

ABSTRACT

Phonon-assisted and zero-phonon radiative transitions in nanoscale silicon quantum dots are studied using a new approach which combines a full calculation of the confined electronic eigenstates and vibration modes. We predict that the confinement, combined with the indirect bandgap of bulk silicon, must have several important consequences on the luminescence of a single silicon dot: i) a large broadening of the peaks, in the range of 10s of meV for a 3 nm dot, in spite of the atomic-like electronic structure of the dot ii) a great sensitivity of the spectrum to the size and the shape of the dot. We obtain that phonon-assisted transitions always dominate, even for size below 2 nm. Finally, we show that the radiative recombination in presence of an oxygen related surface defect (Si=O) is also assisted by optical phonons.

INTRODUCTION

Recent advances in nanofabrication technologies have enabled to fabricate semiconductor quantum dots (QDs) in which electrons are three-dimensionally confined. These QDs are often referred to as artificial atoms since their electronic properties are predicted to have discrete, atomic-like energy levels [1,2]. The spectroscopic studies have mainly concerned ensembles of QDs where structural and environmental inhomogeneities result in a loss of spectral information. This has initiated an effort to study individual QDs [3]. When the QDs are made of direct semiconductors, their photoluminescence (PL) is mainly characterized at low temperature by sharp spikes with linewidth lower than 100s of μeV, confirming the atomic-like nature of the emitting state [3]. For QDs made of silicon, an indirect semiconductor, no such study exists in spite of the recent burst of activity on the luminescence of nanocrystalline silicon [4]. In that case a completely different behavior is expected because in bulk silicon the optical transitions mainly proceed near the fundamental gap with assistance of phonons, with the transverse optic (TO) modes dominant and smaller contributions from the transverse acoustic (TA) modes. In nanocristalline silicon, there are now some experimental [5,6] and theoretical [7] studies showing that transitions remain mainly assisted by phonons in spite of the confinement which partially breaks the selection rules [8]. But the experiments [5,6] have only been performed on ensembles of QDs so that a number of phenomena are hidden in the inhomogeneous linewidth. On the other side, the theoretical study of Ref. [7] is based on the effective-mass approximation (EMA) and neglects the dispersion of the phonon frequencies involved in the transitions. These approximations are likely to break down in the limit of small QDs [9]. Therefore we present a new theoretical approach to study zero-phonon and phonon-assisted optical transitions in QDs using a full calculation of their electronic and vibronic structures. We show that in contrast to direct semiconductors the PL spectra of individual silicon QDs smaller than ~ 3 nm should be dominated by broad lines (~ 10s of meV) due to the coupling to many vibration modes because the confinement breaks the selection rules. We analyze the evolution of this broadening with the size of the QDs and we show that emission spectra could considerably vary from one dot to another even if they have comparable sizes. Finally we examine the recombination process in presence of a surface defect.

THEORY

We consider here the strong confinement regime, i.e. when the diameter of the QDs is smaller than ~4 nm. The confinement of electrons and phonons has several consequences: i) the electronic bandgap exhibits a blue shift (ii) the electron and phonon states are quantized, iii) zero-phonon optical transitions become allowed, as well as phonon-assisted transitions with *all* the phonons of the QD so that a careful calculation of the confined states of both the electrons and the phonons is required. A number of calculations of the electronic states in QDs have been performed over the last few years which show that it is necessary to go beyond the EMA to obtain reliable predictions of the QD bandgap [9]. Therefore, following closely Ref. [10], we use a tight binding method to calculate the electronic structure of silicon QDs with spherical shape and with an hydrogen-passivated surface. The tight binding parameters include interactions up to third nearest neighbors in order to provide an extremely good fit to the bulk silicon band structure [9]. Our prediction of the bandgap versus size is in good agreement (within ~ 0.1 eV) with previous works including non-orthogonal tight binding [8], empirical pseudopotentials [11] and local density calculations [12]. The vibrational modes of the QDs are calculated using a standard Keating model [13], with the parameters α = 42.57 N/m and β = 12.12 N/m slightly modified compared to the original ones [13] to get a better description of the optical modes of bulk silicon (note however that the energy of the TA modes is largely overestimated in the Keating model). The parameters for the Si-H bonds are adjusted to reproduce the vibrations of the SiH_4 molecule. The matrix of the electronic Hamiltonian and the dynamical matrix are diagonalized using an efficient conjugate gradient minimization technique which enable us to treat QDs containing a large number of Si atoms (up to ~ 600). Finally, for each QD, we calculate the radiative recombination rate $W(h\nu)$ of a photon $h\nu$ as follows. We start from the usual expression for the dipolar transition [14]:

$$W(h\nu) = K\sum_{i,f} p(i)\left|\langle\Psi_i|\vec{p}|\Psi_f\rangle\right|^2 \tag{1}$$

with

$$K = \frac{16\pi^2}{3}n\frac{e^2}{h^2 m^2 c^3}h\nu \tag{2}$$

$|\Psi_i\rangle$ and $|\Psi_f\rangle$ are respectively initial and final states of the system with energy E_i and E_f such that $E_f - E_i = h\nu$, $p(i)$ is the probability of occupation of the initial state and n is the refractive index. The wave functions include the coordinates of the electrons and nuclei. Working within the Born-Oppenheimer approximation, the matrix element of the momentum in eqn. (1) becomes:

$$\langle\Psi_i|p|\Psi_f\rangle = \langle\chi_i|\langle\psi_i(\{Q\})|\vec{p}|\psi_f(\{Q\})\rangle|\chi_f\rangle \tag{3}$$

$|\chi_i\rangle$ and $|\chi_f\rangle$ are the vibrational states which in the harmonic approximation are built from the 3N independent harmonic oscillators (N is the number of atoms in the system). $|\psi_i(\{Q\})\rangle$ and $|\psi_f(\{Q\})\rangle$ are the one-electron eigenstates for a given set of positions Q of the nuclei [15]. The optical matrix element can be expanded to first order in atomic displacements:

$$\langle\psi_i(\{Q\})|\vec{p}|\psi_f(\{Q\})\rangle = \langle\psi_i(\{0\})|\vec{p}|\psi_f(\{0\})\rangle + \sum_j \vec{A}_j Q_j \tag{4}$$

where Q_j are the vibrational eigenmodes (j = 1, 3N) of energy $\hbar\omega_j$ and $\{0\}$ means that the nuclei are at their equilibrium positions. As a result of the decomposition $Q_j = \sqrt{\hbar/2\omega_j}(a_j^+ + a_j)$ as function of the creation and annihilation operators, we verify that only one vibrational mode gains or loses a phonon in a particular transition [16]. Thus for each mode j, eqn. (1) reduces to a statistical average over the initial states characterized by the number of phonons in the mode, the final state just differing by one phonon. After straightforward algebra [16], we obtain the following recombination rates:

$$\text{zero-phonon: } W = K\left|\langle\psi_i(\{0\})|\vec{p}|\psi_f(\{0\})\rangle\right|^2 \tag{5}$$

$$\text{one-phonon emission: } W = K\left|\vec{A_j}\right|^2 \frac{\hbar}{2\omega_j}\{\bar{n}_j + 1\} \tag{6}$$

$$\text{one-phonon absorption: } W = K\left|\vec{A_j}\right|^2 \frac{\hbar}{2\omega_j}\bar{n}_j \tag{7}$$

where $\bar{n}_j = (\exp(\hbar\omega_j/kT) - 1)^{-1}$. The heavy part of the work is the evaluation of the coupling coefficients $\vec{A_j} = \partial\langle\psi_i(\{Q\})|\vec{p}|\psi_f(\{Q\})\rangle/\partial Q_j$ which are calculated numerically for *all* the modes j of the QD. For each mode j, it requires to calculate the wavefunctions - therefore the Hamiltonian - and the optical matrix elements when the nuclei are displaced from their equilibrium sites according to Q_j. The optical matrix elements are calculated like in Ref. [17] and the

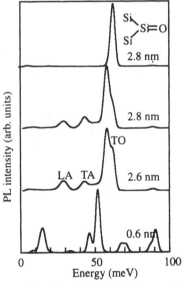

Fig. 1: Intensity of luminescence due to the emission of phonons with respect to the energy of the phonon involved in the transition for various quantum dot.

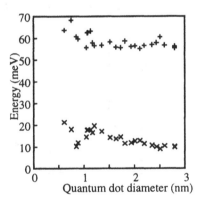

Fig. 2: Difference between the energy of the zero-phonon transition and the mean energy of the luminescence (+). Standard deviation from this mean (×) giving the energy dispersion of the emitted light.

matrix elements of the Hamiltonian are made dependent on the atomic positions following the rules of Ref. [18].

RESULTS

We plot on Fig. 1 the emission spectra at 300K of three QDs with respect to the energy of the phonon involved in the transition. We assume that the PL intensity is directly proportional to the recombination rate. At this temperature, the phonon absorption process remains weak and the recombination mainly proceeds by emission of a phonon. To characterize these spectra with respect to the size, we present on Fig. 2 the mean energy of the emission (with respect to the zero-phonon line) and the standard deviation from this mean which describes the energy dispersion of the emitted light. In agreement with the EMA results of Ref. [7], optical (TO) modes dominate among the phonon-assisted channels. The mean energy of the emitted light is in average ~55 meV below the zero-phonon transition energy (Fig. 2). But our results present a number of interesting new features. The broadening of the peaks is large, of the order of 10s of meV. The energy dispersion of the emission spectra goes from ~20 meV for the smallest QDs to ~10 meV for 3 nm QDs (Fig. 2), meaning that the transitions can be assisted by a large number of vibration modes as a consequence of the strong confinement. Thus the PL of a silicon QD should not be dominated by a sharp spike like for QDs made of direct semiconductors. Peaks corresponding to transitions assisted by LA phonons or low energy acoustic modes are also visible on Fig. 1. Further work is necessary to understand the origin of these peaks which could be due to the breaking of the selection rules (in particular we need to check if they are not due to a bad description of the LA modes by the Keating model).

Fig. 3 shows the evolution of the total recombination rate due to all the phonon-assisted chan-

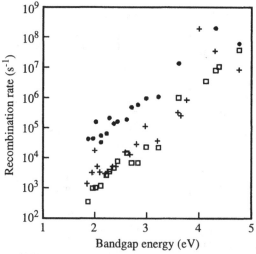

Fig. 3: Recombination rate as a function of the quantum dot bandgap: total of all the phonon-assisted transitions (●); zero-phonon transition (□); phonon-assisted transition with the highest probability (+).

nels with respect to the bandgap of the QDs. We also plot for comparison the recombination rate for the zero-phonon transition and the one for the phonon-assisted transition with the highest probability. Going from small to large bandgaps, the total recombination rate decreases by orders of magnitude. This is a direct consequence of the indirect character of the bulk silicon bandgap [7,8]. We obtain that for the *whole* range of sizes the recombination rate for the phonon-assisted channels is at least one order of magnitude larger than for the zero-phonon transition, in contrast to the EMA results of Ref. [7] which predicts a crossover for bandgap larger than ~ 2.1 eV. Actually we see on Fig. 3 that the most probable phonon-assisted transition and the zero-phonon transition have comparable probabilities when the gap is larger than ~2.2 eV. All these results lead us to conclude that many vibration modes contribute to the total recombination rate and that these contributing modes are more and more dispersed in energy when we go to smaller sizes. In the EMA calculation of Ref. [7], the energy dependence of the phonons involved in the transition was neglected which is only valid in the limit of large sizes.

Another interesting consequence of the strong confinement combined with the indirect bandgap of bulk silicon is the extreme sensitivity of the emission spectrum to the QD size. Fig. 1 shows the calculated spectra for two QDs of 2.1 and 2.2 nm. In spite of the small difference in diameter, the shape of the two spectra presents remarkable variations, in particular below 50 meV. The reason for this is that optical matrix elements and the coefficients $\vec{A}_j$ are somehow related to the reciprocal space overlaps of the electron and hole wave functions which are strongly oscillating functions of the size of the QD. Such differences as shown in Fig. 1 should be observable by spectroscopic experiments on single silicon QDs in this range of size.

Finally, we examine the emission process involving an oxygen related defect. Recently, we have shown that a Si=O bond at the surface of a nanocrystal gives a new electronic state which appears in the bandgap for clusters smaller than ~ 3 nm and which could explain the luminescence of porous silicon [19]. The electronic state is a *p* state on the Si atom. Our calculations show that the zero-phonon transition is also almost forbidden and that phonon assisted transitions again dominate. Fig. 1 shows that the recombination is mainly assisted by optical phonons (TO and LO). Therefore, we conclude that the observation of transitions assisted by optical phonons cannot be used to rule out the influence of defects. However, coupling to acoustic phonons is not obtained in that case which could be used to discriminate between the different types of excitonic recombinations.

CONCLUSIONS

In conclusion, the present calculations show that phonon-assisted transitions dominate radiative recombination in silicon QDs, even for the smaller QDs. When the confinement becomes stronger, the dispersion in energy of the emission spectrum increases due to the contribution of many vibration modes. We predict an important broadening of the PL peaks, of the order of ~10s of meV for a 3 nm QD and we show that the PL spectra should be very sensitive to the size and the shape of the QDs. We show that the recombination on a Si=O bond also involves optical phonons. All these results suggest that spectroscopic studies on single silicon QDs should be particularly interesting. These conclusions can be easily extended to QDs made of other indirect semiconductors. Finally, further works are still in progress to improve the calculation: i) we plan to use a better description of the phonons than in a simple Keating model in order to improve the description of the acoustic modes; ii) we plan to include the effect of the atomic relaxation in presence of an exciton which must be at the origin a substantial electron-phonon coupling [10].

ACKNOWLEDGEMENTS

The "Institut d'Electronique et de Microélectronique du Nord" is "Unité mixte 9929 du Centre National de la Recherche Scientifique".

REFERENCES

1 L. Brus, Appl.Phys. A **53**, 465 (1991), and references therein.
2 Al.L. Efros and A.L. Efros, Sov.Phys.Semicond. **16**, 772 (1982); L. Brus, J. Chem. Phys. 80, 4403 (1984); D.J. Norris and M.G. Bawendi, Phys.Rev. B **53**, 16338 (1996).
3 K. Brunner, G. Abstreiter, G. Bohm, G. Trankle, and G. Weimann, Phys.Rev.Lett. **73**, 1138 (1994); S.A. Empedocles, D.J. Norris, and M.G. Bawendi, Phys.Rev.Lett. **27**, 3873 (1996); D. Gammon, E.S. Snow, B.V. Shanabrook, D.S. Katzer, and D. Park, Phys.Rev.Lett. **76**, 3005 (1996);
4 For a recent review, see A. G. Cullis, L. T. Canham, P. D. J. Calcott **82**, 909 (1997) and in *Theory of Optical Properties and Recombination Processes in Porous Silicon*, edited by G. Amato, C. Delerue and H.-J. von Bardeleben, Vol. 5, Optoelectronic Properties of Semiconductors and Superlattices series, Gordon and Breach Science Publishers (1997).
5 P.D.J. Calcott, K.J. Nash, L.T. Canham, M.J. Kane and D. Brumhead, J.Phys. Condens. Matter 5, **L91** (1993). T. Suemoto, K. Tanaka, A. Nakajima, and T. Itakura, Phys.Rev.Lett. **70**, 3659 (1993).
6 L. Brus, in *Light Emission in Silicon. From Physics to Devices*, edited by D. Lockwood, series Semiconductors and Semimetals, Vol. 49, Academic Press, p. 303 (1998).
7 M.S. Hybertsen, Phys.Rev.Lett. **72**, 1514 (1994).
8 C. Delerue, G. Allan and M. Lannoo, Phys.Rev.B **48**, 11024 (1993).
9 A. Zunger, MRS Bulletin **23**, 35 (1998); C. Delerue, G. Allan, and M. Lannoo, in *Light Emission in Silicon. From Physics to Devices*, edited by D. Lockwood, series Semiconductors and Semimetals, Vol. 49, Academic Press, p. 253 (1998).
10 E. Martin, C. Delerue, G. Allan and M. Lannoo, Phys.Rev.B **50**, 18258 (1994).
11 L.W. Wang and A. Zunger, J.Phys.Chem. **98**, 2158 (1994).
12 B. Delley and E.F. Steigmeier, Appl.Phys.Lett. **67**, 2370 (1995).
13 P.N. Keating, Phys.Rev. **145**, 637 (1966).
14 D.L. Dexter, in Solid State Physics, Advances in Research and Applications, edited by F. Seitz and D. Turnbull (Academic, New York, 1958), Vol. 6, p. 360.
15 The splitting of the lowest excitonic state into singlet and triplet states due to the electron-hole exchange interaction lead to additional complications in the emission spectra [10]. But as the phonons do not mix states with different spins, the phonons sidebands should be relatively unchanged.
16 P.T. Landsberg, in *Recombination in Semiconductors*, Cambridge University Press (1991).
17 J. Petit, G. Allan and M. Lannoo, Phys.Rev.B **33**, 8595 (1986).
18 W.A. Harrison, *Electronic Structure and the Properties of Solids* (Freeman, San Francisco, 1980).
19 M. Wolkin-Vakrat, P.M. Fauchet, G. Allan and C. Delerue, Phys.Rev.Lett., to be published.

LUMINESCENCE STUDY OF SELF-ASSEMBLED, SILICON QUANTUM DOTS

S. Miyazaki, K. Shiba, N. Miyoshi, K. Etoh, A. Kohno, and M. Hirose
Department of Electrical Engineering, Hiroshima University
Kagamiyama 1-4-1, Higashi-Hiroshima 739-8527, Japan

ABSTRACT

Hemispherical silicon quantum dots (QDs) have been self-assembled with an areal density as high as ~2×10^{11} cm^{-2} on SiO$_2$/Si(100) and quartz substrates by controlling the early states of low pressure chemical vapor deposition (LPCVD) of pure silane. It is found that, for the thermally-oxidized Si QDs, when the mean Si dot height is decreased from 6.3 nm to 1~2 nm, the photoluminescence (PL) peak energy is increased from 1.2 to 1.4 eV at room temperature while the optical absorption edge determined by photothermal deflection spectroscopy is shifted from 1.9 to 2.5 eV. In addition to the observed Stokes shift as large as 0.7~1.1 eV, a weak temperature dependence of the broad luminescence band and non-exponential luminescence decay with a mean life time of sub-msec even at room temperature suggest that localized, radiative recombination centers existing presumably in the SiO$_2$/Si dot interface are responsible for the efficient PL from the Si QDs. From the change in room temperature PL by SiO$_2$ thinning and removal in a dilute HF solution, it is demonstrated that the surface passivation of Si QDs plays an important role for the efficient light emission at room temperature.

INTRODUCTION

Physical properties of nanometer-sized silicon structures have been extensively studied to disclose quantum confinement effects and to open up new device applications. For porous silicon layers [1,2] and nanocrystalline silicon films [3-6], a significant blue-shift of the optical absorption edge and the efficient visible light emission at room temperature have been reported so far and the importance of the quantum size effect in the Si nanostructures has often been discussed. However, the origin of the photoluminescence (PL) and its mechanism are still under discussion because the correlation between the nanocrystalline size and the PL properties is not clearly understood yet. In our previous works, we reported that the self-assembling of silicon quantum dots (QDs) with a fairly uniform size distribution and an areal density of the order of 10^{11} cm^{-2} was achieved by controlling the thermal decomposition of silane on SiO$_2$ surfaces [6, 7]. And we have demonstrated a clear negative conductance associated with the resonant tunneling through SiO$_2$/Si QD/SiO$_2$ double barrier structures [8]. More recently, metal-oxide-semiconductor (MOS) devices with Si QDs as a floating gate have been fabricated and their memory operations associated with electron charging to the Si QDs have been reported at room temperature [9-11].
In this work, we have investigated the temperature dependence and the temporal decay of the luminescence from self-assembled Si QDs covered with an ultrathin SiO$_2$ layer to discuss a possible luminescence mechanism. Also, the influence of SiO$_2$ thinning by dilute HF treatment on the luminescence has been examined to reveal the importance of the dot surface passivation for the efficient PL at room temperature.

SAMPLE PREPARETION

P-type Si(100) wafers and quartz plates were used as substrates in this study. After conventional wet-chemical cleaning steps, 15 nm-thick SiO$_2$ was grown on Si (100) at 1000°C in dry O$_2$. Also, precleaned quartz substrates were annealed at the same condition. Subsequently, for some of samples the substrate surface was slightly etched back by 0.3 nm in a 0.1% HF solution to obtain an OH-terminated SiO$_2$ surface which enhances the areal density of Si QDs and improve the uniformity of the dot size distribution [7]. Si QDs were self-assembled with an areal dot density of 2 ~ 5x10¹¹ cm⁻² on substrates so prepared by low pressure chemical vapor deposition (LPCVD) of pure SiH_4 at 560 ~ 580 °C. The dot size was controlled by the deposition temperature and time as already reported elsewhere [7]. The dot surface was successively oxidized at 1000 °C for 60 ~ 90 s in 2% O_2 diluted with nitrogen. Under these oxidation conditions, the QDs surfaces were conformally covered with 1.8 ~ 2.7 nm-thick SiO_2, as confirmed by the angle-resolved Si_{2p} core level spectra for as-deposited QDs and oxidized ones. The formation of single crystalline Si dots with a hemispherical shape was confirmed by high resolution TEM observations [7].

RESULTS AND DISCUSSION

The size distribution of Si QDs was measured by AFM images. Figure 1 shows dot height distributions for thermally-oxidized Si QDs which were deposited on as-grown SiO_2 on Si(100) at 580°C for 60 s or 90 s and then oxidized at 1000 °C for 90 s. The average dot height for each sample was determined as an important dimension to assess the quantum confinement effects by fitting a log-normal function [12] on the measured distribution as indicated by a solid line. Since the measured dot height includes the thickness of the oxide layer on the QDs' surfaces, as illustrated in the inset of the figure. To obtain the true Si-dot height, the oxide thickness of 2.7 nm must be subtracted from the observed dot height. Consequently, the average Si-core heights for the samples shown in Fig. 1 were determined to be 3.3 and 6.3 nm, respectively. The lateral diameter of the oxidized QDs is 2 ~ 4 times larger than the dot height as evaluated from the high-resolution SEM and TEM observations.

For as-grown Si QDs, no photoluminescence (PL) is observable at room temperature. When the as-grown surface is oxidized, the Si QDs start to efficiently luminesce at room temperature. This indicates that the passivation of nonradiative surface states is of importance for light emission from the QDs, which is discussed later. For the dots on SiO₂/Si(100) and quartz substrates at the same preparation condition, no difference in the PL spectral shape was observed. Figure 2 shows room temperature PL spectra for thermally-oxidized QDs with different average Si-core heights prepared on quartz. The PL peak energy shifts toward the higher energy side with decreasing average Si-core size.

Evidently, for the samples deposited on HF-treated SiO₂ at 560 °C, a little short of the deposition time results in a blue-shift of the PL peak and a significant reduction of the PL intensity. The sample obtained by 36 s deposition is no longer luminescent, implying that the Si QDs are fully oxidized during the 1000 °C oxidation for 90 s. In fact, no trace of Si dots in the oxidized sample was confirmed by TEM observations. Notice that the PL peak appears at an energy red-shifted by 0.7~0.9 eV than the optical absorption edge, which was determined from photothermal deflection measurements (PDS) [13], as indicated with the arrow for each sample in the figure. Such a large Stokes shift can not be interpreted in terms of quantum size effects. In the calculation based on the effective mass approximation [14] for a spherical

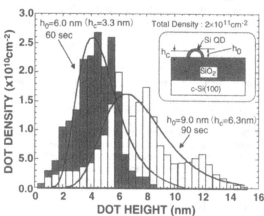

Fig. 1. Size distributions of silicon quantum dots (Si QDs) obtained from AFM images after thermal oxidation at 1000°C for 90 s. The solid curves are log-normal functions well-fitted to the respective observed distributions. The Si QDs were deposited on as-grown SiO₂/Si(100) at 580°C for 60 s or for 90 s.

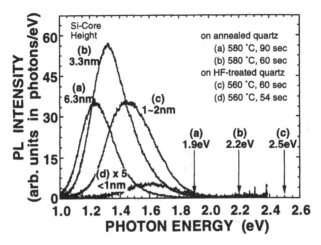

Fig. 2. Photoluminescence (PL) spectra at 293 K for thermally-oxidized Si QDs with different mean Si core heights on quartz. A 488 nm line from an Ar⁺ laser was used as an excitation source. The arrows indicate bandgap energies for the samples determined by photothermal deflection spectroscopy.

Si crystallite with a diameter of 2 nm, the excitonic binding energy is estimated be 0.25 eV. Therefore, the luminescence from the QDs is likely to be caused by radiative recombination through localized states which presumably exists at an interfacial layer between the Si QD core and the SiO_2 surface layer [13]. It is interesting to note that, for the samples (c) which exhibits a similar broad luminescence band to luminescent porous silicon or silicon nanocrystallites embedded in SiO_2 [3-5], the room temperature PL spectrum measured under 325 nm (3.81 eV) excitation is identical in shape to that under 488 nm (2.54 eV) excitation except that the intensity is enhanced by a factor of an increased optical absorption coefficient. This result suggests that a photon energy of 2.54 eV is large enough to excite all the luminescent Si QDs. Namely, it is likely that there exist no significant amount of QDs with an optical bandgap over 2.54 eV in the sample, which is consistent with the optical absorption edge determined from the PDS measurements.

To gain a better understanding of the PL properties from Si QDs, the temperature and excitation power dependences of the luminescence were observed under 488 nm excitation (Figs. 3 and 4). As shown in Fig. 3, the temperature dependence is rather weak. The total PL intensity is increased by a factor of at most 2 around 77 K. Further decrease in the temperature causes an intensity reduction especially for lower energy emissions, which results in the broadening of the spectral shape and the energy shift of the PL peak to the higher energy side. The observed intensity reduction at lower temperatures suggests that the nonradiative recombination due to the Auger effect becomes significant and/or the radiative recombination involves a thermal activation process. The excitation power dependence of the PL intensity confirms that the Auger process plays an important role for carrier recombination processes at a temperature below 77 K as represented in Fig. 4. The PL peak intensity at room temperature is linearly increased with excitation power density up to 320 mW/cm² while at 13 K it tends to be saturated under excitations higher than 40 mW/cm².

In order to get a further insight into the luminescence mechanism, time-resolved PL measurements was carried out by using 337.1 nm light pulse from an N₂ laser. As represented in Fig 5, the luminescence for the sample (c) shows non-exponential decay features which reach to the

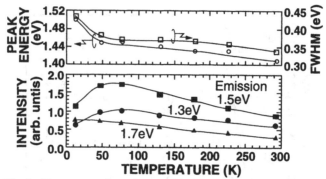

Fig. 3. Temperature dependences of the PL peak energy, the intensities at different emission energies and the full width at half maximum for the sample (c) seen in Fig. 2 measured under 488 nm excitation with a power density of 200 mW/cm².

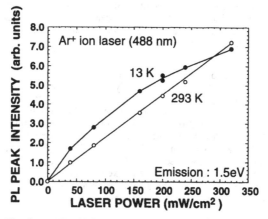

Fig. 4   Excitation intensity dependence of PL peak intensity at 13 and 293 K for the sample (c) seen in Figs. 2 and 3 measured under 488 nm excitation.

sub-msec region even at room temperature. The observed PL decay can be characterized well by using a stretched exponential function as demonstrated in Fig. 5. The PL decay becomes slower for lower energy emissions and at lower temperatures. At 13 K, a decay component faster than 1μsec becomes observable prior to the slow component, being attributable to the Auger recombination. The PL decay features for thermally-oxidized Si QDs suggest that the energy relaxation process of photogenerated carriers involves the carrier trapping through thermally assisted tunneling to localized states, whose structure is not identified yet, and that the Auger effect emerges in the confinement of photogenerated carriers deeply at low temperatures.

For the purpose of an argument for the dot surface passivation deeply concerned in the luminescence, the as-grown 1.8 nm-thick SiO_2 layer on Si QDs with a mean Si-core height of 3.3 nm was thinned stepwise in 0.1% HF solution. At each SiO_2 thinning step, the PL spectrum was measured at room temperature under 325nm excitation (Fig. 6) and Fourier transform infrared (FT-IR) measurements were carried out by contacting the sample surface to a Ge ATR (attenuated total reflection) prism (Fig. 7). The oxide thinning only by ~0.3 nm causes a significant reduction in the PL intensity, implying that photogenerated carriers in Si QDs can tunnel through ultrathin oxide and recombine nonradiatively at the defect generated on and/or near the oxide

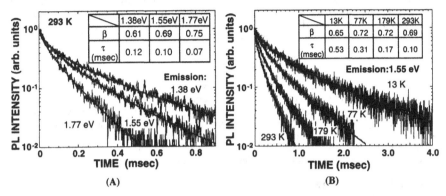

	1.38eV	1.55eV	1.77eV
β	0.61	0.69	0.75
τ (msec)	0.12	0.10	0.07

293 K

Emission: 1.38 eV

1.77 eV   1.55 eV

	13K	77K	179K	293K
β	0.65	0.72	0.72	0.69
τ (msec)	0.53	0.31	0.17	0.10

Emission: 1.55 eV

13 K

77 K

293 K   179 K

(A)   (B)

Fig. 5   PL decay curves at 293 K for emissions at different energies (A) and at different temperatures for a 1.55 emission (B) for the sample (c) seen in Figs. 2 , 3 and 4. The solid lines denote the stretched exponential curves well-fitted to the observed decay curves. The mean life time τ and β values obtained from the stretched exponential function are summarized in the corresponding inset.

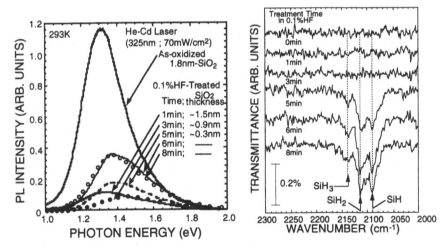

293K   He-Cd Laser
(325nm ; 70mW/cm²)
As-oxidized
1.8nm-SiO₂

0.1%HF-Treated
SiO₂
Time; thickness
1min; ~1.5nm
3min; ~0.9nm
5min; ~0.3nm
6min; ———
8min; ———

Treatment Time
in 0.1%HF

0min
1min
3min
5min
6min
8min

0.2%   SiH₃
SiH₂   SiH

Fig.6   Room temperature PL spectra for Si QDs fabricated on quartz at each SiO_2 thinning step by a 0.1% HF solution. In as-prepared sample, Si QDs with a mean Si-core height of 3.3 nm were covered with 1.8 nm-thick SiO_2. A cw 325 nm line with a power density of 70 mW/cm² from a He-Cd laser was used as an excitation source.

Fig.7   FT-IR-ATR spectra for Si QDs on 15nm-thick SiO_2/Si(100) fabricated and treated under the same conditions as the sample shown in Fig. 6. A Ge crystal prism in contact with the sample surfaces before and after dilute HF etching was used for multiple internal reflection medium.

49

surface by dilute HF etching. The PL peak shift observed after oxide thinning by ~0.9 nm suggests that the emission at lower energy is more sensitive to the defects generated by dilute HF etching because of the slower emission life time (see in Fig. 5 (a)). For surface oxide thickness below ~0.3 nm, the PL intensity is partly recovered and then the surface silicon hydrides becomes observable as indicated in Fig. 7. When the dot surface is fully terminated with hydrogen, the luminescence with an intensity about 30% of the initial value remains almost unchanged by further dilute HF treatment. The result of Figs. 6 and 7 indicates the control of surface passivation is of great importance for efficient light emission from Si QDs.

CONCLUSIONS

The photoluminescence from hemispherical, self-assembled silicon quantum dots (QDs) have been studied as a function of temperature, excitation power and decay time. A Stokes shift more than 0.7 eV, a weak temperature dependence and slow nonexponetial-decay features for the PL band suggests that the radiative recombination of photogenerated carriers through localized states is an important pathway for the emission from the Si QDs rather than the direct transition between quantized states or their excitonic states. The control of the dot surface passivation is a crucial factor for the efficient luminescence at room temperature.

ACKNOWLEDGMENTS

This work was supported in part by Grant-in aids for the "Research for the Future" program by the Japan Society for the Promotion of Science and for the Core Research for Evolutional Science and Technology (CREST) from the Japan Science and Technology Corporation (JST).

REFERENCES

1. L. T. Canham: Appl. Phys. Lett. 57 (1990) 1046.
2. V. Lehmann and U. Gösele: Appl. Phys. Lett 58 (1991) 856.
3. H. Takagi, H. Ogawa, Y. Yamazaki, A. Ishizaki and T. Nakagiri: Appl. Phys. Lett. 56 (1990) 2379.
4. Y. Kanemitsu, T. Ogawa, K. Shiraishi and K. Takeda: Phys. Rev. B 48 (1994) 4883.
5. Y. Yamada, T. Orii, I. Umezu, S. Takeyama and T. Yoshida: Jpn. J. Appl. Lett. 35 (1996) 1361.
6. A. Nakajima, Y. Sugita, K. Kawamura, H. Tomita and N. Yokoyama: Jpn. J. Appl. Phys. 35 (1996) L189.
7. K. Nakagawa, M. Fukuda, S. Miyazaki and M. Hirose: Mat. Res. Soc. Symp. Proc. 452 (1997) 234.
8. M. Fukuda, K. Nakagawa, S. Miyazaki and M. Hirose: Appl. Phys. Lett. 70 (1997) 2291.
9. S. Tiwari, F. Rana, H. Hanafi, H. Hartstein, E. F. Crabbé and K. Chen: Appl. Phys. Lett. 68 (1996) 1377.
10. L. Guo, E. Leobandung and S. Y. Chou: Technical Digest of 1996 Intern. Electron Device Meeing (San Francisco, 1996) p. 955.
11. A. Khono, Murakami, M. Ikeda, S. Miyazaki and M. Hirose: Extended Abstracts of the 1997 Intern. Conf. on Solid State Devices and Materials (Hamamatsu, 1997) p. 566; Extended Abstracts of the 1998 Intern. Conf. on Solid State Devices and Materials (Hiroshima, 1998) p. 174.
12. R. R. Irani and C. F. Callis: Particle Size Measurement: Interpretation and Application (Wiley, New York, 1976) p. 135.
13. K. Shiba, K. Nakagawa, M. Ikeda, A. Kohno, S. Miyazaki and M. Hirose: Jpn. J. Appl. Phys. 36 (1997) L1279.
14. T. Takagahara and K. Takeda: Phys. Rev. B 46 (1992) 15578.

FORMATION PROCESS OF Si NANOPARTICLES
FORMED BY LASER ABLATION METHOD

T. MAKIMURA, T. MIZUTA, T. UEDA and K. MURAKAMI
Institute of Materials Science, University of Tsukuba,
Tsukuba, Ibaraki 305-8573, Japan, makimura@ims.tsukuba.ac.jp

ABSTRACT

Utilizing laser ablation of Si targets, nanoparticles can be cleanly formed in rare gas. In order to fabricate nanoparticles with well-defined structures such as those whose surfaces are chemically modified, it is important to investigate the formation process of the nanoparticles. We have developed a decomposition method for measuring time-resolved spatial distributions of nanoparticles in rare gas. Applying this method, we have investigated formation processes of silicon nanoparticles in 2-Torr argon gas. The nanoparticles are found to grow from 300 μs to 1 ms after the ablation.

INTRODUCTION

Silicon (Si) nanoparticles with sizes of 1-10 nm exhibit bright luminescence and are candidates of opto-electronic materials [1-9]. The nanoparticles with roughly-controlled sizes have been cleanly fabricated by laser ablation of Si targets in rare gas [3-5, 8, 9]. In order to fabricate the nanoparticles with well-defined structures, it is important to investigate the formation process of the nanoparticles: By exposing pulsed reactive gas just after nanopaticles are formed, we could fabricate surface-modified nanoparticles such as Si nanoparticles covered with oxide and those terminatd with hydrogen atoms [10]. We could ionize nanoparticles for size selection after they are formed.

So far, imaging by means of intensified CCD (ICCD) cameras [3, 6] and soft X-ray absorption spectroscopy [9] in addition to traditional spectroscopies [11] have revealed that the laser irradiation causes ejection and formation of a cloud (plume) including light-emitting Si neutrals and ions as well as electrons. Subsequently the plume dissipates its kinetic energy of 1-100 eV to the ambient gas by collisions in a time scale of 10 μs and the plume begins to move in a diffusive way [3, 6, 9, 12]. Recently, Muramoto *et al.* applied an imaging technique utilizing laser-induced fluorescence (LIF) and found that Si dimers are formed in the central region of the plume concomitantly when Si atoms disappear [13]. The dimers exist during limited periods of 400-700 μs and 200-400 μs in helium gas at 5 Torr and 10 Torr, respectively. Finally, relatively large nanoparticles with sizes of order of 10 nm are observed in time scale of 1 ms, applying the Rayleigh Scattering method [6, 12, 13]. Furthermore, Geohegan *et al.* reported gas-phase photoluminescence from Si nanoparticles in the same time scale [5, 6]. The nanoparticles are reported to be probed by detecting light emission from atoms decomposed from the nanoparticles by laser irradiation. [14]

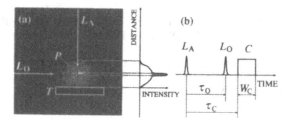

Figure 1: (a) The experimental setup and an obtained image at 400 μs. Enhanced light emission is seen in the track of the decomposition laser beam (L_O) in addition to the plume emission (P) caused by the ablation laser light (L_A). (b) Temporal sequence of the ablation laser pulse (L_A), the OPO laser pulse (L_O) for decomposition and the gate of the intensified CCD camera (C).

This method enables us to detect both a) nanoparticles of our interest with the intermediate sizes between the dimer and the 10-nm nanoparticles and b) those that do not exhibit photoluminescense because of dangling bonds on their surfaces.

In this work, we have developed a technique for measuring time-resolved spatial distributions of Si nanoparticles. Applying the technique, we have investigated the formation processes of Si nanoparticles in argon (Ar) gas. The Ar gas confines the plume most adequately among other rare gas because a Ar atom has mass close to that of a Si atom.

EXPERIMENTAL

The experimental setup is shown in Fig. 1(a). The ablation was performed by irradiation of Si targets (T) with second harmonics of a Q-switched Nd:YAG laser (ablation laser) light (L_A) with a pulse width of 20 ns at approximately 15 J/cm^2, in Ar gas in a vacuum chamber. The base pressure of the chamber was typically 5×10^{-7} Torr. The Ar gas has a purity of 99.999 % and its pressure is either 1 Torr or 2 Torr, which were measured using a thermo-couple gauge. Optical parametric oscillator (OPO: Spectra Physics, MOPO-710) light (L_O) with a pulse width of 10 ns was incident into the plume (P) parallel to the target, 4 mm apart from the target. The focused OPO beam has a beam waist of 200 μm and the energy density is estimated to be approximately 2 J/cm^2 in the center of the plume. The wavelength of the OPO light was varied from 400 nm to 650 nm. The pulse sequence is shown in Fig. 1(b). Temporal delay (τ_O) of the OPO pulse (L_O) from the ablation laser pulse (L_A) was adjusted in a range from 200 μs to 2 ms. Using a gated ICCD camera (Princeton Instruments, ITE/CCD-576-G/RB-E), which is sensitive to light from 200 nm to 850 nm, we observed the time-resolved spatial distributions of light emission induced by both the ablation laser pulse and the subsequent OPO pulse, with gate width (W_C) of either 100 ns or 500 ns. The delay (τ_C) of the ICCD exposure gate (C) is always adjusted to be late from the OPO pulse L_O by 40 ns.

2-Torr Ar

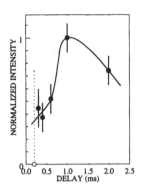

Figure 2: (a) Sections of light emission distributions at given τ_O's. (b) Intensity of the enhanced light emission as a function of τ_O. The intensity is interpreted to be the density of nanoparticles. The density is normalized by that at 1 ms.

In order to determine the origin of the light emission, we measured time-resolved spectrum using a space-resolved spectrometer (Jobin Yvon, HR460) and the gated ICCD camera. The experimental details were given elsewhere [15].

RESULTS

The image in Fig. 1(a) shows a spatial distribution of light emission obtained with 550-nm OPO light at τ_O of 400 μs with W_C of 100 ns. We found that the light emission is enhanced in the track of the OPO beam in contrast with the plume emission(P). The enhanced light emission has its maximum in the central region at all τ_O's investigated. The enhanced light emission could not be observed at 1 Torr. As illustrated in Fig. 1(a), we obtained cross-sectional profiles along the direction perpendicular to the target, averaged over the whole region of the plume. The intensity of the enhanced light emission is plotted as a function of τ_O. It starts growing at 300 μs, increases until 1 ms and is still observed after 2 ms. It can not beidentified at 200 μs. The origin of the enhanced light emission is clearly seen in Fig. 3(a), which shows the spectrum of light emission from the OPO beam region at τ_O of 400 μs with W_C of 500 ns. Broad line widths in spectrum (a) are ascribed to the resolution of the spectrometer with slits widely opened for attaining single-shot measurement [12, 13]. Except for the resolution, the spectrum (a) coincides with that shown in Fig. 3(b), which is observed for the plume formed by only the ablation laser without OPO laser light at τ_C of 100 μs with W_C of 10 μs. As indicated by the markers, the intense lines are due to Ar neutrals and Si neutrals [16].

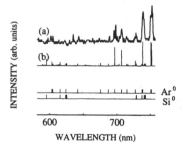

Figure 3: (a) Light emission spectrum for the decomposed region. (b) Spectrum for species formed by only the ablation laser light. The intense lines are assigned to be argon neutrals and silicon neutrals, as indicated by the markers.

DISCUSSION

The following evidence reveals that the enhanced light emission is due to excited Ar and Si neutrals formed through the decomposition of the Si nanoparticles.

Si nanoparticles have an absorption band much broader than atomic species and therefore they can be decomposed by absorption of OPO light like laser ablation of Si wafers. This decomposition can result in light emission from decomposed Si atoms.

The decomposition is also probed by light emission from excited Ar atoms: We confirmed that the intense Ar lines are still prominent in the plume even when a Si target is ablated without the OPO laser at a energy density as low as 3.5 J/cm^2 that is close to the ablation threshold. That is, the laser ablation is always accompanied by the excitation of Ar atoms. The excitation of the Ar atoms is most likely caused by collision with electrons rather than atoms because the cross section is fairly large [17]. The electrons are abundant in Si plasma formed by the laser ablation and are expected to have kinetic energies of order of 10 eV, assuming that they have same energy as ions have in the plasma.

Consequently, we can conclude that Si nanoparticles can be observed by detecting light emission from Si neutrals decomposed by irradiation of OPO light and concomitantly excited Ar neutrals around the decomposed nanoparticles. The decomposition may be caused by heating up of the nanoparticles or by Coulomb explosion due to photo-ionization in analogy to laser ablation [11, 18].

The diameter of the nanoparticles observed in 2-Torr Ar gas is estimated to be larger than 3 nm according to reported diameters for nanoparticles with a band gap of 2.25 eV (= 550 nm) [7]: The nanoparticles larger than 3 nm can absorb the OPO light used here effectively, resulting in the decomposition. On the other hand, the nanoparticles formed in 1-Torr Ar gas may be smaller than 3 nm since the enhanced light emission could not be observed. The estimation above is consistent with our previous result that smaller nanoparticles are formed at lower pressures [3].

As shown in Figs. 1(a) and 2, we found that the intensity of the enhanced light emission saturates at 1 ms in 2-Torr Ar gas. The saturation indicates the end of the formation of the nanoparticles. We also found that the nanoparticles begin to grow at 300 μs. This onset almost coincides with the reported disappearance of the dimers [13]: Our result suggests that nanoparticles grow concomitantly when the dimers disappear. The onset is earlier than most of reported ones measured by Rayleigh scattering method: Muramoto *et al.* measured onsets of 4 ms, 1 ms and 500 μs in He gas at 5 Torr, 10 Torr and 100 Torr, respectively. Geohegan *et al.* reported onsets of 3 ms and 200 μs in 1-Torr Ar gas and 10-Torr He gas, respectively. These may suggest that the decomposition method is more sensitive to smaller nanoparticles than Rayleigh scattering method. Note these onsets can not be compared directly to each other because onset depends on mass and pressure of ambient gas. We should emphasize that we can observe the whole formation processes of nanoparticles using the decomposition method.

CONCLUSION

In conclusion, we have developed a decomposition method in order to investigate formation processes of Si nanoparticles in Ar gas after laser ablation of Si targets. Observing the light emission from atoms decomposed into by laser light, we obtained time-resolved spatial distribution of Si nanoparticles. Applying the technique, we directly observed that the nanoparticles begin to grow at 300 μs after the ablation in the central region of the plume in 2-Torr Ar gas and that they grows till 1 ms. These knowledge will be useful for collecting Si nanoparticles as well as fabricating Si nanoparticles with modified structures such as nanoparticles surrounded by oxide or terminated with hydrogen atoms.

ACKNOWLEDGEMENTS

This work is in part supported by Monbusho International Scientific Research Program: Joint Research. We would like to thank Dr. D. B. Geohegan, Dr. T. Okada, Dr. Y. Nakata, Dr. D. H. Lowndes and Prof. M. J. Aziz for helpful discussions.

REFERENCES

1. L. T. Canham, Appl. Phys. Lett. **57**, 1046 (1990).

2. T. Makimura, Y. Kunii, N. Ono and K. Murakami, Appl. Surf. Sci. **127–129**, 388 (1998).

3. T. Makimura, Y. Kunii and K. Murakami, Jpn. J. Appl. Phys. **35** Part 1, 4780 (1996).

4. T. Yoshida, S. Takeyama, Y. Yamada and K. Mutoh, Appl. Phys. Lett. **68**, 1772 (1996).

5. D. B. Geohegan, A. A. Puretzky, G. Dusher and S. J. Pennycook, Appl. Phys. Lett. **73**, 438 (1998).

6. D. B. Geohegan, A. A. Puretzky, G. Dusher and S. J. Pennycook, Appl. Phys. Lett. **72**, 2987 (1998).

7. Y. Kanemitsu, in *Optical Properties of Low-Dimensional Materials*, edited by T. Ogawa and Y. Kanemitsu (World Scientific, Singapore, 1995), Chap. 5.

8. E. Werwa, A. A. Seraphin, L. A. Chiu, Chuxin Zhou, K.D. Kolenbrander, Appl. Phys. Lett. **64**, 1821, (1994).

9. K. Murakami, T. Makimura, N. Ono, T. Sakuramoto, A. Miyashita and O. Yoda, Appl. Surf. Sci. **127–129** , 368 (1998).

10. These are most possible; We have confirmed that as-deposited nanoparticles formed in either oxygen gas or hydrogen gas diluted with argon gas exhibit photoluminescence.

11. D. B. Geohegan, in *Pulsed Laser Deposition of Thin Films*, edited by D. B. Chrisey and G. K. Hubler (Wiley-Interscience Publisher, 1994), Chap. 5.

12. J. Muramoto, Y. Nakata, T. Okada and M. Maeda, Jpn. J. Appl. Phys. **36** Part 2, L563 (1997).

13. J. Muramoto, Y. Nakata, T. Okada and M. Maeda, Appl. Surf. Sci. **127–129**, 373 (1998).

14. L. Boufendi, J. Hernamm, A. Bouchoule and B. Dubreuil, J. Appl. Phys. **76**, 148 (1994).

15. T. Makimura and K. Murakami, Appl. Surf. Sci. **96–98**, 242 (1996).

16. A.R.Striganov and N.S.Sventitikii: *Tables of Spectral Lines of Neutrals and Ionized Atoms* (IFI-Plenum, New York, 1968).

17. D. Rapp and P. Englander-Golden, J. Chem. Phys. **43**, 1464 (1965).

18. W. Marine, J. M. Scotto d'Aniello, M. Gerri and P. Thomsen-Schmidt, in *Laser Ablation of Electronic Materials, Basic Mechanisms and Applications*, edited by E. Fogarassy and S.Lazare (Elsevier, Amsterdam, 1992), p. 89.

CORRELATION BETWEEN DEFECT STRUCTURES AND LIGHT EMISSION IN Si-NANOCRYSTAL DOPED SiO$_2$ FILMS

K. Sato[*], Y. Sugiyama[*], T. Izumi[*], M. Iwase[**], Y. Show[***], S. Nozaki[***], and H. Morisaki[***]
[*]Department of Electronics, Faculty of Engineering, Tokai University, 1117 Kitakaname, Hiratsuka-shi, Kanagawa 259-1292, JAPAN
[**]Department of Electrical Engineering, Faculty of Engineering, Tokai University, 1117 Kitakaname, Hiratsuka-shi, Kanagawa 259-1292, JAPAN
[***]Department of Communications and Systems, The University of Electro-Communications, 1-5-1 Chofugaoka, Chofu-shi, Tokyo 182-8585, JAPAN

ABSTRACT

Correlation between defect structures and light emission from Si-nanocrystal doped SiO$_2$ films has been studied using electron spin resonance (ESR) and photoluminescence (PL) methods. The ESR analysis revealed the presence of three kinds of ESR centers in the film after annealing at above 900 °C in argon (Ar) atmosphere, i.e. Si dangling bond in amorphous Si cluster (a-center: g=2.006), Si dangling bond at Si-nanocrystal/SiO$_2$ interface (P$_b$-center: g=2.003) and conduction electrons in Si-nanocrystal (P$_{ce}$-center: g=1.998). Moreover, visible light emission was observed in the annealed sample from the PL measurement. Both the PL intensity and the ESR signal intensity of the P$_{ce}$-center were increased with an increase of annealing temperature. These results indicate that the P$_{ce}$-center is strongly associated with the emission center.

INTRODUCTION

Silicon has been thought of as an indirect bandgap semiconductor with a bandgap in infrared region at 1.1 eV for a long time. However, visible light emission spectra at room temperature have recently been reported from various kinds of Si nano-structures, that is, porous Si layer prepared by anodization in HF solution[1], Si ultrafine particles deposited by gas evaporation of Si powders in an argon atmosphere[2,3], Si microcrystals embedded in SiO$_2$ glass films prepared by radio-frequency (RF) magnetron sputtering[4], and Si-nanocrystal doped SiO$_2$ thin film deposited by RF sputtering[5,6]. After that, the studies on the visible light emissions from silicon nanocrystal materials have become of major interest, because of their potential application in Si based light emitting devices. Numerous workers proposed that the quantum size effects caused by the confinement of electrons and holes in the Si crystals of nanometer size[1,7,8], or defects existing interface region between the surface of Si nano-structures and the surface oxide layer[9,10] are responsible for light emission. However, the light emission mechanism has not been completely resolved up to now.

We fabricated Si-nanocrystal doped SiO$_2$ thin films deposited on Si substrate by co-sputtering of Si and SiO$_2$. The visible light emission was obtained from the film after

57

annealing above 900 °C. ESR measurement is one effective technique to characterize defects in the Si-nanocrystal and thereby resolve the mechanism of the light emission. In this paper, we present the defect structures in the Si-nanocrystal doped SiO_2 films by the ESR method. Moreover, we discuss the correlation between the defects and the light emission from the Si-nanocrystal doped SiO_2 films.

EXPERIMENTAL

The Si-nanocrystal doped SiO_2 films were deposited on p-type Si (100) substrates by co-sputtering of p-type Si (100) and SiO_2. The scheme of the RF sputtering apparatus is shown in Fig.1. The typical deposition conditions are listed in Table I. A sputtering target was consisted of 5 mm×5 mm Si chips placed on a SiO_2 substrate with a diameter of 100 mm, and the numbers of Si chips were 16. The target was sputtered in Ar gas at a pressure of 0.1 Torr at a constant RF power of 110W. The film thickness of 3 μm were obtained by the sputtering time of 2 hours.

Thermal treatments were carried out in Ar atmosphere up to 1100 °C for 1 hour.

The paramagnetic defects in the Si-nanocrystal doped SiO_2 films were investigated by the ESR method. The ESR measurements were performed using X-band spectrometer at room temperature. The g-value, the line width (ΔH_{pp}) and the spin density were determined using the signals of Mn^{2+} and 1,1-diphenyl-2-picryl-hydrazyl (DPPH) as the calibration references.

PL spectra were measured using an 488 nm line of an argon-ion laser at room temperature.

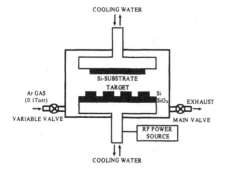

Fig. 1. Schematic diagram of
the RF sputtering apparatus

Table I. Deposition conditions

Substrate	p-type Si (100)
Target	Si chips on SiO_2 glass
Number of Si chips	16
RF power	110 W
Sputtering gas	Ar
Gas pressure	0.1 Torr
Sputtering time	2 hours

RESULTS AND DISCUSSION

Fig. 2 shows the transmission electron microscope (TEM) image of the Si-nanocryatal doped SiO_2 films, which was annealed at 1100 °C. In this micrograph, the presence of Si-nanocrystal was confirmed by the observation of regular lattice image. The nanocrystal was consistent with the (111) plane of diamond structure. We have also observed the sharp rings in the electron-diffraction pattern, which indicated the Si-nanocrystal with diamond structure. The Si-nanocrystals were dispersed uniformly with spherical shape in the SiO_2 films. The average size of the Si-nanocrystal was ~ 2.5 nm.

Fig. 3 shows the typical PL spectra for as-deposited and annealed samples. The films were annealed in Ar atmosphere at 1100 °C for 1 hour. The visible PL spectrum with a peak position of 780 nm was obtained from the films after annealing at 1100 °C, while there was no detectable luminescence from the as-deposited films.

Fig. 4(a) and (b) show the typical ESR spectra for the samples used in Fig. 3. The ESR signals for the as-deposited films were composed of two kinds of ESR centers as shown in Fig. 4(a). One is a ESR signal with g-value of 2.004, which is originated from Si dangling bond in a-SiO_x region[11]. In general, the g-value of ESR spectrum, which is observed in amorphous SiO_x-layers (0<x<2) prepared by the evaporation of silicon at oxygen atmosphere, is usually small than that (g=2.005) of amorphous silicon. If oxygen is introduced into the film, the g-value of this center shifts toward 2.001. The other is a ESR signal with g-value of 2.001 corresponds to E'-center, which is identical with silicon backbonded to three oxygens in SiO_2 region[12].

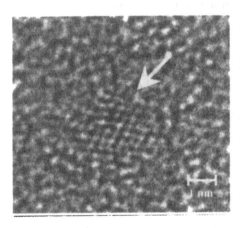

Fig. 2. TEM image for the film annealed at 1100 °C

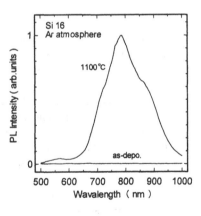

Fig. 3. PL spectra for the as-deposited film and the film annealed at 1100 °C

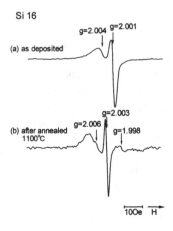

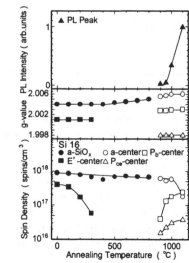

Fig. 4. ESR spectra for (a) the as-deposited film and (b) the film annealed at 1100 °C

Fig. 5. Change in PL intensity, g-value, and the spin density versus annealing temperature (● a-SiOₓ: Si dangling bond in a-SiOₓ region, ■ E'-center: Si backbonded to three oxygens in SiO₂ region, ○ a-center: Si dangling bond in amorphous Si cluster, □ Pb-center: Si dangling bond at the Si-nanocrystal/SiO₂ interface, △ tentatively called the Pce-center: conduction electrons in Si-nanocrystal)

When the film was annealed at 1100 °C, however, the above ESR spectra of Fig. 4(a) disappeared, and then new three kinds of ESR centers appeared, i.e., a-center (Si dangling bond in amorphous Si cluster)[13], P_b-center (Si dangling bond at the Si-nanocrystal/SiO$_2$ interface)[14,15], and the ESR center due to conduction electrons in Si-nanocrystal (tentatively called the P_{ce}-center)[16], as shown in Fig. 4(b).

Fig. 5 shows the change of PL intensity, g-value, and spin density versus annealing temperature. No visible light emission was obtained from the films after annealing up to 800 °C. We only observed the Si dangling bond (g=2.004) in a-SiO$_x$ region and the E'-center from these samples. The g-value of ESR center with 2.004 shifts to 2.005 with rising anneal temperature up to 800 °C. On the other hand, the E'-center disappeared at 400 °C annealing. The visible light emission was obtained when the film was annealed above 900 °C. At the same time, the ESR signals of the a-center, the P_b-center, and the P_{ce}-center were observed. Moreover, the PL intensity, and the ESR signal intensities of the P_b- and the P_{ce}-centers increased with increasing anneal temperature up to 1100 °C. When the annealing temperature increased from 900 °C to 1100 °C, the Si-nanocrystals containing P_b-center in the SiO$_2$ film increased with annealing temperature. The existence of the P_b-center reflect the crystallinity of the Si-nanocrystal. The ESR signal intensity of the P_b-center increased with increasing the number of the Si-nanocrystal. Therefore, the P_b-center increased with PL intensity, even if the P_b-center was an emission killer.

From the results mentioned above, the following relation became evident. When the Si-nanocrystal was formed in the SiO_2 films at annealing temperature above 900 °C, both visible light emission, and the P_b- and the P_{ce}-centers were observed at the same time. We may, therefore, reasonably consider that either of these ESR centers is associated with the visible light emission.

Young et al. observed the ESR signal due to the conduction-band (CB) electrons in porous silicon[16]. Finger et al. also observed it from microcrystalline hydrogenated silicon[17]. However, the visible light emission may or may not reflect the band-to-band transition in the case of Si-nanocrystal doped SiO_2 films, although we also observed the P_{ce}-center, which was identical with CB electrons as shown in Fig. 4(b). Nozaki et al. proposed that the visible light emission was due to the radiative recombination via radiative recombination center[6], that is, the electrons and holes generated by irradiation in a Si-nanocrystal thermally diffuse to the Si-nanocrystal surface, where they recombine via radiative recombination center. Since the radiative recombination center may be associated with interface states in the nanocrystal Si/SiO_2 structure, the light emission energy (1.59 eV) can be smaller than the apparent band gap energy (2.25 eV) of Si-nanocrystal. In addition, the radiative recombination center acts as an oxygen related emission center (OREC)[18]. Therefore, the following process is considered as the light emission mechanism.

First, the CB electron, which excited to the conduction band in the Si-nanocrystal, was trapped in a radiative recombination centers (localized states).

Secondly, the visible light emission was observed as the recombination of electron and hole via the radiative recombination centers.

On the other hand, it is known that the P_b-center is associated with the formation of non-radiative center (emission killer center)[14]. Consequently, it seems reasonably conclude that the P_{ce}-center is strongly associated with the visible light emission. Clearly more work is needed to clarify the mechanism of light emission.

CONCLUSIONS

Correlation between defect structures and light emission in Si-nanocrystal doped SiO_2 films has been investigated by ESR and PL methods. The ESR analysis revealed the presence of two kinds of ESR centers from as-deposited film. One is a defect center with g-value of 2.004, which originated from Si dangling bond in a-SiO_x region. The other is E'-center with g-value of 2.001, which is primarily due to a hole trapped in oxygen vacancy in SiO_2 region. No visible light emission was observed from the as-deposited films. However, when as-deposited film was annealed at annealing temperature above 900 °C, visible light emission was observed. Moreover, the ESR analysis revealed the presence of new three kinds of ESR centers in the annealed sample.

(1) a-center with g-value of 2.006, which originated from Si dangling bond in amorphous Si cluster.

(2) P_b-center with g-value of 2.003, which originated from Si dangling bond at the interface between Si-nanocrystal and SiO_2.

(3) The P_{ce}-center with g-value of 1.998, which is due to the conduction electrons in Si-nanocrystal.

The PL intensity and the ESR signal intensity of the P_{ce}-center increased with increasing anneal temperature up to 1100 °C. It seems reasonable to conclude P_{ce}-center is associated with the radiative recombination center.

ACKNOWLEDGMENTS

The authors gratefully thank Dr. T. Kamino and Ms. T. Yaguchi of Hitachi Techno Research Laboratory for technical assistance on TEM measurement.

REFERENCES

1. L. T. Canham: Appl. Phys. Lett. **57**, 1046 (1990)
2. H. Morisaki, F. W. Ping, H.Ono, and K. Yazawa: J. Appl. Phys. **70**, 1869 (1991)
3. H. Morisaki: Nanotechnology **3**, 196 (1992)
4. Y. Osaka, K. Tsunetomo, F. Toyomura, H. Myoren, and K. Kohno: Jpn. J. Appl. Phys. **31**, L365 (1992)
5. S. Furukawa, and T. Miyasato: Jpn. J. Appl. Phys. **27**, 2207 (1988)
6. S. Nozaki, H. Nakamura, H. Ono, H. Morisaki, and N. Ito: Jpn. J. Appl. Phys. **34**, Suppl. **34-1**, 122 (1995)
7. H. Takagi, H. Ogawa, Y. Yamazaki, A. Ishizaki, and T. Nakagiri: Appl. Phys. Lett. **56**, 2379 (1990)
8. S. S. Iyer, R. T. Collins, and L. T. Canham: Mater. Res. Soc. Symp. Proc. **256**, 1 (1991)
9. S. M. Prokes, and O. J. Glembocki: Phys. Rev. B **49**, 2238 (1994)
10. O. K. Andersen, and E.Veje: Phys. Rev. B **53**, 15643 (1996)
11. E. Holzenkämpfer, F. W. Richter, J. Stuke, and U. Voget-Grote: J. Non-Cryst. Solids. **32**, 327 (1979)
12. D. L. Griscom, and E. J. Friebele: Rad. Effects **65**, 63 (1982)
13. B. L. Crowder, R. S. Title, M. H. Brodsky, and G. D. Pettit: Appl. Phys. Lett. **16**, 205 (1970)
14. M. Shimasaki, Y. Show, M. Iwase, T. Izumi, T. Ichinohe, S. Nozaki, and H. Morisaki: Appl. Surf. Sci. **92**, 617 (1996)
15. Y. Nishi: Jpn. J.Appl. Phys. **10**, 52 (1971)
16. C. F. Young, E. H. Poindexter, and G. J. Gerardi : J. Appl. Phys. **81**, 7468 (1997)
17. F. Finger, C. Malten, P. Hapke, R. Carius, R. Flückiger, and H. Wagner : Phil. Mag. Lett. **70**, 247 (1994)
18. H. Morisaki, and S. Nozaki : J. Vac. Soc. Jpn. **38**, 935 (1995) (in Japanese)

LIGHT EMITTING NANOSTRUCTURES IN IMPLANTED SILICON LAYERS

R. PLUGARU[a], J. PIQUERAS[b], B. MENDEZ[b], G. CRACIUN[a], N. NASTASE[a]
[a]Institute of Microtechnology, 72996 Bucharest, Romania
[b]Dpt. Fisica de Materiales, Facultad de Fisicas, Universidad Complutense, 28040 Madrid, Spain

ABSTRACT

Structural changes in amorphous silicon layers have been investigated as a process route to obtain light emitting silicon nanostructures. The optical emission of the layers was studied by cathodoluminescence (CL). Under 10^{13}-10^{14} boron ions/cm^2 implantation, nanosized crystalline structures grow in the amorphous matrix. A dominant emission band, centered at 400 nm, appears in the catodoluminescence spectrum of the low dose implanted film, while spectra with a 400 nm intense band and 480-500 nm and 650 weak bands are characteristic of higher dose implanted and anodized layers. The structural changes are correlated with the emission properties in the 400-650 nm range.

INTRODUCTION

Different processes have been used in order to obtain light emitting silicon structures. Generally there are two mechanisms involved in their formation: one, is related to disorder effects induced in monocrystalline phase by porosification [1] and the other to the growth of crystallites in an amorphous phase [2,3]. Solid phase crystallization of amorphous silicon films depends on the annealing process parameters and on the initial film structure, e.g. the presence of the nucleation centers. Transmission electron microscopy studies reported by Harbeke *et al* [4] and Voutas *et al* [5] show that depending on the deposition parameters, a crystalline phase is present in the amorphous matrix. Nucleation centers are randomly distributed, with a slight increase of the grain size in the region of the interface between the substrate and the deposited film.

Subsequent annealing processes determine increase of the crystalline grain size starting from the nucleation centers [6,7]. A possible method to control the grain size and orientation is recrystallization of the polycrystalline films by ion irradiation. In this case, depending on the dose used and the orientation of the grains relative to the ion beam direction, the resulting structure consists of crystalline centers. For the high dose irradiation, these centers almost dissapear and an incubation time is necessary for the nucleation and crystallites growth [8]. The implantation processes appear to change the structural properties of amorphous films in a controlled way. The basic process of the amorphous to crystalline phase transition induced by ions implantation has not been established yet. Thermal spikes as temperature assisted processes or increase of the disorder have been invoked for explanation [9,10]. It was proposed that, in highly disordered amorphous silicon the disorder-driven effect causes the formation of crystalline nuclei. The activation energy of grain growth by ion beam irradiation was determined 0.24 or 0.86 eV, as comparing to approximately 2.7 eV reported for thermal growth processes [11]. A contribution to the crystalline phase transition is due to the excited atomic states produced by electronic stopping power loss process of implantation of light ions.

In this work the crystalline phase formation in as-deposited amorphous silicon layers implanted with boron ions has been investigated. The structural changes are correlated with the implantation parameters, dose and energy, and the light emission properties of the implanted films are discussed.

EXPERIMENTAL

Silicon films were obtained by low pressure chemical vapour deposition (LPCVD) at 570_C temperature and 0.4 torr pressure on (100) silicon wafers. Structural changes of the deposited layers were produced by implantation with boron ions at 10^{13} and 10^{14}cm^{-2} doses at an energy of 100 keV. One of the implanted layers was anodized in HF (50wt%), with current densities of 1-2.5mA/cm^2, for 1-5 minutes. Some of the as-deposited and implanted films were annealed at 650_C for 60 min in nitrogen atmosphere. The structural characterization was realized by X-ray diffraction (XRD) and Atomic Force microscopy (AFM). Light emission properties were analysed by cathodoluminescence (CL) in a Hitachi S2500 scanning electron microscope (SEM). The CL measurements were performed at temperatures between 78 and 300 K under different excitation conditions by varying the beam accelerating voltage between 10 and 25 keV or by defocusing the beam on the sample surface.

RESULTS AND DISCUSSION

Nanocrystalline silicon structures in amorphous implanted silicon films.

XRD measurements showed an amorphous structure of the as-deposited films. However, previous transmission electron microscopy studies [4,5] of LPCVD silicon films deposited at 570_C and below revealed the presence of a crystalline phase, with crystallites of 3-10 nm size, in the amorphous matrix. In our samples crystallites with preferred orientation in <111> and <220> directions were revealed by XRD after implantation with boron ions at a dose of 10^{13}cm^{-2} and 100 keV energy, as shown in Figure 1. The estimated grain size was 8-10 nm. Implantation with the 10^{14}cm^{-2} dose cause the increase of the average size of the grains with <111> and <220> orientation and the appearance of a new phase with <211> orientation. The average grain size for the <211> orientation was estimated to be about 8 nm. The XRD spectrum of the 10^{14}cm^{-2} implanted film is shown in Fig. 2.

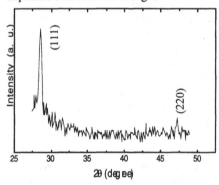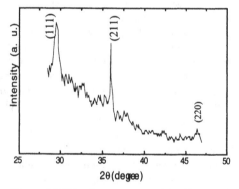

Figure 1 Diffraction spectrum of the film implanted with a dose of 10^{13}cm^{-2}.

Figure 2 Diffraction spectrum of the film implanted with a dose of 10^{14}cm^{-2}.

XRD of the as-deposited film after annealing at 650_C for 60 min, shows the presence of a polycrystalline structure with a strong texture along the <111> direction and a weak texture along the <220> direction. It has been previously reported [11] that thermal annealing of amorphous films at low temperature causes grain growth with preferred <111> orientation.

It appears that the implantation and annealing treatments of this work induce crystalline phase formation with similar structure. The structure is also similar to that obtained after high temperature rapid thermal annealing [12].

CL investigation of the luminescent emission of the silicon layers.

Figure 3 shows the CL spectrum of the as-deposited sample. Emission is observed in the range 350-700 nm with resolved bands centered at 400 nm and about 650 nm as well as a broad band with a peak at 480-500 nm.

The CL emission spectrum of the 10^{13}cm^{-2} implanted sample shows a broad band centered at 400 nm while the emission in the 600-700 nm region appears almost quenched (Fig.4). This is similar to the PL spectrum from nanocrystalline films reported in reference [6].

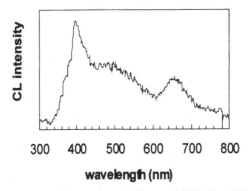

Figure 3 CL spectrum of the as-deposited layer, at 25 keV and focused electron beam.

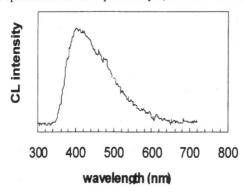

Figure 4 CL spectrum of the 10^{13}cm^{-2} implanted sample.

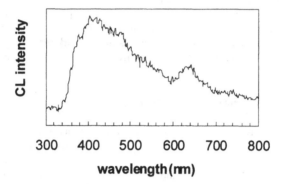

Figure 5 CL spectrum of a film implanted at 10^{14}cm^{-2} dose at 20 keV and defocused electron beam.

The CL spectrum of the film implanted at a dose of 10^{14}cm^{-2} (Fig.5) showed, as in the low dose sample, a band at 400 nm, while the 650 nm emission appeared with lower relative intensity than in the as-deposited sample. The change of the excitation conditions by defocusing the electron beam on the film surface causes the relative increase of the 650 nm band. The total CL intensity was lower in the samples implanted with the high dose.

Anodization produced a slight increase of the emission intensity without spectral changes. This effect was observed in the films after the different implantation or annealing processes. AFM observations at the surfaces of the films after anodization reveal a preferential etching in well defined areas that would correspond to amorphous phase surrounding regions composed of crystalline nanostructures. Figure 6 shows AFM images of the 10^{13}cm^{-2} implanted and annealed film, before and after anodization process.

The main emission band of the as-deposited films appears in the blue region and can be related to the presence of nanocrystallites in the amorphous matrix. In the previous photoluminescence studies [2,16] of amorphous films it was reported that no PL emission was observed from amorphous films while dominant blue emission is characteristic for annealed films. The PL emission was correlated with the presence of the crystallites in the annealed samples and attributed to quantum confinement effects. CL spectra of the implanted layers show enhanced intensity of the blue band. The observed intensity changes can be corelated with the structural evolution of the layers observed in the XRD spectra. Nanocrystallites are present in the low dose as well as in the higer dose implanted layers. The CL spectra in the 400-480 nm region are very similar to the PL spectrum of the nanocrystalline films of ref.2. A similar structure of the layers was observed by XRD investigation.

Annealing produces the decrease of the observed emission intensity that which can be attributed to the reduced quantum size related effects in the larger grains of the annealed sample. The anodization process further increases the blue band intensity. On the other hand, after anodization, green-red bands appear in the CL spectrum of the layer implanted with the high dose. The 650 nm band is also detected in annealed samples. The emission in the range from 600 to 700 nm has been atributed by some authors to states at the interface between nanocrystallites and the amorphous phase [15]. Results of Delgado *et.al* [17] suggest that the observed red emission is related to electronic traps.

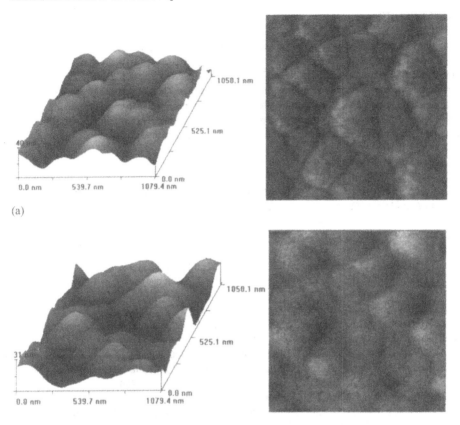

(a)

(b)

Figure 6 AFM images of the layer implanted at 10^{13} cm^{-2} dose after annealing (a) and anodisation at 1mA/cm^{-2} current density for 5 min. (b) processes.

Our CL results indicate that the 400-480 nm emission is present in as-deposited and implanted silicon films. Changes induced in the crystallinity of the samples by implantation or with annealing are correlated variations in the emission intensity which indicate that the luminescence is related to the presence of crystallites. CL emission in the wavelenght region of 530-650 nm was observed in the high dose implanted layers and in the annealed films. The

650 nm band also appears in CL spectra of as-deposited film. Taking into account the structural features of these films, e.g. the presence of cystalline nucleation centers in the as-deposited films and the very narrow grain boundary regions in high dose implanted or annealed samples, the red emission could be related to surface states.

CONCLUSIONS

Structural changes produced by boron ion implantation in amorphous deposited silicon films have been investigated. The growth of nanocrystalline structures with 8-16 nm size can be correlated with the implantation doses. The CL in the SEM reveals the presence of several visible luminescence bands in nanocrystalline silicon films. The dominant blue emission has a complex character with a component at 400 nm which appears related to quantum size effects. We suggest that the 400 nm band is correlated with the nanosized structures, while the 580-600 nm band emission represents the contribution of the defect centers.

ACKNOWLEDGEMENTS

This work was supported by NATO, Grant HTECH.CRG 961392, DGES (Project PB 96-0639) and MRT (Project A39).

REFERENCES

1. A. G. Cullis, L.T. Canham, and P.D. J. Calcott, J.Appl. Phys. **83**, 909 (1997).
2. X. Zaho, O. Schoenfeld, J. Kusano, Y. Aoyagy, and T. Sugano, Jpn. J. Appl. Phys. **33**, L 649 (1994).
3. H.Z. Song, and X. M. Bao, Phys. Rev. B **55**, 6988 (1997).
4. G. Harbeche, L. Krausbauer, E.F. Steigmeier, A. E. Widmer, H.F. Kappert and G. Neuegebaure, J. Electrochem. Soc. **120**, 675 (1984).
5. A. T. Voutas and M.K. Hatalis, J. Electrochem. Soc. **139**, 2659 (1993).
6. X. Zaho, S. Nomura, Y. Aoyagi, T. Sugano, J. Non-Cryst. Solids **198-200**, 847 (1996).
7. T. Asano, K. Makihira, and H. Tsutae, Jpn. J. Appl. Phys.**33**, 659 (1994).
8. N. Yamauchi and R. Reif, J.Appl. Phys. **75**, 3235 (1994).
9. P.A. Stolk, F.W. Saris, A.J.M. Berntsen, W.F.van der Weg, L.T. Sealy, R.C. Barklie, G.Krotz and G. Muller, J.Appl.Phys.**75**, 7266 (1994).
10. A. Yoshinouchi, A. Oda, Y. Murata, T. Morita and S. Tsuchimoto, Jpn. J. Appl. Phys. **33**, 4833 (1994).
11. R.B. Iverson and R. Reif, J. Appl.Phys. **62**, 1675 (1987).
12. S. Nomura, X. Zaho, O. Schoenfeld, K. Misawa, T. Kobayashi, Y. Aoyagi and T. Sugano, Solid State Comm. **92** (8), 665 (1994).
13. N.M. Kalkhoran, F. Namavar and H.P. Maruska, Appl. Phys. Lett. **63**, 2661 (1993).
14. P.G. Han, M.C. Poon, P.K. Ko, J.K.O. Sin, J. Vac. Sci. Technol. B **14**, 824 (1996).
15. X. Zaho, O. Schoenfeld, S. Komuro, Y. Aoyagi and T. Sugano, Phys. Rev.B **50**, 18654 (1994).
16. E. Edelberg, S. Bergh, R. Nanone, M. Hall and E.S. Aydil, J. Appl.Phys.**81**, 2410 (1997).
17. G. R. Delgado, H. W. H. Lee, S. M. Kauzlarich, and R. A. Bley, *Mat. Res. Soc. Symp. Proc.* **452**, 177 (1997).

FORMATION AND LUMINESCENT PROPERTIES OF OXIDIZED POROUS SILICON DOPED WITH ERBIUM BY ELECTROCHEMICAL PROCEDURE

V.Bondarenko*, N.Vorozov*, L.Dolgyi*, V.Yakovtseva*, V.Petrovich*, S.Volchek*,
N.Kazuchits**, G.Grom***, H.A.Lopez***, L.Tsybeskov***, P.M. Fauchet***
*Belarusian State University of Informatics and Radioelectronics, Department of Microelectronics, Minsk, BELARUS 220027, vitaly@cit.org.by
**Belarusian State University, Department of Semiconductor Physics, Minsk, BELARUS 220050
***University of Rochester, Department of Electrical and Computer Engineering, Rochester, NY 14627

ABSTRACT

The present work is concerned with Er-doped oxidized porous silicon (PS). The characteristic feature of the work is that PS doping has been realized by an electrochemical procedure followed by a high temperature treatment. 5-μm thick PS layers were formed on p-type Si of 0.3-Ohm·cm resistivity. Er incorporation was performed by a cathodic polarization of PS in a 0.1M Er(NO$_3$)$_3$ aqueous solution. A high temperature treatment in an oxidizing ambient at 500-1000°C was utilized to provide either partial or total oxidation of PS:Er layers. X-ray microanalysis was used to study chemical composition of the samples. Photoluminescence (PL) and photoluminescence excitation (PLE) spectra were investigated. After the partial oxidation (in the temperature range of 600-800°C), weak Er^{3+}-related PL at 1.53 μm was observed. A high temperature anneal in Ar atmosphere at the temperature of 1100°C caused a significant increase in the Er^{3+}-related PL intensity. Resonant features were observed in PLE spectra of fully oxidized PS. Five peaks at 381, 492, 523, 654, and 980 nm were revealed. The strongest excitation occurred at 381 and 523 nm. The excitation of different Er^{3+} energy levels, cross-relaxation interactions and emission due to the $^4I_{13/2} \rightarrow {}^4I_{15/2}$ transitions were considered. Application of the Er-doped oxidized PS for integrated optical waveguides is presented.

INTRODUCTION

Recently, there has been a growing interest in silicon based integrated optoelectronics [1]. True large-scale integration of optoelectronic components requires integrated optical waveguides to provide optical interconnections. A promising way to form integrated optical waveguides based on oxidized porous silicon (OPS) has been demonstrated [2,3] and found to be compatible with conventional microelectronic silicon technology. OPS-based waveguides buried into a silicon wafer can be formed from a few to 10-15 μm in thickness. An advantage of OPS waveguides is the modification of its optical characteristics by introducing desired dopants into PS before thermal oxidation. OPS waveguides are presently passive devices, whose functionality is to guide and distribute optical signals. The next challenge is to build active devices (amplifiers and lasers) based on OPS waveguides. Incorporating rare-earth elements with strong luminescence into OPS provides a way of realizing integrated optical waveguide amplifiers and lasers. Erbium is of specific interest because its 1.53 μm emission band falls into a principal optical communication window that results in less light absorption. Erbium can be introduced into PS by various methods: ion implantation, diffusion, or electrochemical deposition. Electrochemical deposition offers an advantage over ion implantation and diffusion, because it allows erbium to penetrate into PS to depths of 10 μm [4]. Recent work reported was concerned with Er-doped PS for applications in light-emitting devices [5,6]. In this paper, we present the PL study of partially and fully oxidized erbium doped PS. Er photoluminescence at 1.53 μm in OPS materials has been of our main interest. Excitation spectra that can be helpful to select optimal conditions of active layer pumping in optical waveguide amplifiers and lasers is also studied.

EXPERIMENT

Processing flow to fabricate OPS:Er samples for PL measurements is presented in Fig. 1a. P-type boron doped Si(111) wafers of 0.3 Ohm·cm resistivity were used as initial substrates. Uniform 5 μm thick PS layers were formed by anodization in 48% HF at an anodic current density of 10 mA/cm^2. After anodization, the HF electrolyte was replaced by a 0.1 M Er(NO$_3$)$_3$ solution and an Er-containing film was electrochemically deposited into PS at a cathodic current density of 0.125 mA/cm^2. The samples were then rinsed in de-ionized water, dried at room temperature, and oxidized in a diffusion furnace at 500-1000°C in an O$_2$ atmosphere for 20 min followed by an anneal at 1100°C for 15 min in an Ar environment. Integrated waveguides based on fully oxidized PS were fabricated in a similar manner using a patterned Si$_3$N$_4$ mask (Fig. 1b). Highly doped p$^+$-type Si(111) wafers of 0.03 Ohm·cm resistivity were used as initial substrates.

Room-temperature, 77K, and 4.2K photoluminescence spectra were recorded by a grating spectrometer MDR-23 equipped with a liquid nitrogen-cooled Ge:Cu detector. The 488 nm Ar laser line was used for exciting PL. Photoluminescence excitation (PLE) spectra were recorded at room temperature using a Xe lamp as the excitation source. A chemical composition of the OPS:Er was studied by X-ray microanalysis using a EDX-Spectrometer AN-10000.

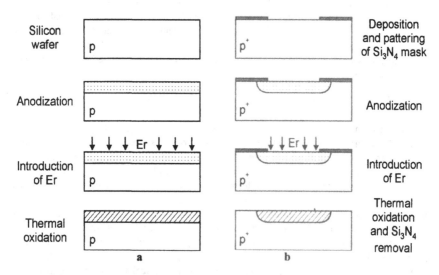

Fig. 1. Processing steps for fabricating (a) OPS:Er samples for PL and PLE measurements and (b) integrated optical waveguides.

RESULTS

Fig. 2 shows the room-temperature PL spectra of PS:Er samples for different processing steps. The PL spectrum of as-prepared PS:Er has an emission band at ~830 nm, which is the familiar S-band, and a broad band between 1600 and 1000 nm, which is attributed to a defect-related radiative transition [7]. The thermal treatment of PS:Er layers causes a significant change in PL spectra. For samples oxidized in oxygen ambient, the intensity of the band related to PS decreases with increased temperature, and a weak Er-related peak at ~1.53 μm shows up. Oxidized PS:Er samples were annealed at 1100°C for 15 min in Ar ambient. Annealing results in a significant increase in intensity of 1.53 μm PL. The most drastic increase is observed for the sample oxidized at 750°C. As for the PL spectrum of this sample, it reveals a weak PS-related band at 750 nm and a weak broad emission band between 1000 and 1400 nm. PL spectra of this type are characteristic of partially oxidized PS:Er. With respect to PS:Er samples oxidized at 950-1000°C followed by annealing, no other emission band in addition to the band at 1.53 μm are revealed in the PL spectra. This strongly suggests that Er ions are the most efficient centers of radiative recombination. PL spectra of this type are characteristic of fully oxidized PS:Er layers.

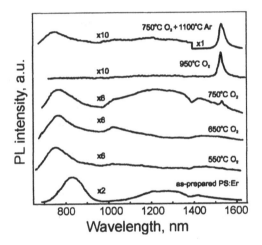

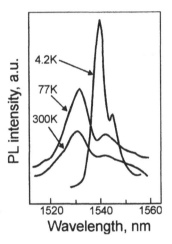

Fig. 2. Room-temperature PL spectra of PS:Er for different processing steps.

Fig. 3. PL spectra of OPS:Er measured at different temperatures.

Fig. 3 shows the PL spectra of Er-doped PS oxidized at 750°C in oxygen and annealed at 1100°C. The spectrum measured at 4.2K shows a sharp peak at 1540 nm and a side (satellite) peak at 1545 nm. These peaks are due to transitions between the first excited manifold $^4I_{13/2}$ and the $^4I_{15/2}$ ground manifold of Er^{3+}-ions [8]. At low temperature (4.2K) the FWHM of the main and side peaks are ~4.8 and ~6-7 nm respectively. This is a result of Stark splitting of the excited and ground states in the host OPS field. The two observed peaks are the only resolved Stark structure at 4.2K as well as at 77K and 300K. We observe a slight difference between the wavelength of the Er^{3+}-related emission peaks measured at different temperatures. The main peak is located at 1540 nm at 4.2K, whereas at 77 and 300K it is observed at 1532 and 1531 nm. A similar shift of the main peak wavelength from 1539 to 1532 nm with increasing temperature from 4.2 to 300K was observed when Er ions were introduced into a silica-like phase of PS [9]. As seen in Fig. 3, the FWHM and the intensity of Er-related emission peaks depend on temperature. The main peak broadening from 4.8 to 21 nm is observed as the temperature increases from 4.2 to 300K. The room-temperature spectral band is broader than that of Er-implanted silica (11 nm FWHM) and soda-lime silicate glass (19 nm FWHM), and narrower than phosphosilicate glass (25 nm FWHM) and aluminum oxide (55 nm FWHM) [8]. The intensity of the main peak decreases by a factor of 3 when the temperature increases from 4.2 to 300K. The spectra presented in Fig. 3 are like those of Er implanted in silica with a high temperature post-annealing [8]. We believe that the PL peaks observed in the present work originate from the Er:O complexes in OPS, which is closely related to silica formed by thermal oxidation of silicon. We did not reveal a peak at 1548 nm that was recently observed at 4.2K in [9]. This peak was attributed to Er ions, which have diffused into the nanocrystallites of PS.

The excitation mechanism of Er in OPS was investigated by measuring the 1.53-μm PL intensity as a function of excitation wavelength. Fig. 4 shows the photoluminescence excitation (PLE) spectra recorded at room temperature for partially and fully oxidized PS:Er.

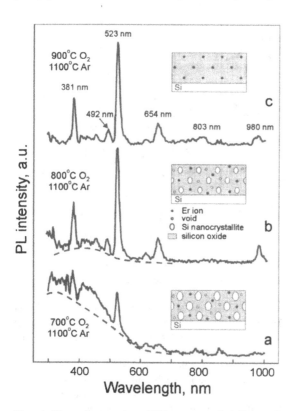

Similar PLE features are observed for both partially and fully oxidized PS:Er samples. They consist of broad band (dashed line) and sharp peaks located at 381, 523, 654, and 980 nm. The broad band PLE spanining from 300 to 600 nm can be attributed to Er ions excitation processes involving silicon nano-crystallites in the partially oxidized PS host. As the oxidation temperature increases this broad excitation band disappears because of oxidation of nanocrystallites. The PLE spectrum of fully oxidized PS:Er is seen to have a pronounced banded structure. Five peaks are exhibited at wavelengths

Fig. 4. Room-temperature PLE spectra of 1.53-μm Er-related emission from (a,b) partially and (c) fully oxidized PS:Er.

of 381, 492, 523, 654, and 980 nm, reflecting the structure of the optical absorption bands of Er^{3+} ions. Five observed excitation peaks are identified with the following transitions within Er^{3+} ions: $^4I_{15/2} \to \,^4G_{11/2}$ (381 nm), $^4I_{15/2} \to \,^4F_{7/2}$ (492 nm), $^4I_{15/2} \to \,^2H_{11/2}$ (523 nm), $^4I_{15/2} \to \,^4F_{9/2}$ (654 nm), and $^4I_{15/2} \to \,^4I_{11/2}$ (980 nm) [8]. A low intensity PLE peak stemming from the transition $^4I_{15/2} \to \,^4I_{9/2}$ is observed at 803 nm. These results indicate that Er ions in fully oxidized PS are excited directly by exciting radiation.

As shown in Fig. 4, the excitation peaks at 381 and 523 nm are clearly more intense than the other peaks. The transitions corresponding to these excitation peaks are generally said to be hypersensitive transitions. Excitation of Er^{3+} ions at wavelengths of 381 and 523 nm are very effective and seem to be the optimal pump wavelengths for Er luminescence at 1532 nm in OPS waveguides. We believe that intense PLE peaks at 381 and 523 nm are due to cross-relaxation of Er^{3+} ions and multiplication of electron excitations. When excited at 523 nm ($^4I_{15/2} \to \,^2H_{11/2}$), Er^{3+} ions relax from the upper $^2H_{11/2}$ level to the $^4S_{3/2}$ level. In this case, the $^4I_{13/2}$ level responsible for PL at 1532 nm is populated as a result of the following cross-relaxation interactions: Er^{3+} ($^4S_{3/2} \to \,^4I_{9/2}) \to Er^{3+}$ ($^4I_{15/2} \to \,^4I_{13/2}$), Er^{3+} ($^4I_{9/2} \to \,^4I_{13/2}) \to Er^{3+}$ ($^4I_{15/2} \to \,^4I_{13/2}$). For every transition, the energy gap does not exceed 580 cm^{-1} [10] and cross-relaxation is very effective. When excited at 381 nm, excitation occurs by transition to the $^2G_{11/2}$ state. In this case, there is a wide selection of cross-relaxation interactions with narrow energy gaps that populate the $^4I_{13/2}$ level. So, excitation at 381 nm is also very effective. It is important to keep in mind that in parallel with the considered cross-relaxation processes, intra-ionic non-radiating multi-phonon relaxation participates in populating all energy levels of Er^{3+} ions, including the $^4I_{13/2}$.

Fig. 5 shows (a) a cross-section of the waveguide based on OPS:Er, and profiles of Er distribution (b) over the surface and (c) in depth of the waveguide channel. Fig. 5d shows the PL spectra from the surface of the waveguide channel and from OPS sample doped with Er.

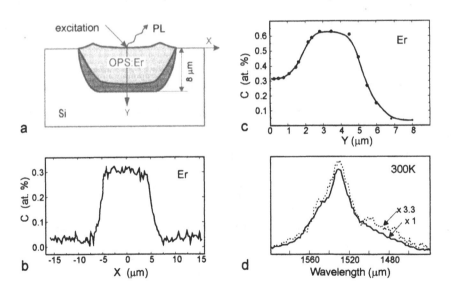

Fig. 5. Cross-section of waveguide (a), profiles of Er distribution (b, c), and PL spectra (d) measured from waveguide (solid line) and OPS sample doped with Er (dotted line).

As Fig. 5a shows, the channel waveguide has dimensions of 8-μm thick and 20-μm width. Scanning over the surface (Fig. 5b) reveals Er at the waveguide core that is 11-12 μm wide. In this area, Er concentration was constant around 0.3 atomic percent. With scanning in depth of the waveguide (Fig. 5c), at first the Er concentration increases from 0.3 atomic percent at the surface to 0.6 atomic percent at a depth range of 2.5-4.5 μm, and then decreases to under 0.1 atomic percent at a depth of over 7 μm. A correlation between Fig. 5a, Figs. 5b, and 5c discloses that Er is introduced mainly into the core part of the channel, through which waveguide modes propagate [2-5]. Outer regions of the waveguide were doped with Er to a lesser extent. Fig. 5d shows that the PL spectra of OPS:Er waveguiding channel is quite similar to that of reference continuous OPS:Er layer.

CONCLUSIONS

In order to make an effective fiber-optic amplifier, Er concentration in the fiber should be no less than 0.001 atomic percent for fiber lengths of a few tenths of a meter [8]. For integrated waveguiding structures, waveguide lengths are only a few centimeters. So, to attain the amplification effect, Er concentration should be no less than 0.1 atomic percent [8]. Er concentration reported in this work fulfils this requirement. The results obtained in this report can be helpful in designing integrated optical waveguides amplifiers.

ACKNOWLEDGMENTS

The authors would like to thank V.V.Kuznetzova for fruitful discussions. The research described in this publication was made possible in part by Award No BE2-108 of the U.S. Civilian Research and Development Foundation for the Independent States of the Former Soviet Union (CRDF).

REFERENCES

1. R.Soref, MRS Bulletin, **23,** 20 (1998).
2. V.Bondarenko, V.Varichenko, A.Dorofeev, N.Kazuchits, V.Labunov, V.Stelmah, Tech. Phys. Lett., **19,** 463 (1993).
3. V.Bondarenko, A.Dorofeev, N.Kazuchits, Microelectron. Engineering, **28**, 447 (1995).
4. T.Kimura, A.Yokoi, H.Horiguchi, R.Saito, T.Ikoma, A. Sato, Appl. Phys. Lett., **65,** 983 1994).
5. L.Tsybeskov, S.Duttagupta, K.Hirschman, P.Fauchet, K.Moore, D.Hall, Appl.Phys.Lett., **70,** 1790 (1997).
6. H.Lopez, S.Chan, L.Tsybeskov, P.Fauchet, and V.Bondarenko, this volume.
7. A.Cullis, L.Canham, and P.Calcott, J.Appl.Phys., **82,** 909 (1997).
8. A.Polman, J.Appl.Phys., **82,** 1 (1997).
9. M.Stepikhova, W.Jantsch, G.Kocher, L.Palmetshofer, M.Schoisswohl, H.Bardeleben, Appl. Phys.Lett., **71,** 2975 (1997).
10. T.Kojan, V.Kuznetzova, P.Pershukevich, I.Sergeev, V.Homenko, J.Appl.Spectroscopy, **63,** 992 (1996).

LONG LUMINESCENCE LIFETIME OF 1.54 μ m Er^{3+} LUMINESCENCE FROM ERBIUM DOPED SILICON RICH SILICON OXIDE AND ITS ORIGIN

SE-YOUNG SEO, JUNG H. SHIN AND CHOOCHON LEE
Department of Physics, Korea Advanced Institute of Science and Technology (KAIST), 373-1~Kusung-dong, Yusung-gu, Taejon, Korea

ABSTRACT

The photoluminescent properties of erbium doped silicon rich silicon oxide (SRSO) is investigated. The silicon content of SRSO was varied from 43 to 33 at. %, and Er concentration was 0.4-0.7 at. % in all cases. We observe strong 1.54 μ m luminescence due to $^4I_{13/2} \Rightarrow {}^4I_{15/2}$ Er^{3+} 4f transition, excited via energy transfer from carrier recombination in silicon nanoclusters to Er 4f shells. The luminescent lifetimes at the room temperature are found to be 4-7 msec, which is longer than that reported from Er in any semiconducting host material, and comparable to that of Er doped SiO$_2$ and Al$_2$O$_3$. The dependence of the Er^{3+} luminescent intensities and lifetimes on temperature, pump power and on background illumination shows that by using SRSO, almost all non-radiative decay paths of excited Er^{3+} can be effectively suppressed, and that such suppression is more important than increasing excitation rate of Er^{3+}. A planar waveguide using Er doped SRSO is also demonstrated.

INTRODUCTION

Erbium is a rare earth atom which can luminesce at 1.54 μm, corresponding to an intra-4f shell transition from the first excited $^4I_{13/2}$ to the ground state $^4I_{15/2}$. This wavelength coincides with the minimum absorption window of silica optical fibers. Thus, Er doped Si offers the possibility of Si-based light source for an all-silicon integrated optoelectronics. Since the initial study [1], intense research has resulted in Si:Er based LEDs operating at room temperature [2,3].

However, Si:Er based LEDs suffer such a tremendous quenching of luminescence intensities and efficiencies as either the temperature or the pump power is increased that so far, they remain unpractical. By now, there is a general consensus that such activation of non-radiative decay paths is intrinsic to the excitation mechanism of Er in semiconductors including Si. [4] This is shown in more detail in Fig. 1, which shows, schematically, the excitation and de-excitation mechanisms of Er in Si. The dark arrows show the excitation mechanisms, and the light arrows show the de-excitation mechanisms. Er in Si is excited via Auger-type interaction with carriers, either by bound e-h pairs as on the left side of Fig. 1, or by hot electrons as shown on the right side of Fig. 1. De-excitation mechanism is in essence the reverse of the excitation mechanism.

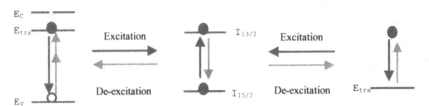

Fig. 1: Schematic description of excitation and de-excitation mechanisms of Er in Si

Er 4f electrons can decay by creating another e-h pair. In this case (usually referred to as back-transfer), phonons are needed to overcome the energy mismatch between Er 4f transition and the formation energy of e-h pair. Er 4f electrons can also decay non-radiatively by Auger exciting carriers, both bound and free, as they become more numerous, either at higher temperatures or during photo- and electrical excitation of Si:Er.

As these non-radiative mechanisms are intrinsic to the excitation mechanisms of Er, they can be alleviated, but not eliminated. One known way of suppressing such non-radiative decay paths of Er is using a wide-bandgap material. This increases the energy mismatch between the 4f transition and the formation energy of e-h pair, while at the same time reducing the number of free carriers. The benefits of using wide-bandgap materials have been demonstrated by Er doped GaN [5] and SiC [6]. Therefore, by using Si quantum dots, whose bandgap can be engineered by reducing its size, we may be able to suppress the activation of non-radiative decay paths of Er while still using Si. Recently, we have shown that silicon–rich–silicon–oxide (SRSO), which consists of Si nanoclusters embedded in SiO_2 matrix, is an excellent host material for Er [7]. In this paper, we show that by using SRSO, nearly all non-radiative decay paths of Er can be effectively suppressed, leading to a very efficient Er^{3+} luminescence. We also demonstrate the feasibility of waveguides using using Er doped SRSO.

EXPERIMENT

Er doped SRSO films of varying Si contents were deposited by ECR-PECVD of SiH_4 and O_2 with concurrent sputtering of Er and subsequent rapid thermal anneal. Details of the deposition process can be found in Ref. [7]. Compositions of deposited films were determined by RBS (not shown). The Si content ranged from 43 to 33 %, and the Er content was 0.7-0.4 at. %. Visible photoluminescence (PL) from SRSO measured using the 325 nm line of He-Cd laser and GaAs PMT. 1.54 μm Er^{3+} PL measured using 488 or 515 nm line of Ar laser and TE cooled InGaAs detector. All PL measured using grating monochromators and lock-in technique. Low temperature 1.54 μm Er^{3+} PL measured using a closed-cycle He cryostat.

RESULTS AND DISCUSSION

Figure 2 shows the visible luminescence spectra of SRSO. Broad luminescence peaks in the 600-800 nm range typical of silicon nanoclusters are observed. From the compiled data on the relationship between PL peak position and cluster sizes [8], we estimate the cluster sizes to be <2 nm, 2 nm, 3 nm, and 3 nm for SRSO films with Si content of 33%, 35%, 41%, and 45%, respectively. From these values, we estimate the cluster density to be ≈ 1×10^{18} cm^{-3} for all films.

Figure 3 shows the room temperature infrared spectra from these SRSO films, showing the typical Er^{3+} luminescence spectra. The inset shows the dependence of

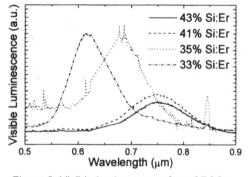

Figure 2. Visible luminescence from SRSO

Er³⁺ luminescence intensity upon the pump wavelengths. We observe Er³⁺ luminescence at all excitation wavelengths and not just those corresponding to Er³⁺ absorption bands. This indicates that in all cases, excitation of Er³⁺ atoms is dominated by carriers in Si nanoclusters even when the composition of SRSO films is nearly that of stoichiometric SiO₂, and shows that such energy transfer mechanism is very efficient. It should be noted that SRSO films with different Si content absorb the excitation beam differently. Thus, caution should be used when comparing Er³⁺ luminescence intensities from different films.

Figures 4 and 5 show the temperature dependence of the Er³⁺ luminescence intensities and luminescence decay times, respectively. The PL intensities are normalized to the values at 25 K. Two things become immediately clear. First, the luminescence decay times are very long, especially for 35% Si film. It's nearly 9 msec at 25 K, and even at room temperature, is nearly 7 msec. This is, to our best knowledge, the longest Er³⁺ luminescence decay time observed from a non-glass host, and is in fact is very close to that observed from Er doped glasses. Second, both the Er³⁺ luminescence intensities and lifetimes undergo virtually no quenching as the temperature is raised from 25K to 300 K except for a small onset of quenching above 250 K for samples with Si contents above 40%. The lack of temperature quenching of Er³⁺ luminescence decay times indicates that both the "back-transfer" and Auger-excitation of equilibrium free carriers are absent in Er doped SRSO. This demonstrates that bandgap-engineering by using silicon nanostructures is

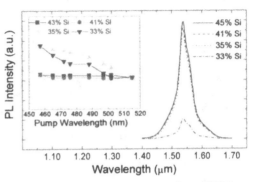

Figure 3. Er luminescence from SRSO

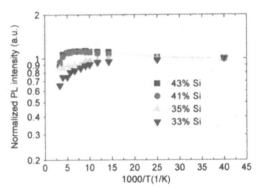

Figure 4: Er³⁺ PL intensity vs temp.

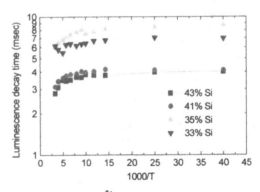

Figure 5: Er³⁺ PL decay time vs temp.

effective suppressing thermal activation of non-radiative decay paths for excited Er^{3+} atoms if clusters are smaller than 3 nm. And the lack of temperature quenching of Er^{3+} luminescence intensity further indicates that we also have suppressed thermally activated reduction of Er^{3+} excitation efficiency, which was shown to be important for Er doped bulk Si. [9]

Now we investigate the effect of excess carriers generated by excitation beam on the Er^{3+} luminescence. Figure 6 shows the dependence of Er^{3+} luminescence decay times on the excitation power. In case of bulk Si, it is known that the luminescence decay times are affected by the excitation power, becoming significantly shorter as the excitation power is increased [10]. As Fig. 6 shows, however, Er doped SRSO films show only very little such quenching. It should be noted that in this case, as we are observing the decay of Er^{3+} luminescence after the excitation beam has been turned off, excited Er^{3+} 4f electrons can interact only with bound carriers. Thus, in this case, the lack of quenching of Er^{3+} luminescence by high-power excitation indicates that Auger-excitation of bound carriers by excited 4f electrons of Er^{3+} has been suppressed. The reason for such suppression is not clear, as Auger-excitation of bound carriers does not involve the bandgap of the material. Furthermore, as the exciton lifetimes in Si nanoclusters can be as long as 50 μsec [11], interaction with bound carriers should be observable in the time scale used in the experiments. One possible reason may be that since carriers are confined in Si nanoclusters, Er atoms in SRSO simply "see" less bound carriers than they do in bulk Si. This also may explain why SRSO samples with high Si contents (41 and 43% Si) do show a small but definite (15%) quenching by high power excitation.

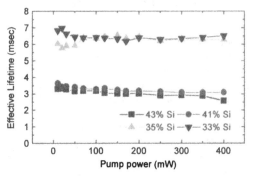

Figure 6: Er^{3+} decay time vs. excitation power

To investigate the effects of excess free carriers on the Er^{3+} luminescence, two-beam experiments were performed. Using a variable beam splitter, the excitation beam was split into two beams, one which illuminated the sample continuously, and the other which was chopped by the mechanical chopper. Figure 7 shows the decay times of the Er^{3+} luminescence as functions of the power of the continuous beam. In this case, a clear quenching of the Er^{3+} luminescence lifetime by the excitation beam is observed. It should be noted, however, that the degree of quenching is much less the case with Er doped bulk Si. In that case, background

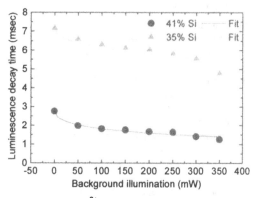

Figure 7: Er^{3+} decay time vs. background illumination

illumination of 80 mW results in a more than three fold reduction of the Er^{3+} luminescence decay time [4]. One important corollary to this result is that the Er^{3+} luminescence decay times usually quoted in the literature are not the actual decay times during excitation.

In bulk Si, the excess free carriers recombine bimolecularly. Thus, the excess free carrier density is proportional to $\sqrt{(P_{background})}$. We assume that excess free carriers in Si nanoclusters, too, recombine bimolecularly. From Fig. 4-6, we know that other non-radiative decay paths of Er^{3+} 4f electrons such as back-transfer, Auger-excitation of equilibrium free carriers, and Auger-excitation of bound carriers are negligible. Thus, we may take the Auger-excitation of excess free carriers to be the dominant non-radiative decay paths for excited Er^{3+} 4f electrons. In such a case, we can model the dependence of Er^{3+} decay time on the power of background illumination as

$$\tau = \frac{1}{1/\tau_o + c\sqrt{(P_{background})}} \tag{1}$$

where τ_o is the luminescence decay time in the absence of background illumination, and c is a constant to be determined by fitting that includes the absorption coefficient as well as the strength of Er^{4f} electron-free carrier Auger interaction. The solid curves in Fig. 7 are the results of such fit. Satisfactory fits are obtained for c values of 0.02 and 0.002 for SRSO films with Si contents of 41% and 35%, respectively, indicating a much stronger suppression of Auger-excitation of excess free carriers in case of SRSO film with 35% Si.

The effect of such suppression of non-radiative decay paths of excited Er^{3+} on the absolute Er^{3+} luminescence is shown in Fig. 8. At high excitation powers, the SRSO film with 35% Si shows a much stronger Er^{3+} intensity than the SRSO film with 41% Si even though it should have a smaller absorption coefficient. Furthermore, at 800 mW, the Er^{3+} luminescence intensity of 43% Si SRSO film has nearly saturated, while that from 35% Si SRSO film still is still increasing. This indicates that for strong Er^{3+} luminescence, it is much more important to suppress the non-radiative decay paths of excited Er^{3+} atoms by using as little of smaller Si clusters as possible than to increase the excitation rate by using more of larger Si clusters.

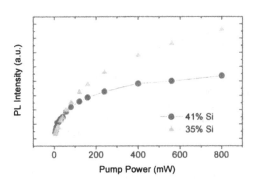

Figure 8: Er^{3+} PL Intensity vs. excitation power

One possible application for such SRSO films would be an Er doped planar waveguide. The advantage of SRSO-based waveguides over glass-based waveguides would be the possibility to excite Er atoms in the waveguide with any broad-band light source instead of a laser tuned to an absorption band of Er^{3+}. To test such a possibility, a slab-waveguide using a 1.3 μm thick Er doped SRSO film deposited on Si substrate with a 5 μm thick oxide layer was fabricated. The refractive index of SRSO layer was measured to be 2.4. Figure 9 shows a CCD camera image of 1.3 μm light confined in such a waveguide. Good confinement of light is seen, confirming the feasibility of such waveguides.

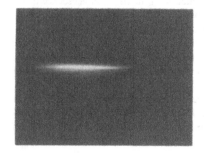

Figure 9: CCD camera image of 1.3 μm light confined in a SRSO/SiO₂/Si slab waveguide

CONCLUSION

In conclusion, we have demonstrated Er^{3+} luminescence with room temperature luminescence lifetimes of nearly 7 msec from Er doped silicon-rich-silicon-oxide. Dependence of Er^{3+} luminescence intensities and lifetimes on the temperature, pump power, and background illumination show that by using SRSO, we have suppressed nearly all possible non-radiative decay paths of excited Er^{3+} 4f electrons while still maintaining efficient excitation via carriers. Using such Er doped SRSO films, slab-waveguide in the infrared region has been demonstrated.

CONCLUSION

It is a pleasure to acknowledge Dr. J. H. Song and Prof. S. Y. Shin for expert technical assistance. This work was supported by STEPI.

REFERENCES

[1] H. Ennen, J. Schneider, G. Pomrenke, and A. Axmann, Appl. Phys. Lett., **43**, 943 (1983).

[2] B. Zheng, J. Michel, F. Y. G. Ren, L. C. Kimerling, D. C. Jacobson, and J. M. Poate, Appl. Phys. Lett. **64** 2842 (1994)

[3] G. Franzó, F. Priolo, S. Coffa, A. Polman, and A. Carnera, Appl. Phys. Lett. **64** 2235 (1993).

[4] J. Palm, F. Gan, B. Zheng, J. Michel, and L. C. Kimerling, Phys. Rev. B **54** 17603 (1996)

[5] J. T. Torvik, C. H. Qui, R. J. Feuerstein, J. I. Pankove, and F. Namavar, J. Appl. Phys., **81** 6343 (1997)

[6] W. J. Choyke, R. P. Devaty, L. L. Clemen, M. Yoganathan, G. Pensi and Ch. Hässler, Appl. Phys. Lett. **65** 1668 (1994)

[7] J. H. Shin, M. Kim, S. Seo, and C. Lee, Appl. Phys. Lett. **72** 1092 (1998)

[8] C. Delerue, G. Allan, and M. Lannoo, *Light Emission in Silicon: From Physics to Devices* Semiconductors and Semimetals **49** 272 (Academic Press, 1998)

[9] F. Priolo, G. Franzò, S. Coffa, and A. Carnera, Phys. Rev. B **57** 4443 (1998)

[10] S. Coffa, G. Franzò F. Priolo, A. Polman, and R. Serna, Phys. Rev. B **49** 16313 (1994)

[11] M. L. Brongersma, A. Polman, K. S. Min, E. Boer, T. Tambo, and H. A. Atwater, Appl. Phys. Lett. **72** 2577 (1998)

LOCAL STRUCTURE AND Er³⁺ EMISSION FROM PSEUDO-AMORPHOUS GaN:Er THIN FILMS

S.B.ALDABERGENOVA*, M.ALBRECHT*, A.A.ANDREEV***, C.INGLEFIELD**,
J.VINER**, P.C.TAYLOR** and H.P.STRUNK*
* Institut für Werkstoffwissenschaften, Universität Erlangen-Nürnberg, Cauerstr.6
91058 Erlangen, Germany
** University of Utah, Salt Lake City, UT 84112, USA
*** A.F.Ioffe Physical-Technical Institute, St.-Petersburg, 194021, Russia

ABSTRACT

We report on strong Er³⁺ luminescence in the visible and infra-red regions at room temperature in amorphous GaN:Er thin films prepared by DC magnetron co-sputtering. The intensity of the Er³⁺ luminescence at 1.535 µm corresponding to $^4I_{13/2} \rightarrow {}^4I_{15/2}$ transitions is greatly enhanced after annealing at 750°C. In this material GaN crystallites have formed and embedded in the continuous amorphous matrix. The crystallites are 4 to 7 nm in diameter as analyzed by high resolution transmission electron microscopy. The absorption edge, extending three orders of magnitude in absorption coefficient in the spectral range from 0.5 to 3.5 eV, is superimposed on resonant absorption bands of Er³⁺ ions.The total photoluminescence spectrum consists of well-defined Er³⁺ luminescence peaks imposed on a broad band edge luminescence from the amorphous GaN host matrix.

INTRODUCTION

It is well-known that Er³⁺ luminescence is observable in most III-V semiconductors at room temperature. Favennec et al.[1] showed a strong dependence of Er³⁺ emission intensity both on the band gap of the different hosts and temperature: the wider the band gap the less the temperature quenching. Therefore, it is advisable to choose a semiconductor host with a larger optical gap, for example, III-V nitrides. The Er³⁺ luminescence in GaN (Eg=3.3eV) and AlN (Eg=6.3eV) was reported by several groups [2-6]. Because of the small mass of the nitrogen atom, the shorter and more ionic Er ligand bonds can be associated with more intense Er³⁺ luminescence. The thermal stability of Er³⁺ emission in wide gap host is expected because of the relatively large binding energy of the exciton with subsequent transfer of the recombination energy to the nearby Er³⁺ ion. There are two separate problems: (i) the limited fraction of optically active Er³⁺ ions even at the highest levels of doping, (ii) the high concentration of dislocations and other extended defects (more than $10^9 cm^{-3}$), where excitons can non-radiatively recombine. In essence, all that is important is a high concentration of Er atoms to make them close in the host: the more the exciton passes on, the more likely it looses its energy to the radiative centers Er³⁺.

In this paper we report on strong Er³⁺ luminescence at room temperature in a pseudo-amorphous GaN:Er system at 1.54µm and in the visible region. One of the prime advantages as opposed to their crystalline analogues is that an amorphous material is much easier to fabricate. As the main advantage, being amorphous, GaN and AlN hosts can support larger quantities of active Er³⁺ ions and other impurities such as O. On the whole, the geometric structure and the absence of long-range order in the amorphous state broadens the absorption and the luminescence spectra . We shall show that the broader absorption in the nitride matrices allows a better coupling of the excitation energy.

EXPERIMENT AND RESULTS

Pseudoamorphous GaN thin films doped by Er were prepared by (DC+HF) magnetron co-sputtering of Ga targets with additional pellets of metallic Er. A gas mixture of 20% N_2 and 80% Ar was used to deposit nitrides at a substrate temperature of about 350°C. The GaN films were grown 1μm thick on glass and Si substrates. Secondary ion mass spectroscopy (SIMS) measurements show that both Er and O concentrations are about 10^{20} atoms/cm³.

To activate the Er, the nitrides were annealed in pure N_2 flow (under a pressure of about 1 Torr) at a given temperature in a quartz tube furnace. The annealing temperatures were choosen at intervals of 100°C for temperatures up to 500°C and at 50°C intervals beyond 500°C (550°C, 600°C ...). The samples were kept at each annealing temperature for 15 minutes and allowed to cool down to room temperature to make measurements. Photoluminescence (PL) measurements were repeated after each annealing cycle. The highest PL intensities of Er^{3+} luminescence in GaN:Er were observed for annealing at around 750°C. The rapid decrease in PL intensity above 750°C can be associated with precipitation of Er atoms. Optical absorption in the spectral range up to 3.5 eV was measured using photothermal deflection spectroscopy (PDS). We measured photoluminescence spectra from Er doped GaN at 1.54μm and in the visible region at room temperature and 77K. The PL spectra were excited by incident light of wavelengh 357, 488 and 514.5 nm from an argon ion laser. The spectra were measured by using a 0.85m Spex 1404 double monochromator with a 600 nm grating and detected with a cooled Ge pin detector.

In order to optimize the Er^{3+} PL it is essential to clarify the exact microstructure of the host nitride and its development with annealing. A high resolution transmission electron microscope (HRTEM) with a point resolution of 1.7Å is used to directly observe the structure. Figure 1 shows a HRTEM micrograph of GaN sample annealed at optimal temperature. This image clearly shows

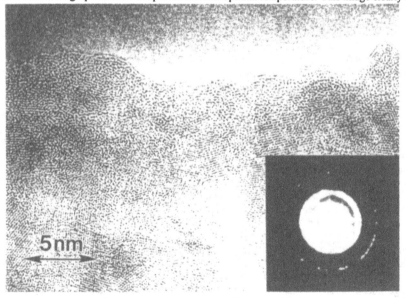

Fig.1.HRTEM micrograph reveals tiny GaN nanocrystallites embedded in an amorphous matrix. The insert shows the diffraction pattern of this region.

that the host nitride has a two-phase microstructure. It can be described by small GaN crystallites embedded in a continuous amorphous matrix. Mainly GaN particles with a diameter of 4 to 7 nm can be seen. The structure of these crystallites is first examined by electron diffraction. Several non-continuous rings show that the film is composed of randomly distributed GaN crystallites. Two rings are seen corresponding to approximate lattice plane spacings d=0.24 and d=0.27 nm. These values can be attributed to the calculated spacings $d_{\{0111\}}$= 0.244 nm and $d_{\{0110\}}$ = 0.276 nm of the hexagonal structure of GaN. Annealing increases the volume fraction of crystalline phase from 10% in as prepared a-GaN to 35% in an annealed sample. Tendency to crystallization with the annealing procedures is confirmed by the Raman data (not shown here, will be discussed elsewhere).

Photothermal deflection spectroscopy yields a broad absorption edge reminiscent of the Urbach behaviour in other amorphous systems. The absorption edge extends over three orders of magnitude in the spectral range from 0.5 to 3.4 eV (Fig.2) and results from contributions of all electronic transitions between localized states in the optical gap. Over a whole range the smooth slope deviates markedly at all exactly defined energetical positions of Er^{3+} states. Absorption bands due to electronic transitions in the Er^{3+} ion peak at 1.535µm ($^4I_{15/2} \rightarrow {}^4I_{13/2}$), 920nm ($^4I_{15/2} \rightarrow {}^4I_{11/2}$), 800nm ($^4I_{15/2} \rightarrow {}^4I_{9/2}$), 660nm ($^4I_{15/2} \rightarrow {}^4F_{9/2}$) and 540nm ($^4I_{15/2} \rightarrow {}^4S_{3/2}$) are superimposed with the background absorption. The broadening of the Er^{3+} peaks can be explained by the Stark splitting due to the electric field in the host and the distributions of the oscillator strengths of optical transitions from the ground state to the higher energy levels of the Er^{3+} ion. The possible optical transitions from the initial state of the absorbing atom can be classified in four types: 1) to continuum states, 2) to the localized states in the gap, which are consistent with the amorphous nature of the material, 3) to well-defined states of Er^{3+} ions and 4) to excitonic states in small GaN crystallites. The real advantage of the amorphous and crystalline structure of our samples comes from the broad background absorption which may create a large number of (e-h) pairs localized deep in the gap with subsequent energy transfer to Er^{3+} ions and an additional excitation of (e-h) pairs in GaN particles which then effectively transfer their energy to Er^{3+} ions where then recombination occurs.

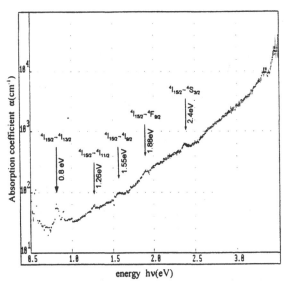

Fig.2. Optical absorption spectrum Er-doped a-GaN sample annealed at 750°C. The absorption edge extends up to 3.5 eV and resonant absorption bands of Er^{3+} ions are superimposed.

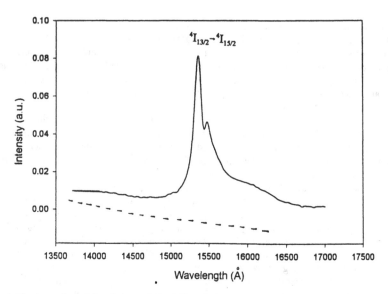

Fig.3.The intensity of the characteristic Er^{3+} emission at 1.535 μm at room temperature from Er-doped a-GaN is enhanced by a factor of 25 after annealing at optimal 750°C(continuous line) in comparison to that of as prepared sample (dashed line). Excitation wavelength is 514.5nm.

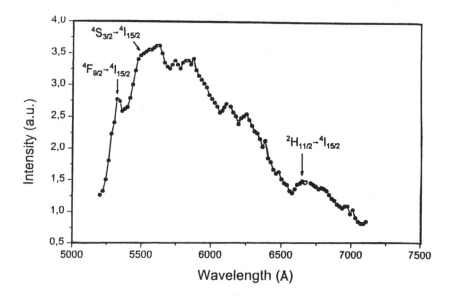

Fig.4. Photoluminescence spectrum of Er-doped a-GaN sample annealed at 750°C as measured at 77K with excitation by the 488nm line of the Ar laser.

Characteristic Er^{3+} emission peaking at 1.535 μm due to $^4I_{13/2} \rightarrow {}^4I_{15/2}$ intra-shell transitions was observed under excitation at 488 and 514.5 nm and always shows nearly identical shape in GaN:Er. Figure 3 shows the PL spectrum of Er^{3+} ion in GaN:Er measured at room temperature after optimal annealing at 750°C. The Er^{3+} luminescence intensity at 1.535 μm is strongly enhanced by more than 25 times in comparison to that of an as prepared sample. We note that when temperature of measurements is raised from 77 K to 300 K, the total PL intensity at 1.535 μm decreases only by a factor of 3-4.

We now go into details about strong visible (green-yellow colored) room temperature luminescence from Er doped a-GaN observed when different excitation wavelengths of 514.5, 488 and 357 nm applied. The broad luminescence spectrum consists of several emission bands peaking at 534, 550, 580 and 610nm. The broad luminescence coming from wide band gap amorphous matrix masks the transitions from intra-shell states of Er^{3+} ions in the spectrum we measured. The PL peaks at 534 and 550nm we can be tentatively identified with the $^2H_{11/2} \rightarrow {}^4I_{15/2}$ and $^4S_{/2} \rightarrow {}^4I_{5/2}$ optical transitions, respectively and a band peaking at 660nm with the $^4F_{9/2} \rightarrow {}^4I_{15/2}$ optical transitions of Er^{3+} ion. The PL bands in between green (550nm) and red (660nm) Er^{3+} emission peaks in the PL spectrum at 77K (Fig.4.) comes very probably from radiative recombination between tail or defect states in the amorphous host. The PL spectrum at room temperature is slightly different (not shown), but the intensity of the Er^{3+} emission wavelengths deacreases only by a factor of 3. Much stronger green light emission comes from the a-GaN:Er sample when 488nm line of the Ar+ laser is used as a pumping wavelength. This photon energy is in resonance with the $^4F_{7/2}$ energy level. Once effectively excited it can promote the light emission from the underlying core levels to the ground states. Thus, we can state that the shape of the emission spectrum in a-GaN:Er in the visible region is determined by the superposition of intra-shell optical transitions of Er^{3+} ions and the edge photoluminescence of the semiconductor host. In our amorphous films, free carriers photo-excited in the host are captured in the localized states of conduction and valence band tails. As a result, by tunneling localized (e-h) pairs occupy every localized tail state, providing better coupling between tail-to-tail recombination and the Er core excitation energy. In this case, both radiative recombination of band tail electron and holes in a-GaN matrix and Er^{3+} intra-shell emission can be observed. Mechanisms of energy transfer from the absorbing sites in the matrix to the Er^{3+} ions can be changed after annealing, when new small GaN crystallites appear, as TEM analysis shows. The contribution of (e-h) pairs excited in small GaN crystallites can be significant as the strong intensity increase of the luminescence after nanocrystallites formation shows. Because of the large exciton binding energy in small crystallites, the (e-h) pairs can store energy effectively, thus with a high probability transfer to nearby Er^{3+} ions.

CONCLUSIONS

In summary, we observe room temperature green-yellow light emission which can be easily seen by the naked eye from Er doped GaN films prepared by magnetron co-sputtering. Characteristic Er^{3+} PL at 1.535 μm wavelength and in the visible region are drastically enhanced after annealing at 750°C. The HRTEM analysis shows the appearance of new small GaN crystallites in the continuous amorphous matrix after annealing, which is paralleled by a strong luminescence increase (factor 25 at 1.535μm and 8 in the visible range). The broadening of the optical spectra occurs due to formation of the band tail states which lead to strong localization of the free carriers by amorphous host. Transitions from these localized states contribute to the broad edge luminescence from the amorphous host, then the total emission spectrum consists of broad edge luminescenscence superimposed with strong Er^{3+} photoluminescence peaks.

ACKNOWLEDGMENTS

This work is supported by National Research Council, National Academy of Science (COBASE Grant).

REFERENCES

1. P.N.Favennec, H.L'Haridon, D.Moutonnet, M.Salvi, and M.Gauneau in Rare Earth Doped Semiconductors, edited by G.S.Pomrenke, P.B.Klein, and D.W.Langer, Mater.Res.Soc.Symp.Proc., vol.31 (Materials Research Society,Pittsburg, PA,1993), p.181.
2. R.G.Wilson, R.N.Schwartz, C.R.Abernathy, S.J.Pearton, N.Newman, M.Rubin, T.Fu, and J.M.Zavada, Appl.Phys.Lett.65, 992 (1994).
3. S.Kim, S.J.Rhee, D.A.Turnbull, E.E.Reuter, X.Li, J.J.Coleman, and S.G.Bishop, Appl.Phys.Lett. 71, 231 (1997).
4. X.Wu, U.Hommerich, J.D.Mackenzie, C.R.Abernathy, S.J.Pearton, R.N.Schwartz, R.G.Wilson and J.M.Zavada, Appl.Phys.Lett.70, 2126 (1997).
5. Myo Thaik, U.Hommerich, R.N.Schwartz, R.G.Wilson, and J.M.Zavada, Appl.Phys. Lett.71, 2641 (1997).
6. S.Kim, S.J.Rhee, D.A.Turnbull, X.Li, J.J.Coleman, S.G.Bishop, and P.B.Klein, Appl. Phys.Lett.71, 2662 (1997).

Part II

Properties of Nanocrystalline Semiconductors and Periodic Structures

PERSPECTIVES OF POROUS SILICON MULTILAYER TECHNOLOGY

M. Thönissen
Research center Jülich, D-52425 Jülich, Germany
E-Mail: m.thoenissen@fz-juelich.de

ABSTRACT

The exact control and adjustment of the etch parameters is important for the fabrication of well defined and reproducible devices based on porous silicon (PS) multilayers. In this paper the etch parameters "electrolyte volume" and "diffusion in the electrolyte" will be discussed as specific problems during a commercial fabrication of multilayer applications. Additionally the stability of the filter characteristics of multilayer systems will be analyzed by changing the climatic environments. New perspectives of multilayer applications will be given, e. g. the electrical control of filters, spatially graded interference filters and multiple interference filters on one sample.

INTRODUCTION

The formation process of porous silicon multilayers is known very well and the resulting multilayer structures have been intensively studied [1-7]. Possible applications of these structures are for example interference filters [8], waveguides [9], photodiodes [10] and microcavities [11]. For all these applications well defined reflectance/transmittance characteristics are required. This makes high demands on the control of all etch parameters taking part in the formation process. Starting with the adjustment of the etch rate and current density [12] and a calibration of refractive index/current density [13] also further etch parameters need to be controlled for a professional fabrication of PS multilayers. Whereas in [14] the adjustment of the current density with depth is discussed in order to produce layer stacks with homogeneous optical thickness, in this paper the role of electrolyte volume and diffusion problems in the electrolyte will be demonstrated. Following these issues relating to reproducible fabrication, multilayer stabilization and to the control of the depth homogeneity new applications of PS multilayers will be discussed, like the electrical control of filter reflectance, gradual changed reflectance and multiple interference filters on one sample.

EXPERIMENTAL

PS layers were formed by anodization of Czochralski grown boron doped silicon (100) wafers. PS layers on substrates with resistivities of 10 mΩcm (p$^+$) and 200 mΩcm (p) were investigated. For the p-doped substrates a metal contact was evaporated on the back side of the wafer to allow a homogeneous current flow. Immediately before anodization the substrates were cleaned in propanol in an ultrasonic bath and rinsed in deionized water. Anodization was performed under standard galvanostatic conditions using a mixture of $H_2O:HF:C_2H_5OH = 1:1:2$. The anodization current was supplied by a Keithley 238 high precision constant current source which is controlled by a computer to allow the formation of PS multilayers. To prevent the photogeneration of carriers, the anodization is performed in the dark. After formation the samples are rinsed with pure ethanol and dried with nitrogen gas. Reflectance spectra of the samples were taken with a Perkin Elmer $\lambda 2$ photospectrometer in the range of 9000-50000 cm^{-1} (1111-200 nm). The reflectance was measured under an incidence angle of 8° using s-polarized light.

RESULTS AND DISCUSSION

I. Influence of electrolyte volume on the optical properties

For a well defined formation of PS multilayers in a process line the influence of the electrolyte volume on the optical properties has to be analyzed. Assuming the formation of a 5μm layer with a current density of 100 mA/cm² (valence v = 2.3 from gravimetrical data) in a electrolyte volume of 5 ml, the molecular/atomic relation of HF/Si is given by

$$n(HF)/n(Si) \approx 315$$

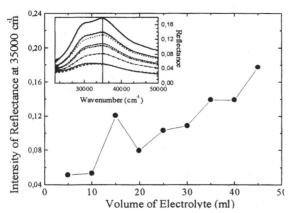

Fig. 1: Change of reflection in the UV for different electrolyte volumes for the maximum of the reflectance at 35.000cm^{-1}. The samples has been formed with a current density of 100mA/cm² on p⁺ doped Si, the volume of electrolyte was changed from 5ml to 45ml in steps of 5ml. In the inset the increase of reflection with increasing electrolyte volume can be seen.

Furthermore a reduction of a factor of 6 has to be taken into account because of the formation of SiF$_6$. On the whole volume this is an effect in the range of 2% so that the consumption of HF during the chemical reaction can not cause diffusion problems in the whole electrolyte unless it takes place in the pores and the HF exchange between the pores and the reservoir is drastically reduced. A possible explanation for changes in the chemical reaction can be caused by a restricted transport in the electrolyte. This restricted transport of HF into the pores can be explained by two phenomena: By the small and decreasing cross section of a pore and/or by a intermediate layer between the electrolyte in the pores and the electrolyte in the etch cell, the so called Prandtl- layer [15].

The following experiment should show the effects of this consumption of HF in the pores on the optical properties. In Fig. 1 the reflectance spectra of a layer system described in the figure caption was measured for different electrolyte volumes starting from 5 ml up to 45 ml in steps of 5 ml. Especially in the UV range changes of the reflectance as a function of the electrolyte volume can be found. The increasing reflectance corresponds to a decrease in the porosity in the near surface region. This is in agreement with a possible increase of the HF concentration caused by the increasing electrolyte volume. Fitting the spectra with the Looyenga model assuming a two-layer-system of a small top layer and a large main layer below, the porosity and the refractive index can be calculated:

Electrolyte volume [ml]	Porosity [%]	Surface layer [μm]	Main layer [μm]	Refractive index at 15.000 cm^{-1}
5	66	0.25	5.2	2.045
25	62	0.21	5.0	2.128
45	60	0.19	4.9	2.176

Table 1: Porosity, thickness and refractive index for selected samples of Fig. 1

In table 1 it is shown that the porosity and layer thickness decreases with increasing electrolyte volume corresponding to an increased electrolyte concentration. All changes in the refractive index are obviously larger than probably caused by the 2% changes in the whole electrolyte. If the diffusion problems are restricted within the pores by the decrease of the concentration of the electrolyte, the

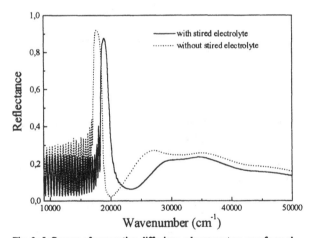

Fig. 2: Influence of convective diffusion; a layer system was formed with current densities of 30/120mA/cm^2 and etching times of 2.92/1.13s respectively. The layer system was repeated 20 times. After formation the reflectance was measured. The reflectance is different if electrolyte is stirred.

whole volume in the etch cell should have almost no influence on the porosity or etch rate. The problems might be caused by changes in the thickness of a diffusion layer between the reservoir and the porous structure - the so called Prandtl layer - which will be investigated in the next subsection.

II.   *Diffusion in the electrolyte*

The so called Prandtl layer is known very well from other electrochemical systems [15]. The calculation of the thickness and the experiments can be found elsewhere, especially the experiments to determine the thickness of this layer by using a rotating disc electrode (RDE) [16]. Important for the following experiments is the thickness of the Prandtl layer which depends on the viscosity of the liquid and of a possible rotation of the electrode: A magnet embedded in Teflon was put in the electrolyte and a rotating magnetic field was switched on during the etch process. A layer system was formed as is described in the figure caption of fig. 2. After the formation the reflectance was measured for a sample with and without stirred electrolyte. In Fig. 2 the shift of the reflectance maximum can be observed. It should be remarked that the spectra of the sample without magnet but with rotating magnetic field is the same as the one without rotating magnetic field. The characteristic in the reflectance shows that possibly the surface layer has a higher porosity because of a lower reflectance. To observe changes in the transport of HF the time t to etch a layer with thickness d was first calculated by the relation [see 17,18]

$$t = a \cdot d + b \cdot d^2$$

where a and b are fitting parameters. The diffusion D can then be calculated using the parameter b by the equation [see 18]

$$D = (\alpha \cdot N \cdot n)/(2 \cdot b \cdot c_0)$$

N are the silicon atoms per volume, n the solved HF molecules (per Si atom), c_0 the concentration at the beginning, c_d the concentration at the depth d and α the exponent of the concentration/current density formula $I_d = I_0 \cdot (c_d/c_0)^{\alpha}$. In the literature $\alpha = 1.35$ can be found [19]. Calculating now the diffusion D for different samples from the measurement of the thickness for a certain etching time the following data can be obtained:

Doping	a [s/µm]	B [s/µm²]	D [10^{-5} cm²/s]
p	88.605	0.0187	0.124
p (stired)	14.150	0.0087	0.266
p+	16.246	0.0023	1.0

Table 2: Diffusion for calculated for different samples by measurements of the depth profiler. The p-doped samples was formed with a current density of 96.64 mA/cm² for 1min, the p+ doped sample was formed with 245mA/cm² for 30 s. The rotation frequency of the magnet was about 300rpm.

In the last column of Table 2 the values for the diffusion can be found. It can be seen that the diffusion is about two times higher for the stirred electrolyte. In this case the Prandtl layer is assumed to be lower than without stirring. Also for the p+ doped sample the diffusion is much higher in comparison to the p doped sample. So it can be concluded that the diffusion depends drastically not only on the morphology but also on the electrolyte reservoir and an additional mechanical stirring of the electrolyte. For a reproducible formation of PS multilayers the conditions in the environment are extremely important. The electrolyte volume should be kept constant, if the electrolyte is used many times, an on-line control of the concentration is very important.

III. Aging in natural environment

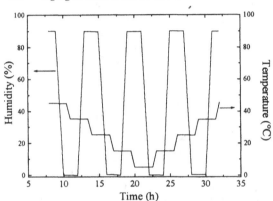

Fig. 3: Profile of temperature and humidity in a climatic chamber which was repeated periodically for 10 days. Additionally the samples have been illuminated.

In [20] and [21] the stabilization of PS single layer and multilayers respectively via thermal oxidation was demonstrated. Also in a high temperature environment the samples remain stable in their reflectance characteristics. For the daily use of reflectance filters it is important not only to investigate the thermal stability but also to analyze this in combination with the stability against environmental influences e.g. the humidity. We have investigated changes of temperature, humidity and illumination after the formation on as prepared samples (layer systems on p-PS with 30/120 mA/cm² and 20 repetitions) and oxidized samples. The preoxidized samples have been oxidized for 60 min at 450°C and 20 min at 950°C. The reflectance spectra have been measured before the 10 day cycle and afterwards (fig. 4). Shifts of the reflectance maximum for all samples can be observed. For the as prepared samples the filter frequency shifts to higher wavenumbers whereas for the oxidized samples the filter shifts to lower ones. In the first case the increase can be explained by the oxidation so that the refractive index of the silicon component in the effective medium is reduced due to the oxidation. A detailed analysis with IR spectroscopy shows the reason for this shift: Between 2700 cm^{-1} and 2900 cm^{-1} Si-C-O modes and at 3000 cm^{-1} H$_2$O embedded in the porous structure can be found [22]. The frequency shift now can be explained by the embedded H$_2$O so that there is an effective medium consisting of the three components water, air and Si. The refractive index of the air component now increases so that the ratio of refractive indices

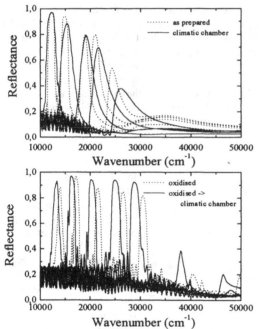

Fig. 4: Reflectance spectra of Bragg reflectors formed for different wavelengths. Top: as prepared samples, with and without storage in the climatic chamber for 10 days. Bottom: oxidized samples before and after storage for 10 days in the climatic chamber.

between the high and low porous layer is decreased. From these experiments it can be seen, that oxidized PS multilayers are hydrophilic. For applications where high accuracy in the filter frequency is required this has to be taken into account. To avoid this either the samples have to be dried supercritically, they have to be used in a stable environment with low humidity or they have to be passivated with a top layer that does not affect the optical properties strongly.

IV. New applications and perspectives

In the following subsections an overview and perspectives of interesting new features of PS multilayer applications should be given. For details see the references given in the sections.

A) Electrical control of multilayers

Electrical control of multilayers after their formation opens a broad market for possible applications. In "as prepared" samples neither the structure of PS nor the reflectance can be subsequently controlled (e. g. electrically), but the large surface area of PS can be used to be filled with components like polymers or liquid crystals (LC) which can be controlled by environmental parameters: Molecular chains of liquid crystals change their refractive index e.g. in an electric field. Their properties are well known from PDLC (Polymer Dispersed Liquid Crystals) Displays [23]. The electrical orientation of the molecules will cause a change of the effective refractive index of the porous layer as a function of the volume of the LC in the pores. If the refractive index of the LC fits to the refractive index of the silicon skeleton in the porous silicon then there is no difference between the medium in the voids and the skeleton itself. Especially for multilayer structures there is no difference in the optical properties for different porosities (fig. 5).

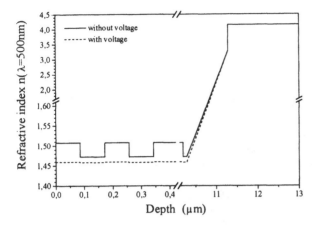

Fig. 5: Simulation of refractive index of a PS multilayer. The filling factors are assumed to be 0.3 and 0.8 respectively, the thickness of a single layer is 0.85 µm. The refractive index of the LC is 1.46. Starting with a thickness of 10.2 µm the refractive index is assumed to increase linearly and is adapted to the one of the substrate.

If the molecule chains are oriented by an external voltage, the refractive index of the filled pores is slightly different from the one of the porous skeleton. As a result different layers in a porous multilayer system become visible and the reflectance is determined by the filter structure. First investigations [24] show that oxidized PS single layers can be filled with liquid crystals and that the reflectance can be changed by the applied voltage. At the moment the switching time is in the range of a few seconds which means that it is not fast enough.

B) Spatially graded interference filters

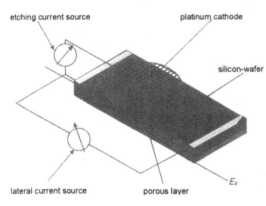

Fig 6: Schematical view of the experimental setup for etching lateral gradually changed interference filters.

Spatially graded interference filters are formed using an additional current through the substrate perpendicular to the etching current (lateral current, see Fig. 6). Details are given in [25]. These types of interference filter can be used for photodiodes with lateral changing reflectance characteristics. Important for the formation is the control of the lateral current which is depending on the anodization current density. This means also that the lateral current has to be adapted to the anodization current and that it has to be changed if the anodization current is changed in a multilayer system.

C) Multiple filters

From the literature, problems during the formation of patterned PS and PS multilayers are known very well [26], e.g. influence of photresist and structure size. To avoid these problems neighbouring filters for filter arrays should be formed using a technology presented here: Stacks of filters can be formed cheaply and easily by etching different filters one after the other at the same location. To improve the

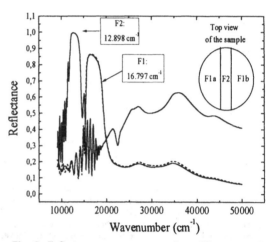

Fig. 7: **Reflectance spectra measured at different sample positions. In the area F2 the topmost filter is removed by RIE.**

quality with respect to the absorption filters with high absorption (e.g., filters which reflect the blue spectral range) should be etched at first. In Fig. 7 a schematical top view of a porous sample with two filters below each other is shown. The filters are etched on a p$^+$-doped substrate, filter 1 has a thickness of 3 μm and is optimized for a reflectance of 16666 cm^{-1} (repetitions of the layer system: 24), filter 2 has a thickness of 2 μm and is optimized for an reflectance of 12500 cm^{-1} (repetitions: 12). After anodisation, a 100 nm Ti layer was evaporated on the PS layer. Using lithography steps a stripe of 3 mm is opened in the middle of the sample. With reactive ion etching (RIE) (for RIE etching rates of PS see [21]) the topmost filter 1 is dissolved and the Ti layer is dissolved in a mixture of $H_2O_2:H_2SO_4$. The resulting reflectance spectra are also shown in Fig. 7. The reflectance maxima fit very well with the wavenumbers the sample is designed for. In the different areas the filter characteristic can be seen clearly.

This method shows the formation of neighbouring filters with sharp frontiers between the different characteristics. The method has to be improved by using e. g. illumination to dissolve partially the topmost filter in order to save costs.

CONCLUSION

In this paper problems of diffusion in the electrolyte and electrolyte volume have been discussed. It can be concluded, that these parameters can drastically influence the reproducibility and homogeneity of the reflectance on the whole sample area. The diffusion can be increased by stirring the electrolyte which is important for high current densities and large layer stacks. The electrolyte volume has to be controlled and the concentration be measured if the electrolyte will be recycled.

Also environmental parameters like the humidity have an influence on the stability of the reflectance peak. The samples have to be capped for applications in high humidity locations or they have to be heated slightly. Another possibility to minimize the physisorption of water is to dry the samples supercritically.

New applications have been demonstrated like the electrical control of PS multilayers, the reflectance filter with spatially graded reflectance and the formation of closely packed filter arrays of different optical characteristics.

REFERENCES

1. M.G. Berger et al., J. Phys. D: Appl. Phys. **27**, 1333 (1994)

2. M.G. Berger et al., Mat. Res. Soc. Symp. Proc. **358**, 327 (1995)

3. G. Vincent, Appl. Phys. Lett. **64**, 2367 (1994)

4. D. Buttard et al., Thin Solid Films **276,** 69-72 (1996)

5. C. Mazzoleni, L. Pavesi, Appl. Phys. Lett. **67**, 2983 (1995)

6. Xing-Long Wu et al., Appl. Phys. Lett. **68**, 611 (1996)

7. M. Araki et al., Jap. J. Appl. Phys. **35** (2B), 1041 (1996)

8. M.G. Berger et al., "Optical interference coatings", Florin Abeles ed., Proc. SPIE **2253**, 865 (1994)

9. A. Loni et al., Proceedings of the IEE colloquium "Microengineering Applications in Optoelectronics", 27th February, 1996, London, (Digest No.96/39)

10. M.Krüger et. al., Proceedings of the 26th European solid state device research conference, ISBN 2-86332-196-X, 1996

11. L. Pavesi, *Porous silicon dielectric multilayers and microcavities*, La Rivista del Nuovo Cimento, 1997

12. M.Thönissen, M.G. Berger, W. Theiß, S. Hilbrich, M. Krüger, R. Arens-Fischer, S. Billat, G. Lerondel, H. Lüth, "Depth gradients in porous silicon: how to measure them and how to avoid them" Mat. Res. Soc. Symp. Proc. 1996

13. Michael Berger, PhD thesis, University of Aachen, Germany, 1996

14. M. Thönissen, M.G. Berger, S. Billat, S. Hilbrich, G. Lerondel, M. Krüger, P. Grosse, H. Lüth, „Analysis of the depth homogeneity of p-PS by reflectance spectroscopy", Thin Solid Films, 1996

15. J. Albery ed., *Electrode kinetics*, Oxford Chemistry series, 1975

16. A.C. Riddiford ed., *Advances in electrochemistry and electrochemical engineering*, McGraw-Hill, 1965

17. Volker Lehmann, PhD thesis, University of Erlangen, Germany, 1988

18. Markus Thönissen, PhD thesis, University of Aachen, Germany, 1998

19. X.G. Zhang, S. D. Collins, R.L. Smith, J. Electrochem. Soc., 136, 1561 (1989)

20. J.J. Yon, K. Barla, R. Herino, G. Bomchil, J. Appl. Phys., 62, 1042-1048 (1987)

21. M. Krüger, PhD thesis, University of Aachen, Germany, 1997

22. W. Theiss, Optical properties of porous silicon, RWTH Aachen 1995

23. Iam-Chon Khoo ed., Liquid Crystals, John Wiley & Sons Inc., 1995

24. M. Thönissen et. al., Proceedings of the PSST98, Mallorca, accepted

25. D. Hunkel et al., EMRS 98, Strasbourg, Proceedings in "Thin solid films", accepted

POROSITY-INDUCED OPTICAL PHONON ENGINEERING IN III-V COMPOUNDS

I.M. TIGINYANU*, G. IRMER**, J. MONECKE**, H.L. HARTNAGEL***,
A. VOGT***, C. SCHWAB****, J.-J. GROB****
* Technical University of Moldova, MD-2004 Chisinau, Moldova, hfmwe103@hrzpub.tu-darmstadt.de
** TU Bergakademie Freiberg, D-09596 Freiberg, Germany
***Technische Universität Darmstadt, D-64283 Darmstadt, Germany
****CNRS/PHASE, BP-20, F-67037 Strasbourg Cedex 2, France

ABSTRACT

New possibilities for modifying the phonon spectra of III-V compounds are evidenced by micro-Raman analysis of porous layers prepared by electrochemical anodization of (111)A-oriented n-GaP substrates. In particular, a surface-related vibrational mode along with a porosity-induced decoupling between the longitudinal optical (LO) phonon and plasmon are observed. We prove that filling in the pores with other materials (aniline as a first approach) is a promising tool for controlling the surface phonon frequency.

INTRODUCTION

Increasing attention is paid nowadays to the manufacturing of nanostructured materials with size-tunable optical properties. As an inexpensive variant, porosity has emerged as a versatile approach for micro- and nanostructuring semiconductors with the potential of controlling their optical characteristics.[1-6] With regard to elemental silicon, where the porous structure actually comparable with other man-made quantum structures was first investigated, porosity in compound semiconductors could offer additional advantages for devices. Apart from those related to the possibility of changing the chemical composition, the shift from element to compound entails a major crystallographic change. Although the overall tetrahedral sp³ bonding between atoms is retained, the centro-symmetrical lattice of the column IV element of diamond type becomes a non-centro-symmetrical lattice of sphalerite-type for the derived I_B-VII, II-VI and III-V binary compounds. This paves the way for new physical properties specific to acentricity to occur in these polar materials. For instance, Fröhlich-type surface vibrational modes are expected in the spontaneous Raman scattering (RS) spectra of III-V compounds and are the more easily observed as the surface-to-volume ratio is considerably enhanced by porosity. These modes are predicted to occur whenever the wavelength of the incident radiation becomes greater than the average size of the crystallites.[7] They have been reported for both GaP and GaAs porous films obtained by anodic etching of n-type substrates[5,8] and the dependence of the surface-phonon frequency upon the degree of porosity could be accounted for by the effective-dielectric-function approach.[5] These findings may enable one to elaborate effective porosity-based methods for optical phonon engineering in III-V compounds and alloys as well as for controlling the strength of the interaction between vibrational and charge carrier excitations.

In this work, we present the results of a micro-Raman study of the first-order modes, with a particular emphasis on the Fröhlich vibrational mode, in porous GaP layers prepared by electrochemical anodization of bulk GaP substrates.

EXPERIMENT

The GaP substrates used in the present study were cut from liquid-encapsulation-Czochralsky-grown n-type ingots moderately or highly Te-doped. The values of the electron concentration in the as-grown crystals were $n_1 = 3 \times 10^{17}$ cm^{-3} and $n_2 = 5 \times 10^{18}$ cm^{-3} (T = 300 K) according to the manufacturer. Porous layers were produced by anodic etching of (111)A-oriented GaP substrates in a 0.5 M H_2SO_4 electrolyte for 30 min. The dissolution was carried out potentiostatically using a Pt metal cathode with an applied voltage ranging from 4 to 10 V. During anodization, the samples were kept in the dark. The porous layers obtained were about 10 µm thick and exhibited a light-yellow color, easily distinguishable from the orange one of the underlying GaP substrate. According to the images obtained by scanning electron and scanning tunneling microscopes, anodic etching of (111)A-oriented GaP surface leads to a top layer with a pillar structure[9] characterized by isolated columns with transverse sizes of about 50 and 200 nm in highly and moderately doped crystals respectively. The analysis of the cross sectional views of the anodized samples showed the pillar structure to stretch perpendicularly to the initial surface. The transverse dimension of the pores was found to be spatially modulated. This spatial modulation is connected probably with dopant and other defects distribution in relation with crystal growth conditions.

The micro-Raman spectra were excited with the 514.5 nm line of an Ar$^+$ laser. To avoid thermal effects,[5] the laser beam power was limited to 1 mW for a spot diameter of 2 µm. The scattered light was analyzed in backscattering geometry using a Jobin -Yvon triple monochromator with a spectral resolution set at about 0.5 cm^{-1}.

RESULTS

Fröhlich Character of the Surface-Related Phonon

A Fröhlich mode may be identified by analyzing the dependence of its frequency on the shapes and sizes of the microstructure forming the porous layers as well as on the dielectric constant of the surrounding material. In fact, a porous layer under ambient conditions may be considered as a nanocomposite solid/air. The dielectric constant of the medium surrounding the microstructures can be changed by filling in the pores with liquids. In this work, advantage has been taken of this fact to throw additional light on the Fröhlich character of the surface-related vibrational mode in porous GaP.

To exclude the contribution of the GaP substrate to the RS signal measured from the porous layer, a sample was cleaved and the incident laser beam was oriented perpendicular to the freshly prepared (110)-surface. The RS spectrum taken from the substrate exhibits only the TO-phonon at 366 cm^{-1} (Fig. 1, dotted curve). The q = 0 LO phonon is forbidden due to the selection rules in (110) backscattering geometry and, therefore, it is absent in the spectrum of bulk GaP. At the same time, when the laser beam propagates along the porous layer, one can observe the emergence of an intense LO-peak accompanied by a well defined shoulder on the low-energy side. The occurrence of a strong LO-signal in (110) geometry is indicative of the breakdown of the selection rules.

The stoichiometric composition of the GaP skeleton remaining after dissolution was checked by electron microprobe analysis. A further analysis by the second harmonic generation technique[10] confirmed the good crystal perfection of the porous GaP as derived from the RS data. Taking this into account the enhancement of RS in porous layers can be attributed only to an increase of the optical path length due to multiple reflections of the light in the porous network. In other words, as in powders, the increase of the RS signal is due to inelastic scattering taking place after many elastic scattering events. Furthermore, since internal reflections of the light occur in all directions, the information about the initial scattering geometry is lost, resulting in a breakdown of the polarization selection rules.

So, from the optical point of view, the effective volume of the material in a porous layer is increased. One can note, in this regard, that the effect under consideration was found recently to cause a strongly enhanced photoresponse of porous GaP for light absorbed via indirect optical transitions.[11]

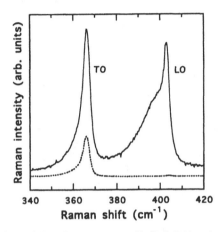

FIG. 1. Micro-Raman spectra of bulk GaP (dotted curve) and porous layer (solid curve) measured in (110) backscattering geometry.

To prove the Fröhlich character of the porosity-induced RS peak centered in the gap between the bulk optical phonons (Fig. 1), a Raman analysis of a porous layer was carried out before and after filling in the pores with aniline. Fig. 2 presents the RS spectra of a GaP porous layer measured before (a) and after (b) filling in the pores with liquid. According to the results of a spectral decomposition, the position of the LO-phonon band is nearly the same in both cases. At the same time, the frequency of the peak centered in the gap between the TO and LO phonons was found to decrease from 396.4 to 394.7 cm^{-1} when filling in the pores with aniline. This downward frequency shift is a result of changing the dielectric constant of the surrounding medium. This is strong evidence for the involved RS peak to correspond to a Fröhlich-type surface-related vibrational mode.

Although the area of the Raman band, corresponding to the surface-related phonon, practically does not change when filling in the pores with aniline, a visible broadening of the band occurs. The full widths at half height of the band are 13.0 and 16.0 cm^{-1} before and after filling in the pores, respectively. This broadening may be explained by assuming an only partial filling of the pores with aniline. We did not succeed with filling in the pores in case of porous layers produced on (100)-oriented n-GaP substrates (such layers are known to exhibit a catacomb-like porosity[4,12]). In contrast, the pillar-like structure inherent to as-anodized (111)-oriented GaP favors the filling of the pores with liquid, but we expect a part of them to remain empty. In this case the contribution of the regions with filled in and empty pores to the RS spectrum will consist of two strongly overlapping surface related phonon bands, leading to the observation of a single broad band. The bulk optical phonons are practically not sensitive to the dielectric constant of the surrounding medium and, therefore, the corresponding Raman bands exhibit neither a shift nor a broadening.

The effective dielectric constant of a nanocomposite GaP/air or GaP/aniline in a sphere approximation (relevant when the transverse dimensions of the pores are spatially modulated) may be calculated using the formula[13]

$$\varepsilon_{eff} = \varepsilon_1 \left(1 - \frac{f - \beta}{s} - \frac{\beta}{s - s_0} \right) \qquad (1)$$

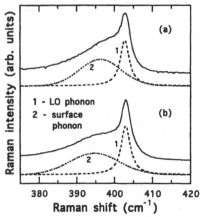

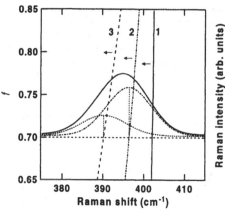

FIG. 2. Micro-Raman spectra of a porous gallium phosphide layer measured in (110) backscattering geometry before (a) and after (b) filling in the pores with aniline.

FIG. 3. The relationship between the volume fraction of GaP skeleton and the frequencies of LO-phonon (curve 1) and Fröhlich phonon when $\varepsilon_1 = 1$ (curve 2) and $\varepsilon_1 = 2.56$ (curve 3). The deconvolution of the surface related Raman band takes into account the effect of partial filling in of the pores with aniline.

where $s = \varepsilon_1/(\varepsilon_1 - \varepsilon_2)$, $\beta = (s_0/3)(f - f^2)$, and $s_0 = (1 - f + 6f^2 - 4f^3)/3$. Here f denotes the volume fraction of GaP in the porous layer, ε_1 equals 1 and 2.56 (Ref. 14) respectively for air and aniline as surrounding mediums, and $\varepsilon_2(\omega)$ is the dielectric function of bulk GaP at the phonon frequencies. From the zeros of $\varepsilon_{eff}(\omega)$ one can deduce a downward frequency shift of both LO and Fröhlich phonons with decreasing f. Fig. 3 illustrates the frequency shifts of the surface phonon in porous GaP assuming the pores to be filled in with air (curve 2) and aniline (curve 3). Using the frequency of the surface-related phonon for $\varepsilon_1 = 1$ (Fig. 2a), we found from Eq. (1) the value of the volume fraction of GaP in the porous layer studied to be $f = 0.7$. According to theory,[13] for $\varepsilon_1 = 2.56$ and $f = 0.7$ the frequency of the Fröhlich mode should be 390.2 cm^{-1}. These data enabled us to carry out a decomposition of the surface-related phonon band (curve 2 in Fig. 2b), taking into consideration two contributions coming from regions with empty ($\varepsilon_1 = 1$) and filled ($\varepsilon_1 = 2.56$) pores. The result of the spectral decomposition is illustrated in Fig. 3, the numerical data are summarized in Table 1. The analysis of the relative areas of the two component-peaks shows that the degree of filling in the porous layer with

TABLE 1. Numerical results of the decomposition of surface-related Raman band for porous GaP layer filled in with aniline: shift (ω), full width at half height (b), and areas (A) of the constituent peaks normalized to the area of the initial band

Raman band	ω (cm^{-1})	b (cm^{-1})	A (%)
Initial band	394.7	16	100
Constituent	390.2	13	30
Peaks	396.4	13	70

aniline equals 30 %. So, Raman spectroscopy seems to be an useful approach for controlling the process of filling in pores in polar compounds with other materials.

Decoupling between LO-Phonon and Plasmon in Porous Layers

The dotted curves in Fig. 4 illustrate the Raman spectra of (a) moderately and (b) highly doped bulk (111)-GaP. The reference spectrum of moderately doped bulk n-GaP exhibits the Brillouin-zone-center LO and TO phonons, in accordance with the selection rules of crystals with the zinc-blende structure. As to the highly doped bulk material, it exhibits a strong TO-phonon and a weak and relatively broad RS band at high frequencies (see dotted curve in Fig. 1b). The absence of pure or „unscreened" LO-phonon can be explained as follows. In highly doped crystals of polar semiconductors, the plasmon interacts with the LO-phonon and the spectrum is dominated by LO-phonon plasmon coupled modes (LOPC).[8,15] The broad band in Fig. 1b (dotted curve) corresponds to the high-frequency L_+ LOPC.[15]

It is to be noticed that the wavelength $\lambda = 514.5$ nm of the laser radiation provided a bulk excitation of the as-grown crystal. Under these conditions one can ignore the contribution to the RS of the surface depleted layer where the LO vibration is decoupled from the plasmon and exhibits an „unscreened" LO signal.[8,15]

For the porous layers, the anodic etching of moderately doped GaP leads to a downward frequency shift of the LO-phonon and to a considerable RS signal intensification (up to 3-4 times). Furthermore, the anodization caused the emergence of the surface-related phonon band positioned in the frequency gap between the TO and LO phonons. A similar peak can be observed in the RS spectrum of a porous layer obtained from highly doped GaP (Fig. 1b, solid curve). The analysis of high-frequency regions in Fig. 1b shows that porosity induces a decoupling between the LO-phonon and plasmon. From the emergence of an „unscreened" LO-phonon a relatively low carrier concentration in the porous GaP network could be deduced. The most probable reason for this is the surface space-charge effect which results in a depletion of the skeleton, the free carriers being transferred to the substrate. For small enough skeleton sizes, a depletion may be caused also by spatial confinement effects on free carriers.

Thus, these results suggest that porosity could be an effective means for controlling the strength of the interaction between LO-phonon and plasmon. Since the LO-phonon disappears at high degrees of porosity,[5] an important task of further investigation is to explore

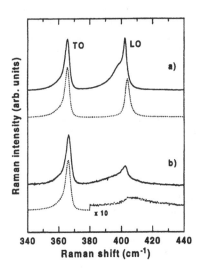

FIG. 4. Micro-Raman spectra of bulk GaP (dotted curves) and porous layers (solid curves). The curves are normalized to the intensity of the TO peak. The free carrier concentration in bulk samples, cm^{-3}: a) $n_1 = 3 \times 10^{17}$; b) $n_2 = 5 \times 10^{18}$.

the peculiarities of interaction between free carriers and the surface vibrational mode.

CONCLUSION

A porosity-induced intensification of Raman scattering accompanied by the breakdown of the polarization selection rules was observed in n-GaP and explained taking into account multiple reflections of the light in the porous layers. The ability of the pillar structure to support surface-related phonon mode at 396.4 cm^{-1} was established. We have studied the behaviour of this surface phonon when filling in the pores with aniline. A downward frequency shift was observed, proving the Fröhlich character of this excitation.

The possibility to excite Fröhlich vibrational modes in porous layers may find applications in quantum structures working on phonon resonance. Besides of changing the shape and size of microstructures forming the porous layers, filling in the pores with other materials (electrochemically, for example) seems to be a promising tool for controlling the surface phonon frequency, i.e., for optical phonon engineering. Porosity proves to be also an effective means for modifying the strength of the interaction between LO-phonon and plasmon.

ACKNOWLEDGMENTS

This work was supported by the NATO Scientific and Environmental Affairs Division under Grant No. HTECH.LG 961399. One of us (I.M.T.) gratefully acknowledges the support of the Alexander von Humboldt Foundation.

REFERENCES

1. A. G. Cullis, L. T. Canham, and P. D. J. Calcott, J. Appl. Phys. **82**, 909 (1997).
2. A. Anedda, A. Serpi, V. A. Karavanskii, I. M. Tiginyanu, and V. M. Ichizli, Appl. Phys. Lett. **67**, 3316 (1995).
3. E. Kikuno, M. Amiotti, T. Takizawa, and S. Arai, Jpn. J. Appl. Phys. **34**, 177 (1995).
4. B. H. Erne, D. Vanmeakelbergh, and J. J. Kelly, J. Electrochem. Soc. **143**, 305 (1996).
5. I. M. Tiginyanu, G. Irmer, J. Monecke, and H. L. Hartnagel, Phys. Rev. B **55**, 6739 (1997).
6. P. Schmuki, D. J. Lockwood, H. J. Labbe, and J. W. Fraser, Appl. Phys. Lett. **69**, 1620 (1996).
7. R. Ruppin and R. Englman, Rep. Prog. Phys. **33**, 144 (1970).
8. I. M. Tiginyanu, G. Irmer, J. Monecke, A. Vogt, and H. L. Hartnagel, Semicond. Sci. & Technol. **12**, 491 (1997).
9. V. M. Ichizli, I. M. Tiginyanu, and H. L. Hartnagel, J. Scanning Microscopies **20**, 156 (1998).
10. I. V. Kravetsky, private communication.
11. F. Iranzo Marin, M. A. Hamstra, and D. Vanmaekelbergh, J. Electrochem. Soc. **143**, 1137 (1996).
12. I. M.Tiginyanu, C. Schwab, J.-J. Grob, B. Prevot, H. L. Hartnagel, A. Vogt, G. Irmer, and J. Monecke, Appl. Phys. Lett. **71**, 3829 (1997).
13. J. Monecke, Phys. Status Solidi b **155**, 437 (1989).
14. S. Hayashi and H. Kanamori, Phys. Rev. B **26**, 7097 (1982).
15. B. Prevot and J. Wagner, Prog. Crystal Growth and Charact. **22**, 245 (1991).

ULTRASOUND EMISSION FROM POROUS SILICON: EFFICIENT THERMO-ACOUSTIC FUNCTION AS A DEPLETED NANOCRYSTALLINE SYSTEM

N. KOSHIDA, T. NAKAJIMA, M. YOSHIYAMA, K. UENO, T. NAKAGAWA, H. SHINODA
Faculty of Technology, Tokyo University of A&T, Naka-cho, Koganei, Tokyo, 184, Japan

ABSTRACT

It is demonstrated that luminescent porous silicon (PS) exhibits an efficient thermo-acoustic effect owing to its extremely low thermal conductivity. The experimental device is composed of a patterned thin Al film electrode (30 nm thick), a microporous PS layer (10-50 μm thick), and a single-crystalline Si (c-Si) wafer. The PS layer was formed by a conventional anodization technique. When an electrical input is provided to the Al electrode as a sinusoidal current followed by Joule's heating, a significant acoustic pressure is produced in front of the device as a result of an efficient heat exchange between PS and air. The output amplitude is in inverse proportion to the square root of the input frequency (0.1-100 kHz) as predicted by a theoretical analysis. The observed effect is a novel useful function of PS as a completely depleted nanocrystalline system.

INTRODUCTION

Luminescent porous silicon (PS) consists of highly packed silicon nanocrystallites with the same band dispersions as those of single-crystal silicon (c-Si) [1]. Since the mean size of silicon nanocrystallites in micro PS is below the criterion for quantum confinement (e.g., 4.3 nm corresponding to Bohr radius of exciton), various activities appear in the optical, electrical, and related physical properties.

One of the remarkable features is an optical bandgap widening induced by strong confinement which results in the visible photo- [2] and electro-luminescence [3]. It also leads to a significant decrease in both the dielectric constant [4] and the refractive index [5]. The next one is detected as the nonlinear electrical behavior such as negative-resistance [6] and nonvolatile optoelectonic bistable memory [7], in which carrier injection into silicon nanocrystallites produces a large field distortion in the PS layer through significant charging effects. As a related issue, electron emission phenomenon is also a characteristic function of well-controlled PS layers [8,9], where hot or quasiballistic electrons are efficiently produced under high electric field by multitunneling through nanocrystallites.

Another important one relates to the thermal property. Because of a complete carrier depletion in silicon nanocrystallites, the thermal conductivity of PS is a few orders of magnitude lower than that of c-Si [10,11]. This would make it possible to promote further application studies of PS. We report here that PS is very useful for a key component of a new-type ultrasonic generator owing to its efficient thermo-acoustic transmission capability.

EXPERIMENTS

In contrast to the conventional methods based on transmission of a mechanical solid vibration to air, the ultrasound generation presented here simply utilizes a direct thermo-acoustic heat exchange at the still solid surface without any vibration systems. As described in the next section, the key issue for implementation of ultrasound devices based on this mechanism is the formation of solid-state layer with an extremely low thermal conductivity on a substrate with a sufficiently high thermal conductivity.

Mat. Res. Soc. Symp. Proc. Vol. 536 © 1999 Materials Research Society

The experimental device is composed of a patterned thin Al film electrode (30 nm thick), a microporous PS layer (10-50 μm thick), and a p- or n-type c-Si wafer as shown in Fig. 1 (a) and (b). The PS layer with a porosity of about 70% was formed by conventional anodization technique at a current density of 20 mA/cm^2 for 8 to 40 min. The electrical input (1 mW/cm^2) was provided to the Al electrode as a sinusoidal current followed by Joule's heating. The emitted acoustic pressure was measured as a function of output frequency (0.1-100 kHz) using a microphone in a closed air space. The spacing between the PS device and the microphone was 0.1 mm.

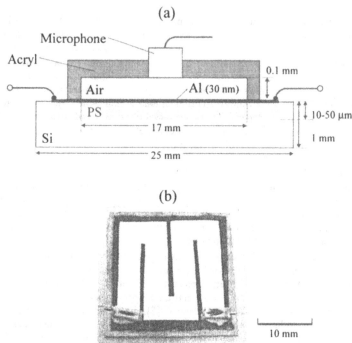

Fig. 1. Schematic illustration of a cross sectional view of the fabricated device with the experimental configuration for detecting the thermo-acoustic pressure (a) and photograph of a top view of the device (b).

THEORETICAL BACKGROUND

In Fig. 2 is shown a schematic illustration of the device and coordinate configuration. Suppose that a thermal power density $q(\omega)$ [W/cm^2] with an angular frequency ω is provided from sufficiently thin metal film to the unit area on the PS layer surface. We can assume that the power diffusion into the air is negligible, and that the c-Si region at x>l has a sufficiently high thermal conductivity and heat capacity.

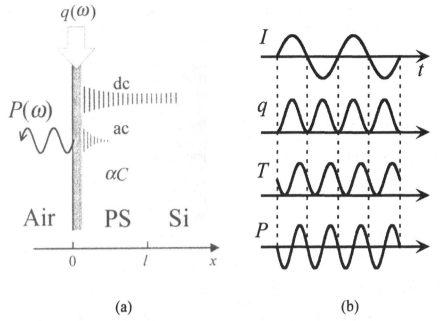

(a)   (b)

Fig. 2. Schematic diagram of the air/thin Al film/PS/c-Si structure and
the corresponding one dimensional coordinate configuration (a).
When an electrical power I is introduced into the top electrode,
the surface temperature T should change in response to the induced
Joule's heating q, and then the acoustic pressure P is generated (b).

Solving the thermal conduction equation, the surface temperature $T(\omega)$ is given by

$$T(\omega) = \frac{1-j}{\sqrt{2}} \frac{1}{\sqrt{\omega\alpha C}} \tanh\left(l\sqrt{\frac{j\omega C}{\alpha}} \right) q(\omega), \qquad (1)$$

Where α and C are the thermal conductivity and the heat capacity per unit volume of the PS
layer which corresponds to the region $0<x<l$ in Fig. 2..

According to Eq. (1), $T(\omega)$ should depended strongly on l. If l is adjusted to

$$L = \sqrt{\frac{2\alpha}{\omega C}}, \qquad (2)$$

which corresponds to the charactersitic range of optimum heat transfer, $T(\omega)$ should become a
maximum, since the stationary dc component can be almost completely removed away from
PS into the silicon substrate [12]. In either case of $l<L$ or $l>L$, $T(\omega)$ cannot respond well to
input power because of a mismatched heat transfer.

Under further simplified assumption that the induced temperature change is transfered to air within a thickness h in front of the surface, an acoustic pressure amplitude $|P(\omega)|$ of the generated sound wave is represented by

$$|P(\omega)| = \frac{P_0}{\left|1 - j\cot(\sqrt{\gamma}kh)/\sqrt{\gamma}\right|} \frac{|T(\omega)|}{T_0},\qquad(3)$$

where P_0, T_0, k, and γ are ambient pressure, room temperature, the wave number of the sound at a certain value of ω, and ratio of specific heats, respectively. If h is nearly equal to $\lambda/(4\gamma^{1/2})$, where λ is the wavelength of sound, Eq. (3) is simplified as follows:

$$P(\omega) = P_0 \frac{T(\omega)}{T_0}.\qquad(4)$$

From these results, we can see that it is important to lower the value of the product αC for efficient sound emission, and that the output acoustic power should be in inverse proportion to the square root of the electrical input frequency.

Above-mentioned theoretical analyses suggest that an appropriate combination of PS and c-Si produces a desirable thermo-acoustic activity, since both the depth of PS and the porosity can be easily and accurately controlled by the anodization conditions. Of especial importance is the αC value. Table 1 shows typical thermal data of PS in comparison to that of c-Si. The data of SiO_2 are also shown for reference. It is clear that the αC value is about 1/400 of c-Si. This enables an efficient heat exchange between PS and air.

The α and C values are also important parameters in another sense, because they determine the required minimum PS layer thickness L mentioned above. According to the data in Table 1, $L=2$ and 6 μm at frequencies of 100 and 10 kHz, respectively.

Table 1. Thermal conductivity α and heat capacity per unit volume C of c-Si in comparison to those typical values of PS. Data for SiO_2 are also shown for reference. Note that the relative value of the product αC of PS is extremely low.

	α [W/m/K]	C [10^6 J/K/m^3]	$\alpha C/(\alpha C)_{\text{c-Si}}$
c-Si	168	1.67	1
PS	1	0.7	1/400
SiO_2	1.4	2.27	1/88

EXPERIMENTAL RESULTS

The measured acoustic pressure amplitudes for various PS layer thicknesses are plotted in Fig. 3 as a function of frequency for sinusoidal Joule's heating power of 1 mW/cm². In this case, the PS layer was formed on a p-type c-Si substrate. It is observed that a significantly high acoustic pressure is observed in a wide band of frequency, and that it is increased with increasing the PS layer thickness due to a possible error in the α and C values. In addition, the output pressure is approximately in inverse proportion to the square root of frequency as predicted from the theoretical analysis. Similar results were also obtained from the device fabricated on n-type c-Si substrates.

This Si-based thermo-acoustic device shows a high reproducibility and stability. Mechanical toughness against the external pressure is another advantage over conventional ultrasound sources, since it doesn't need any fragile structure for vibration such as a diaphragm. Furthermore, the employment of fine segment array electrode structure would make a drastic improvement in the efficiency and the output pressure. As suggested from the data in Table 1, the use of oxidized PS would also be promising and practical approach for enhancing the efficiency.

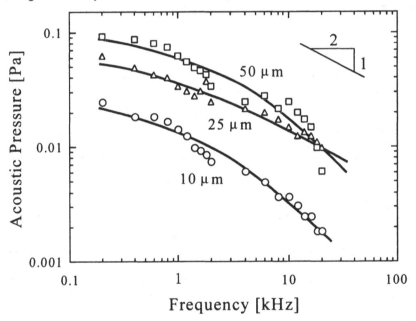

Fig. 3. The detected thermo-acoustic pressure amplitudes for three
devices with different PS layer thicknesses as a function of
output frequency. The electrical power input in this case was
1 mW/cm². Since the output signal responds to Joule's heating,
the acoustic frequency is twice the input current frequency.
The acoustic output is inversely proportional to the square
root of input fequency as expected from a simplified theoretical
analysis.

CONCLUSION

It has been demonstrated that an appropriate combination of PS and c-Si with quite different thermal properties enables a significant heat exchange even at a solid surface, and that a simple metal/PS/c-Si device operates as an efficient ultrasonic emitter. The output acoustic pressure behaves in a way as expected from theoretical analyses. The observed strong thermo-acoustic effect occurs since PS has extremely low values of thermal conductivity and heat capacity per unit volume, reflecting the completely depleted electronic property of Si nanocrystallites in PS.

A novel function of PS related to quantum confinement is now made clear. The PS device has some advantageous features for development of functional integrated ultrasonic emitters with a sufficient output intensity in a wide band width.

ACKNOWLEDGEMENTS

This work was partially supported by a Grant-in-Aid for Scientific Research from the Ministry of Education, Science, Sports, and Culture of Japan (#10355015) and Japan Society for the Promotion of Science (JSPS-RFTF 96P00801).

REFERENCES

1. Y. Suda, K. Obata, and N. Koshida, Phys. Rev. Lett. **80**, 3559 (1998).
2. L. Canham, Appl. Phys. Lett. **57**, 1046 (1990).
3. N. Koshida and H. Koyama, Appl. Phys. Lett. **60**, 347 (1992).
4. N. Koshida, Dielectric constant of porous silicon, in EMIS Data Review Series No. **18** "Properties of Porous Silicon", Ed. L. Canham (IEE, London, 1997) pp. 234-237.
5. W. Theiss and S. Hilbrich, Refractive Index of porous silicon, in EMIS Data Review Series No. **18** "Properties of Porous Silicon", Ed. L. Canham (IEE, London, 1997) pp. 223-228.
6. K. Ueno and N. Koshida, Jpn. J. Appl. Phys. **37**, 1096 (1998).
7. K. Ueno and N. Koshida, Appl. Phys. Lett. **73** (1998) (in press).
8. X. Sheng, T. Ozaki, H. Koyama, and N. Koshida, J. Vac. Sci. & Technol. B **16**, 793 (1998).
9. X. Sheng and N. Koshida, Mat. Res. Soc. Symp. Proc. **509**, 193 (1998).
10. W. Lang, A. Drost, P. Steiner, and H. Sandmaier, Mat. Res. Soc. Symp. Proc. **358**, 561 (1995).
11. W. Lang, Thermal conductivity of porous silicon, in EMIS Data Review Series No. **18** "Properties of Porous Silicon", Ed. L. Canham (IEE, London, 1997) pp. 138-141.
12. T. Nakajima, M. Yoshiyama, K. Ueno, N. Koshida, and H. Shinoda, Proc. IEEJ 16th Sensor Symp. (IEEJ, 1998, Tokyo) pp. 185-188.

PRODUCTION OF SILICON NANOCRYSTALS BY THERMAL ANNEALING OF SILICON-OXYGEN AND SILICON-OXYGEN-CARBON ALLOYS: MODEL SYSTEMS FOR CHEMICAL AND STRUCTURAL RELAXATION AT Si-SiO$_2$ and SiC-SiO$_2$ INTERFACES

Gerald Lucovsky[1,2,3], David Wolfe[2] and Bruce Hinds[1]
Deptartments of Physics[1], Materials Science and Engineering[2], and Electrical and Computer Engineering[3], North Carolina State University, Raleigh, NC 27695-8202

ABSTRACT

This paper discusses the formation of silicon nanocrystals, nc-Si, by thermal annealing of hydrogenated amorphous thin films of SiO$_x$, $x<2$ and (Si,C)O$_x$, $x<2$. Comparisons are made with SiC$_x$ films, providing additional insights into pathways for generation of nc-Si. These alloys are used as model systems for understanding chemical and structural relaxations occurring at Si-SiO$_2$ and SiC-SiO$_2$ interfaces during post-oxidation thermal annealing. This then provides important information for optimized processing of Si-SiO$_2$ and SiC-SiO$_2$ interfaces for device applications.

INTRODUCTION

The formation of Si-SiO$_2$ interfaces for advanced CMOS devices with ultra thin gate oxides (~2-4 nm thick) has been achieved in several different ways: i) by thermal oxidation and rapid thermal oxidation, RTO, of Si(100) substrates at temperatures of 800-1000°C [1], and ii) by remote plasma-assisted oxidation at 300°C [2]. In the thermal oxidation and RTO processes, the interface is continuously regenerated as oxidation progresses. In contrast, for preparation of ultra thin gate oxides by remote plasma techniques, interface formation and film deposition are separately and independently controlled in a two-step process comprised of remote plasma-assisted oxidation, RPAO, followed by remote plasma-enhanced chemical vapor deposition, RPECVD [2,3]. Optimization of device properties is achieved for both types of processing after a post-oxidation or post-deposition anneal at ~900°C for 30 s or more [4]. Studies of interface bonding chemistry and structure have shown that improvements after annealing derive from reduction of suboxide bonding groups in interfacial transition regions, effectively smoothing the Si-SiO$_2$ interface [4]. Optimized SiC-SiO$_2$ interfaces have been formed by thermal oxidation at temperatures greater than about 1025°C, also followed by post-oxidation annealing, but at *lower* temperatures in the range of 900 to 950°C [5]. SiC-SiO$_2$ interfaces have not been studied as extensively as Si-SiO$_2$ interfaces, and the effects of the post-deposition anneal, or the requirement for a higher temperature oxidation step are not as well understood microscopically.

This paper extends studies of the thermal-stability of plasma-deposited SiO$_x$ thin films [6, and references therein] to suboxides of Si and C, hereafter (Si,C)O$_x$, $x<2$. The research identifies changes in structure and bonding as a function of rapid thermal annealing temperatures that are of importance for understanding changes in interface bonding and chemistry occurring during the thermal oxidation of SiC and subsequent annealing of the SiC-SiO$_2$ interfaces. The results of the study thereby provide important information underpinning SiC device processing technology.

EXPERIMENTAL PROCEDURES

Thin films of hydrogenated SiO$_x$ and (Si,C)O$_x$, $x<2$, were deposited by RPECVD, at 250°C using i) excited species (e.g., O$_2$* and O atoms) transported out of an upstream He/O$_2$ RF plasma, and ii) neutral species from down-stream injected SiH$_4$, and SiH$_4$/CH$_4$ mixtures, respectively. As a control, thin films of hydrogenated SiC$_x$, x ~0.2, were also prepared by similar RPECVD processes. The bonding of hydrogen in these three different types of alloy films was studied by Fourier transform infrared spectroscopy, FTIR. Alloy compositions for the films were determined by X-ray photoelectron spectroscopy, XPS, and Rutherford backscattering, RBS, with O and C compositions in the range of 5 to 20 at.% with an uncertainty of ±0.02 at.%

Mat. Res. Soc. Symp. Proc. Vol. 536 © 1999 Materials Research Society

Significant changes in bonding and morphology occurred during rapid thermal annealing in an inert Ar environment. The chemical bonding and morphology were studied as a function of annealing temperature for temperatures up to about 1150°C by FTIR, Raman scattering, and high resolution TEM imaging. Experimental results are discussed in the next section providing a basis for correlating these basic materials studies with important aspects of $Si-SiO_2$ and SiC-SiO_2 interface processing and characterization.

EXPERIMENTAL RESULTS

Figures 1 and 2 show, respectively, FTIR and Raman scattering results for SiO_x thin films, x ~0.15±0.02. FTIR absorption spectra are included for as-deposited films, and for films annealed at 600°C, 900°C and 1050°C; Raman spectra are shown for as-deposited films, and for films annealed at 600°C, 950°C and 1050°C. Analysis of FTIR and Raman spectra show that: i) substantial amounts of H evolve from the films at temperatures below 600°C, and ii) the films undergo a significant morphology change after annealing at temperatures >900°C, resulting in the formation of nc-Si in an amorphous SiO_2 matrix. Loss of bonded H was determined by changes in the FTIR spectra, in particular the reductions in the absorption strength of the of Si-H bond-bending feature at 630 cm^{-1} and a coupled Si-O-Si-H vibration at ~ 780 cm^{-2} [7]. Separation of the alloy material at temperatures up to about 900°C into Si-rich and O-rich regions was inferred from systematic shifts in the frequency and full-width-at-half-maximum, FWHM, of the Si-O-Si bond-stretching feature at ~1000-1100 cm^{-1}. The as-deposited films were not compositionally homogeneous; comparisons between characterizations of O content by FTIR [8] and RBS, coupled with photoluminesce studies indicated the as-deposited films were compositionally inhomogeneous with SiO_x and Si rich regions [9]. The onset of crystallization for nc-Si was observed in the Raman spectra by the emergence of a feature at ~515 cm^{-1}, and coincided with observation of Si-crystallites in the high resolution TEM images. These in turn correlated with changes in Si-O-Si features at 1080, 800 and 460 cm^{-1} that indicated that stoichiometric SiO_2 had been formed. Fig. 1 shows that for annealing up to about 900°C, there is a decrease in peak spectral absorbance of the Si-O-Si stretch mode; as the annealing temperature is increased above 900°C, the amplitude trend reverses, and the FWHM decreases. The morphology and bonding changes that occur between as-deposited films and annealing temperatures of ~900°C correspond to the following reaction: SiO_x ---> nc-Si + a-SiO_2.

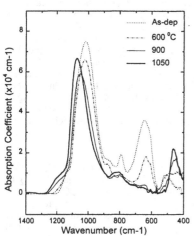

Fig. 1. FTIR Spectra of SiO_x as a function of temperature.

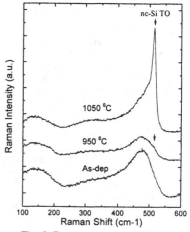

Fig. 2. Raman spectra of SiO_x as a function of temperature.

112

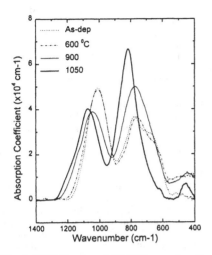

Fig. 3. FTIR Spectra of $(Si,C)O_x$ as a function of temperature.

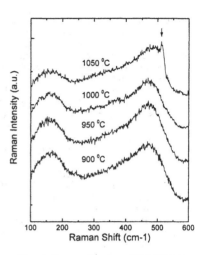

Fig. 4. Raman spectra of $(Si,C)O_x$ as a function of temperature.

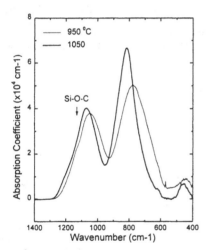

Fig. 5. FTIR Spectra of $(Si,C)O_x$ in the Si-O-Si and Si-O-C regime.

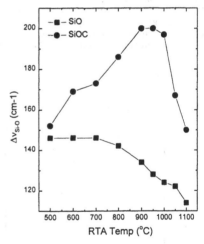

Fig. 6. Δv in the Si-O-Si and Si-O-C regime as a function of annealing.

Finally XPS measurements of as-deposited and annealed films indicated no measurable changes in the Si to O atom ratio after the high temperature anneals.

Figures 3 and 4 show similar FTIR and Raman spectra for the $(Si,C)O_x$ films, x ~0.15±0.02 and C ~0.15±0.02. As in the case of the SiO_x films, elimination of bonded H is determined from changes in the FTIR absorption spectra, the formation of nc-Si from changes in the Raman spectra, and the end-member amorphous phases from the FTIR. There are significant differences between the changes in bonding and morphology for the annealed $(Si,C)O_x$ films when compared with SiO_x films: i) H is evolved from *both* Si-H and C-H groups (the 600 cm^{-1} shoulder on the broad feature centered at about 800 cm^{-1}, and the very weak feature at ~ 1340 cm^{-1} in the as-deposited films), ii) nc-Si formation is not observed in the Raman spectrum until the annealing temperature is increased to ~1050°C, compared to ~900°C in SiO_x, and finally iii) there are two end-member amorphous phases: SiO_2 (spectral peak at ~1080 cm^{-1}), and an a-Si-C alloy (spectral peak at 820 cm^{-1}). Similar to the SiO_x spectra, the absorption strength in the Si-O-Si stretch mode initially decreases, and then after the onset of crystallization at ~1050 cm^{-1}, it increases. In contrast, during annealing the Si-C feature shifts monotonically to higher wavenumber, the peak absorption increases and the FWHM decreases. The decrease in the FWHM is most pronounced at the onset of crystallization. Additionally, there is no spectrocopic evidence for Si-O-C groups in the as-deposited films. However, Si-O-C bonding is clearly evident at annealing temperatures of ~900°C, but is not detected once the annealing temperature is raised to 1050°C (see Figs. 5 and 6).

The FTIR feature at about 1150 cm^{-1} in Fig. 5 has been assigned to Si-O-C bonding groups on the basis of its frequency relative to the Si-O-Si stretch. The temperature range in which this vibration is evident is shown in Fig. 6, which displays the FWHM, Δv, of the Si-O-Si and Si-O-C bands. It is significant to note that Δv increases by more than 20 cm^{-1} when the annealing temperature is increased from about 800 to 950°C, and then drops sharply for annealing temperatures above about 1000°C. The increase in Δv coincides with detection of the Si-O-C feature in the FTIR spectrum, and the sharp decrease in Δv above 1000°C occurs at about the same temperature at which a nc-Si feature becomes evident in the Raman spectrum (see Fig. 4).

Figures 7 and 8 show FTIR and Raman spectra for the SiC_x films, x ~ 0.18±0.02 as a function of annealing temperature. As in the case of the SiO_x and $(Si,C)O_x$ films, the elimination of bonded H is determined from changes in the FTIR absorption spectra, the formation of nc-Si from changes in the Raman spectra, and the end-member amorphous phases from the FTIR. There are several significant differences between the changes in bonding and morphology for the SiC_x films when compared with the SiO_x and $(Si,C)O_x$ films: i) H elimination is essentially complete at a higher temperatue, 800°C, ii) nc-Si becomes evident at 950°C, a temperature intermediate between the

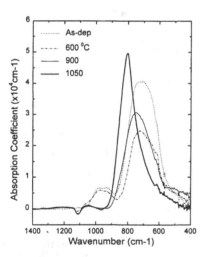

Fig. 7. FTIR Spectra of SiC_x as a function of temperature.

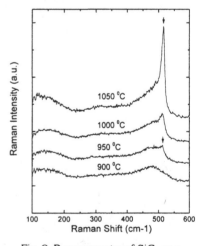

Fig. 8. Raman spectra of SiC_x as a function of temperature

respective 900°C and ~1050°C crystallization temperatures of SiO_x and $(Si,C)O_x$ films, and iii) the end-member amorphous phase is a SiC alloy. The degree of chemical ordering in the a-SiC alloy and a-SiO_2 are assumed to be similar from the sharpness of their spectroscopic feaures. To a good approximation, the FTIR spectra of the $(Si,C)O_x$ films are a *linear combination* of the FTIR spectra of the SiO_x and Si,C_x films. The only significant differences are additional spectral features in the SiC_x films associated with an increased concentration CH bonding groups.

DISCUSSION

This section is separated into two parts: the first deals with bonding interpretations of the FTIR and Raman spectra, and the second with the relevance of the alloy studies to device processing of Si-SiO_2 and SiC-SiO_2 interfaces. One focus of the second section is on the chemistry of the oxidation process, and a second is on changes in the chemical bonding at interface after annealing.

Microscopic interpretation of the alloy bonding as a function of annealing

SiO_x films have been subjected to numerous studies and these have been in part reviewed in a recent publication [6]. The as-deposited films are amorphous and can either be chemically homogeneous [8], or chemically inhomogeneous as for the films deposited in this study [9]. The FTIR and Raman spectra establish that the dominant local bonding groups within the film are: i) Si-Si_4 tetrahedra in the Si-rich regions, and ii) Si-H,O,Si_2 tetrahedra in the O rich regions. The bonding of H and O to a common Si has been noted in previous publications and is associated with the sharp FTIR feature at ~780 cm^{-1} in the as-deposited films (Fig. 1) [7]. Upon annealing H is evolved, and the inhomogeniety is increased. This evidenced by the loss of Si-H features, and shifts of the Si-O-Si stretch frequency to higher wavenumber. Separation into stoichiometric SiO_2 and nc-Si occurs at about 900°C, and annealing beyond that temperature provides additional structural relaxation of the oxide phase and the internal nc-Si-SiO_2 interfaces.

$(Si,C)O_x$ films have received much less attention, and this study represents the first attempt to make comparisons with SiO_x and SiC_x deposited films. The as-deposited films are amorphous and as expected show spectroscopic features associated with Si-H, C-H, Si-O and Si-C bonding. Changes in the spectral character of the Si-O-Si stretch vibration with annealing are similar to those of the SiO_x films indicating increases in inhomogeneous bonding into O and Si rich regions. Changes in the Si-C stretch band indicate that Si-C rich regions form concurrently with the changes in Si and O bonding noted above. The development of Si-O-C bonding arrangements at about 900°C is assumed to be the cause of the increase in the temperature at which nc-Si is observed in the Raman spectra, and by high resolution TEM. The spectral peak frequency of the Si-C feature is higher than in the SiC_x alloys, suggesting that the Si-C bonding groups in the $(Si,C)O_x$ system after crystallization have C and O atoms bonded to a common Si atom; i.e., that the frequency is increased to chemical induction effects [10].

Finally, as noted above, the $(Si,C)O_x$ FTIR spectra are essentially a linear combination of the SiO_x and SiC_x spectra suggesting that the bonding of the SiC_x films parallels that of the SiO_x films with Si-H, C-H and Si-C arrangements. Of particular interest are the relatively high frequencies of the Si-C vibrations after annealing compared to those observed in the as-deposited films. The increases in frequency in both the $(Si,C)Ox$ and $SiCx$ films are consistent with; i) the loss of H from C-H bond groups, and ii) the formation of local bonding arrangements with strong chemical ordering, i.e., based on filamentary C-Si_4 structures interconnected through the four-fold coordinated Si atoms atoms.

Application of alloy studies to processing of Si-SiO_2 and SiC-SiO_2 interfaces

Consider first the application to Si-SiO_2 interfaces. It has been shown that post-oxidation annealing of thermally-oxidized interfaces and plasma-oxidized interfaces [3,4] improves device performance and reliability. This discussion focuses only on chemically-pristine Si-SiO_2 interfaces, e.g., the incorporation of monolayer nitride bonding at interfaces has been studied extensively, and is known to provide additional improvements as well [3].

Suboxide bonding at Si-SiO$_2$ interfaces has been studied by XPS [11] and more recently XPS studies using monochromatic synchrotron radiation have confirmed the reduction of excess suboxide bonding by annealing in an inert ambient at 900°C [12]. One layer of suboxide bonding is required at Si-SiO$_2$ interfaces to bridge the bonding of Si in the substrate (Si-Si$_4$) to Si in the oxide (Si-O$_4$). In addition, the experiments described in Ref. 13 establish that growing an oxide at 900°C is not equivalent to growth at 900°C followed by annealing at the same temperature. The apparent saturation of annealing effects in Si-SiO$_2$interfaces after 900°C annealing and downstream processing exposures suggests that the phase separation reaction observed in the deposited thin films at 900°C; i.e., SiO$_x$ ---> SiO$_2$ + nc-Si, also occurs at the crystal interface. However, there are two important differences in the effects that take place at the Si-SiO$_2$ interface. First, the separation is incomplete at the monolayer level, i.e., the annealed Si-SiO$_2$ interfaces have a residual suboxide bonding of the order of a monolayer. Second, there is no evidence for the formation of nc-Si. The separation reaction results in increased crystallization at the Si-SiO$_2$ interface, so that in place of the formation of nc-Si crystallites, Si derived from SiO$_x$ regions is incorporated into the Si substrate.

Consider now the SiC-SiO$_2$ interfaces. These have been the many studies of the SiC-SiO$_2$ interfaces; however a complete review of these is beyond the scope of this paper. Instead the focus is on two observations: i) the occurrence of Si-oxycarbide groups in the interfacial region by XPS [14], and ii) significant improvements in device performance by an oxidation/annealing sequence in which the oxidation is performed at a temperature in excess of 1025°C, and the post-oxidation annealing at 900°C [5]. Based on the alloy studies, the requirement for a high temperature oxidation is to remove Si-O-C bonding groups that form at the boundary between Si surface atoms and the growing oxide layer. At temperatures > 1025°C , the oxidation process is assumed to eliminate residual C from the growth surface by forming volatile species such as CO and/or CO$_2$. Since device fabrication is usually done on Si faces of 6H or 4H wafers, the annealing step at ~900°C presumably performs the same function as at Si-SiO$_2$ interfaces, reducing suboxide bonding in interfacial transition regions. XPS studies should be performed on SiC-SiO$_2$ interfaces with ultra thin oxide layer. While the newly develop oxidation/annealing sequence of Ref. 5 has yielded significant improvements of electron channel mobilities, values are still considerably below values that are anticipated on the basis of electron mobilities in bulk SiC.

ACKNOWLEDGEMENTS

Supported in part by ONR and AFOSR.

REFERENCES

[1] J. Ahn, J. Kim, G.Q. Lo and D.-L. Kwong, Appl. Phys. Lett. **60**, 2089 (1992).
[2] T. Yasuda, Y. Ma, S. Habermehl and G. Lucovsky, Appl. Phys. Lett. **60**, 434 (1992).
[3] G. Lucovsky, H, Niimi, K. Koh, D.R. Lee and Z. Jing, in *The Physics and Chemistry of SiO$_2$ and the Si-SiO$_2$ Interface*, Ed. by H.Z. Massoud, E.H. Poindexter and C.R. Helms (Electrochemical Soc., Pennington, 1996), p. 441.
[4] G. Lucovsky, A. Banerjee, B. Hinds, B. Claflin, K. Koh and H. Yang, J. Vac. Sci. Technol. **B 15**, 1074 (1997).
[5] L. Lipkin and J.W. Palmour, J. Electronic Materials **25**, 909 (1996).
[6] G. Lucovsky, J. Non-Cryst. Solids **227**, 1 (1998).
[7] G. Lucovsky and W.B. Pollard, J. Vac. Sci. Technol. **A 1**, 313 (1983).
[8] D.V. Tsu, G. Lucovsky and B.N. Davidson, Phys. Rev. **B 40**, 1795 (1989).
[9] F. Wang et al., these proceedings.
[10] G. Lucovsky, Solid State Commun. **29**, 571 (1979).
[11] G. Hollinger and F.J. Himpsel, Appl. Phys. Lett. **44**, 93 (1984).
[12] J. Keister et al., submitted to J. Vac. Sci. Technol. (1999).
[13] M X. Chen and J.M. Gibson, Appl. Physics Lett. **70**, 1462 (1997).

[14] B. Hornetz, H-J. Michel, J. Halbritter, J. Mater. Res **9**, 3088 (1994).

POROUS SILICON MULTILAYER MIRRORS AND MICROCAVITY RESONATORS FOR OPTOELECTRONIC APPLICATIONS

S. CHAN, L. TSYBESKOV, AND P. M. FAUCHET
Department of Electrical and Computer Engineering, University of Rochester, Rochester, NY 14627

ABSTRACT

Porous silicon multilayer structures are easily manufactured using a periodic current density square pulse during the electrochemical dissolution process. The difference in porosity profile, corresponding to a variation in current density, is attributed to a difference in refractive index. Manipulating the difference in refractive index, high quality optical filters can be made with a maximum reflectivity peak ~ 100%. The next logical step to further exploit these optical mirrors is to incorporate them into an LED device. The benefit of adding a multilayer mirror below a luminescent film of porous silicon is to significantly reduce the amount of light loss to the silicon substrate and increase the light output. However, oxidation is required to stabilize the as-anodized porous silicon film. This disrupts the overall index profile of the multilayer stack, causing the peak reflectance to blue shift. This phenomenon must be quantified and accounted before device implementation. We present a detailed study on the effects of oxidation temperature, gas environment, and annealing time of porous silicon multilayer structures in a device configuration.

INTRODUCTION

Since 1990, a huge amount of interest has been directed to a new optical material that emits visible light at room temperature, porous silicon (PSi) [1]. Various structures have been engineered since then, such as optical waveguides, light emitting diodes (LEDs), Bragg reflectors or multilayer mirrors, and microcavity resonators [2-4]. The advantage of creating such a dielectric multilayer is the multiple interference a light beam can undergo when it is reflected at each interface. By correctly choosing the appropriate thickness and refractive index of each layer, it is possible to control the multilayer structure's reflectivity spectrum. However, the as-anodized structures are unstable and mechanically fragile, which render them useless for optoelectronic devices. To rectify this problem, we subject all multilayer mirrors and microcavity resonators to partial oxidation. Silicon nanoclusters surrounded by an oxide matrix result after oxidation, which have excellent light emitting and carrier transport properties [5]. These properties aid in the realizablity of the first oxidized multilayer structure device.

EXPERIMENTAL

PSi multilayer mirrors and microcavity resonators are electrochemically formed by anodic etching in an hydrofluoric electrolyte. Table I details the experimental conditions used in the formation of multilayer structures. A 2 sec/layer regeneration time at 0.0 mA is incorporated into

117

the formation scheme to allow the system to reach equilibrium before etching another layer. The porosity values are estimated using a porosity dependence on current density plot for various concentrations of HF [6,7]. Thickness values are obtained through SEM photographs. The microcavity resonator contains an additional active layer embedded between two multilayer mirrors, formed with a current density of 80-150 mA/cm^2 and anodization time of 1-20 seconds.

Table I. Formation of PSi Multilayer Structures

	p+ (0.008 – 0.012)	p- (~1.0)
Substrate resistivity (Ω-cm)		
Electrolyte (HF:H$_2$O:EtOH)	3:3:14	1:1:2
Regeneration time (sec/layer)	2	2
Periods	5 - 10	18
J$_1$ (mA/cm^2)	5	5
t$_1$ (sec)	20	30
Porosity$_1$ (%)	43	63
Thickness$_1$ (nm)	80	60
J$_2$ (mA/cm^2)	30	50
t$_2$ (sec)	10	5
Porosity$_2$ (%)	62	72
Thickness$_2$ (nm)	160	190

Oxidation of the multilayer structures is performed in both oxygen and nitrogen environments, at temperatures ranging from 800-1100° C for a time duration of 10 minutes.

A Perkin-Elmer Lambda 9 UV/VIS spectrometer with an integrating sphere accessory is utilized to obtain reflection data from 200-2000 nm. Photoluminescence (PL) is obtained with an OMA detector using the 514 nm excitation wavelength of an Ar$^+$ ion laser. Lifetime measurements are obtained using an excimer pumped dye laser at 343 nm. To probe the electrical properties of these structures, ~15 nm of gold is sputtered onto the PSi with a circular area of 0.079 cm^2.

RESULTS AND DISCUSSION

Optical Properties

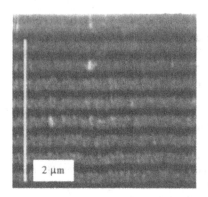

Figure 1. Cross-sectional SEM micrograph of a 10 period p+ oxidized PSi multilayer mirror with overall thickness of ~2.4 μm. The thickness of the low porosity layers is 80 nm, and that of the high porosity layers is 160 nm.

It has been established that the dissolution of silicon occurs at the solid-liquid interface, which makes anodic etching of PSi multilayers possible [8]. The quality of the layers can be seen from figure 1, which is a cross-sectional SEM micrograph of an oxidized PSi multilayer mirror taken at 30kV. The thicknesses of the low and high porosity layers are 80 nm and 160 nm, respectively. To simulate the reflection spectrum, the dielectric function of the PSi layers is described by an effective medium approximation [9]. The porosities of the two layers used to simulate the multilayer reflection spectrum were 43% and 62% for the low and high porosity, respectively. We accounted for the dependence of the refractive index of silicon on wavelength, but absorption is neglected.

The quality of a PSi multilayer mirror can be determined by examining its reflection spectrum. It is desirable to position the mirror's reflection peak at the precise location corresponding to the peak luminescence of an active light-emitting PSi layer. Upon oxidation, the reflection peak will shift, and if not accounted for, the luminescent peak of the active PSi layer and the reflection peak of the multilayer mirror will not coincide. Therefore, it is necessary to factor in the effects of thermal oxidation. A blue-shift occurs when the PSi multilayer mirrors are thermally oxidized for 10 minutes in both a nitrogen and oxygen ambient, due to a change in refractive index of the layers, as shown in figure 2. In addition to the change in reflection peak position, it is important to determine how oxidation affects the maximum peak reflectance. A good quality mirror requires a peak reflectance to be ~100% even after oxidation. Figure 3 demonstrates that the peak reflectance does not decrease upon thermal treatment.

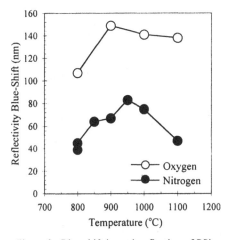

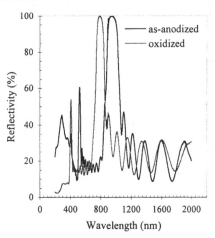

Figure 2. Blue shift in peak reflection of PSi multilayer mirrors. The structures are thermally treated in nitrogen and oxygen at various temperatures for 10 minutes.

Figure 3. Reflection spectra of 10 period as-anodized and oxidized PSi multilayer mirrors.

The oxidized sample was annealed at 900° C in oxygen for 10 minutes. A 150 nm blue shift in peak reflection is observed, while the peak reflectance is maintained at ~100%.

When an active layer is sandwiched between two highly reflecting mirrors, a microcavity resonator structure is formed. The top and bottom multilayer mirrors are composed of 10 periods of alternating porosities, 43% and 62%, while the active layer possesses a higher porosity, to ensure high luminescence. Higher porosity active layers can also be used (up to 95% without subsequent treatment, such as supercritical drying), because the top multilayer mirror prevents the highly porous layer from flaking. All PL spectra are obtained in a direction perpendicular to the sample. Because the stop-band of the multilayer mirror is narrower than the broad emission band of PSi, leaky modes from both sides of the resonating wavelength are observed. Thermal treatment of the samples further enhances the overall luminescence of the structures. Figure 4a shows a typical reflection spectrum of a microcavity resonator and figure 4b displays the PL spectra of PSi microcavity resonators having different resonating wavelengths. The minimum of the reflection dip is approximately 80%. This value can be significantly decreased by increasing the thickness of the active layer [10].

Because transport through a thick, highly porous layer is difficult, a compromise between the minimum reflection dip and active layer thickness is needed. All active layers in the microcavity resonator structures have thicknesses ~ 100-150 nm, and minimum reflection dip of ~80%. The resonant wavelength of the microcavity resonator can be tuned by manipulating the porosities of the active PSi layer. All porosity values are estimated using the same porosity dependence on current density plot as stated in the experimental section. As figure 4b shows, the peak PL wavelength can be shifted from 848-712 nm.

The lifetime of PSi can be extracted using a stretch exponential function, defined as

$$I_{PL}(t) = I_{PL}(0) exp\left[-\left(\frac{t}{\tau}\right)^{\beta}\right] \quad (1)$$

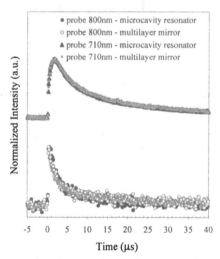

Figure 5. Room temperature luminescence decay of a p- PSi multilayer mirror and microcavity at the leaky modes corresponding to 710 and 800 nm. At these two wavelengths, the decays of the mirror and resonator are identical, indicating that the non-resonating luminescence that arises in the microcavity structure is due to the luminescence of the multilayer mirror.

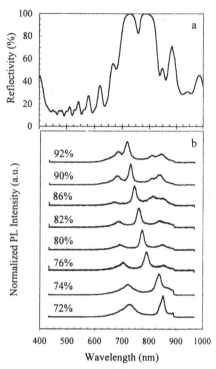

Figure 4. (a) Reflection spectrum of a p+ PSi microcavity resonator at the 750 nm resonant wavelength. The reflection dip is ~80%. (b) PL of several p+ PSi microcavity resonators with varying active layer porosities. The average porosity of each active layer is labeled adjacent to their corresponding spectra. The PL features that appear at fixed wavelengths ~700nm and ~850nm are produced by leaky modes.

where τ = lifetime, I = intensity, and β is the dispersion coefficient. Generally, $\beta < 1$ signifies the presence of a broad distribution of lifetimes, which is valid for PSi [11]. The lifetimes of a p- PSi multilayer mirror at the wavelengths of 710 and 800 nm are equivalent to the lifetimes of the p- PSi microcavity resonator at the same wavelengths. The lifetime of the mirror and resonator structures probed at 710 nm is ~13 μs (β = 0.77) and ~4 μs when probed at

800nm ($\beta = 0.53$) This confirms the fact that the non-resonating luminescence of the leaky modes observed in the microcavity resonator is due to the luminescence of the multilayer mirror. Figure 5 displays the luminescent lifetimes probed at 710 and 800 nm for the multilayer mirror and microcavity resonator.

Electrical Properties

Electrical measurements are conducted on both oxidized PSi multilayer mirrors and microcavity resonators. A thin semi-transparent layer (~15 nm) of Au suffices as the top contact, with a diameter of 3.175 mm. The transmission of the Au layer is approximately 50% in the 750 nm wavelength of interest. PL spectra are detected from samples through the gold contact in order to make a fair comparison between PL and electroluminescence (EL).

The current-voltage characteristics of these structures are shown in figure 6, where the Au contact is biased positive. A threshold voltage of 3V is observed for both the microcavity resonator and multilayer mirror, however the current density increases faster for the multilayer mirror. EL detection is possible at initial current levels of 5mA/cm^2 and increases with higher current injection.

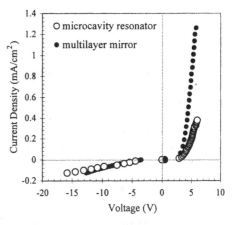

Figure 6. Current density - voltage relationship for a PSi multilayer mirror and a microcavity resonator. A faster current increase is observed for the multilayer mirror. Threshold voltage is ~3V, and Au contact area is 0.079 cm^2.

Typical EL spectra from these devices are displayed in figure 7, where the PL of the samples is also shown for comparison. Two peaks are observed in the multilayer mirror

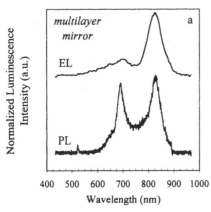
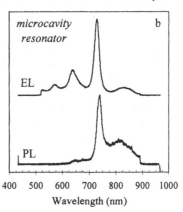

Figure 7. Electroluminescence from an oxidized PSi multilayer mirror (a) and microcavity resonator (b). Voltage applied was ~100V at a current density ~30 mA/cm^2.

device, at 700 nm and 810 nm. They correspond exactly to the leaky modes in the reflection stopband. The designed resonating wavelength for the microcavity resonator in figure 7b is 730 nm. A narrow peak is clearly observed at this wavelength in both the EL and PL spectra. This reduction in line shape gives evidence to the fact that the luminescence of the active layer is altered due to the effects of confinement by two highly reflecting mirrors (Fabry-Perot cavity). Through thermal oxidation, both the multilayer mirror and microcavity resonator devices are well-passivated, thus maintaining a stable structure.

CONCLUSIONS

The feasibility of an all PSi multilayer device has been successfully demonstrated. To overcome the inherent instability in PSi, thermal oxidation has been incorporated as a necessary and vital step. After that step, the multilayer stack remains intact and the peak reflectance does not decrease. Constructing a microcavity resonating device has tremendous advantages, such as offering the possibility of obtaining high spectral purity by narrowing the emission band of a single PSi layer. Tunable and narrow PL spectra have been observed in PSi microcavity resonators from 848-712 nm, and more recently, EL from these structures has been obtained.

ACKNOWLEDGEMENTS

We gratefully acknowledge support from the US Army Research Office.

REFERENCES

[1]   L.T. Canham, Appl. Phys. Lett. **57**, 1046 (1990).
[2]   K.D. Hirschman, L. Tsybeskov, S.P. Duttagupta, and P.M. Fauchet, Nature **384**, 338 (1996).
[3]   S. Frohnhoff and M.G. Berger, Advanced Materials **6**, 963 (1994).
[4]   A. Loni, L.T. Canham, M.G. Berger, R. Arens-Fischer, H. Munder, H. Luth, H.F. Arrand, and T.M. Benson, Thin Solid Films **276**, 143, (1996).
[5]   L. Tsybeskov, S.P. Duttagupta, KD. Hirschman, and P.M. Fauchet, Appl. Phys. Lett. **68**, 2058 (1996).
[6]   L. Pavesi, La Rivista del Nuovo Cimento **20**, (10), 1-76 (1997).
[7]   W. Theiβ, Surface Science Reports **29**, 91 (1997).
[8]   G. Vincent, Appl. Phys. Lett. **64**, 2367, (1994).
[9]   J. von Behren, PhD thesis, Technische Universitat Munchen, 1998.
[10]   M.G. Berger, M. Thönissen, R. Arens-Fischer, H. Münder, H. Lüth, M. Arntzen, and W. Theiβ, *Thin Solid Films* **255**, 313, (1995).
[11]   L. Pavesi, *J. Appl. Phys.* **80**, 216 (1996).

SILICON-BASED UV DETECTOR PROTOTYPES USING LUMINESCENT POROUS SILICON FILMS.

Limarix Peraza [*], Madeline Cruz [*], Angel Estrada [*], Carlos Navarro [**], Javier Avalos [*], Luis F. Fonseca [**], Oscar Resto[**], and S. Z. Weisz[**].
[*]Dept. of Sciences and Technology, Metropolitan University, San Juan, PR.
[**]Dept. of Physics, University of Puerto Rico, San Juan, PR 00931.

ABSTRACT

The luminescent properties of porous silicon (PSi) films in the visible region were used to improve the photoresponse of PSi/Si-wafer and PSi/Si p-n junctions UV detector prototypes in the region below 500nm. A luminescent PSi overlayer was formed on top of the Si wafers and p-n junctions by electrochemical anodization. These overlayers have emission spectra peaking close to 690nm. In the case of the PSi/Si wafer, the PSi film was produced with a high optical transparency above 600nm and highly absorbent below this value. With such characteristics, the incident UV radiation is partially absorbed and converted into visible radiation that can be highly transmitted through the PSi film and efficiently absorbed by the wafer or the junction. The UV measurements show enhancement of the photoresponse at 366nm as compared with control prototypes without PSi. Details about the enhancement process are discussed.

INTRODUCTION

The use of silicon for UV-photodetection applications faces the problem of the silicon surface UV response that reduces significantly the interband light absorption in this region of the EM spectrum. In a typical commercial Si-based detector, the photosensitivity at 400nm is of the order of 5 times smaller than at 800nm around which the detector has its maximum sensitivity. Methods have been developed to overcome such difficulty and produce the so-called UV-enhanced silicon detectors. Today, a new possibility to enhance the UV-photoresponse of silicon-based detectors is the inclusion of photoluminescent PSi layers into the design of the device.

After the report of efficient photoluminescence from electrochemically etched silicon wafers [1], considerable effort has been done on the study and the development of electrooptical applications of PSi [2,3]. The maximum of the photoluminescence emission of PSi changes with the characteristics of the silicon wafer and the preparation method but it is in the range between 800nm and 650nm for a typical freshly dried and exposed to air sample. These emission spectra have broad bands. The main idea to improve the sensitivity of a silicon-based UV detector using PSi is to exploit the efficient UV-to-visible conversion mechanism of PSi layers. A prototype of such a PSi-based new detector will operate under the assumption that UV incoming radiation is absorbed by the PSi layer and re-emitted as visible radiation according to the above discussed photoluminescence properties of the PSi layer. The converted energy then is in the frequency range of maximum photosensitivity of the bottom part of the device. Several prototypes have been developed in recent years using PSi to improve the photosensitivity of the Si-based UV detectors. Some of the studied photodiode structures are: metal/PSi/c-Si, and n-PSi/p-PSi/c-Si [4-7]. The performance of those structures is based on the I-V response for electric charge transport along the direction perpendicular to the layers. Characteristic band bending and an increasing photosensitivity with strong dependence on the reverse bias voltage were found at the PSi/c-Si interface[8].

We studied the photoresponse of two types of structures under UV illumination: PSi/p-Si/Al and PSi/n-Si/p-Si/Al. The first structure was studied by measuring the change in the voltage drop, with and without illumination, for an electric current passing through the p-Si layer. In this case

Mat. Res. Soc. Symp. Proc. Vol. 536 © 1999 Materials Research Society

PSi overlayer was used as a UV-to-visible conversion interface. The second heterostructure was studied by measuring the photovoltage across the p-n junction/PSi structure.

EXPERIMENT AND RESULTS

We prepare PSi by electrochemical anodization of crystalline Si using an acid solution containing HF, water, and ethanol. The Si wafer, as well as the p-n structure, are previously prepared to have an aluminum ohmic contact at the bottom (at the p-Si side) deposited by sputtering and annealed at 440 ^{0}C during 10min. The wafer is then mounted at the bottom of a teflon electrolytic cell with the crystalline Si surface facing towards the cell where the acid solution is poured once the surface is cleaned with 20%HF-water solution during 20min. A platinum cathode in contact with the acid solution and a computer controlled constant current source complete the circuit. With this procedure, the PSi layer grows from the top towards the bottom of the wafer as the electric current is applied.

It is generally accepted that the visible photoluminescence of PSi is related to the quantum confinement of the free carriers in nanocrystalline Si-structures formed by the electrochemical etching process [1,2]. Depending on the resistivity of the silicon wafer and the anodization parameters, one can get luminescent PSi with quite different microstructures. SEM and TEM of PSi samples show that it can indeed be prepared to yield nanoporous or microporous films. Samples in which a columnar or coral-like structure of pores with diameters of the order of microns also exhibit PL indicating the existence of nanoporous structure on the pore-walls but this kind of microstructure produces large light scattering at the visible wavelengths. Optically, nanoporous-Si behaves as a homogenous material in the visible and longer wavelength regions, as shown by ellipsometry, reflectivity, and transmitance measurements, and effective medium calculations for the dielectric function of such samples. Considering that our application requires high transparency above 600nm, the formation parameters were set to produce nanoporous PSi layers.

A prototype was prepared from a 250μm-thick p-type Si wafer with 20Ωcm resistivity and (100) surface orientation. After the corresponding sputtering and annealing processes that produces the bottom aluminum ohmic contact, the disk was mounted on the electrolytic cell and a central circular area of 0.5cm^2 was reduced in thickness to 60μm by dissolving the wafer during 105s by pouring a mixture of hydrofluoric, nitric and acetic acids (CP-4) in proportions 3:5:3 into the cell. After this step, the cell and the wafer were washed with deionized water and the cell filled up with a 1:1:2 (HF:water:ethanol) electrolytic solution. The exposed area was then anodized for 120s by passing a current density of 100mA/cm^2 between the aluminum contact and the platinum cathode. With this current density, a PSi film of 10μm was formed. After anodization, the wafer was washed with ethanol and left under ethanol during 3 hours. Finally, the prototype was dried using the supercritical point drying method [9]. After the drying process, two bottom ohmic contacts were prepared by partially dissolving the aluminum bottom layer with HCl. Figure 1 shows the schematic configuration of the prototype.

Figure 2 shows the photoluminescence spectrum of the PSi overlayer. The spectrum was obtained by exciting the layer with a 442nm defocused laser beam. The figure shows a broad band emission peaking at 670nm which is within the radiation energy range of high sensitivity for Si-based photodetection. Figure 3 shows the transmission spectrum of a free standing PSi film prepared from a p-type Si wafer and with similar characteristics to the one used in the prototype. The figure shows high transparency above 600nm. The significant reduction of the transmission coefficient below 500nm is associated with a large absorption of the incident radiation at this frequency range. This absorption is related with the conversion of the UV incoming radiation to visible as is shown in the photoluminescence response of the PSi film (figure 2).

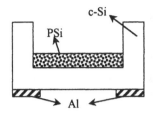

Figure 1. Schematic configuration of the first prototype.

The UV photoresponse of the prototype was determined and compared with a control device similarly prepared but without the PSi cover layer. We studied the photoresponse by passing a lateral current through the c-Si layer between the two aluminum ohmic contacts. The change in the voltage drop across the devices (ΔV), with and without UV illumination, as a function of the lateral current was measured using an exciting 360nm UV radiation with an intensity of $7\mu W/cm^2$. Figure 4 shows the photoresponse of the prototype as compared with the control.

The second studied prototype was prepared from a 250μm p-type Si wafer of ~1Ωcm resistivity and with a 20μm thick n-type Si overlayer, epitaxially grown on top forming a p-n junction. The resistivity of the epitaxial n-type layer was ~4.6Ωcm. This wafer was treated in similar way as the previous one. First, the aluminum ohmic contact was formed at the bottom side, after that, the n-type epitaxial layer was cleaned with 20% HF solution and exposed to the 1:1:2 electrolytic solution and anodized during 40s using a current density of 100mA/cm². Because of the hole-deficiency of the n-type layer, the surface was illuminated during anodization with a quartz white lamp with an intensity of 100mW/cm². The process, as describe above, results in the formation of a PSi layer of ~3μm. After the anodization, the prototype was dried using the same procedure as explained for the first prototype. In this case the photoresponse was tested on the photovoltage across the p-Si/n-Si/PSi structure. For such measurements, the contacts were the

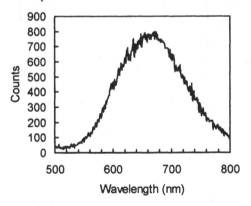

Figure 2. Photoluminescence spectrum of the PSi overlayer.

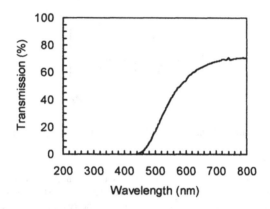

Figure 3. The transmission spectrum of a PSi free standing film.

aluminum bottom contact deposited at the p-side and a sputtered gold thick film of 1mm in diameter deposited on top of the PSi overlayer and close to the border of the anodized region, as shown in figure 5. To determine the UV photoresponse of this prototype, we measured the voltage between the gold and the aluminum contacts when the PSi layer was illuminated with the exciting 360nm UV radiation with an intensity of $7\mu W/cm^2$. The measured voltage was 12.3mV. The measured voltage for a control prototype prepared in similar way but without the PSi layer was 1.2mV under similar conditions.

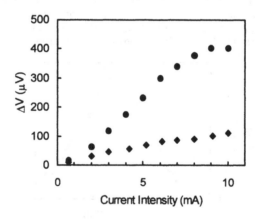

Figure 4. Photoresponse of the first prototype (circles) as compared with the control (diamonds).

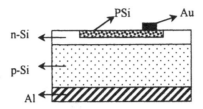

Figure 5. Schematic configuration of the second prototype

CONCLUSIONS

Two Si-based light detector prototypes using PSi and with enhanced UV response were prepared and studied. The supporting idea for using PSi to produce such enhancement rests on the fact that Si have reduced UV photosensitivity and PSi can be used as an interface that converts UV radiation into visible radiation which is efficiently used by Si for the photogeneration of free carriers. The application of such procedure requires the preparation of a PSi overlayer with enough optical quality that will allow the transmission of the re-emitted visible photons to the crystalline Si region. We succeeded on the preparation of high optical quality PSi films, as is evident from figure 3. The enhancement of the photoresponse of the first prototype, measured as the increasing of the slope of the (ΔV)-I curve, was confirmed from the graph of figure 4 which shows an increasing of 354% as compared with the corresponding slope for the control. It is worth noting that, in the case of the first prototype, the electric current flows in a lateral configuration instead of along the different layers. In that case, it is reasonable to assume that the electric current is passing mainly through the c-Si layer that has less resistivity than the PSi layer which acts as a visible photons provider.

The second prototype based on a p-n junction structure also showed a significant enhancement of the UV photoresponse in the near UV region. The electric charge transport in this case transverses through the different layers, including the PSi region. It is worth mentioning that the SEM image of the PSi structure for the case of the p-n junction points to the existence of porous structures with diameters of the order of 4µm. The size of these pores are large enough that scattering of visible light is expected. However, for that kind of prototype, the thickness of the PSi film is small, as compared with the first prototype, to reduce the scattering losses.

Both prototypes have shown that PSi can be used to increase the photosensitivity of Si-based detectors in the UV spectral region. The operational principle is based on the ability of PSi to absorb UV photons and reemit in the visible region where the Si shows better photosensitivity. The expected enhancement will depend on the efficiency of the energy conversion process related with the fraction of visible photons that are re-emitted from the total absorbed UV photons. From the presented results, one can conclude that the re-emission efficiency is large enough to overpass the small fraction of UV photons that causes direct free carriers photogeneration in c-Si. As a final remark, due to the geometry of our prototypes, there is a possibility that free carriers generated at the PSi layer can diffuse into the active region. This other mechanism will also contribute to the enhanced photoresponse of the detector.

ACKNOWLEDGMENTS

Authors at the Metropolitan University acknowledge support from NSF-MIE grant Authors from University of Puerto Rico acknowledge partial support from NASA grant No. NCCW-0088 and US ARO grant No. DAAHO4-96-1-0405.

REFERENCES

1. L.T. Canham, Appl. Phys. Lett., 57, 1046 (1990).
2. A.G. Cullis, L.T. Canham, and P. D. Calcott, J. Appl. Phys. 82, 909 (1997).
3. See contributions to MRS Bulletin 23, (1998).
4. J.P. Zheng, K. L. Jiao, W. P. Shen, W. A. Anderson, and H.S. Kwok, Appl. Phys. Lett. 61, 459 (1992).
5. L. V. Belyakov, D. N. Goryachev, O. M. Sreseli, and I. D. Yorashestkii, Semiconductors 27, 758 (1993).
6. T. Ozaki, M. Araki, S. Yoshimura, H. Koyama, and N. Koshida, J. Appl. Phys. 76, 1986 (1994).
7. B. Unal and S. C. Bayliss, J. Phys. D: Appl. Phys. 30, 2763 (1997).
8. L. A. Balagurov, D. G. Yarkin, G. A. Petrovicheva, E. A. Petrova, A. F. Orlov, and S. Ya. Andryushin, J. Appl. Phys. 82, 4647 (1997).
9. J.Von Beheren, P.M. Fauchet, E.H. Chimowitz, and C.T. Lira, Mat. Res. Soc. Symp. Proc. 452, 565 (1997).

EFFECTS OF VARIOUS HYDROGEN DILUTION RATIOS ON THE PERFORMANCE OF THIN FILM NANOCRYSTALLINE/CRYSTALLINE SILICON SOLAR CELLS

Y. J. SONG AND W. A. ANDERSON
Department of Electrical Engineering, State University of New York at Buffalo, Amherst, NY 14260

ABSTRACT

Low temperature growth of hydrogenated nanocrystalline silicon film (nc-Si:H) by microwave electron cyclotron resonance chemical vapor deposition has been performed employing a double dilution of silane, using a He carrier for SiH_4 and its subsequent dilution by H_2. A series of Raman spectra and AFM pictures has shown that a very thin (<100Å) nc-Si:H layer initially grown with high H_2 dilution on a glass substrate can serve as a seed layer for the subsequent growth of the film with lower H_2 dilution, which results in a higher crystallinity of the whole film. The role of this thin layer in low temperature junction formation has been examined by the insertion of the layer between the interface of both nc-Si:H (deposited with lower H_2 dilution)/c-Si and a-Si:H/c-Si heterojunction type photovoltaic cells. This is to address the knowledge that the device's performance is strongly influenced by the quality of the thin film silicon/crystalline silicon interface. Various thicknesses and H_2 dilution ratios have been used to find the optimized condition providing the best performance of the cells. The maximum efficiency of 10.5% (J_{sc}=35.1mA/cm^2, V_{oc}=0.51V and FF=0.59) has been obtained, without an AR coating, by the successive deposition of nc-Si:H film with four different H_2 dilution ratios on a crystalline silicon substrate. This is potentially a low-temperature, low-cost solar cell fabrication process.

INTRODUCTION

Hydrogenated nanocrystalline silicon (nc-Si:H) films, obtained by plasma-assisted decomposition of silane diluted in H_2 at low temperatures, has become an important material for the application of solar cells and thin film transistors (TFTs). Although the nc-Si:H film generally reveals higher mobility, stronger light absorption in the infrared and improved stability, compared to hydrogenated amorphous silicon (a-Si:H) film, the as-deposited nc-Si:H film (without passivation and/or recrystallization) still needs improvements in grain size, passivation of dangling bonds and deposition rate. Recently, the use of high density plasmas generated in electron cyclotron resonance (ECR) condition has become one of the standard techniques for the growth of epitaxial-like silicon thin films because it provides a better control of energetic ion bombardment (which is believed to enhance the etching effect and promote the growth of amorphous phase in the film), together with a relatively high deposition rate (typically, 20~50 Å/min), even at lower temperatures (≤ 400 °C) [1-2]. Furthermore, it is known that the control of initial growth of nc-Si:H is essential to obtain the high degree of crystallinity of thin films (<2000 Å) since the initial thin layer on the top of a substrate tends to be amorphous. The common technique for this includes the utilization of a seed layer, grown in highly H_2 diluted condition with very low deposition rate (<20 Å/min), resulting in a higher degree of crystallinity of the following nc-Si:H film deposited with low H_2 dilutions [3-4].

In this paper, we first demonstrate the optimized growth condition of nc-Si:H films, providing both a high crystallinity and a reasonable throughput, by utilization of multi-layer structures having different H_2 dilution ratios and layer thicknesses. Finally, in order to examine

129

the nc-Si:H film as an effective solar cell material, we investigate the dependence of various H_2 dilution ratios and film thicknesses on the photovoltaic properties in single or multi-layer thin film nc-Si:H/crystalline Si heterojunction type solar cells.

EXPERIMENT

The nc-Si:H films were deposited by microwave (2.45 GHz) ECR-CVD, whose schematic diagram is described elsewhere [5]. The source gas was hydrogen diluted 2% SiH_4/H_2 mixture, while pure Ar formed the background gas for plasma generation. All the depositions, including single or multi-layer nc-Si:H films prepared at different H_2 dilution levels in the discharge, were performed at a substrate temperature of 400 °C, a chamber pressure of 1~5 mTorr and an input power of 300 W. Both flat Corning glass (for AFM and Raman analysis) and HF treated p-type 1 ~ 5 Ω-cm (100) float-zone silicon wafers (for solar cells) were used as substrates. In particular, the total thickness of all films on glass substrates was adjusted to around 1000 Å, in order to eliminate the thickness dependence on film's properties. The structural properties of films were investigated by both atomic force microscopy (AFM) and Raman spectroscopy. The deposition rate of the films obtained in various H_2 dilution ratios was determined by the low angle irradiation of synchrotron X-ray. The photovoltaic cell structure utilized in this work was undoped nc-Si:H/p-type c-Si heterojunction with Mg/Al (100 Å/1000 Å) double layer grid contact on the top (0.28 cm^2 cell area), while 1000 Å thick evaporated and sintered (at 600 °C) Al served as an ohmic contact on the back. Photovoltaic response test of solar cells was done under 100 mW/cm^2 AM1.0 spectrum from a tungsten halogen lamp calibrated with a cell previously tested at NREL.

RESULTS AND DISCUSSION

Film Properties

Figure 1 demonstrates the dependence of H_2 dilution ratios in single or multi-layer nc-Si:H films on Raman shift. As revealed in the figure, the single layer film deposited at $R_H = 0.33$ (R_H is defined as $H_2/(H_2+Ar+SiH_4+He)$) shows the peak at 508 cm^{-1} (the peak position of crystalline silicon is at around 521 cm^{-1}) with the broadest full width at half maximum (FWHM) of 32 cm^{-1} (Fig.1 (a)), indicating the lowest degree of crystallinity of the four samples shown here. By comparison, the Raman peak of another single layer film of $R_H = 0.58$ (Fig.1 (b)) exhibits the highest degree of crystallinity (peak position at 517 cm^{-1} with the narrowest FWHM of 11 cm^{-1}). However, despite the superior crystallinity of the film grown at high R_H, this suffers from its slow deposition rate (due to the etching effect of hydrogen atom at the surface) of only 10 Å/min for $R_H = 0.58$ in this study (other deposition rates are 35, 56 and 68 Å/min for R_H of 0.33, 0.11 and 0, respectively). This low throughput may not be suitable for commercial low-cost processes.

To increase the throughput without losing the high degree of crystallinity, the film growth at lower R_H has been followed by the growth of a very thin film deposited at higher R_H (called 'seeded growth'). From the figure, the Raman peak of the film deposited at three different H_2 dilution ratios (i.e., successive depositions in the conditions of $R_H = 0.58$ (90 Å thick), $R_H = 0.33$ (800 Å thick) and $R_H = 0$ (100 Å thick)) presents a pretty comparable crystallinity (peak position at 516 cm^{-1} with FWHM of 14 cm^{-1}, as in Fig.1 (c)) to the film deposited at $R_H = 0.58$ alone. Moreover, the nc-Si:H film deposited by the successive growth at four different H_2 dilution ratios still exhibits good crystallinity, in spite of the shorter total deposition time (30 mins).

130

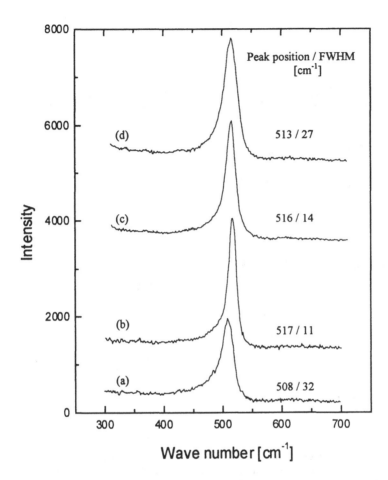

Fig. 1 Raman spectra of the ECR-CVD nc-Si:H films on glass substrates deposited at four different conditions. The peak position and its FWHM are given in the figure. (a) $R_H = 0.33$, thickness = 1000 Å, total dep. time = 28 mins (b) $R_H = 0.58$, thickness = 1000 Å, total dep. time = 100 mins (c) $R_H = 0.58/0.33/0$, thickness = 90/800/100 Å, total dep. time = 33 mins (d) $R_H = 0.58/0.33/0.11/0$, thickness = 90/400/400/100 Å, total dep. time = 30 mins.

Figure 2 shows the AFM images of the four samples, which were previously used for Raman spectroscopy. As expected, the largest grains with a diameter up to 700 Å can be observed in the film grown at $R_H = 0.58$ (Fig. 2 (b)), which showed the crystallinity in Raman spectrum. In addition, large grains of around 500 Å are also found in seeded grown nc-Si:H films ((c) and (d) in Fig.2). On the other hand, the smaller grains (200 ~300 Å) are observed in the film grown at $R_H = 0.33$ (Fig. 2 (a)).

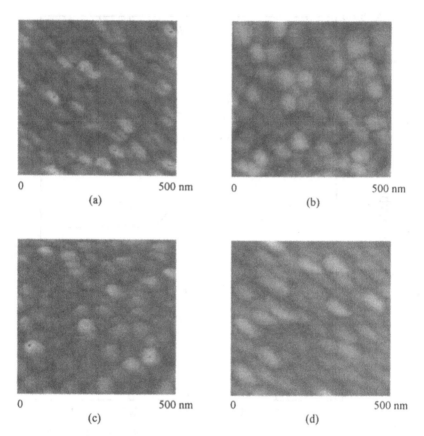

Fig. 2   AFM images of ECR-CVD nc-Si:H films on glass substrates grown at four different conditions (the same samples as used in Raman spectroscopy). (a) R_H = 0.33, thickness = 1000 Å (b) R_H = 0.58, thickness = 1000 Å (c) R_H = 0.58/0.33/0, thickness = 90/800/100 Å (d) R_H = 0.58/0.33/0.11/0, thickness = 90/400/400/100 Å.

Photovoltaic Properties

As illustrated in Table I, several different combinations of R_H and layer thickness have been examined to investigate their effects on thin film nc-Si:H/c-Si type solar cells. Since the performance of heterojunction type cells is strongly dependent on the interface quality [6], it can be said that the thin seed layer (deposited at high R_H) might play a key role in photovoltaic properties of the cell, in addition to its dominant role in structural properties. The solar cell (Cell #20), fabricated with a single layer nc-Si:H film grown at R_H = 0.58 onto crystalline silicon substrate, reveals the lowest efficiency (2.0 %, see also J-V curves in Figure 3) of all, in spite of

Table I. Dependence of H_2 dilution ratio and film thickness on photovoltaic properties of thin film nc-Si:H/c-Si heterojunction solar cells.

Sample #	R_H	film thickness [Å]	J_{sc} [mA/cm^2]	V_{oc} [V]	FF	η^* [%]	total dep. time [mins]
Cell #15	0.33	800	24.9	0.49	0.57	7.0	23
Cell #20	0.58	800	6.9	0.41	0.71	2.0	80
Cell #13	0.58/0.33/0	70/500/200	31.9	0.51	0.52	8.3	24
Cell #14	0.58/0.33/0	70/500/100	27.2	0.50	0.64	8.7	23
Cell #8	0.58/0.33/0.11/0	70/300/200/100	35.1	0.51	0.59	10.5	21

* Solar cell efficiency without an anti-reflective coating.

its high crystalline nature. This result could be explained by the low hydrogen content in the film, leaving dangling bonds unpassivated at grain boundaries which creates defect states within the bandgap. Its relatively low open circuit voltage (V_{oc} = 0.41 V) implies the existence of states near the interface. Its low short circuit density (J_{sc} = 6.9 mA/cm^2) is due to less light absorption since there is less amorphous phase in the film. The Cell #15 shows the improved photovoltaic properties compared to Cell #20. Since the film in Cell #15 has been grown at lower H_2 dilution level (R_H = 0.33), it is apparent that the enhanced amorphous phase with larger amount of hydrogen content in the film has contributed to more light trapping (higher J_{sc}) and better passivation of dangling bonds (higher V_{oc}), simultaneously.

In the fabrication of Cell #13 and #14, a very thin nc-Si:H layer ($\sim$ 70 Å) deposited at R_H = 0.58 has been inserted into the interface (between nc-Si:H and c-Si) of Cell #15. In addition, another thin layer (200 Å for Cell #13 and 100 Å for Cell #14) has been grown on the top without H_2 dilution. These photovoltaic cell configurations have increased the efficiency by 15 % (8.3 % efficiency for Cell #13) and 24 % (8.7 % efficiency for Cell #14), even though comparable deposition times (23 $\sim$ 24 mins) have still been used. Now, it can

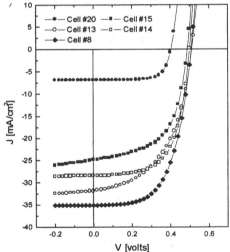

Fig. 3 J-V photovoltaic response of the nc-Si:H/c-Si solar cells described in Table I.

be considered that the interface states existing in Cell #20 have been removed by the subsequent deposition of low H_2 diluted films, which contains a higher amount of hydrogen (the hydrogen atom may easily diffuse through the inserted layer because the layer is very thin, and then, it passivates the dangling bonds effectively). The elevated V_{oc} (0.50 ~ 0.51 V) and J_{sc} (27.2 ~ 31.9 mA/cm^2) are reflecting the fact that the inserted layer with high crystallinity acts as an effective buffer layer, which improves the interface quality of the cell, as long as the defects in the layer are passivated sufficiently.

Moreover, it is clear that the thin layer on the top (R_H = 0) provides not only a further light absorption, but also a better surface passivation, according to its amorphous-like properties. However, although the thicker top layer in Cell #13 seems to increase J_{sc} by more light trapping, it has reduced the fill factor (FF) significantly, due to its lower conductivity and shorter diffusion length of carriers. Finally, the maximum efficiency of 10.5 % (Cell #8) has been obtained, without an anti-reflective coating, by the successive deposition of nc-Si:H film at four different H_2 dilution ratios, with a relatively high throughput (21 minutes).

CONCLUSIONS

In summary, we have investigated the role of H_2 dilution ratio in the low temperature growth of ECR-CVD nc-Si:H films, in terms of structural and photovoltaic properties. It is found that a very thin initial layer deposited at high H_2 dilution improves both the crystallinity of the entire film (as a seed layer) and the interface quality of nc-Si:H/c-Si heterojunction type solar cells (as a buffer layer), resulting in an elevated solar cell efficiency. Furthermore, this seeded growth process reduces the processing time significantly, maintaining high crystallinity and high solar cell efficiency. This optimized process is potentially attractive for a low-cost, low-temperature solar cell fabrication.

ACKNOWLEDGMENT

The authers greatly appreciate the technical support of En-hwa Lee in AFM analysis, and also want to thank Hae-du Chung and Sun-suk Kim for their help in Raman spectroscopy.

REFERENCES

1. J. L. Rogers, P. S. Andry, W. J. Varhue, P. McGaughnea, E. Adams, and R. Kontra, Appl. Phys. Lett. **67**, 971 (1995).

2. K. Erickson, V. L. Dalal, and G. Chumanov, Mat. Res. Soc. Symp. Proc. **467**, 409 (1997).

3. J. Zhou, K. Ikuta, T. Yasuda, T. Umeda, S. Yamasaki, and K. Tanaka, Appl. Phys. Lett. **71**, 1534 (1997).

4. W. S. Park, Y. H. Jang, M. Takeya, G. S. Jong, and T. Ohmi, Mat. Res. Soc. Symp. Proc. **467**, 403 (1997).

5. B. B. Jagannathan, R. L. Wallace, and W. A. Anderson, J. Vac. Sci. and Tech. A **16**, 2751 (1998).

6. B. Jagannathan and W. A. Anderson, Solar Energy Material and Solar Cells **44**, 165 (1996).

INTEGRATION OF MULTILAYERS IN Er-DOPED POROUS SILICON STRUCTURES AND ADVANCES IN 1.5 μm OPTOELECTRONIC DEVICES

H. A. LOPEZ, S. CHAN*, L. TSYBESKOV*, H. KOYAMA*, V. P. BONDARENKO**, and P. M. FAUCHET*
Materials Science Program, University of Rochester, Rochester, NY 14627
*Department of Electrical and Computer Engineering, University of Rochester, Rochester, NY 14627
**Belarusian State University of Informatics and Radioelectronics, Minsk, Belarus

ABSTRACT

Infrared photoluminescence (PL) and electroluminescence (EL) from erbium-doped porous silicon (PSi) structures are studied. The PL and EL from the Er-doped PSi structures and the absence of silicon band edge recombination, point defect, and dislocation luminescence bands suggest that the Er-complex centers are the most efficient recombination sites. PSi multilayers with very high reflectivity (R ≥ 90%) in the 1.5 μm range have been incorporated in the structures resulting in a PL enhancement of over 100%. Stable and intense EL is obtained from the Er-doped structures. The EL spectrum is similar to that of the PL, but shifted towards higher energy. The unexpected shift in emission opens up the possibility for erbium related luminescence to encompass a larger part of the optimal wavelength window for fiber optic communications.

INTRODUCTION

Erbium-doped silicon has received a lot of attention for its importance to fiber optics and the possibilities of intra-chip and chip-to-chip optical interconnect communication. After the initial report of weak $4f$-shell luminescence at ~1.54 μm from Er-doped silicon [1], the discovery of strong room temperature luminescence obtained from Er-doped silicon by co-doping with oxygen, nitrogen, and fluorine sparked great interest [2-3]. From that point on, numerous silicon based materials have been doped by erbium. Among them is porous silicon, a material that has been widely studied because of its efficient visible PL [4], compatibility with standard silicon processes in making integrated silicon based optoelectronic devices [5-6], and control over the index of refraction in making PSi multilayes [7]. Its very large surface to volume ratio makes the PSi matrix very accessible for Er-doping, as well as a host for large concentrations of oxygen. PSi can be easily doped by ion implantation [8] and electrochemical deposition [9]. Cathodic electrochemical deposition is preferred over ion implantation because it offers the advantages of deeper erbium penetration (10-20 μm), lower price, and simplicity of processing.

EXPERIMENTAL

PSi was produced by anodic etching of highly boron-doped c-Si wafers (0.008-0.012 Ω·cm) in a hydrofluoric acid-ethanol electrolyte (3:2). The PSi active layer was produced under a current density of 90 mA/cm^2 over a time of 5-150 seconds depending on the desired thickness.

Mat. Res. Soc. Symp. Proc. Vol. 536 © 1999 Materials Research Society

The thickness of the PSi active layer varied from 0.5 to 15 μm and its porosity was ~46 % (calculated gravimetrically). The PSi multilayers consisted of 10 periods each containing two layers of different porosities. They were produced under current densities of 35-165 mA/cm^2 over times of 2-3 seconds per layer. After etching, the samples were rinsed with ethanol and placed in deionized water for about 10 minutes. Later, the samples were boiled in a hydrogen peroxide/hydrochloric acid/water (1:3:15) solution (RCA solution) for 10 minutes in order to clean the structure. During this time the RCA solution was magnetically stirred to remove any bubbles on the surface of the PSi. The samples were rinsed and placed in water for 10 minutes. After a rinse in ethanol, the samples were immersed in an ErCl$_3$.6H$_2$O/ethanol solution and negatively biased relative to the Pt electrode in order to introduce the erbium ions into the PSi matrix. The incorporated erbium concentration is estimated to be ≤ 10^{-19} cm^{-3} by taking 10% of the total charge of the electrochemical process [9]. After doping, the samples were oxidized for 20 minutes at 950°C in a dilute oxygen environment and densified for 10 minutes at 1100°C in a nitrogen environment. For electrical measurements semi-transparent (~13 nm) gold contacts were sputtered with a Denton Vacuum Desk II cold sputter. EL and PL were measured with a nitrogen cooled Ge detector using the 514 nm Ar$^+$ line as the excitation for PL. Reflectivity was taken using a Perkin-Elmer Lambda 9 UV/VIS Spectrometer. SEM, AFM, and electron fluorescence were also used in order to evaluate the quality of the PSi surface prior to erbium doping.

RESULTS

PSi RCA CLEANING

It is important to get large concentrations of erbium ions into the PSi matrix in order to obtain more efficient emission. There are reports, discussed in reference 10, which mention that for certain glasses the PL intensity levels off for high erbium concentrations as a result of ion precipitation. To our knowledge, this has not been reported for PSi. In order to facilitate the dopant infiltration and increase the erbium concentration, it is important to have a PSi surface and matrix free of blocking species. Removing surface species and obtaining a more atomically flat surface is also desired for improved optical properties and better adhesion of the top contact. This can be achieved by performing an RCA cleaning step to PSi prior to doping. Figure 1 shows SEM micrographs of similarly prepared PSi samples with and without cleaning. Figure 1a shows a uniform surface for the cleaned sample and figure 1b shows a surface that is being covered by some surface species. These species are probably blocking and capping pores and are responsible for the non-uniform profile of erbium reported in PSi [8]. Electron fluorescence shows that the sample that was not cleaned contains a large oxygen concentration (2.09%) with respect to the sample that was cleaned (0.10%). The oxygen is probably incorporated in the form of SiO$_x$ and is possibly represented as the light structures covering the PSi surface in the SEM picture. AFM of both samples was also performed and it was determined that the surface roughness decreased by ~25 % by cleaning the sample. The decrease of surface roughness will improve the optical and electrical characteristics of the structure. Both samples were also electrochemically doped by erbium under the same conditions, and as expected the sample that was cleaned showed an increase in ~1.54 μm emission by a factor of 4-5, which supports the hypothesis that more erbium is incorporated in the structure.

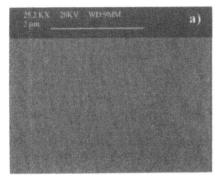
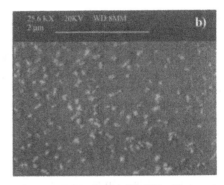

Figure 1. (a) SEM micrograph of a PSi surface that was RCA cleaned (99.87% silicon and 0.10% oxygen). (b) SEM micrograph of a PSi surface that was not cleaned (97.77% silicon and 2.09% oxygen by electron fluorescence).

PSi MULTILAYER INCORPORATION

In order to increase the output PL from Er-doped PSi, a multilayer structure was incorporated below the PSi active layer. The multilayer structure has a high reflectivity in the erbium emission range, so it acts as a mirror reflecting any emission that would be lost into the silicon substrate. Since the erbium goes through two annealing processes, it is important to establish what happens to the reflectivity peak during the thermal steps. Figure 2 shows what happens to the position and intensity of the reflectivity peak during 20 minutes of activation and 10 minutes of densification as a function of activation temperature. It is clear from the figure that larger changes are observed for higher temperatures. Similar results in position and intensity of the reflectivity peak are observed when the annealing time is increased. The annealing parameters were chosen because they produce strong Er-related PL while still maintaining high reflectivities from the multilayers. The temperatures chosen (950°C and 1100°C) result in a decrease in reflectivity of about 10 % and a shift in peak position of about 256 nm towards higher energy. Figure 3 shows the PL comparison of a 1 μm active layer + 4 μm multilayer structure and a 5 μm active layer structure. The PL of the multilayer structure shows a ~188% increase with respect to the 5 μm active layer control. Theoretically, a 100% increase should be the maximum increase if the multilayer reflects the

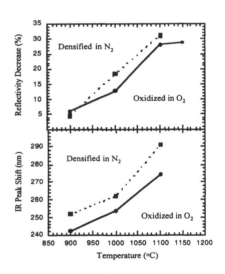

Figure 2. Percent decrease and shift to higher energy of the IR reflectivity peak as a function of activation temperature (20 minutes activation in O_2 and 10 minutes densification in N_2 at 1100 °C)

emission that would otherwise be lost to the substrate. The larger increase is possibly due to a change of reflectivity of the 514 nm excitation source as well as the PSi luminescence (600-800 nm) which further excites more erbium ions increasing the enhancement to higher than 100%. Figure 3 also shows the reflectivity of the 1 μm active layer + 4 μm multilayer structure with peaks at the erbium and PSi emission, as well as the laser excitation wavelength.

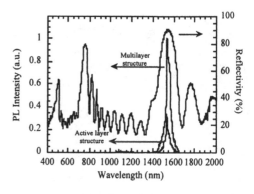

Figure 3. PL comparison of a 1 μm active layer + 4 μm multilayer and a 5 μm active layer structure. The reflectivity of the multilayer structure is also shown

We have also shown that spontaneous emission from erbium can be controlled by placing an active layer between two multilayers such that resonance occurs at a new wavelength of choice (resonator). Preliminary results show that the erbium emission can be controlled to occur at different wavelengths, where the natural spectrum of erbium is very weak. The location of the peak is easily controlled by changing the annealing temperature and duration since the reflectivity peak varies as a function of both parameters.

ADVANCES IN Er-DOPED PSi EL

The driving force behind this work is to obtain efficient electrically excited ~1.5 μm emission from silicon based materials for integration in optoelectronic devices. Since the first report of EL from Er-doped PSi in 1997 [11], we have been working on making a more efficient and stable LED structure. The inset in figure 4 shows a schematic of the device structure used for this study, where the Er-doped active layer is varied from 0.5-7 μm. Figure 4 also shows the I-V characteristics of a 1 and 5 μm active layer device at room temperature. The high resistance observed in the I-V plot is due to the poor transport through the oxidized PSi, but as the thickness of the PSi layer is decreased from 5 to 1 μm, improvements in transport are observed. A lower turn on voltage is observed for the thinner active layer device.

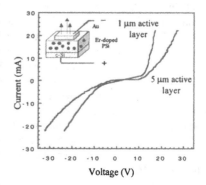

Figure 4. I-V characteristics of a 1 and 5 μm Er-doped PSi active layer device. The inset shows a schematic of the device.

Figure 5 shows room temperature EL from the 5 μm active layer device. The spectra was collected under a forward bias of 68 V and a current density of ~77.7 mA/cm^2. The EL turn on voltage for the device is about 35 volts. Thinner active layer devices exhibit better voltage characteristics, but the intensity of the EL peak is weaker due to lower erbium concentrations. Taking into account the high voltage, excitation of erbium probably occurs by impact of hot electrons as is the case for Er-doped silicon devices under reverse bias [12]. Figure 5 also compares the room temperature EL and PL spectra taken from the same sample. The EL emission is stable and the intensity is lower by a factor of 35-40 than that of PL excited with a laser intensity of ~ 4 W/cm². The shape of the EL spectrum is similar to that of the PL spectrum, with two differences: it is slightly broader and especially it is shifted towards lower

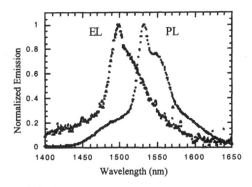

Figure 5. EL and PL comparison. EL is under a forward bias of 68 V and a current density of ~ 77.7 mA/cm² and PL is exited by the 514 nm line of an Ar⁺ laser.

wavelengths. To our knowledge such a large shift in erbium emission has not been reported. We have observed similar shifts for different LED's under forward and reverse bias, making this effect reproducible. Heating is always a possible cause for modification of the spectra since the devices are under large voltages, but no shifts were observed in the PL spectrum by exciting the same structure through the gold contact with a high laser intensity (40 W/cm²), suggesting that heating may not be responsible for the shift. Some very small spectral shifts in EL have been reported in the literature [11, 12], but the usual explanation is that different subsets of erbium centers that are in different environments are being excited. In this case, the shift is too large to be attributed to excitation of different erbium centers. Further studies (EL, PL, absorption, lifetime, and low temperature) are needed to better understand the origin of the new position of the peak. The new EL peak centered at ~1.50 μm is of great importance because it opens the door to larger usable bandwidths still in the window of maximum transmission for optical fiber communication.

CONCLUSION

We demonstrated enhancements in ~1.54 μm PL of over 100% by incorporating highly reflecting PSi multilayers into the PSi structure. The incorporation of the multilayer increases the PL intensity by reflecting the erbium emission that would be lost into the substrate, as well as increasing the number of excited Er-centers by reflecting the laser excitation and PSi emission. Control over the PL peak to new positions where the natural spectrum of erbium is weak is also observed. Stable and intense room temperature EL form Er-doped PSi is achieved. The observed EL exhibits an unexpected shift towards higher energy from the normal erbium emission, which enables erbium related emission to encompass a larger part of the high transmission window of optical fibers. It was also shown that a PSi pre-doping cleaning step is necessary because it cleans and unplugs the pores making erbium infiltration more efficient, improving the quality of the surface, and leading to better devices.

ACKNOWLEDGEMENTS

This research was made possible by support from the NSF Science and Technology Center (grant # CHE-9120001), the US Army Research Office, and by award No. BE2-108 of the US Civilian Research and Development Foundation for the Independent States of the Former Soviet Union.

REFERENCES

1. H. Ennen, J. Scheider, G. Pomrenke, and A. Axman, Appl. Phys. Lett. **43**, 943 (1983).
2. S. Coffa, F. Priolo, G. Franzo, V. Bellani, A. Carnera, and C. Spinella, Phys. Rev. B **48**, 11782 (1993).
3. J. Michael, L. Benton, R.F. Ferrante, D.C. Jacobson, D.J. Eaglesham, E.A. Fitzgerald, Y.-H. Xie, J.M. Poate, and L.C. Kimerling, J. Appl. Phys. **70**, 2762 (1991).
4. L.T. Canham, Appl. Phys. Lett. **57**, 1046 (1990).
5. K.D. Hirschman, L.Tsybeskov, S.P. Duttagupta, and P.M. Fauchet, Nature **384**, 338 (1996).
6. R.T. Collins, P.M. Fauchet, and M.A. Tischler, Physics Today **50**, 24 (1997).
7. L. Pavesi, La Revista del Nuovo Cimento **20** (10), 1-76 (1997).
8. F. Namavar, F. Lu, C.H. Perry, A. Cremins, N.M. Kalkhoran, J.T. Daly, and R.A. Soref, Mat. Res. Soc. Symp. Proc. **358**, 375 (1995).
9. T. Kimura, A. Yokoi, H. Horiguchi, R. Saito, T. Ikoma, and A. Sato, Appl. Phys. Lett. **65**, 983 (1994).
10. A. Polman, J. Appl. Phys. **82**, 1 (1997).
11. L. Tsybeskov, S.P. Duttagupta, K.D. Hirschman, K.L. Moore, D.G. Hall, and P.M. Fauchet, Appl. Phys. Lett. **70**, 1790 (1997).
12. S. Coffa, G. Franzo, and F. Priolo, MRS Bulletin **23** (4), 25-32 (1998).

STRUCTURAL CHARACTERIZATION OF nc-Si/a-SiO$_2$ SUPERLATTICES SUBJECTED TO THERMAL TREATMENT

G. F. Grom[*], L. Tsybeskov[**], K. D. Hirschman[**], and P. M. Fauchet[**]
[*] Materials Science Program, Department of Mechanical Engineering
[**] Department of Electrical and Computer Engineering
University of Rochester, Rochester, NY 14627

J. P. McCaffrey, H. J. Labbé, and D. J. Lockwood,
Institute for Microstructural Sciences, National Research Council, Ottawa K1A OR6, Canada

ABSTRACT

The morphology of nanocrystalline (nc)-Si/amorphous (a)-SiO$_2$ superlattices (SLs) is studied using Raman spectroscopy in the acoustic and optical phonon ranges, transmission electron microscopy (TEM), and atomic force microscopy (AFM). It is demonstrated that high temperature annealing (up to 1100°C) and oxidation in O$_2$/H$_2$O ambient do not destroy the SL structure, which retains its original periodicity and nc-Si/a-SiO$_2$ interface abruptness. It is found that oxidation at high temperatures reduces the defect density in nc-Si/a-SiO$_2$ SLs and induces the lateral coalescence of Si nanocrystals (NCs). The size, shape, packing density, and crystallographic orientation of the Si nanocrystals are studied as a function of the oxidation time.

INTRODUCTION

Nano-sized Si structures have been attracting considerable attention due to their potential use in devices based on single-electron phenomena (SEDs).[1] The quest for higher integration and better performance of VLSI devices pushes their sizes into the sub-micrometer range, where unwanted quantum mechanical effects and the high cost of manufacturing will eventually set the limit to further downscaling. SEDs are promising candidates to achieve higher integration with lower power consumption compared to VLSI-type devices. To realize the technological potential of SEDs, inexpensive techniques compatible with microelectronics processing need to be developed that allow fabrication and precise positioning of nanometer-sized structures with controllable size and shape, and minimal defect density. In recent years impressive progress has been made toward the fabrication of 1D, 2D, and 3D confined Si nanostructures[2] although more work is needed to improve the quality of Si nanomaterials and to gain more control over their parameters.

Among the different approaches for making 3D confined Si nanostructures, the fabrication of Si quantum dots (QDs) by crystallization of thin a-Si layers sandwiched between higher band-gap materials (e.g. Si$_3$N$_4$, SiN:H, SiO$_2$), offers several advantages.[3] One of them is that the size of the Si nanocrystals can be controlled precisely in the growth direction by varying the thickness of the a-Si layer. Recently, we reported the formation of Si nanocrystals after crystallization of a-Si/a-SiO$_2$ SLs.[4] The crystallized SLs contain layers of densely packed Si nanocrystals with a characteristic size equal to the thickness of the initially deposited a-Si layers. In this paper we show that nc-Si/a-SiO$_2$ SLs can withstand high temperature (up to 1100°C) processing. We present conclusive evidence that thermal treatments such as furnace annealing and oxidation can

141

change the lateral size of the NCs and improve their surface passivation with no degradation in interface sharpness in the SL direction.

SAMPLE PREPARATION

The a-Si/a-SiO$_2$ SLs were grown by radio-frequency magnetron sputtering and plasma oxidation in a Perkin-Elmer 2400 sputtering system with film deposition rates of 1.1±0.1 nm/min and 2.7±0.2 nm/min for a-Si and SiO$_2$ respectively. In some samples a SiO$_2$ buffer layer of several tens of nanometers in thickness was thermally grown on the Si substrate before deposition of the SL. The layer thicknesses of both Si and SiO$_2$ were kept constant throughout each SL and their values were determined by TEM. SLs with a-Si layer thicknesses in the 2-14 nm range were fabricated with up to 60 periods.

The a-Si layers were crystallized by rapid thermal annealing (RTA) at 900°C for 60 s with subsequent quasi-equilibrium (10°C/min temperature ramp-up) furnace annealing at 1100°C for 15 min. Rapid thermal and furnace annealing were performed in a nitrogen atmosphere. An optional wet oxidation step was done for some SLs by keeping them in O$_2$/H$_2$O vapor ambient at 700-900°C for periods of 10 min up to 4 h. Using X-ray diffraction and Raman spectroscopy, we have already shown that these deposition and annealing conditions result in the crystallization of the a-Si layers for thicknesses down to ~2 nm.[4,5] It was also demonstrated that the RTA step allows a significant reduction of the crystallization temperature for thin a-Si layers (less than 10 nm).[6]

RESULTS AND DISCUSSION

Transmission Electron Microscopy

Figure 1 shows cross-sectional TEM micrographs of a crystallized nc-Si/a-SiO$_2$ SL before and after wet oxidation. The insets show "representative" crystallites at each stage. The different crystallographic orientations of the Si NCs relative to the electron beam produce the different brightnesses of the Si crystallites in the TEM images. This interpretation was confirmed by X-

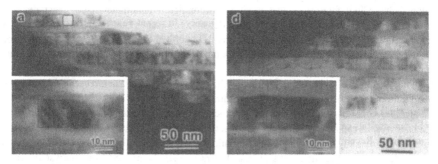

Fig. 1. Cross-sectional TEM micrographs of the (15nm nc-Si/7.5nm SiO$_2$)$_{15}$ SL (a) before wet oxidation and (d) after wet oxidation at 910°C for 30 min.

ray diffraction analysis.[5] The as-crystallized "average" Si crystallite (Fig. 1a) appears cubic in shape with an average size equal to the thickness of the initial a-Si layer. Neither annealing nor wet oxidation disturb the layered structure and the sharp, flat nc-Si/a-SiO$_2$ interfaces are preserved. We believe that, during wet oxidation, oxygen diffuses into the multilayer structure through the SiO$_2$ layers and grain boundaries in the nc-Si layers. It penetrates the whole multilayer structure since the SiO$_2$ layer thickness increases regardless of the layer's distance from the surface. Surprisingly, oxygen does not oxidize all of the Si in its diffusion path. With longer wet oxidation times, the size of an "average" crystallite decreases slightly in the growth direction of the SL but expands laterally, acquiring a rectangular shape. According to the TEM images, some NCs with clearly different crystallographic orientations (different brightness on Fig. 1a) transform into a single crystallite (uniform brightness on Fig. 1d). That might be interpreted as a result of coalescence between the adjacent crystallites. It is unclear whether the NCs are separated from each other laterally by SiO$_x$ or by Si grain boundaries after the wet oxidation step. The defects seen on the micrograph are typical of polycrystalline Si growth, i.e. dislocations, twins etc. In addition, it should be remembered that the TEM image is a 2D picture of a 3D volume filled with Si nanocrystals, that have different crystallographic orientations and no spatial ordering.

Atomic Force Microscopy

Atomic force microscopy was used to study the shape and density of Si NCs in the nc-Si

layers. The presence of Si NCs was revealed by dipping the SLs in dilute HF (25%) for a short period of time (from 1 to 10 s), in order to etch the SiO$_2$ on top of the SL and between NCs. Figure 2 is a lateral view of Si NCs in the top layer of a (21nm nc-Si/90nm SiO$_2$)$_5$ SL, obtained using the tapping mode of AFM. On this image the Si NCs are the bright spots. The maximum shadow contrast (between black and white) corresponds to 25 nm in height change, which was dependent on the etching time. Qualitatively, the Si NCs are densely packed, have similar in-plane sizes, and rounded shapes. No spatial ordering was observed. This result was confirmed from plan view TEM images.

Fig. 2 AFM lateral image of the top nc-Si layer in a (21nm nc-Si/90nm SiO$_2$)$_5$ SL .

Raman Spectroscopy: Optical Phonons

Figure 3 shows the Raman spectra from a (14nm nc-Si/5nm SiO$_2$)$_{60}$ SL after it was subjected to wet oxidation of different durations (spectra 1-3). We made sure that the Raman signal originated from the SL rather than the substrate by selecting a laser wavelength (λ=514 nm, Ar$^+$ laser) that has an absorption length ($\sim$0.7 μm) smaller than the total thickness of the nc-Si layers ($\sim$0.8 μm). The spectrum from crystalline Si (c-Si) is shown for reference (spectrum 4). In the crystallized sample (spectrum 1), there is no detectable signal from a-Si (~480 cm^{-1}), indicating complete crystallization of the a-Si layers. The Raman peak from nc-Si is red shifted by ~3 cm^{-1} relative to that of c-Si (~521 cm^{-1}), possibly due to tensile strain in Si NCs induced by the high density of Si/SiO$_2$ interfaces. For the Si NCs under study (d = 14 nm), the effect of the NC size on the position of the Raman peak is negligible.[7]

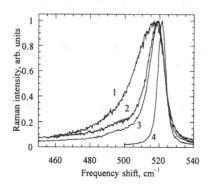

Fig. 3. Raman spectra of a (14nm nc-Si/5nm SiO$_2$)$_{60}$ SL obtained with Ar$^+$ laser (λ=514 nm): (1) as-crystallized, (2) after wet thermal oxidation at 950°C for 60 min, (3) after wet thermal oxidation at 1050°C for ~4h, and (4) Raman spectrum of c-Si.

Surprisingly, the high wavenumber slopes in spectra 1-4 are nearly identical. The low wavenumber slopes of spectra 1-3 are stretched and have considerable tails compared to that of spectrum 4. With longer wet oxidation time (going from 1 to 3) the spectra tend to narrow down and shift toward spectrum 4 (c-Si). This behavior might be explained as follows. High temperature annealing (at 1100°C) and wet oxidation do not dramatically reduce the size of Si NCs (no red shift), but rather lead to (i) a strain release (shift toward the c-Si peak position) and (ii) a reduction in defect density, possibly via defect ejection onto the surface of NCs accompanied by their oxygen passivation (narrowing of the peak). This interpretation is consistent with the TEM analysis and shows that high temperature annealing and wet oxidation can be used to improve the quality of Si NCs by lowering their defect density.

Raman Spectroscopy: Folded Acoustic Phonon Modes

The dispersion curve of acoustic phonons propagating in the growth direction of a SL is obtained by folding the dispersion curve of an average bulk compound into a reduced Brillouin zone that is defined by the SL period. In other words, the periodicity of the SL causes additional phonon modes to appear at the Brillouin zone center. The frequencies and Raman intensities of folded longitudinal acoustic phonons (FLAP) provide information on the compositional profile (i.e. layer thickness and interface sharpness) of the SL.[8] The FLAP in the nc-Si/SiO$_2$ SLs were studied in the quasi-backscattering geometry with an Ar$^+$ laser (λ=457.9 nm) as the excitation source. The spectra were measured with a SOPRA DMDP2000 instrument. The details of the experimental setup are given elsewhere.[9]

Figure 4 shows the Raman spectrum in the acoustic phonon range (2.5-40 cm^{-1}) of an as-crystallized (15nm Si/7.5nm SiO$_2$)$_{20}$ SL. The arrows indicate the positions of FLAP peaks calculated using Ritov's model.[10] The relative intensities of the peaks were calculated assuming a photoelastic scattering mechanism.[10] The spectrum is very rich in structure. Since the experimental data were obtained with fixed incident light wavelength and for SLs with identical periodicity, the assignment of peaks was not obvious. However, the agreement between calculated and observed peak positions is good (the details of the calculations will be presented

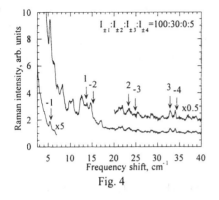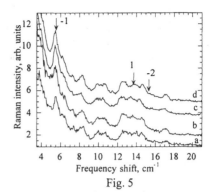

Fig. 4   Fig. 5

Fig. 4-5. Raman spectra of a (17nm nc-Si/8nm SiO_2)$_{20}$ SL excited by an Ar^+ laser (λ=457.9 nm). The arrows indicate the theoretical positions of the Raman peaks of folded longitudinal acoustic phonons (FLAP) propagating in the [100] direction. Fig. 4 and 5a: as-crystallized SL. Fig. 5b-5d: after wet oxidation at 910°C for 10, 20 and, 30 min, respectively. The spectra are shifted along the ordinate for clarity.

elsewhere[11]). In the calculations, all required parameters for Si NCs (i.e. photoelastic coefficients, density, velocities of acoustic waves) were assumed identical to those of c-Si. Scattering from phonons with folding index m=±1, ±2, ±3, and -4 was identified. The observation of a Brillouin peak (m=0) at 4 cm^{-1} was obscured by a large background of Rayleigh scattering. As shown on Fig. 4, not all of the peaks were identified. Qualitatively, unassigned peaks might originate from (i) phonons at the zone center (at 10 cm^{-1}), (ii) confined quasi-longitudinal and quasi-transverse acoustic phonons in Si NCs, which might be present due to the finite size and random crystallographic orientations of NCs, (iii) confined surface phonons in Si NCs, and (iv) additional FLAP that appear due to the finite number of SL periods, accompanied by some variation in the thickness of both the Si and SiO$_2$ layers (more recent samples with better thickness control and spatial uniformity show no additional FLAP peaks). No peaks were expected from folded transverse acoustic phonons (FTAP), since the intensity of scattering from FTAP was found to be four orders of magnitude smaller than that from FLAP regardless of the NC crystallographic orientation.[11] The fact that Raman peaks due to scattering from FLAP were observed up to high index values provides conclusive evidence that the periodicity of the SLs under consideration was good. The relative intensities of the peaks assigned to the FLAP are in a good agreement with theoretically calculated intensities, proving that the Si/SiO$_2$ interfaces are very sharp. In Fig. 5 the Raman spectra are compared after the sample was subjected to wet oxidation of different durations. The spectra are similar, as expected, since no dramatic changes occur in the periodicity of the SL and the Si/SiO$_2$ interfaces remain abrupt after wet oxidation.

CONCLUSIONS

We have demonstrated the fabrication of nc-Si SLs by controlled thermal crystallization. TEM and AFM show the formation of Si nanocrystals, that are densely packed and have comparable sizes. The nc-Si/SiO$_2$ SL structure can withstand high temperature annealing (up to 1100°C) and aggressive wet oxidation. It has excellent periodicity and abrupt interfaces. As a result, Raman scattering from folded acoustic phonon modes is observed. In addition, it is found that high temperature treatment reduces the defect density in nc-Si SLs and induces Si nanocrystal lateral coalescence.

ACKNOWLEDGMENTS

This work was supported in part by the US Army Research Office and Motorola.

REFERENCES

1. T. Haricot in *Mesoscopic Physics and Electronics*, edited by T. Ando, Y. Arakawa, K. Furuya, S. Komiyama and H. Nakashima, (Springer-Verlag, 1998), p.213-219
2. L. Guo, E. Leobandung, S. Y. Chou, Science **275**, 649-651 (1997); Y. Ishikawa, N. Shibata, S. Fukatsu, Appl. Phys. Lett. **72**, 2592-2594 (1998); Y. Kanemitsu, S. Okamoto, Phys. Rev B **56**, R15561-R15564 (1997)
3. D. A. Grützmacher, E. F. Steigmeier, H. Auderset, M. Morf, B Delley, and R. Wessicken in *Microcrystalline and Nanocrystalline Semiconductors*, edited by R. W. Collins, C. C. Tsai, M. Hirose, F. Koch, L. Brus (Mater. Res. Soc. Proc. **358**, Pittsburgh, PA, 1995) pp. 833-838; X. Huang, W. Shi, K. Chen, S. Yu, and D. Feng, ibid. pp. 839-844
4. L. Tsybeskov, K. D. Hirschman, S. P. Duttagupta, M. Zacharias, P. M. Fauchet, J. P. McCaffrey and D. J. Lockwood, Appl. Phys. Lett. **72**, 43-45 (1998)
5. P. M. Fauchet, L. Tsybeskov, M. Zacharias and K. D. Hirschman in *Thin-Films Structures for Photovoltaics*, edited by E. D. Jones, J. Kalejs, R. Noufi and B. Sopori (Mater. Res. Soc. Proc. **485**, Warrendale, PA, 1998) pp. 49-59
6. M. Zacharias, L. Tsybeskov, K. D. Hirschman, P. M. Fauchet, J. Bläsing, P. Kohlert, P. Veit, J. of Non-Crys. Solids, **227-230**, 1132-1136 (1998)
7. I. H. Campbell and P. M. Fauchet, Solid State Commun., **58**, 739 (1986)
8. B. Jusserand and M. Cardona, in *Light Scattering in Solids* V, edited by M. Cardona and G. Güntherodt (Springer, Berlin, 1989), p. 49.
9. D. J. Lockwood, M. W. C. Dharma-wardana, J. -M. Baribeau, and D. C. Houghton, Phys. Rev. **B 35**, 2243-2251 (1986)
10. C. Colvard, T. A. Gant, M. V. Klein, R. Merlin, R. Fischer, H. Morkoc, A. C. Gossard, Phys. Rev. **B 31**, 2080-2091 (1985)
11. G. Grom, L. Tsybeskov, K. D. Hirschman, P. M. Fauchet, J. P. McCaffrey, J.-M. Baribeau, H. J. Labbé and D. J. Lockwood, to be published

Part III

Biological Applications and Surface Chemistry of Nanocrystalline Semiconductors

IN-VIVO ASSESSMENT OF TISSUE COMPATIBILITY AND CALCIFICATION OF BULK AND POROUS SILICON

A P Bowditch, K Waters, H Gale, P Rice, E A M Scott.
Biomedical Sciences Dept, CBD, DERA Porton Down, Wiltshire, SP4 0JQ.
L T Canham, C L Reeves, A Loni , T I Cox.
Electronics Sector, DERA Malvern, St Andrews Road, Malvern, Worcestershire, WR14 3PS

ABSTRACT

The compatibility of both bulk and porous silicon at the subcutaneous site has been assessed for the first time, following ISO standard procedures. The in-vivo responses to implantation were monitored in the guinea pig and histopathological reactions evaluated at 1, 4, 12 and 26 weeks. Attention is focused here on the histological assessment protocols used, and the results demonstrating in-vivo evidence for good tissue compatibility, and porous Si bioactivity with regards calcification.

INTRODUCTION

Silicon technology, with its associated range of miniaturised intelligent devices, has in principle a lot to offer the medical implant industry. It is hence surprising that the dominant semiconductor has not been thoroughly assessed, nor developed, as a potential biomaterial in its own right. However, in 1995 it was claimed that silicon could be made much more 'biocompatible' in specific sites of the body, than had been previously thought [1]. In-vitro evidence was presented, demonstrating that nanostructured silicon surfaces were bioactive with regard promoting hydroxyapatite growth in simulated human plasma [1].

The aim of this study was to thoroughly assess the compatibility of silicon structures at the subcutaneous site, in a guinea pig model. The in-vivo assessment of potential biomaterials and medical devices is a critical element of the development and implementation of prosthetic implants for human use [2]. The only prior in-vivo studies of silicon biocompatibility of which we are aware concern bulk silicon in cortical tissue sites [3-5] and polycrystalline silicon structures at various body sites [6]. The latter small study involved no control materials for histological comparison [6]. The present work was conducted in accordance with the methods specified in the appropriate International Standard for biological evaluation of medical devices, and designed to use the minimum number of animals to allow valid statistical comparisons [2, 7].

SUMMARY OF TRIAL PROTOCOLS

We summarise here the key aspects of experimental design used in the trial; a more detailed description of test material manufacture, trial protocols and results will be published elsewhere [8]. Four classes of material were studied: bulk monocrystalline silicon (BSi), porous silicon of the type shown to exhibit calcification in-vitro (PSi) [9], porous silicon pre-treated in simulated plasma (CoPSi) [10] and bulk titanium (Ti) as a negative control material. Each test specimen was in the form of a disc (11 ± 0.5 mm diameter, 0.40 ± 0.05 mm thickness), whose surface topography was the same on all faces, and stored in air prior to sterilisation by autoclaving at 134°C for 10 minutes in pressurised steam. Four discs were evaluated in each animal and a total of 140 discs were studied. The clinical response to implantation was monitored daily by visual inspection of the implant sites, assessment of body weight and measurements of body temperature using a commercial electronic transponder (Biomedic Data Systems Inc. Model IPTT100). Histopathology was conducted at 1, 4 , 12 and 26 weeks post-implant. The specimen type at each implant site was randomised and the experiments and histology conducted blind.

HISTOPATHOLOGY OF IMPLANT SITES

Subcutaneous pockets were made by blunt dissection on the back of the animal and closed with appropriate suture material. Following explant, standard tissue sections of each site were stained with haematoxylin and eosin, for evaluation using a Zeiss Axioplan Photomicroscope. Pathological features reflecting the local tissue response, were graded by assigning a numerical score to each feature, as detailed in Table 1.

Table 1 Score grade criteria for the effects on the surrounding tissue of implanted specimens.

Grade.	Acute Inflammatory Reaction.
0.	No histological evidence of any acute inflammatory reaction.
1.	Small discrete clusters of inflammatory cells consisting predominantly of neutrophils and activated macrophages with occasional eosinophils and lymphocytes.
2.	Continuous sheets of acute inflammatory cells showing invasion of connective tissues in the immediate vicinity of the implanted material.
3.	Similar features to 2. above but associated with either necrosis of connective tissues and/or extension of cellular infiltrate beyond the vicinity of the implant.

Grade.	Tissue Degeneration, Oedema, Haemorrhage and Necrosis.
0.	No histological evidence of oedema, haemorrhage or tissue necrosis.
1.	Mild oedema of the connective tissues in the immediate vicinity of the implant.
2.	Significant oedema associated with either haemorrhage and/or necrosis in the vicinity of the implant.
3.	Similar features to 2. above but extending beyond the implant and involving adjacent connective tissues and muscle.

Grade.	New Vessel and Granulation Tissue Formation.
0.	No histological evidence of new vessel formation.
1.	Focal formation of isolated capillary loops in regions of tissue haemorrhage and/or necrosis.
2.	Continuous sheets of new vessel formation in association with local accumulations of fibroblasts to form loose granulation tissue limited to the vicinity of the implant.
3.	Similar features to 2. above but extending beyond the implant and involving adjacent connective tissues and muscle.

Grade.	Persistent (Chronic) Inflammation and Tissue Fibrosis.
0.	No histological evidence of either persistent (chronic) inflammation or early deposition of collagen (fibrosis).
1.	Small discrete foci of macrophages and lymphocytes which may or may not be associated with small populations of fibroblasts and new collagen deposition.
2.	Obvious sheets of chronic inflammatory cells and/or discrete granulomata associated with fibrous scar tissue in the vicinity of the implant.
3.	Similar features to 2. above but extending beyond the implant and involving adjacent connective tissues and muscle.

An objective comparison of the silicon materials with the tissue-compatible control (titanium) was obtained by comparing score grades with respect to implant site and time. Following histological appraisal the discs' identification were decoded and the scores for each time point compared using non-parametric tests. Multi-factorial analysis of variance was used with post hoc tests for differences between groups. Table 2 summarises the average rank scores for each implant type, by histology category, for each time period. The small differences observed in chronic inflammation/ fibrosis observed at weeks 4 and 12 had resolved by week 26. The data will be discussed in more detail elsewhere [8] but in summary demonstrate little or no reaction to the test or control materials beyond that expected from the surgical procedure employed, and the natural processes of wound repair. Both bulk and porous Si have comparable biocompatibility to titanium at the subcutaneous site.

Table 2   Median (lower-upper quartile) score grades for each implant type by histology category for each time period.

*Denotes a significant difference between score grade for that implant type in comparison to the titanium control ($p<0.05$).

Group	Implant type	Inflammation	Tissue Degeneration	New vessels/ Granulation	Chronic Inflammation/ Fibrosis
Week 1	Titanium	1 (1-1)	0 (0-0)	1 (1-1)	0 (0-0)
	BSi	1 (1-1)	0 (0-0)	1 (1-1)	0 (0-0)
	PSi	1 (1-1.75)	0 (0-0.75)	1 (1-1)	0 (0-0)
	CoPSi	1 (1-1)	0 (0-0)	1 (1-1)	0 (0-0)
Week 4	Titanium	0 (0-0)	0 (0-0)	1 (0-1)	1 (1-1.75)
	BSi	0 (0-0)	0 (0-0)	1 (1-1)	2 (1.25-2)
	PSi	0 (0-0)	0 (0-0)	1 (1-1)	2 (2-2)*
	CoPSi	0 (0-0)	0.5 (0-1)	1 (0.25-1)	1 (1-2)
Week 12	Titanium	0 (0-0)	0 (0-0)	0 (0-0)	0 (0-0)
	BSi	0 (0-0)	0 (0-0)	0 (0-0)	0 (0-0)
	PSi	0 (0-0)	0 (0-0)	0 (0-1)	2 (1-2)*
	CoPSi	0 (0-0)	0 (0-0)	0 (0-0)	0 (0-1)
Week 26	Titanium	0 (0-0)	0 (0-0)	0 (0-0)	0 (0-0)
	PSi	0 (0-0)	0 (0-0)	0 (0-0)	0 0-1)

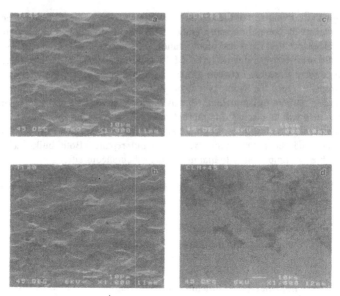

Figure 1   Plan view SEM images of bulk titanium and silicon surfaces before (a, c) and
after (b, d) subcutaneous implantation for 12 weeks.

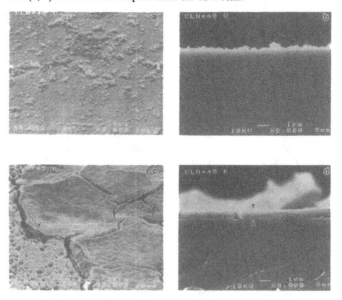

Figure 2   Plan view (a, c) and cross sectional (b, d) SEM images of SBF treated PSi
before (a, b) and after (c, d) subcutaneous implantation for 12 weeks.
A calcium phosphate overlayer (arrowed) is evident on explanted material.

152

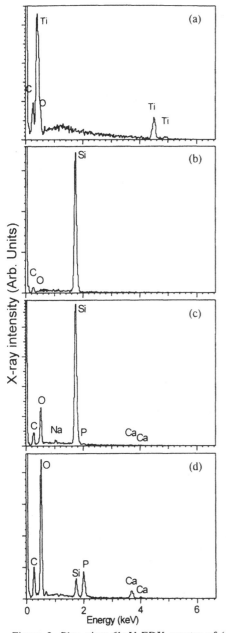

EVIDENCE FOR IN-VIVO CALCIFICATION

Explanted test specimens were rinsed in de-ionised water and examined in plan view and cross section using a JEOL 6400F scanning electron microscope equipped with EDX for compositional analysis. Figures 1, 2 and 3 contrast the effects that in-vivo implantation has on bulk Si and Ti, with that on bioactive porous Si. No gross changes in surface topography (figure 1), or surface composition (figure 3 a, b) are seen for the bulk materials, aside from that introduced by organic debris contamination. For porous Si however, seeded with a thin and discontinuous array of calcium phosphate spherulites [10], a substantial level of further calcification has clearly occurred in-vivo (figures 2 and 3 c,d). The smooth morphology of this calcium phosphate deposit formed at the subcutaneous site differs substantially from that seen in-vitro in simulated plasma, in analogy with similar studies on bioactive glasses [11]. Indeed, a number of porous bioceramics used in orthopaedics are reported to be capable of eventually inducing osteogenesis when implanted in non-bony sites [12]. Whilst there seems to be little controversy regarding the use of simulated plasma as the first test for bioactivity [1], such in-vivo evidence of calcification of porous Si clearly provides additional evidence that the semiconductor can be developed to bond to bone and perhaps even soft tissue [1].

Figure 3. Plan view 6keV EDX spectra of (a) bulk titanium after 12wks in-vivo *cf* Figure 1b, (b) bulk Si after 12wks in-vivo *cf* Figure 1d, (c) porous Si after SBF treatment *cf* Figure 2a, (d) SBF treated porous Si after 12wks in-vivo *cf* Figure 2c.

CONCLUSIONS

This in-vivo assessment of silicon structures at a subcutaneous site in the guinea pig has revealed that both bulk and porous Si exhibit tissue compatibility comparable to that of pure titanium, an established biomaterial. In addition, we have provided evidence for porous Si promoting calcification in-vivo. This, together with the micromachining capabilities of Si technology [13] suggests that there is considerable potential for exploiting this material, including areas such as tissue engineering and orthopaedics.

ACKNOWLEDGEMENTS

We are grateful to R Knight and J Platt for assistance with histopathology, R Gwyther for statistical analysis, J Bannister for assistance with disc manufacture and experimental animal house staff at CBD for technical support.

REFERENCES

[1]   L T Canham. Adv Mater 7, 1033 (1995).

[2]   J Black in Biological Performance of Materials - Fundamentals of Biocompatibility. 2nd Edition. Marcel Dekker Inc (1992).

[3]   S S Stensas, L J Stensas. Acta Neuropathologica 41, 145-155 (1978).

[4]   D J Edell, V V Toi, V M McNeill, L D Clark. IEEE Trans BME 39, 635 (1992).

[5]   S Schmitt, K Horch, R Normann. J Biomed Mater Res 27, 1393 (1993).

[6]   T Desai, M Ferrari, G Mazzoni. Trans ASME Vol PD-71, p97-101 (1995).

[7]   Biological Evaluation of Medical Devices. Part 6 - Tests for Local Effects after Implantation. ISO standard 10993-6 (1994).

[8]   A P Bowditch et al. Biocompatibility of bulk and porous silicon at a subcutaneous site in the guinea pig. To be published.

[9]   L T Canham, C L Reeves in Thin Films and Surfaces for Bioactivity and Biomedical Applications edited by C M Cotell et al. MRS Proc Vol 414, 189-194 (1996).

[10] L T Canham et al. Adv Mater 8, 847 (1996).

[11] O H Andersson et al in Bioceramics, Vol 7, p67-72 (1994).

[12] Z Yang, H Yuan, W Tong, P Zhou, W Chen, X Zhang. Biomaterials 17, 2131 (1996).

[13] L T Canham et al in Advances in Microcrystalline and Nanocrystalline Semiconductors - 1996. Edited by R W Collins et al. MRS Proc Vol 452, p579-590 (1997).

SURFACE CHEMISTRY OF POROUS SILICON

J.-N. CHAZALVIEL, F. OZANAM
Laboratoire PMC, CNRS-École Polytechnique, 91128 Palaiseau, FRANCE.

ABSTRACT

As-prepared porous silicon comes out covered with covalently bonded hydrogen. This hydrogen coating provides a good electronic passivation of the surface, but it exhibits limited stability, being removed by thermal desorption or converted into an oxide upon prolonged storage in air. Starting from the hydrogenated surface, an oxide layer with good electronic properties is also obtained by anodic oxidation or rapid thermal oxidation.

The hydrogenated surface may be nitridized using thermal treatments in nitrogen or ammonia. Fast halogenation of the surface may be obtained at room temperature, but the resulting coating is rapidly converted to an oxide in the presence of moisture. Many metals have been incorporated into the pores, using chemical or vacuum techniques, or even direct incorporation during porous silicon formation.

More interestingly, organic derivatization may increase surface stability or provide chemical functionalities. The poor reactivity of the hydrogenated surface can be remedied by using various methods: thermal desorption of hydrogen, hydroxylation or halogenation of the surface, thermal or UV assisted reaction. However, most promising results have been obtained through either Lewis-acid catalyzed grafting or electrochemical activation of the surface. The latter method has been used for grafting formate, alkoxy, and recently methyl groups. In most of these methods, oxidation is present as a parallel path, and care must be taken if it is not desired. Also, 100% substitution of the hydrogens by organic groups has never been attained, due to steric hindrance problems. The electrochemical method appears especially fast, and has led to 80% substitution of the hydrogens by methyl groups, with no photoluminescence loss and a chemical stability increased by one order of magnitude.

INTRODUCTION

The surface of freshly prepared porous silicon is known to be coated by a covalently attached layer of hydrogen, under the form of SiH, SiH_2 and SiH_3 surface groups, as evidenced by infrared spectroscopy [1-3] (deuterium is found instead of hydrogen, if preparation is carried out in DF electrolyte [4]). This hydrogen coating provides the surface with good electronic passivation properties [5]. It is now about accepted that such simple surface species cannot play an active role as chromophores associated with the photoluminescence of porous silicon [6]. However, surface passivation still stands as a key ingredient for reaching high photoluminescence efficiencies: any dangling bond left on the surface will result in an electron state near midgap, which acts as an efficient nonradiative recombination center. Eventhough the monoatomic hydrogen coating may not fully protect the luminescence, e.g., against the quenching effect of surface adsorbates or a liquid in the pores [7], the absence of surface midgap states is a prerequisite, which can be fulfilled only by a proper (covalent) termination of the silicon lattice. However, the hydrogen coating exhibits limited stability. Upon storage in air, the surface gets slowly oxidized, a process which may be speeded up under illumination, e.g., during photoluminescence measurements [8]. The resulting surfaces generally exhibit a degraded electronic quality, hence a loss in photoluminescence intensity, so that controlled substitution of surface hydrogen by another chemical species has appeared desirable. Here we will review our knowledge about surface oxidation, either spontaneous or purposely induced, together with the methods of alternate surface modifications which have been worked out in order to improve surface stability while preserving the photoluminescence. Most of the information for these modifications has been obtained using infrared spectroscopy.

OXIDATION

Spontaneous oxidation

In some cases, the surface of freshly prepared porous silicon may already appear contaminated with oxide, especially when prepared under illumination [9,10]. Recent investigations have shown that the oxidation actually takes place when the sample is removed from the liquid and exposed to air [11]. Also, the detail of the rinsing procedure may play a crucial role in the surface chemistry of as-prepared samples: though the hydrogenated surface is hydrophobic, exceedingly short rinses may result in water and fluoride being left inside the pores [12,13]. Alternately, exceedingly long rinses may possibly lead to some surface oxidation.

Controlled oxidation of the surface may be realized by slow ageing in the dark under atmosphere during weeks or months [14]. Using dark conditions preserves the good electronic properties of the surface. Visible or ultraviolet illumination speeds up oxidation, either through thermal effects or more plausibly through photoactivation of the surface Si-H bonds, but the resulting oxide exhibits poor electronic properties, leading to a loss of photoluminescence intensity [8].

Wet chemical oxidation

A number of authors have reported on chemical oxidation of the porous silicon surface with hydrogen peroxide [15,16] or nitric acid [15,17]. The obtained samples exhibit luminescence blue shifts, luminescence intensity is generally preserved and possibly improved, while stability is increased. Oxidation in boiling water has also been reported, and it gives an especially large blue shift of the luminescence [18]. Recently, a chemical modification using tetraethoxysilane has been described [19]. It involves partial oxidation of the surface associated with formation of an SiO_2 gel inside the pores, which leads to a strengthening of the layer, while the luminescence is preserved and even increased.

Anodic oxidation

Anodic oxidation has been realized in non-fluoride electrolytes (HCl, KNO_3, ...). In order to insure removal of fluoride ions which might be left from porous silicon preparation, $CaCl_2$ has been used as a fluoride-ion scavenger [20]. Also, care must be taken that wetting of the surface is preserved when transferring the sample from the HF electrolyte to the oxidation bath [21]. Anodic oxidation is typically realized at current densities of 1-10 mA/cm^2, and durations of $10-10^3$ s. The electrochemical process occurs preferentially near the bottom of the pores, but this geometrical effect tends to diminish upon increasing the current density, and layers of a few micrometer thickness may be oxidized almost homogeneously [22]. Oxidation stops when electrical contact to the substrate is broken, leaving an oxide layer on the order of a monolayer [22]. Several authors have observed that hydrogen is still present at the anodically oxidized surface in the form of SiH bonds [20,23]. However, a controversy is still present on this point [24]. Still more controversial is the mechanism of anodic oxidation. According to many authors, the oxidation can only start through electrochemical attack of the SiH bonds [25-28]. According to many others, the Si-Si bonds may be attacked first [23,29]. The anodically oxidized samples have been found to exhibit improved photoluminescence efficiency with respect to the hydrogenated surface, and the photoluminescence intensity further increases upon ageing [22].

Thermal oxidation

Thermal oxidation must be realized in a controlled way. Oxidation at temperatures from 300°C to 900°C results in a loss of hydrogen and the formation of an Si/SiO_2 interface of poor electronic quality. On the other hand, oxidation at high temperatures may result in a full oxidation of the porous layer. Rapid thermal oxidation (typically 900°C in O_2 for a few minutes) appears to give optimum results, in providing samples with good electronic surface passivation and improved stability to ageing [30]. Using controlled oxidation (800-900°C in nitrogen + 10% oxygen), Hirschman et al. [31] made efficient light-emitting devices with improved stability.

UHV controlled oxidation

Annealing porous silicon under UHV conditions (4 min at 360°C + 4 min at 520°C) leads to removal of the surface hydrogen coating [32]. On the clean surface, water adsorbs dissociatively (20 min of water vapor exposure at ambient temperature under 10^{-5} Torr), which produces a silicon surface exclusively coated with SiH and SiOH species, as shown by infrared spectroscopy [32] (Fig.1). Annealing this surface above 300°C promotes hydroxyl conversion into SiOSi bridges and produces a partially oxidized porous silicon surface. The optical properties of these materials have not been investigated.

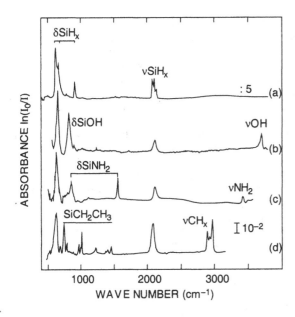

Fig.1: Infrared absorption of porous silicon (a) freshly prepared, (b-d) after hydrogen desorption and dissociative adsorption of various chemicals: (b) water (after [32]), (c) ammonia (after [34]), (d) diethylsilane (after [48]).

157

NITRIDATION

Rapid thermal annealing (30s at 1100°C) of porous silicon in N_2 or NH_3 promotes nitridation of the surface, as evidenced by the build-up of a SiNSi vibrational peak at 915 cm^{-1} in infrared spectra [33]. The reaction appears more efficient in ammonia than in nitrogen. Dillon et al. have also obtained nitridized surfaces under UHV conditions [34]. After hydrogen desorption, and much in the same way as for water, ammonia adsorbs dissociatively on the porous silicon surface, covering it with SiH and SiNH$_2$ species (Fig.1). Annealing the surface above 400°C converts the surface coating into a nitride. Finally, oxynitridation of the surface has been achieved by annealing porous silicon first at 300°C in O_2, then at 1000°C in NH_3 [35]. The resulting oxynitride is hydrogen free and stable upon annealing during several hours at 600°C.

HALOGENATION

Lauerhaas and Sailor [36] showed that porous silicon is modified by exposure to halogen vapor for durations of the order of minutes. The photoluminescence is quenched, and is hardly restored upon pumping over the sample for hours. After air exposure, a significant fraction of the original luminescence is recovered, but the surface appears oxidized. However, halogenation is interesting as an intermediate step for other surface modifications.

Treatments in chlorinated solvents (hot CCl_4 [23,37] or trichloroethylene [38]) have been found to promote hydrogen removal but do not lead to halogenation. Oxidation of the surface is found instead. This illustrates the inherent fragility of the halogenated surface and the difficulty of the halogenation route.

MODIFICATIONS INVOLVING METALS

Weakly electropositive metals, such as Cu and Ag, have been deposited on porous silicon by contact-displacement deposition from solution, using Cu^{2+} and Ag^+ ions, respectively [39]. However, such modifications result in a strong quenching of the photoluminescence.

A porous silicon light-emitting device has been modified by mild electrodeposition of indium from indium chloride (1.7 mA/cm^2, 30 s) [40]. On a partially oxidized layer, the deposit did not quench the luminescence and allowed for increased electroluminescence efficiency.

Porous silicon has been modified by soaking in a solution of aluminum isopropoxide in toluene for durations of several days [41]. The modified samples retained their luminescence properties and exhibited improved stability. It seems that the modification largely consists in an oxidation, with Si-O-Al groups present at the surface. Similarly, SnO_x and TiO_2 have been deposited in porous silicon, using the respective tetrachlorides as precursors [42,43].

A modified CVD technique has been used in order to deposit cobalt into the pores [44]. With a $HCo(CO)_4$ precursor, an actual surface modification consisting in H desorption from the surface and formation of Si-Co bonds was achieved.

Recently, porous silicon was prepared by chemical etching of a Si wafer in a bath composed of 40% HF + 0.3M Fe^{3+} at 140°C for 50 min [45]. The surface of the resulting porous material appeared passivated with iron. It exhibited an improved photoluminescence as compared to standard porous silicon, together with a good stability to ageing.

ORGANIC CHEMICAL DERIVATIZATION

For stabilization of the porous silicon surface, derivatization by organic groups offers an alternate possibility to oxidation, since the process is expected to stop at a monolayer at most. Also, the variety of groups which may be attached to the surface should allow for different applications (e.g., sensors). However, the tendency of silicon to irreversibly form SiO bonds makes oxidation a competing process for any other surface modification.

The effect of various organic molecules on luminescence quenching, especially organoamines [46], has been discussed in terms of "addition" of the molecule to the surface. The amine molecules were dissolved into heptane and the porous-silicon samples were immersed in that solution. However, there is no direct evidence for new bonds associated with a direct attachment of the molecule to the surface, and it is not clear whether the effect should be discussed in terms of surface modification. The impregnation of porous silicon by dye molecules [47] even more clearly belongs to this class of "physical" bonding.

Unambiguous chemical bonding of various organic groups to the porous silicon surface has been obtained by many groups using various methods. Here again, a first strategy consists in exposing the bare surface to reactants under UHV conditions, after thermal desorption of the initial hydrogen coating. In this way, Dillon et al. obtained a surface covered with SiH and $SiCH_2CH_3$ upon dissociative adsorption of diethylsilane [48] (Fig.1). A similar behavior is obtained with methanol, leading to SiH and $SiOCH_3$ [49].

Among the other methods, one may distinguish those aiming at the attachment of an organic group through an Si-O bond, and those aiming at the realization of a direct Si-C bond.

Si-O-R bonding

Anderson et al. [33] and Dubin et al. [50] have grafted trimethylsiloxy groups on the porous silicon surface. Substitution of -H by $-OSi(CH_3)_3$ was realized by exposing the porous silicon layer to vapors of trimethylchlorosilane or hexamethyldisilazane in a wet atmosphere (Fig.2). The reaction scheme was (here written for hexamethyldisilazane)

$$\equiv SiH + H_2O \rightarrow \equiv SiOH + H_2\uparrow \quad (slow),$$
$$2 \equiv SiOH + HN[Si(CH_3)_3]_2 \rightarrow 2 \equiv SiOSi(CH_3)_3 + NH_3\uparrow \quad (fast).$$

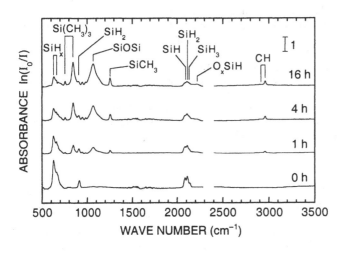

Fig.2: Infrared absorption of porous silicon when exposed to vapors of hexamethyldisilazane, leading to grafting of trimethylsiloxy groups (after [50]).

An equilibrium state was reached after an exposure time of a few hours. The photoluminescence was preserved upon the substitution. However, only a small fraction of the surface hydrogens were substituted, a limitation attributed to steric hindrance among the large trimethylsiloxy groups.

Lee et al. [51] brominated the porous silicon surface by exposure to Br_2 vapor for 5 min. The brominated surface was reactive toward a variety of alcohols, and ethoxy, triethylsiloxy, or 3-thiophene methoxy groups could be attached to the surface upon exposing the sample to the corresponding alcohol for ~10 min. According to the authors, the reaction may be written

$$\equiv Si\text{-}Si\equiv + Br_2 \rightarrow \equiv SiBr + Br\text{-}Si\equiv,$$
$$\equiv SiBr + HOR \rightarrow \equiv SiOR + HBr.$$

10-40% of the initial luminescence was retained upon derivatization.

Thermal derivatization of porous silicon with alcohols has been investigated by a number of groups. Hory et al. [23] have reported partial methoxylation of the surface, accompanied by significant oxidation, upon immersing porous silicon in boiling methanol. Glass et al. [49] reported that modification of the surface in the presence of methanol vapor occurs only at temperatures promoting partial hydride desorption (>300°C). At these temperatures, the C-O bond itself may be thermally broken, and SiH, $SiOCH_3$, $SiCH_3$ and SiOSi species are found on the surface. Kim and Laibinis [52] have reported successful grafting of many alkoxy groups (including ethoxy, undecoxy, 3-phenylpropoxy, and various substituted alcohols) upon exposing the surface to the pure alcohol or its solution in anhydrous dioxane for durations of the order of hours, the luminescence being essentially preserved. As in [23], the modification is accompanied by some oxidation, and the surface concentration by alkoxy groups is limited by steric hindrance. The authors suggest that the reaction proceeds through cleavage of Si-Si bonds by the alcohol:

$$\equiv Si\text{-}Si\equiv + HOR \rightarrow \equiv SiH + ROSi\equiv,$$

with formation of new SiH groups. Such a mechanism has also been invoked by Vieillard et al. [53] in a slightly different context. At room temperature, direct reaction of porous silicon with alcohols in the vapor phase takes place on a time scale of days [54].

An attractive alternative has been used by Li et al. in order to graft alkoxy groups on the porous Si surface [55]. They perform ultraviolet irradiation of fresh luminescent porous Si samples immersed in methanol, ethanol or propanol (irradiation of a few minutes using a 6W UV lamp located 2 cm away from the samples). This photochemical treatment results in a slight decrease and a red shift of the luminescence. After the treatment, infrared spectra demonstrate that alkoxy species covalently bind to the surface, which is also probably oxidized to some extent.

Finally, in the same class of modifications, one may also mention the recent report by VanderKam et al. about the use of zirconium alkoxides for the grafting of copolymers to the porous silicon surface through Si-O-Zr-O-C bridges [56].

Direct Si-C bonding

Direct grafting of organic groups to the PS surface through a Si-C bond is of considerable interest, because Si-C bonds are expected to be more stable than Si-O-C bridges, and the resulting modification should be much more resistant to ageing, even in aggressive environments.

Chemical derivatization of flat hydrogenated silicon surfaces with direct Si-C bonding has been realized by different methods. Among these methods, are substitution from the halogenated surface [57], according to various schemes, such as

$$\equiv SiH + 2\ Cl\cdot \rightarrow\ \equiv SiCl + HCl,$$
$$\equiv SiCl + RMgCl \rightarrow\ \equiv SiR + MgCl_2\,,$$

grafting of alkyl radicals [58]

$$\equiv SiH + 2\ R\cdot \rightarrow\ \equiv Si\cdot + R\cdot + RH \rightarrow\ \equiv SiR + RH,$$

or addition of olefins using a platinum complex as a catalyst [59]

$$\equiv SiH + H_2C{=}CH{-}R \rightarrow\ \equiv Si{-}CH_2{-}CH_2{-}R.$$

However, none of these methods has as yet been transposed to the case of porous silicon, and it is not clear whether this could be done while preserving the photoluminescence. For the case of a platinum-complex catalyzed methoxylation, luminescence is actually lost [53]: platinum is plausibly left on the surface, leading to fast surface recombination.

Recently, the possibility of alkene or alkyne addition on porous silicon using the Lewis acid EtAlCl$_2$ as a catalyst has been demonstrated [60] (Fig.3). The reactions may be written

$$\equiv SiH + R{-}CH{=}CR'R'' \rightarrow\ \equiv Si{-}(CRH){-}(CR'R''H)$$
$$\text{or} \qquad \equiv SiH + R{-}C{\equiv}C{-}R' \rightarrow\ \equiv Si{-}CR{=}(CR'H).$$

The addition takes place at room temperature in a typical time of one hour, and the unwanted oxide formation is essentially avoided. When simple alkyl chains (from 1-dodecene and 1-dodecyne) are grafted, the photoluminescence is essentially preserved, and the modified surface is stable in diluted boiling KOH solution. Chains terminated with various functional end groups can also be grafted. These results clearly open the way to applications of porous silicon in the field of sensors.

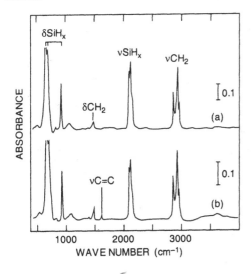

Fig.3: Infrared absorption of porous silicon after grafting of (a) 1-dodecyl and (b) 1-dodecenyl groups, using the Lewis-acid catalyzed route (after [60]).

ELECTROCHEMICAL DERIVATIZATION

The poor reactivity of the hydrogenated silicon surface has led people to explore alternate techniques such as electrochemical activation of the surface. Here again we will distinguish between grafting through an oxygen bridge and direct Si-C bonding.

Si-O-C bonding

Lee et al. [51,61,62] have derivatized the porous silicon surface with formate groups, by anodization of porous silicon in a HCOOH + 1M HCOONa electrolyte. In these experiments, porous silicon was made from n-Si and the modification required illumination, allowing for a possibility of defining the modified surface by patterning. Infrared spectroscopy shows that formate attachment occurs through a single SiOC bonding (Fig.4). Some oxidation is also present, especially when the duration of the treatment is increased. Trifluoroacetate groups were grafted in a similar way starting from a CF_3COOH + 1M CF_3COONa electrolyte. These modifications are detrimental to the photoluminescence. The absence of SiH loss, together with the results of deuteration experiments, suggest that the mechanism involves breaking of Si-Si bonds rather than SiH bonds, and formation of new SiH's from solution. According to the authors, the reaction can be written

$$\equiv Si-Si\equiv + HCOO^- + h^+ \rightarrow \equiv SiO(CO)H + \equiv Si\cdot,$$

the Si dangling bond then abstracts an H· from solution, and the backbonds of the esterified silicon atom are oxidized [62]. The grafted surface-carboxylate groups can be taken as a starting point for further modification. For example, they can react with an alcohol [51], with formation of a surface alkoxide group, or with an organo-methoxysilane $RSiOCH_3$ in toluene [62] (reaction time ~1 hour), allowing for substitution of the carboxylate by the RSiO- group. The modified surfaces exhibit different wetting properties, whence different sensitivities to the photoluminescence quenching by chemical vapors [61].

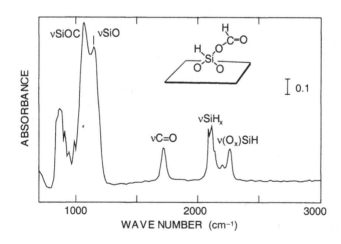

Fig.4: Infrared absorption of porous silicon after electrochemical modification in HCOOH + HCOONa electrolyte, showing grafting of formate groups (after [62]).

Warntjes et al. [53,63] have reported methoxy derivatization of the porous silicon surface by anodization of the porous layer in anhydrous methanol electrolyte (Fig.5). Here again, the modification probably proceeds through electrochemical activation of the surface:

$$\equiv SiH + h^+ \rightarrow \equiv SiH^{\cdot+},$$
$$\equiv SiH^{\cdot+} + CH_3OH \rightarrow \equiv SiOCH_3 + 2\ H^+ + e^-.$$

The modification is very fast (10 mA/cm^2, 20 s) but prolonged treatment results in dissolution of the layer. Infrared investigations with deuterated methanol prove that the dissolution occurs by chemical attack of the Si-Si backbonds at a methoxylated silicon site. The silicon ends dissolved as $Si(OCH_3)_4$ and the surface is rehydrogenated behind [53]. This problem arises from the electronegativity of the methoxy group and the induced polarization of the Si-Si backbonds. It limits the substitution efficiency to about one half of the initial hydrogens. The luminescence intensity is preserved, and stability to ageing is improved by about a factor of 2.

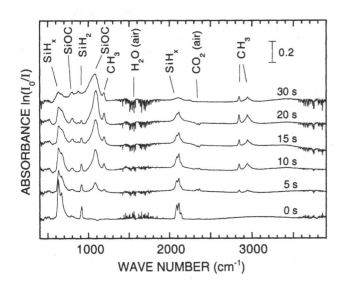

Fig.5: Infrared absorption of porous silicon, during anodization in anhydrous methanol electrolyte, showing fast grafting of methoxy groups. Notice that the layer is partly destroyed for times in excess of 20 s (after [63]).

Direct Si-C bonding

Direct electrochemical formation of a Si-C bond is known to occur on a flat silicon surface, for example by cathodic reduction of a diazo precursor. For example, reduction of $(C_6H_5)N_2^+$ in acetonitrile at a hydrogenated atomically flat (111) silicon surface results in substitution of half of the hydrogens by phenyl groups (with elimination of a N_2 molecule) [64]. Up to now, this reaction has not been realized at a porous silicon surface. However, Dubois et al. have obtained direct grafting of methyl groups by an anodic route [65]. The methyl group is an ideal candidate for surface stabilization, because of its small size and the resulting possibility of substituting most of the hydrogens. Note that methyl grafting is not possible by any of the other mentioned schemes. Here, the poor reactivity of the hydrogenated silicon surface has been remedied by using anodization of porous silicon in an electrolyte consisting of diethylether with $CH_3Li.LiI$ or

CH$_3$MgI. The latter reagent is advantageous, in that it provides good electrolyte properties without addition of a supporting salt. Typical conditions are 1-10 mA/cm^2, 10 min. The reaction scheme for the modification represents a minority reaction pathway, the dominant process at the electrode being the decomposition of the electrolyte, e.g., in the case of the Grignard, the dominant global reactions may be written

$$2\ CH_3MgI + 2\ h^+ \rightarrow C_2H_6 + 2\ MgI^+,$$
and
$$2\ CH_3MgI + 2\ h^+ \rightarrow CH_3I + MgI^+ + CH_3Mg^+.$$

The modification scheme is not clear, and many different routes are possible. A plausible one involves the role of I· radicals formed as intermediates in the decomposition reaction:

$$\equiv SiH + I\cdot \rightarrow\ \equiv Si\cdot + HI,$$
$$\equiv Si\cdot + CH_3MgI \rightarrow\ \equiv SiCH_3 + MgI^+ + e^-.$$

The substitution efficiency of -H by -CH$_3$ is over 80%, and only a few percent of the surface Si atoms get oxidized. A typical infrared spectrum of such a methylated surface is shown in Fig.6. The luminescence intensity is preserved, and the stability to ageing is improved by about an order of magnitude. Modification of porous silicon made from n-Si is possible under illumination.

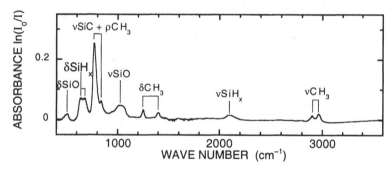

Fig.6: Infrared absorption of porous silicon after anodization in CH$_3$MgI electrolyte, showing grafting of methyl groups, with less than 20% of the original SiH$_x$'s left on the surface (after [65]).

CONCLUSION

Much progress has been achieved in the chemical engineering of the porous silicon surface. Most interesting appears rapid thermal oxidation, which turns porous silicon into a very stable material. Among the organic derivatization techniques, methylation appears promising as it preserves the luminescence and also improves stability significantly. Also, the possibility has emerged of derivatizing the surface and grafting virtually any organic groups, as allowed for by either the carboxylate route or Lewis-acid catalyzed direct Si-C grafting. This possibility appears promising for applications outside the domain of luminescence, e.g., sensors or biotechnical applications.

REFERENCES

1. Y. Kato, T. Ito, and A. Hiraki, Jpn. J. Appl. Phys. **27**, L1406 (1988).
2. P. Gupta, V.L. Colvin and S.M. George, Phys. Rev. B **37**, 8234 (1988).
3. A. Venkateswara Rao, F. Ozanam and J.-N. Chazalviel, J. Electrochem. Soc. **138**, 153 (1991).
4. T. Matsumoto, Y. Masumoto, S. Nakashima and N. Koshida, Thin Solid Films **297**, 31 (1997).
5. E. Yablonovitch, D.L. Allara, C.C. Tsang, T. Gmitter and T.B. Bright, Phys. Rev. Lett. **57**, 249 (1986).
6. V.M. Dubin, F. Ozanam and J.-N. Chazalviel, MRS Symp. Proc. **358**, 519 (1995).
7. J.-N. Chazalviel and F. Ozanam, in *Structural and Optical Properties of Porous Silicon Nanostructures*, edited by G. Amato, C. Delerue and H.-J. von Bardeleben (Gordon and Breach, Amsterdam, 1997) pp. 53-71.
8. M.A. Tischler, R.T. Collins, J.H. Stathis and J.C. Tsang, Appl. Phys. Lett. **60**, 639 (1992).
9. L. Tsybeskov, C. Peng, S.P. Duttagupta, E. Ettedgui, Y. Gao, P.M. Fauchet and G.E. Carver, Mat. Res. Soc. Symp. Proc. **298**, 307 (1993).
10. I. Suemune, N. Noguchi and M. Yamanishi, Jpn. J. Appl. Phys. **31**, L494 (1992).
11. M. Wolkin-Vakrat, P.M. Fauchet, G. Allan and C. Delerue, Phys. Rev. Lett. (under press).
12. D. Petit, J.-N. Chazalviel, F. Ozanam and F. Devreux, Appl. Phys. Lett. **70**, 191 (1997).
13. N. Hadj Zoubir and M. Vergnat, Appl. Surf. Sci. **89**, 35 (1995).
14. A. Kux, F. Müller and F. Koch, MRS Symp. Proc. **283**, 311 (1993).
15. A. Nakajima, T. Ikatura, S. Watanabe and N. Nakayama, Appl. Phys. Lett. **61**, 46 (1992).
16. F. Kozlowski, W. Wagenseil, P. Steiner and W. Lang, MRS Symp. Proc. **358**, 677 (1995).
17. G. Mauckner, T. Walter, T. Baier, K. Thonke and R. Sauer, MRS Symp. Proc. **283**, 109 (1993).
18. X.Y. Hou, G. Shi, W. Wang, F.L. Zhang, P.H. Hao, D.M. Huang and X. Wang, Appl. Phys. Lett. **62**, 1097 (1993).
19. J. Linsmeier, K. Wüst, H. Schenk, U. Hilpert, W. Ossau, J. Fricke and R. Arens-Fischer, Thin Solid Films **297**, 26 (1997).
20. V.M. Dubin, F. Ozanam and J.-N. Chazalviel, Vibr. Spec. **8**, 159 (1995).
21. A. Halimaoui, NATO ASI Ser. E **244** (Kluwer, Dordrecht, 1993) p.11.
22. A. Bsiesy, J.C. Vial, F. Gaspard, R. Hérino, M. Ligeon, F. Muller, R. Romestain, A. Wasiela, A. Halimaoui and G. Bomchil, Surf. Sci. **254**, 195 (1991).
23. M.A. Hory, R. Hérino, M. Ligeon, F. Muller, F. Gaspard, I. Mihalcescu and J.C. Vial, Thin Solid Films **255**, 200 (1995).
24. K. Uosaki, T. Kondo, H. Noguchi, K. Murakoshi and Y.Y. Kim, J. Phys. Chem. **100**, 4564 (1996).
25. V. Lehmann and U. Gösele, Appl. Phys. Lett. **58**, 856 (1991).
26. H. Gerischer, P. Allongue and V. Costa-Kieling, Ber. Bunsenges. Phys. Chem. **97**, 753 (1993).
27. J.-N. Chazalviel and F. Ozanam, MRS Symp. Proc. **283**, 359 (1993).
28. J.-N. Chazalviel, in *Porous Silicon Science and Technology*, edited by J.C. Vial and J. Derrien (Les Éditions de Physique, les Ulis, 1995) pp. 17-32.
29. E.S. Kooij, A.R. Rama and J.J. Kelly, Surf. Sci. **370**, 125 (1997).
30. V. Petrova-Koch, T. Muschik, A. Kux, B.K. Meyer, F. Koch and V. Lehmann, Appl. Phys. Lett. **61**, 943 (1992).
31. K.D. Hirschman, L. Tsybeskov, S.P. Duttagupta and P.M. Fauchet, Nature **384**, 338 (1996).
32. P. Gupta, A.C. Dillon, A.S. Bracker and S.M. George, Surf. Sci. **245**, 360 (1991).
33. R.C. Anderson, R.S. Muller and C.W. Tobias, J. Electrochem. Soc. **140**, 1393 (1993).
34. A.C. Dillon, P. Gupta, M.B. Robinson, A.S. Bracker and S.M. George, J. Vac. Sci.

Technol. A **9**, 2222 (1991).
35. V. Morazzani , J.-L. Cantin, C. Ortega, B. Pajot, R. Rahbi, M. Rosenbauer, H.J. von Bardeleben and E. Vazsonyi, Thin Solid Films **276**, 32 (1996).
36. J.M. Lauerhaas and M.J. Sailor, MRS Symp. Proc. **298**, 259 (1993).
37. J.M. Lavine, S.P. Sawan, Y.T. Shieh and A.J. Bellezza, Appl. Phys. Lett. **62**, 1099 (1993).
38. Y.H. Seo, H.-J. Lee, H.I. Jeon, D.H. Oh, K.S. Nahm, Y.H. Lee, E.-K. Suh, H.J. Lee and Y.G. Kwang, Appl. Phys. Lett. **62**, 1812 (1993).
39. D. Andsager, J. Hilliard and M.H. Nayfeh, Appl. Phys. Lett. **64**, 1141 (1994).
40. P. Steiner, F. Kozlowski, M. Wielunski and W. Lang, Jpn. J. Appl. Phys. **33**, 6075 (1994).
41. L. Zhang, J.L. Coffer, D. Xu and R.F. Pinizzotto, J. Electrochem. Soc. **143**, 1390 (1996).
42. C. Dücsö, N.Q. Khanh, Z. Horváth, I. Bársony, M. Utriainen, S. Lehto, M. Nieminen and L. Niinistö, J. Electrochem. Soc. **143**, 683 (1996).
43. R.B. Bjorklund, S. Zangooie and H. Arwin, Langmuir **13**, 1440 (1997).
44. B.J. Aylett, I.S. Harding, L.G. Earwaker, K. Forcey and T. Giaddui, Thin Solid Films **276**, 253 (1996).
45. Y.H. Zhang, X.J. Li, L. Zheng and Q.W. Chen, Phys. Rev. Lett. **81**, 1710 (1998).
46. J.L. Coffer, S.C. Lilley, R.A. Martin and L.A. Files-Sesler, J. Appl. Phys. **74**, 2094 (1993).
47. L.T. Canham, Appl. Phys. Lett. **63**, 337 (1993).
48. A.C. Dillon, M.B. Robinson, M.Y. Han and S.M. George, J. Electrochem. Soc. **139**, 537 (1992).
49. J.A. Glass Jr., E.A. Wovchko and J.T. Yates, Jr., Surf. Sci. **338**, 125 (1995).
50. V.M. Dubin, C. Vieillard, F. Ozanam and J.-N. Chazalviel, Phys. Stat. Sol. (b) **190**, 47 (1995).
51. E.J. Lee, J.S. Ha and M.J. Sailor, MRS Symp. Proc. **358**, 387 (1995).
52. N.Y. Kim and P.E. Laibinis, J. Am. Chem. Soc. **119**, 2297 (1997).
53. C. Vieillard, M. Warntjes, F. Ozanam and J.-N. Chazalviel, ECS Conf. Proc. **95-25**, 350 (1996).
54. S. Fellah, N. Gabouze, F. Ozanam, J.-N. Chazalviel, A. Dakhia and Y. Belkacem, EMRS Meeting (Strasbourg, June 1998) symposium B, abstract I/P.10 (to be published).
55. K.-H. Li, C. Tsai, J.C. Campbell, M. Kovar and J.M. White, J. Electronic Mat. **23**, 409 (1994).
56. S.K. VanderKam, A.B. Bocarsly and J. Schwartz, Chem. Mater. **10**, 685 (1998).
57. A. Bansal, X. Li, I. Lauermann, N.S. Lewis, S.I. Yi and W.H. Weinberg, J. Am. Chem. Soc. **118**, 7225 (1996).
58. M.R. Linford, P. Fenter, P.M. Eisenberger and C.E.D. Chidsey, J. Am. Chem. Soc. **117**, 3145 (1996).
59. L.A. Zazzera, J.F. Evans, M. Deruelle, M. Tirrell, C.R. Kessel and P. Mckeown, J. Electrochem. Soc. **144**, 2184 (1997).
60. J.M. Buriak and M.J. Allen, J. Am. Chem. Soc. **120,** 1339 (1998).
61. E.J. Lee, J.S. Ha and M.J. Sailor, J. Am. Chem. Soc. **117**, 8295 (1995).
62. E.J. Lee, T.W. Bitner, J.S. Ha, M.J. Shane and M.J. Sailor, J. Am. Chem. Soc. **118**, 5375 (1996).
63. M. Warntjes, C. Vieillard, F. Ozanam and J.-N. Chazalviel, J. Electrochem. Soc. **142**, 4138 (1995).
64. C. Henry de Villeneuve, J. Pinson, M.C. Bernard and P. Allongue, J. Phys. Chem. B **101**, 2415 (1997).
65. T. Dubois, F. Ozanam and J.-N. Chazalviel, ECS Conf. Proc. **97-7**, 296 (1997).

FUNCTIONALIZATION OF POROUS SILICON SURFACES BY SOLUTION-PHASE REACTIONS WITH ALCOHOLS AND GRIGNARD REAGENTS

N.Y. KIM, P.E. LAIBINIS
Departments of Chemistry and Chemical Engineering, Massachusetts Institute of Technology, Cambridge, MA 02139

ABSTRACT

This paper describes the covalent attachment of various organic molecules to the hydrogen-terminated surface of porous silicon using alcohols and Grignard reagents. With alcohols, the chemical reaction forms Si-O-C attachments to the silicon substrate and requires modest heating (40-70 °C). With Grignard reagents, the reaction proceeds at room temperature and forms a covalent film that is attached by Si-C bonds to the silicon support. Evidence for these reactions is provided by infrared and x-ray photoelectron spectroscopies.

INTRODUCTION

The discovery of photoluminescence from porous silicon earlier this decade [1] sparked a tremendous interest in the properties of this material and in its chemistry that continues today [2–9]. The possibility of producing optoelectronic and photonic devices from silicon (with the added fabrication advantage of likely compatibility with current processing schemes [10]) has provided a need to understand both practical and fundamental issues associated with the reactivity, stability, and overall performance of porous silicon. This material is most often prepared as a supported film on silicon by the anodic oxidation of crystalline silicon in a HF(aq)/ethanol mixture. This process produces a porous network of columnar silicon nanostructures that are terminated at their surface by Si-H bonds. The resulting hydrogen-terminated porous silicon is photoluminescent; however, in air and in contact with water, surface oxidation and other processes effect loss of the Si-H terminations and degrade the performance of the material. Methods that would stabilize porous silicon against these deleterious processes or provide a direct physical or electrical connection to other materials would aid in the development of porous silicon for optoelectronic applications.

To accomplish the above goals, the chemical reactivity of porous silicon has been investigated both to provide an understanding of the degradation mechanisms that cause loss of luminescence as well as for methods that would modify its surface and properties. Various electro- and photochemical reactions have been reported to date that allow attachment of organic species to the porous silicon surface [2-5]. In the present work, we describe two solution-phase reactions that produce covalent attachment to the silicon framework of this porous material without requiring photo- or electrochemical input. These reactions use either alcohols or Grignard reagents and have differences in their conditions, abilities, limitations, and stabilities. This paper describes our present understanding of these two reactions and their demonstrated abilities as well as contrasting their identified differences in behavior and scope. Together, the two reactions provide a flexible approach for modifying the surface properties of porous silicon with covalently attached organic films as is detailed below.

EXPERIMENT

Porous silicon was prepared by anodically etching a p-type 1-10 Ω-cm Si(100) wafer (Silicon Sense, Nashua, NH) in 1:1 48% HF(aq)-ethanol in a Teflon cell with a platinum mesh counter electrode; the p-Si(100) wafers were coated on their unpolished face with an evaporated film of aluminum (1000 Å) to produce an ohmic contact.. The etch conditions to produce porous silicon were either 24 mA/cm^2 for 20 minutes or 10 mA/cm^2 for 5 minutes followed by 2.5 hours of aging in 48% HF(aq) solution; these conditions produce a H-terminated layer of porous silicon on the silicon substrate. After etching, the samples were rinsed with ethanol, dried in a stream of N$_2$, and cut into ~1 x 1 cm^2 pieces.

Derivatizations were performed on silicon slides in glass vials under an atmosphere of N_2. With alcohols, the silicon surface was contacted with neat reagent and heated at a specified temperature for reaction; the required reaction time for optimal results was dependent on the alcohol, with typical times being 0.5 to 12 hours. With Grignard reagents, the slide was placed in a 1 M solution in THF or ether. After 1-2 hours at room temperature, the reaction was quenched by addition of 1 M anhydrous HCl in ether at room temperature. For both reactions, derivatized samples were rinsed with ethanol and dried under a stream of N_2 before analysis.

Diffuse reflectance infrared spectroscopy were performed at a resolution of 2 cm^{-1} using a Bio-Rad FTS 175 spectrometer equipped with an MCT detector and a Universal Reflectance accessory (4 scans). X-ray photoelectron spectra were obtained on a Surface Science X-100 spectrometer at a take-off angle of 55° using a monochromatized Al Kα x-ray source and a concentric hemispherical analyzer.

RESULTS

Derivatization of Porous Silicon with Alcohols

Figure 1 displays diffuse reflectance infrared spectra for porous silicon before and after derivatization with a representative alcohol. The spectrum in Figure 1a for the as-prepared porous silicon displays characteristic Si-H$_x$ stretching peaks at 2080-2150 cm^{-1} and Si-H$_2$ bending at 914 cm^{-1}, where the former contains three identifiable peaks (2139, 2115, and 2089 cm^{-1}) that can be assigned to silicon tri-, di-, and monohydride species, respectively. The native porous silicon samples contained oxygenated species as evidenced by the Si-O stretching peak at 1031 cm^{-1}. Exposure of porous silicon to an alcohol (usually neat or as a concentrated solution in tetrahydrofuran or dioxane) at temperatures of 40 to 90 °C resulted in spectroscopic changes that were indicative of chemical modification.

Figure 1b shows the IR spectrum for porous silicon after derivatization with neat ethyl 6-hydroxy-hexanoate [HO(CH$_2$)$_5$CO$_2$CH$_2$CH$_3$]; this alcohol provides an illustrative example for the

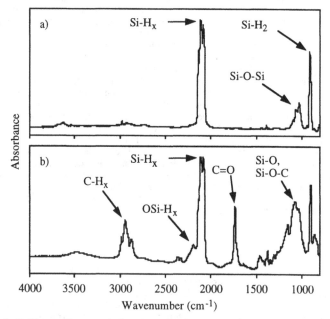

Figure 1. Diffuse reflectance infrared spectra for porous silicon a) before and b) after derivatization with ethyl-6-hexanoate at 87 °C for 20 minutes.

more general reaction as the ester carbonyl has a characteristic infrared absorption that allows easy detection. The spectrum in Figure 1b displays peaks at 2850-2960 and 1743 cm^{-1} for the C-H and C=O stretching modes, respectively. Notably, the relative intensity of the O-H stretching absorption at ~3400 cm^{-1} to that for the CO peak in the infrared spectrum for the parent compound was greatly diminished in the spectrum of the derivatized sample suggesting that the attachment to the surface occurs with loss of the OH moiety. We infer that the alcohol attaches to the silicon surface by an Si-O bond as evidenced by the increased intensity at ~1080 cm^{-1} for Si-O modes and the appearance of O-Si-H peaks at 2150-2260 cm^{-1}. The spectrum of the derivatized sample exhibited no change after rinsing with various solvents, sonication, or exposure to vacuum, demonstrating that the reaction covalently grafts the molecule to the porous silicon surface.

Further proof of the covalent attachment of the compound comes from x-ray photoelectron spectroscopy (XPS). In Figure 2, the silicon signal exhibits the typical doublet for the $2p_{3/2}$ and $2p_{1/2}$ emissions of silicon as well as a small amount of oxidized material as noted by peaks at 102 to 103 eV, compatible with the Si-O peaks observed for the parent material in Figure 1a. The C(1s) region in Figure 2b for the derivatized substrate displays peaks between 284 and 290 eV, with a signal at ~289 eV indicative of a highly oxidized carbon species (RCO_2R'), a shoulder at 286 to 287 eV for less shifted carbon atoms, and a primary peak at ~285 eV for carbon atoms within a hydrocarbon chain. The ratio of the different carbon peaks in Figure 2b is compatible with the stoichiometry of the parent compound and confirms its attachment to the surface and the presence of little introduced adventitious carbonaceous material by the procedure. As the conditions for XPS (10^{-9} torr) would cause evaporation of the parent compound from the surface if present in a physisorbed state, the spectrum also confirms that the reaction produces a robust chemisorbed attachment to the porous silicon surface.

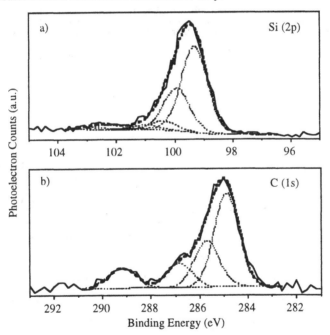

Figure 2. X-ray photoelectron spectra of the a) Si(2p) and b) C(1s) regions for porous silicon derivatized with ethyl-6-hydroxy-hexanoate. The dotted lines result from a deconvolution of the spectra (solid lines) into their component peaks.

Using this thermally-driven reaction, we have attached both alkyl and aromatic alcohols to the porous silicon surface covalently and produced films that expose tail groups including CH_3, $CH=CH_2$, CO_2H, CO_2R, and various halogens [7]. The method appears to be general; however, the exposure time and temperature required for different molecules of interest exhibited some variation. In all cases, extended exposure resulted in loss of the porous silicon layer and etching of the crystalline substrate; however, covalently attached molecular films could be produced easily by control of the reaction conditions. We note that the derivatization reaction appears to proceed by cleavage of Si-Si bonds to form Si-OR and Si-H bonds upon reaction with the alcohol (ROH) rather than by replacement of Si-H bonds [7].

$$H_xSi\text{-}SiH_y + ROH \rightarrow H_xSiOR + SiH_yH \tag{1}$$

Support for this reaction mechanism (where x and y = 0 to 3 and the remaining coordination sites on silicon are to lattice silicon atoms) is noted by the observation that samples derivatized with deuterium-labelled alcohols (ROD) exhibited Si-D peaks in their infrared spectra [7].

<u>Derivatization of Porous Silicon with Grignard Reagents</u>

Figure 3 displays the infrared spectrum for porous silicon after exposure at room temperature to 4-fluorophenylmagnesium bromide and a subsequent quench with HCl in ether. After reaction with the Grignard reagent, the spectrum displays the characteristic peaks (1165, 1245, 1498, and 1591 cm^{-1}) for a para-substituted aromatic species. In contrast to the spectrum in Figure 1b for reaction with an alcohol, the SiH stretching modes are similar to those for the original porous silicon (Figure 1a) as they do not display the O-Si-H peaks at 2150-2260 cm^{-1} and the Si-O peak at ~1050 cm^{-1} remains relatively unchanged in intensity. For the sample in Figure 3, its infrared spectrum exhibited no change after rinsing the sample with various solvents, sonication, exposure to vacuum, and notably even after exposure to HF(aq). The cumulative spectral evidence suggests that the reaction proceeds by attachment of the Grignard reagent to the silicon support directly rather than by Si-O-C linkages that would produce spectral changes such as those presented in Figure 1 (but not observed in Figure 4) and attached species that would not survive exposure to HF. Similar results demonstrating robust covalent attachment were also obtained using alkyl Grignard reagents at room temperature and similar reaction times [9].

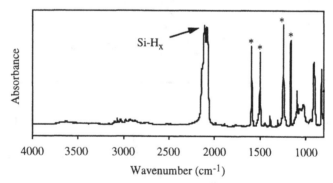

Figure 3. Diffuse reflectance infrared spectrum for porous silicon derivatized with 1 M 4-fluorophenylMgBr in THF at room temperature for 2 hours and quench with 1 M HCl in ether. The astericks note characteristic absorptions for the 4-fluorophenyl species.

Figure 4 shows various XPS spectra for porous silicon before and after derivatization with 4-fluorophenylmagnesium bromide. The survey spectrum for as-prepared porous silicon contains peaks for silicon, oxygen, and adventitious carbon. After reaction with 4-fluorophenylmagnesium bromide, the survey spectrum exhibits a peak for fluorine and additional carbon intensity from the attached fluorophenyl species. Notably, we observed no peaks for magnesium or bromine indicating that the fluorine and carbon peaks are not simply due to the presence of unreacted Grignard reagent on the surface. High resolution spectra of the fluorine region revealed the presence of a small amount of fluorine on the as-prepared porous silicon sample, with increased intensity being present after derivatization. The difference in binding energies for the two fluorine species reflects their bonding characteristics. For porous silicon (a), the fluorine is attached to electropositive silicon and appears at a lower binding energy than for fluorine bonded to carbon in the phenyl ring (b). This shift with bonding to silicon is also observed in the C(1s) spectrum where this region shows a peak at ~287 eV for carbon bonded to an electronegative atom (presumably fluorine), a primary peak at ~285 eV for aromatic carbon (and also adventitous carbonaceous material), and a peak at ~284 eV for carbon bonded to an electropositive atom (presumably silicon). The intensity ratio of the carbon peaks at 284 and 287 eV is ~1:1 suggesting that the presence of each carbon species attached to fluorine requires the presence of a carbon species attached to silicon. At the low pressures of XPS, the volatility of fluorobenzene (the product of 4-fluorophenylmagnesium bromide and HCl) requires that the fluorine signal from the Grignard reagent is from an attached species, where the spectral evidence indicates that the reaction with hydrogen-terminated silicon produces a SiC bond. The observation that the film remains stable during exposure to HF is compatible with the chemical inertness of Si-C bonds to these conditions.

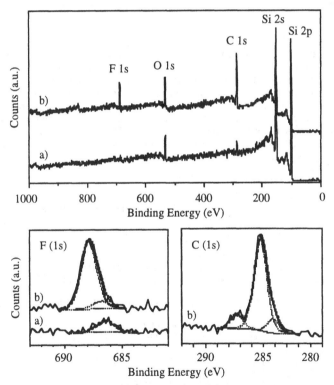

Figure 4. X-ray photoelectron spectra for porous silicon a) before and b) after derivatization with 4-fluorophenylMgBr as noted in Figure 3. The dotted lines result from a deconvolution of the spectra (solid lines) into their component peaks.

In contrast to observations with the alcohol reagents, extended exposure of porous silicon to the Grignard reagents did not produce the catastrophic etching noted above for the alcohols. Despite this difference, the reaction process is likely to be similar and operate by a mechanism that does not involve the simple substitution of SiH bonds by Si-OR or Si-C attachments. By analogy to our proposed mechanism for the reaction of alcohols with the hydrogen-terminated porous silicon surface (equation 1), we similarly hypothesize that the Grignard reagent attaches to the surface by breaking lattice Si-Si bonds (equation 2). This reaction would proceed with

$$H_xSi\text{-}SiH_y + RMgBr \rightarrow H_xSi\text{-}R + SiH_y\text{-}MgBr \qquad (2)$$
$$SiH_y\text{-}MgBr + HCl \rightarrow SiH_y\text{-}H \qquad (3)$$

formation of surfacial Grignard species that reach a limiting surface concentration and then limit further reaction. This asymptotic behavior is suggested by the observation that extended exposure of porous silicon to the Grignard reagent does not effect catastrophic etching of the substrate not increase the surface coverage of attached species [9]. Higher coverages are possible by quenching a reaction with HCl (or other protic sources, as in equation 3) and reexposing the sample to the Grignard solution (equation 2). When a quenched surface is exposed again to the Grignard reagent, we note that additional derivatization occurs [9].

The mechanisms for these reactions and their continued development are present targets of our ongoing work.

CONCLUSIONS

Porous silicon surfaces react with alcohols and Grignard reagents and form covalently attached films. With alcohols, the reaction occurs at temperatures slightly above room temperature and proceeds with concurrent etching of the silicon support, where some level of reaction control is required to functionalize porous silicon and maintain the structure of the parent substrate. With Grignard reagents, the reaction proceeds at room temperature and is less sensitive to reaction conditions. This latter method forms Si-C bonds to the surface that provide a notable stability to the films, including exposure to HF. The developed reactions with alcohols and Grignard reagents provide a general strategy for attaching organic species covalently and directly to the silicon framework of porous silicon by solution-phase means.

ACKNOWLEDGMENTS

We gratefully acknowledge the financial support of the Office of Naval Research.

REFERENCES

1. L.T. Canham, Appl. Phys. Lett. 57, 1046 (1990).
2. E.J. Lee, J.S. Ha, and M.J. Sailor, J. Am. Chem. Soc. 117, 8295 (1995).
3. E.J. Lee, T.W. Bitner, J.S. Ha, M.J. Shane, and M.J. Sailor, J. Am. Chem. Soc. 118, 5375 (1996).
4. J.-N. Chazalviel, C. Vieillard, M. Warntjes, and F. Ozanam, Proc. Electrochem. Soc. 95, 249 (1996).
5. M.J. Sailor and E.J. Lee, Adv. Mater. 9, 783 (1997) and references therein.
6. L.A. Zazzera, J.F. Evans; M. Dereulle, M. Tirrell, C.R. Kessel, and P.J. McKeown, J. Electrochem. Soc. 144, 2184 (1997).
7. N.Y. Kim and P.E. Laibinis, J. Am. Chem. Soc. 119, 2297 (1997).
8. J.M. Buriak and M.J. Allen, J. Am. Chem. Soc. 120, 1339 (1998).
9. N.Y. Kim and P.E. Laibinis, J. Am. Chem. Soc. 120, 4516 (1998).
10. B. Hamilton, Semicond. Sci. Technol. 10, 1187 (1995).

FUNCTIONALIZATION OF POROUS SILICON SURFACES THROUGH HYDROSILYLATION REACTIONS

J. M. BURIAK*, M. P. STEWART, M. J. ALLEN
Department of Chemistry, Purdue University, West Lafayette, IN 47907-1393, E-mail: buriak@purdue.edu

ABSTRACT

Hydrosilylation of alkynes and alkenes on silicon surfaces utilizing the native Si-H termination can be smoothly and rapidly carried out (30 s to 24 h) at room temperature through hydrosilylation mediated by Lewis acid catalysts or photoinduction with white light. Insertion of alkynes and alkenes into surface silicon hydride bonds yields covalently bound alkenyl and alkyl groups, respectively. Different chemical functionalities can be incorporated through these Hydrosilylation reactions, including ester, hydroxy, chloro, nitrile and chiral groups. Hydrophobic porous silicon surfaces demonstrate remarkable stability with respect to boiling aqueous aerated pH 1 to 10 solutions, and protect the bulk silicon from attack. Modification and tailoring of surface properties through this series of reactions induce wide variations in photoluminescent behavior of porous silicon, leading to almost complete quenching in the case of substituted and unsubstituted styrenyl termination, and minor decreases for alkyl and alkenyl functionalization. Because of the broad range of stable, modified surfaces produced using this chemistry, the work described here represents an important step towards technological applications of silicon surfaces.

INTRODUCTION

Because the native silicon hydride terminated surface of porous silicon is only metastable with respect to oxidation under ambient conditions (air, room temperature) [1], there has been intense interest in functionalizing the surface with the hopes of preventing its degradation [2]. Using the silicon-hydride groups as chemical handles, we have examined a broad range of reactions on porous silicon in order to provide a more stable organic monolayer. Hydrosilylation of alkynes or alkenes, shown in scheme 1, has been found to provide a very simple and efficient strategy to produce a wide variety of functionalized porous silicon surfaces

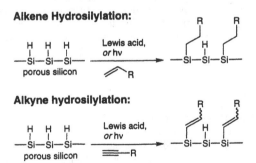

Scheme 1. Hydrosilylation of alkenes and alkynes on silicon hydride terminated porous silicon surfaces, resulting in alkyl or alkenyl termination, respectively. The reaction is Lewis acid mediated or induced by white light illumination.

(figure 1) [3]. The hydrophobic surfaces, especially those with extended alkyl chains, demonstrate dramatically increased stability under both ambient and demanding chemical conditions such as boiling alkali and acid (pH 1-10). The hydrosilylation reaction can be Lewis acid mediated, or photoinduced with white light at room temperature. The effects of different reaction strategies and terminations on the photoluminescence (PL) of the material are discussed.

EXPERIMENT

Photoluminescent silicon samples were prepared according to previously published procedures through a galvanostatic etching procedure [4]. For n-type samples: a 0.28 cm^2 exposed area of a polished, crystalline n-type silicon wafer (prime grade, P-doped, 0.70 Ω·cm, (100) orientation) was etched for 3 minutes in a Teflon cell at 75 mA/cm^2 (positive bias) under 10 mW/cm^2 white light illumination, derived from a 300 W tungsten filament ELH bulb using 1:1 49% HF (aq)/EtOH as the electrolyte/etchant. For p-type samples: a 0.28 cm^2 exposed area of a polished, crystalline p-type silicon wafer (prime grade, B-doped, 7.5-8.5 Ω·cm, (100) orientation) was etched for 30 minutes in a Teflon cell at 2 mA/cm^2 (positive bias) using 1:1 49% HF (aq)/EtOH as the electrolyte/etchant. After etching, the porous silicon samples were rinsed copiously with ethanol, blown dry under a stream of nitrogen and dried *in vacuo* for 1 h prior to use.

Transmission Fourier transform infrared (FTIR) spectra of the porous silicon samples were obtained with a Perkin Elmer 2000 FTIR spectrometer in absorbance mode. 16 to 32 scans were accumulated with a usual spectral resolution of 4 cm^{-1} and the sample chamber was purged with dry nitrogen.

Steady state PL measurements were taken in air using an Acton Research Spectra Pro 275 0.275 m monochromator and a Princeton Instruments liquid nitrogen cooled CCD detector, model LN/CCD-1024-E/1. The illumination source was an Oriel 250 W mercury arc lamp, filtered through a Bausch and Lomb monochromator set to 440 nm, and a short wave pass filter (CVI SPF 450). The luminescence is filtered by a long pass filter (CVI LP 490) before it is dispersed through the SP-275. Measured light intensity on the sample was 0.2 mW/cm^2. The sample mounting set-up was configured in order to reproducibly replace the sample before and after chemical functionalization reactions to ensure accurate comparisons.

The procedures for carrying out the hydrosilylation of alkenes and alkynes are described in references [6] (Lewis acid), [7] (thermal) and [8] (white light). WARNING: EtAlCl$_2$ is pyrophoric in air and should only be handled under an inert atmosphere until quenched.

RESULTS

Lewis Acid Mediated Hydrosilylation

Hydrosilylation on silicon hydride terminated porous silicon involves insertion of an alkene or alkyne into the Si-H group, yielding an alkyl or alkenyl functionalized surface, respectively, as shown in scheme 1 [5]. We previously reported that the inexpensive, commercially available Lewis acid, EtAlCl$_2$, effectively mediates the hydrosilylation reaction on porous silicon at room temperature [6]. Late transition metal complexes known to catalyze hydrosilylation of unsaturates with soluble, molecular silanes were also tested but resulted in considerable darkening of the surface and concomitant PL quenching, presumably through metal deposition

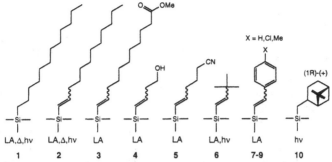

Figure 1. Representative surfaces produced through hydrosilylation reactions of alkenes or alkynes. LA = Lewis acid mediated, Δ = thermally promoted, hv = white light induced.

[7]. The Lewis acid, on the other hand, is not prone to such side reactions and can be removed simply by rinsing with THF and water upon completion of the reaction. By addition of an excess of Lewis acid, substrates containing Lewis basic substituents such as esters, hydroxy and nitrile groups can be cleanly incorporated (figure 1, surfaces 3-5). The reaction takes place smoothly, regardless of initial doping of the crystalline silicon precursor material. The FTIR spectra in figure 2 demonstrate that hydrosilylation of 1-dodecene (surface 1) proceeds well on both n-type, P-doped (n$^+$) and p-type, B-doped (p$^-$) porous silicon samples, for example. The spectra clearly indicate that hydrosilylation has taken place through the appearance of the ν (C-H) stretches of the dodecyl chain between 2960-2850 cm^{-1} and δ (C-H) methylene and methyl bending modes at 1466 and 1387 cm^{-1}, the disappearance of the olefin ν (C=C) at 1640 cm^{-1}, and decrease in ν (Si-H) by ~25%. The hydrophobic surfaces terminated with dodecyl (surface 1), dodecenyl (surface 2), and various styrenyl derivatives (surfaces 7-9) all demonstrate excellent chemical stability to boiling in aerated pH 1 (H$_2$SO$_4$) and pH 10 (KOH) solutions with and without added EtOH.

White Light Promoted Hydrosilylation

In order to avoid the use of any catalyst, a new, white light mediated reaction has been developed in our laboratory [8]. While the Lewis acid EtAlCl$_2$ is relatively innocuous with

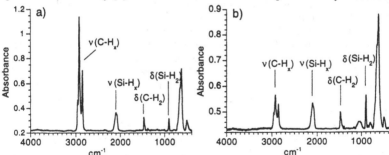

Figure 2. Transmission FTIR spectra of dodecyl terminated porous silicon (surface 1) prepared through Lewis acid mediated 1-dodecene hydrosilylation. a) n-type, P-doped silicon (n$^+$), b) p-type, B-doped silicon (p$^-$).

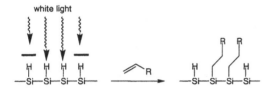

white light

Figure 3. Photopatterned hydrosilylation of alkenes on porous silicon utilizing white light (22-44 mW/cm^2). The reaction is also effective for alkynes.

respect to porous silicon, avoidance of this air and water sensitive organometallic reagent would be preferable for large scale production. We, and others, have observed that hydrosilylation of alkenes and alkynes may be thermally induced (T>100°C) on both flat crystal [9][10] and porous Si-H terminated surfaces [7][11] in absence of any additives. The high temperatures, however, can damage the fragile nanoscale architecture of porous silicon and contribute unfavorably to the thermal budget in an IC manufacturing process. White light of moderate intensity (22-44 mW/cm^2, 1 h, 22°C), however, can quickly and efficiently induce hydrosilylation on n-type porous silicon as shown schematically for alkenes in figure 3. A wide range of alkynes and alkenes react under these conditions. While the reaction with white light is similar to that of the Lewis acid mediated hydrosilylation, the two methodologies are highly complementary. Certain substrates are incompatible with the Lewis acid reaction, in which case the light promoted reaction is preferred. For instance, alkyl fluoride and chloride containing alkenes/alkynes are polymerized in the presence of EtAlCl$_2$ due to carbocation mediated polymerization [12]. With the white light hydrosilylation, alkyl chloride and fluoride groups are tolerated and can be successfully incorporated. In other cases, a substrate may attack the surface through a Lewis basic pendant group (hydroxy groups, amines) and thus be inapplicable with the light chemistry; the Lewis acid chemistry is highly effective for such substrates since it coordinates to the Lewis base and prevents it from damaging the porous silicon [6]. Hydrophobic surfaces produced through white light photoinduction also demonstrate excellent stability with respect to aerated boiling pH 1-10 solutions. Because the reaction is light promoted, the porous silicon surfaces can be easily photopatterned, allowing for the preparation of spatially defined arrays of chemical functionalities.

Effects of Hydrosilylation on Photoluminescence (PL)

While the hydrosilylation reactions are important for the stabilization of porous silicon surfaces, the ability to rationally incorporate any functional group upon demand may also allow for a deeper understanding of the effects of surface states on PL. It is widely accepted that surface states play major roles in the electronics of nanocrystallites since a very large fraction of the atoms reside on or near the surface [2]. In order to carry out such a study, it is important to verify that the reaction itself does not damage the porous silicon, thus having negative consequences on the PL, irrespective of surface termination. Figure 4 demonstrates the effects of 1-dodecene hydrosilylation (surface 1) on the PL n-type porous silicon samples. Lewis acid mediated and thermally induced hydrosilylation result in quenching of over 75%, whereas the white light promoted reaction has little effect on the PL intensity - only a small red shift is observed. This suggests that the white light promoted reaction is the most gentle of these three hydrosilylation approaches since almost no quenching is observed.

We have observed a significant dependence of PL on surface termination. Conjugated vinyl aromatic substituents (surfaces 7-9), most notably, induce complete quenching of the PL, regardless of synthetic procedure. Earlier work by Sailor and Song noted that phenyl acetylide termination, a closely related conjugated alkynyl aromatic substituent, also effectively

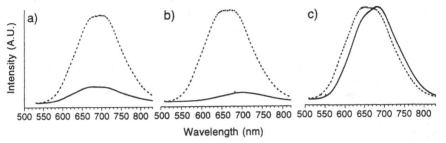

Figure 4. PL spectra of 1-dodecene hydrosilylation (surface **1**) on n-type porous silicon (440 nm excitation). The dotted spectrum is that of freshly etched porous silicon, and the solid that of the resulting dodecyl terminated surface. **a)** Lewis acid mediated, **b)** thermally induced (2 h reflux in neat 1-dodecene), **c)** white light promoted (60 minute reaction at 22 mW cm^{-2} white light).

quenches the PL of porous silicon [13] which suggests that this class of conjugated substrates generally eliminates the PL of porous silicon. An isolated vinyl group, as in surface **2**, however, only weakly quenches the PL by approximately 35% using light chemistry [8], and 80% with the Lewis acid reaction. Direct substitution of the surface with a phenyl [13] or 4-fluorophenyl [14] has been shown to only weakly quench the intrinsic PL of porous silicon, suggesting that the extended conjugation of surfaces **7-9** and the phenyl acetylide termination act as especially effective non-radiative traps. By combining these trends with those in the literature, an *approximate* order of quenching ability of various functionalities can be formulated as shown in figure 5. This ordering can only be approximate due to the different synthetic approaches used to prepare these surfaces.

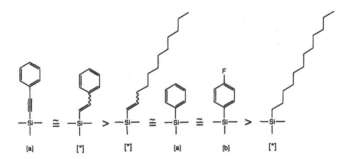

Figure 5. *Approximate* order of quenching efficiencies of various surface functionalities on porous silicon, in order of greatest quenching to least. The data for surfaces labeled [a] and [b] are found in references [13] and [14], respectively.

CONCLUSIONS

Hydrosilylation reactions on porous silicon surfaces can be easily and efficiently mediated at room temperature with the assistance of either a Lewis acid or white light. The latter reaction is particularly promising due to the simplicity of the reaction, the potential for conditions. Because of the complementarity of the chemistries of these two reactions, a wide variety of alkenes and alkynes with different substituents may be incorporated with ease.

ACKNOWLEDGMENTS

Financial support from Purdue University, the Purdue Research Foundation, the Camille and Henry Dreyfus Foundation in the form of a New Faculty Award (J.M.B.), a Link Foundation Energy Fellowship (M.P.S.), and an Eli Lilly-Purdue Alumni Scholarship (M.J.A.) is gratefully acknowledged.

REFERENCES

1 A.G. Cullis, L.T. Canham, P.D.J. Calcott, J. Appl. Phys. 82, 909 (1997).

2. M.J. Sailor, E.J. Lee, Adv. Mater. 9, 783 (1997).

3. J.M. Buriak, Advanced Materials (in press).

4. E.J. Lee, E. T.W. Bitner, J.S. Ha, M.J. Shane, M.J. Sailor, J. Am. Chem. Soc. 118, 1390 (1996).

5. F.A. Cotton, G. Wilkinson, *Advanced Inorganic Chemistry*, 5th ed. (John Wiley and Sons, New York, 1988), pp. 1255-1256.

6. J.M. Buriak, M.A. Allen, J. Am. Chem. Soc., 120, 1339 (1998).

7. J.M. Holland, M.P. Stewart, M.J. Allen, J.M. Buriak, J. Solid State Chem. (submitted).

8. M.P. Stewart, J.M. Buriak, Angew. Chem. Int. Ed. Engl. (in press).

9. On (111) Si-H: M.R. Linford, P. Fenter, P.M. Eisenberger, C.E.D. Chidsey, J. Am. Chem. Soc. 117, 3145 (1995).

10. On (100) Si-H: A.B. Sieval, A.L. Demirel, J.W.M. Nissink, M.R. Linford, J.H. van der Maas, W.H. de Jeu, H. Zuilhof, E.J.R. Sudhölter, Langmuir, 14, 1759 (1998).

11. J.E. Bateman, R.D. Eagling, D.R. Worrall, B.R. Horrocks, A. Houlton, Angew. Chem. Int. Ed. Engl. 37, 2683 (1998).

12. *Cationic Polymerization*, volume 665, edited by R. Faust and T. D. Shaffer (American Chemical Society, Washington DC, 1997).

13. J. H. Song, M. J. Sailor, J. Am. Chem. Soc. 120, 2376 (1998).

14. N. Y. Kim, P. E. Laibinis, J. Am. Chem. Soc. 120, 4516 (1998).

ENHANCEMENT IN EFFICIENCY AND STABILITY OF OXIDE-FREE BLUE EMISSION FROM POROUS SILICON BY SURFACE PASSIVATION

H. MIZUNO and N. KOSHIDA
Division of Electronic and Information Engineering, Faculty of Technology,
Tokyo University of Agriculture and Technology, Koganei, Tokyo 184, Japan

ABSTRACT

The establishment of tuning techniques of visible emission from porous silicon (PS) is very important from both physical and technological viewpoints. As previously reported that the photoluminescence (PL) spectra of PS can continuously be controlled from red to blue simply by postanodization illumination method without any growth of the surface oxide. Details of this oxide-free blue emission have been studied in terms of the PL decay dynamics and surface chemistry. We report here that post-preparation exposure of PS to hydrogen gas is very useful for improvement in efficiency and stability of blue PL. Based on this technique, it has become possible to get blue electroluminescence even at low bias voltages.

INTRODUCTION

Since room-temperature visible photoluminescence (PL) [1] and electroluminescence (EL) [2] from porous silicon (PS) were reported, many studies have been conducted toward PS-based optoelectronic integration. Because PS consists of a great number of Si nanocrystallites, visible light-emission from PS has been discussed in relation to quantum confinement effects. Recently-reported blue PL without any growth of the surface oxide [3, 4] and blue-green electroluminescence (EL) [5] leads to possibilities of silicon-based optoelectronic systems. More detailed information about dynamics of blue emissive PS and improvement of efficiency and stability are required to apply the properties of PS to optoelectronic devices.

In this study, we show the PL life time of oxide-free blue emission in comparison to that of red and green ones. Simple surface passivation technique of PS for improvement of the blue luminescence characteristics is presented. The experimental results show that the additional hydrogen terminations play an important role for suppressing the deterioration of PL against surface oxidation and enhancing the EL efficiency.

EXPERIMENTAL

Sample Preparation

The PS samples are formed on 8-12 Ωcm p-type (100) Si wafers. The anodization is carried out in the dark in a solution of HF (20%) : ethanol (99.5%)=1:1 at a constant current density of 50 mAcm^{-2} for 5 min. Immediately after this anodization, the samples were illuminated by a 500 W tungsten lamp from a distance of 20 cm, for 0-15 min under the open-circuit condition in order to tune the emission energy. Prepared samples were rinsed with ethanol in the dark, and then immediately transferred into a H$_2$ gas ambient without exposure to air to avoid possible surface oxidation. The samples were kept for 0-12 h at room temperature. After this treatment, the samples were transferred to the optical measurement system.

Measurements

Three measurements were carried out.
Measurement A : PL decay dynamics of the three tuned PS samples with red, green, and blue emission were measured in vacuum chamber ($\sim$10^{-5} Torr), using a 337 nm N$_2$ pulse laser

(pulse width : 2 ns) as an excitation source. The power density of this laser was varied in the range of 12-120 μJcm^{-2}. The overall experimental time resolution was about 10 ns.

Measurement B : PL spectra of the H_2-treated samples were measured by multichannel Si detector for 0-60 min in a N_2 gas ambient. The excitation wavelength is 325 nm of He-Cd laser. To investigate the relation between the PL characteristics and the surface composition, fourier transform infrared (FTIR) absorption spectra were measured in a N_2 gas atmosphere.

Measurement C : After the H_2 gas treatment, experimental PS light-emitting diodes were fabricated in the form of Au electrode/PS/c-Si/ohmic contact (see inset in Fig. 5). The current–voltage (I–V) characteristics of PS diodes were measured under the forward and reverse bias conditions in a N_2 gas ambient at room temperature. The EL intensities and spectra are also measured at the same time. The forward bias condition corresponds to the case in which a negative voltage is applied to the Au electrode with respect to the substrate.

RESULTS

PL decay dynamics (Measurement A)

PL life time τ and δ values were obtained by fitting the following stretched exponential function;

$$I_{PL} = I_0 \exp[-(t/\tau)^{\delta}]$$

In Fig. 1 (a) and (b), the measured PL life time and δ value at 300 K and 10 K for three tuned samples are plotted as a function of the respective emission energies. At higher emission energies, the PL life time and δ value decrease in each band. It should be noted that the behavior of δ is independent of temperature. Especially of importance is that there seems to be a universal relationship between the PL decay dynamics and the emission energy, independent of the PL emission band. This is a strong indication that the PL emission from PS is determined by a common mechanism throughout the visible region.

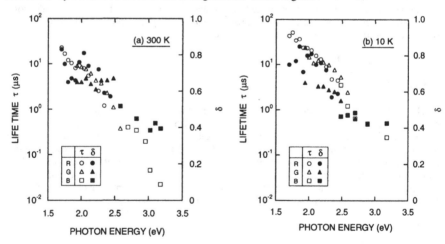

Fig. 1 The emission energy dependence of the PL decay time τ and δ value for three tuned PS samples with red(R), green(G), and blue(B) emission. Temperature of left and right side figure is 300 K and 10 K, respectively.

The plots of life time at 10 K are about one order longer than those at 300 K. According to the result of pico-second measurements which was carried out separately, the fast decay time was determined only in the green-blue band (>2.5 eV). These characteristics imply that the oxide-free blue emission should be regarded as a part of a series of visible luminescence bands all based on confinement.

The PL characteristics of H_2 treated samples (Measurement B)

The PL spectra of H_2 treated and as-anodized sample are shown in Fig. 2. At the initial stage, PL intensity of H_2 treated sample is twice as that of as-anodized sample with the same peak position. After continuous excitation for 60 min, PL intensity of the as-anodized sample show a significant decrease, and peak position shifts to longer wavelength side. On the other hand, in H_2 treated sample there is no change in the peak position while the PL intensity somewhat degrades.

PL peak wavelength and intensity at 430 nm are plotted against the laser excitation time in Fig. 3. PL peak wavelength of as-anodized sample shift toward longer side over 70 nm from the initial position (430 nm). In contrast, the PL peak of H_2 treated sample remained unchanged even after 60 min UV illumination. Similar desirable effects are also observed with the PL intensity.

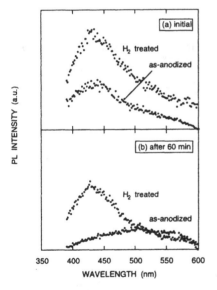

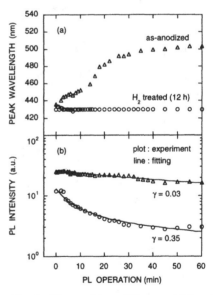

Fig. 2   PL spectra of H_2 treated (for 12 h) and as-anodized sample in a N_2 gas ambient. (a) As-illuminated by He-Cd laser. (b) after 60 min illumination by laser.

Fig. 3   PL transition of H_2 treated (o) and as-anodized (Δ) sample. (a) PL peak wavelength. (b) PL intensity at 430 nm. Plots and lines are experiment data and fitting curve, respectively.

The degradation process has been analyzed by the photoinduced oxidation model [6]. Under the assumption that the PL efficiency deteriorates in accordance with the progress of surface oxidation, the PL intensity I_{PL} given by

$$I_{PL} \propto I_0(\gamma t)^{-\theta} \int_{\gamma\mu t}^{\gamma t} d\xi \cdot \xi^{\theta-1} \cdot \exp[-\xi]$$

where I_0 is the excitation intensity, $\theta=(\alpha/\alpha+\beta)$, $\mu=\exp[-(\alpha+\beta)d]$, $\xi=\gamma t\exp[-(\alpha+\beta)x]$ and γ is a fitting parameter which is proportional to the oxidation rate. Here, we used $\gamma=0.35$ for as-anodized PS and $\gamma=0.03$ for H_2 treated PS, and $\alpha=2\times10^5 cm^{-1}$, $\beta=2\times10^4 cm^{-1}$ and $d=1\mu m$ for both to obtain the best fit to the experimental results. It is evident that H_2-treated PS produces more complete surface passivation against oxidation.

To confirm that the differences in PL degradation arise from differences in the surface composition, FTIR spectra were investigated for a few samples with different times of exposure to H_2 gas after anodization(Fig. 4). As the exposure time is increased, hydrogen termination signal corresponding to stretching mode, in particular, becomes apparent. On the other hand, the Si-O-Si features which originate from residual oxygen in the substrate do not show any changes in both position and intensity. We can conclude that blue emission is activated and stabilized only by additional hydrogen termination.

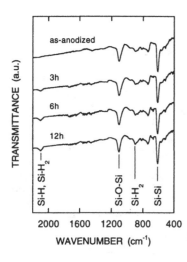

Fig. 4 FTIR spectra of H_2 treated samples were measured in a N_2 gas ambient. The H_2 gas exposure time for each spectrum are 0, 3 h, 6 h, 12 h, respectively.

EL characteristics (Measurement C)

The I-V and EL characteristics for PS diodes with H_2 treated and as-anodized sample are shown in Fig. 5. The time of H_2 exposure in this case was 12 h. In comparison to as-anodized sample, the rectifying ratio of H_2 treated sample is improved significantly. The n factor of H_2 treated and as-anodized sample are about 2.1 and 8.2, respectively. Thus additional hydrogen termination is useful for decreasing the density of generation-recombination centers. With increasing the diode current density, EL intensity of both samples increase rapidly under the forward and reverse bias condition. The current dependence of EL intensity is shown in Fig. 6. It is obvious that in both side of the bias direction, quantum efficiency of H_2 treated sample is higher than that of as-anodized sample. Threshold points of EL emission at reverse bias conditions are lower than that at forward bias condition. Since threshold voltages of EL emission at reverse bias condition are higher than that at forward bias condition, EL emission occur from a high electrical field region in the PS layer. The power efficiency of H_2 treated diode, which is important at technological viewpoint, is also one order higher than that of the as-anodized one.

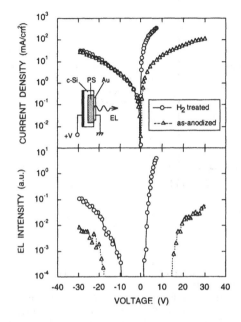

Fig. 5 The I-V curve and EL characteristics of H_2 treated (for 12 h) and as-anodized sample. The inset is a schematic illustration of the experimental PS diode under the forward bias condition.

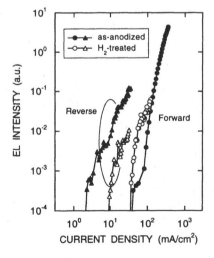

Fig.6 EL intensity of a hydrogen terminated PS diode and a conventionally diode as a function of the diode current.

To determine the EL emission energy, an EL spectrum was measured at the highest emission point using the H$_2$-treated sample. The experimental result is shown in Fig. 7. The peak wavelength is about 430 nm which is the same as that of the PL peak. The EL peak position was independent of both the bias voltage and its polarity.

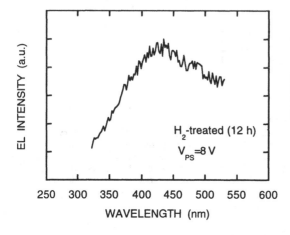

Fig. 7   An EL spectrum from a hydrogen terminated PS diode. The diode current was about 300 mAcm^{-2}.

CONCLUSIONS

The following three results from oxide-free blue emissive PS sample have been observed.

(a) There is a common PL decay dynamics throughout the visible region.

(b) A long-term exposure of the blue-emissive sample to hydrogen gas at room temperature enhances the surface passivation, and consequently improves the PL efficiency and stability.

(c) This surface termination also produces a desirable effect on the EL emission through an improvement of junction characteristics.

It appears that complementary effects of confinement and surface termination in silicon nanocrystallites become more important as the emission band enters into shorter wavelength region.

ACKNOWLEDGEMENTS

This work was partially supported by the Research Foundation Opto-Sience and technology and a Grant-in-Aid from the Ministry of Education, Science, Sports and Culture of Japan.

REFERENCES

1. L. T. Canham, Appl. Phys. Lett. **57**, 1046 (1990).
2. N. Koshida and H. Koyama, Appl. Phys. Lett. **60**, 347 (1992).
3. H. Mizuno, H. Koyama, and N. Koshida, Appl. Phys. Lett. **69**, 3379 (1996).
4. H. Mizuno, H. Koyama, and N. Koshida, Thin Solid Films **297**, 314 (1997).
5. V. A. Kuznetsov, I. Andrienko, and D. Haneman, Appl. Phys. Lett. **72**, 3323 (1998).
6. T. Matsumoto, Y. Masumoto, T. Nakagawa, M. Hashimoto, K. Ueno, and N. Koshida, Jpn. J. Appl Phys. **36**, 1089 (1997).

MODIFICATION OF VISIBLE LIGHT EMISSION FROM SILICON NANOCRYSTALS AS A FUNCTION OF SIZE, ELECTRONIC STRUCTURE, AND SURFACE PASSIVATION

M.V. WOLKIN, J. JORNE* and P.M. FAUCHET**
Materials Science Program, Department of Chemical Engineering*, Department of Electrical and Computer Engineering**, University of Rochester, Rochester NY 14627, USA

G. ALLAN, C. DELERUE
IEMN-ISEN, Lille, France

ABSTRACT

The effect of surface passivation and crystallite size on the photoluminescence of porous silicon is reported. Oxygen-free porous silicon samples with medium to ultra high porosities have been prepared by using electrochemical etching followed by photoassisted stain etching. As long as the samples were hydrogen-passivated the PL could be tuned from the red (750nm) to the blue (400nm) by increasing the porosity. We show that when surface oxidation occurred, the photoluminescence was red-shifted. For sizes smaller than 2.8nm, the red shift can be as large as 1eV but for larger sizes no shift has been observed. Comparing the experimental results with theoretical calculations, we suggest that the decrease in PL energy upon exposure to oxygen is related to recombination involving an electron or an exciton trapped in Si=O double bonds. This result clarifies the recombination mechanisms in porous silicon.

INTRODUCTION

The strong demand for optoelectronic devices based on silicon has led to a great interest in light emitting porous silicon (PSi) [1-2]. Modifying the photoluminescence (PL) peak in the visible range is very important from the theoretical and technological points of view. Several models have tried to establish the mechanism of the visible PL. The quantum confinement model [3-4] suggests that when the silicon particle size is reduced to few nanometer the energy bandgap is enlarged, resulting in a blue shift of the PL. The surface states model [5] proposes that since the surface to volume ratio increases as the crystallite size decreases, the surface bonds play a significant role in the recombination mechanism. Other models such as defects in the oxide [6] and specific molecules [7] have also been considered. Presently, in spite of all the past work, the PL of PSi is not completely understood. In this work we investigate the effect of different factors such as surface passivation, crystallite size and electronic states in order to evaluate their role in modifying the PL [8].

EXPERIMENTAL

Psi samples with medium to ultra-high porosities were produced by electrochemical etching and subsequently dried as reported in reference 8. All samples were formed on 6 Ω·cm <100> p-type silicon wafers. A thin layer of aluminum was deposited on the back side of the wafers to provide a good electrical contact. To prevent IR absorption by the aluminum during FTIR measurements, thick aluminum stripes (1mm) were etched away using a sequence of photolithography steps.

Mat. Res. Soc. Symp. Proc. Vol. 536 © 1999 Materials Research Society

Solutions of HF/ethanol/H$_2$O with an HF volume concentration of 10%-25% were used to achieved the desired porosities. The anodization was carried out in the dark at current densities of 8-50 mA/cm^2 and followed by photo-assisted stain etching using a 500 W halogen lamp, to further increase the porosity [9]. The samples were illuminated for 0-20 minutes, depending on the required porosity. Since the infrared radiation of the halogen lamp heats the etching solution, resulting in inhomogeneties, illumination was performed using a 650 nm cutoff filter.

Immediately after etching, each sample was rinsed in ethanol for few minutes and then dried using one of the subsequent procedures. A first set of samples was directly transferred from the ethanol solution into an Ar environment, without exposure to air. A second set of similar samples was exposed to air for 24 hours. All measurements were done at room temperature.

RESULTS AND DISCUSSION

1. SURFACE PASSIVATION:

The PL response to surface chemical coverage was investigated in the following way. An ultra-high porosity sample was prepared and kept in an Ar environment. As shown in Fig.1 its PL lies in the green (530nm). The PL peak position and intensity were stable as long as the sample was kept in Ar. However, after exposure to air, the PL spectrum was modified and a marked red shift was observed in less than 3 minutes. The evolution of the surface chemistry of a similar sample is depicted in Fig.2. The FTIR transmission spectrum of the as-anodized sample shows a strong absorption band near 2100 cm^{-1} and no oxygen peaks. This fresh samples are passivated by hydrogen and contain no oxygen. After 3 minutes of exposure to air, a small peak appears at 1070 cm^{-1}, associated with Si-O-Si bond. While

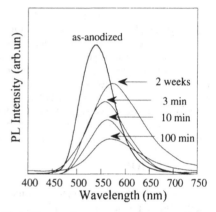

Fig.1 Photoluminescence spectra of a high porosity PSi sample kept under Ar (as-anodized) and as it shifts when exposed to air.

this peak increases with time, the hydrogen feature decreases progressively until it disappears. From these results, it seems logical to assume that the large PL shift is related to surface oxidation.

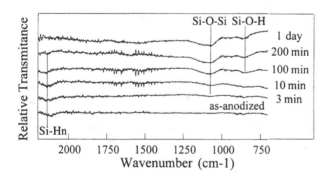

Fig.2 Evolution of the FTIR transmission spectra of a green-emitting sample as a function of the time it was exposed to air.

Fig.3a correlates the integrated absorption of the Si-O-Si bonds with the PL peak wavelength. Immediately after exposure, a large red shift of 125 meV is obtained with a slight change in the Si-O-Si integrated absorption. Further exposure significantly increases the Si-O-Si peak area but red shifts the PL by only 40 meV. This result suggests that only few surface bonds are responsible for the red shift. Fig.3b quantifies the ratio between the Si-O-Si and Si-H$_n$ bonds during air exposure, and proves that the surface coverage is completely transformed from hydrogen to oxygen in one day.

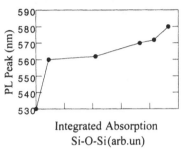
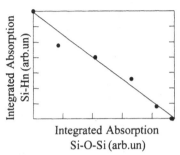

Integrated Absorption
Si-O-Si(arb.un)

Integrated Absorption
Si-O-Si (arb.un)

Fig.3a Variation of the peak position with the Si-O-Si integrated absorption when the green emitting sample is exposed to air.

Fig.3b Si-H$_n$ vs. Si-O-Si integrated absorption at different times during air exposure.

2. SIZE: It has been reported in the PSi literature that when the particle size is reduced to several nanometers, the PL in air is blue shifted up to a limit of 2.1eV [2,10]. Moreover, the Stokes-shift between bandgap and PL energies increases with decreasing crystallite size, indicating that the PL energy becomes different from the free excitonic bandgap [11]. We now examine the correlation between size, PL peak, PL intensity and time-resolved PL [8].

Fig.4 shows how the PL spectra of oxygen-free PSi samples can be tuned from the blue (400nm) [12] to the red (750nm) as a result of decreasing porosities. All PL spectra were stable as long as the samples were hydrogen-passivated. However, after exposure to air, a red shift of the PL was observed specially if the initial PL was in the blue-green-yellow region. The magnitude of the red shifts decreased with decreasing porosities, and was 1eV, 380meV, 200meV, 95meV, 60meV and 0meV for samples originally emitting at the blue, blue-green, green, yellow, orange and red respectively. It is important to note that there is

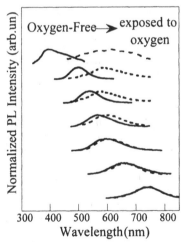

Fig.4 Photoluminescence spectra from PSi samples with different porosities before (———) and after (- - - -) they were exposed to air for 24 hours.

an upper limit to the PL in air (2.14 eV) and that the PL of samples which originally emitted in the red was not shifted after exposure to air. These observations suggest that a full tuning of the PL spectrum by varying the crystalline size is only possible for oxygen-free samples.

3. ENERGY STATES: The decrease in PL energy upon exposure to air may be related to recombination of trapped carriers via oxygen-related localized states which are stabilized due to quantum confinement. We have applied electronic band structure calculations to Si-O-Si bonds and shown that they produce no localized states in the gap even for the smallest crystallites. On the other hand, the Si=O double bond, which may stabilize the Si/SiO$_2$ interface [13], has electronic states within the PSi band gap as shown by Fig.3 in reference 8. Fig.5 presents the calculated free excitonic bandgap (upper line) and the lowest transition energy calculated in the presence of a Si=O bond (lower line), as a function of the nanocrystal diameter. On that same graph, we plot the experimental PL energies (taken from Fig.4) of the oxide-free (full dots) and oxidized (empty dots) samples. Fig.5 shows that theory and experimental results coincide.

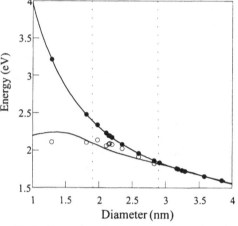

Fig.5 Comparison between experimental and theoretical PL energies as a function of crystallite size. The calculated free excitonic bandgap (upper line) and the lowest transition energy calculated in the presence of a Si=O bond (lower line) are plotted, as well as the experimental PL energy (taken from Fig.4) of the oxide-free (●) and oxidized (○) samples.

4. COMPREHENSIVE MODEL: Based on the experimental and calculated results, we have developed a comprehensive model [8] for the recombination mechanisms as a function of surface passivation and crystallite size. Fig.6 shows silicon crystallites, with different sizes and surface passivations. The first row shows oxygen-free crystallites, in which electrons and holes are free, and recombination occurs via free exciton states for all sizes. The PL energy is therefore equal to the free excitonic bandgap, and can be tuned by changing the crystallite size as seen in Fig.4. The second row shows Si crystallites that have been exposed to air. The carriers can be trapped on the Si=O covalent bond and three regions, corresponding to different recombination mechanisms, are depicted in Fig.5. For sizes larger than 2.8nm, the Si=O localized states are yet to be stabilized. The electrons and the holes remain free, and recombination is via free exciton states. As expected, no red shift was observed after exposure to oxygen since the recombination mechanisms are identical. For sizes smaller than 2.8nm and larger than 1.8nm, recombination is via a free hole and an electron trapped at the Si=O bond. As the size of the crystallite decreases, the PL energy still increases, but slower than for the oxygen-free samples. Moreover, after exposing the samples to air, the PL red shift increases from 60meV up to 200meV when the size decreases from 2.8nm down to 1.8nm (Fig.4). Last, when the crystallites are smaller than 1.8nm, both electrons and holes are trapped at the Si=O bond, and recombination is via this trapped exciton. In this case, the PL energy is constant for all sizes, and upon exposure to air very large red shifts (of up to 1eV) are observed.

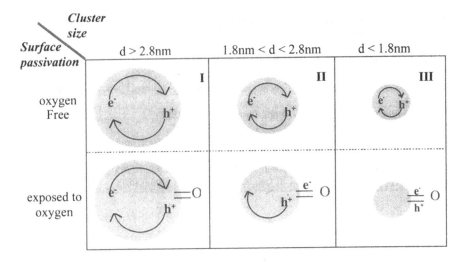

Fig.6 Schematic of six silicon crystallites explains the PL in PSi. To simplify the pictures, the Si-H bonds are not included, and only one Si=O double bond appears in each crystallite exposed to air.

CONCLUSIONS

The PL of PSi samples kept under Ar environment and hydrogen passivated is determined entirely by quantum confinement and can be tuned from the blue (400nm) to the red (750nm) by varying the crystallite size. After exposure to air, we observe a red shift of the PL that can be as large as 1 eV for initially blue emitting samples. During the red shift the hydrogen present in the surface can be replaced by oxygen. This indicates that modifying the PL energy by varying the crystalline size is only possible for oxygen-free samples.

We propose a comprehensive model that correlates the PL with crystallite size and surface passivation, and explains how the PL can be modified by both of these factors. In this model, Si=O surface bonds trap the electron for crystallite smaller than ~2.8nm and both the electron and the hole for crystallite smaller than ~1.8nm. This model is in quantitative agreement with the experimental data.

ACKNOWLEDGMENT

Support from the US Army Research Office and the Electric Power Research Institute is gratefully acknowledged. We thank Leonid Tsybeskov for technical support. The "Institut d'Electronique et de Microélectronique du Nord" is "Unite mixte 9929 du Centre National de la Recherche Scientifique".

REFERENCES

[1]   A. G. Cullis, L. T. Canham and P. D. J. Calcott, J. Appl. Phys. **82**, 909 (1997).

[2]   P. M. Fauchet, J. Lumin. **70**, 294 (1996).

[3]   L. T. Canham, Appl. Phys. Lett. **57**, 1046 (1990).

[4]   J.P. Proot, C. Delerue, and G. Allan, Appl. Phys. Lett. **61**, 1948 (1992).

[5]   F. Koch, V. Petrova-Koch, T. Muschik, A. Nikolov and V. Gavrilenko, Mat. Res. Soc. Symp. Proc. **283**, 197 (1993).

[6]   G. G. Qin and Y. Q. Jia, Solid state Commun. **86**, 559 (1993)

[7]   Z. Y. Xu, M. Gal, and M. Gross, Appl. Phys. Lett. **60**, 1375 (1992)

[8]   M. V. Wolkin, J. Jorne, P. M. Fauchet, G. Allan and C. Delerue, Phys. Rev. Lett. **82**, 197 (1999)

[9]   H. Koyama and N. Koshida, J. Appl. Phys. **74**, 6365 (1993)

[10]   C. Malone and J. Jorne, Appl. Phys. Lett. **70**, 3537 (1997)

[11]   J. von Behren, T. Van Buuren, M. Zacharias, E. H. Chimowitz and P. M. Fauchet, Solid State Comm. **105**, 317 (1998).

[12]   H. Mizuno, H. Koyama and N. Koshida, Appl. Phys. Lett. **69**, 3779 (1996).

[13]   F. Herman and R. V. Kasowski, J. Vac. Sci. Technol. **19**, 395 (1981).

NANOSTRUCTURE OF POROUS SILICON USING TRANSMISSION MICROSCOPY

M. H. NAYFEH, Z. YAMANI, O. GURDAL, A. AlAQL*
Department of Physics, University of Illinois at Urbana-Champaign, Urbana, IL, USA
* Department of Physics, King Saud University, Riyadh, Saudi Arabia

ABSTRACT

We use high resolution transmission electron microscopy (XTEM) to image the nanostructure of (100) p-type porous Si. A network of pore tracks subdivide the material into nanoislands and nanocrystallites are resolved through out the material. With distance from the substrate, electron diffraction develops, in addition to coherent diffraction, amorphous-like patterns that dominates the coherent scattering in the topmost luminescent layer. Also, with distance from the substrate, crystalline island size diminshes to as small as 1 nm in the topmost luminescence material. Although their uppermost layer has the most resolved nano crystallites, it has the strongest diffuse scattering of all regions. This suggests that the diffuse scattering is due to a size reduction effects rather than to an amorphous state. We discuss the relevance of a new dimer restructuring model in ultra small nanocrystallites to the loss of crystalline effects.

INTRODUCTION

The status of the precise structure of porous silicon [1] remains open [2]. There is evidence that the material is composed of interconnected nanocrystallites as small as 2 to 3 nm. But there is also evidence from x-ray photoelectron spectroscopy (XPS) [3], Raman spectroscopy [4], and transmission electron microscopy and diffraction (XTEM) [5-6] of an amorphous component. In reflectance measurements there was evidence for a loss of crystallinity, and the appearance of features similar to an amorphous response. However, it was attributed to a reduction in crystalline size and not to an amorphous phase. However, there was no suggestion as to the nature of the changes that reduce the interband resonance structure. In this paper we use XTEM and electron diffraction to directly observe the structure and its development through the sample. TEM has been successfully used to provide some of the most detailed information on the internal structure of porous silicon, down to the atomic scale [2]. However, the procedure is limited by difficulties in producing thin enough samples for electron transparency without damage and introduction of extraneous structure. Nanocrystalline islands and nano-pore networks have been observed [7-8], and have become characteristic of the material. But there have been no systematic studies of the evolution with depth, especially towards the top most luminescent layers. Our results indicate that with distance from the substrate, and in addition to coherent diffraction, electron diffraction develops diffuse patterns that dominate the coherent scattering in the topmost luminescent layer. Correspondingly, the crystallite size drop to as small as 1 nm in the topmost layer. Although, the topmost layer has the most resolved nanocrystallites, it has the strongest diffuse scattering of all regions. This suggests that diffuse scattering may be due to a size reduction effects rather than to an amorphous state. We discuss the relevance of a new restructuring model in ultra small crystallites to the loss of crystallinity effects.

SAMPLE PREPARATION AND MICROSCOPY DATA

The samples were (100) oriented, 1-10 ohm-cm resistivity; p-type boron-doped silicon is used as it has been known to exhibit the finest pore structure of all doping [8]. They were laterally anodized [9] in an HF solution with a brief UV irradiation, resulting in luminescence in the visible. TEM images or etching with KOH [10] show that the samples are 1.75 µm thick. Specimens were prepared by cutting a vertical section and thinning by mechanical grinding. Final thinning to electron transparency (~100 A°) was done by Ar^+ ion milling in a liquid N_2-cooled stage in which the beam angle and energy were progressively reduced from 15° to 9° and 5 keV to 500 eV to minimize radiation damage [11]. This procedure reproduces the structure of samples prepared by Cullis et al [7] and is an alternative to almost damage-free cleavage [2, 11]. Some oxide contribution to the amorphous background will be present[12]. Imaging was performed using Hitachi H9000 microscope operated at 300 kV, with alignment along the <110> direction.

Figure 1 gives an XTEM image near the bottom of the sample. In the substrate, Si (111) atomic planes are resolved, with the correct spacing of 3.14 A°. We see pores whose width is ~1 nm. It is known that p-type silicon develops very fine pores [8] with propagation in the <100>, forming complicated networks [2,8]. Tracks generally run from the top to the substrate where they visibly terminate slightly inside, with some forming sharp points. Some crystalline islands are resolved easily and appear dark. Much of the unresolved structure may be due to thickness variation (roughness). Fig. 2 shows a close up of the resolved crystalline islands whose size, measured by counting the number of atomic planes within, ranges from 3 to 5 nm. The islands have a varying degree of connection. We took (110) electron diffraction at different spots. In the substrate, we see the standard two dimensional pattern of crystalline Si, composed of sharp crisp diffraction spots. The patterns, just inside the film, and at the top of this image (inset in Figure 1),

Figure 1 A high resolution XTEM image in the (110) zone showing crystalline islands. The insets are (110) diffraction pattern near the top of the image. It shows diffused rings with strong dots, representing the crystalline islands. The directions [001] and [110] are in the plane of the page, vertically up and horizontal respectively.

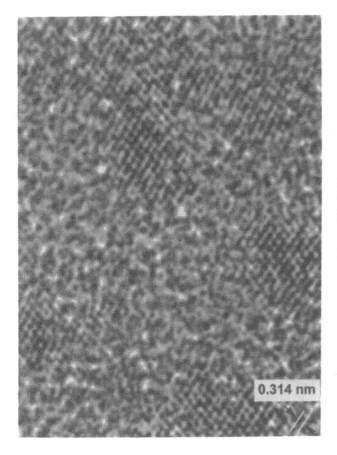

Figure 2 Close up of nano-crystallites from the region near the substrate interface. The directions [001] and [110] are in the plane of page, vertically up and horizontal respectively.

0.314 nm

show strong crystalline dots but also some weak diffused rings. Figure 3, taken from the middle region, shows crystallites, appearing dark. They are 2-4 nm across, and close ups show atomic planes. Electron diffraction, given in the inset, confirms presence of crystallinity, but shows stronger diffuse scatter than the lower region.

We now present the topmost layer (Figure 4). First, the density of pores is higher, and the the island size is the smallest. Second, most of the islands are crystalline, 6 to 10 atomic spacing across (2 to 3 nm). Third, pore tracks often get to within 3 to 4 lattice spacings, forming constrictions or bottle necks, effectively creating islands of ultra small widths that are 4 -5 nm apart. The diffraction pattern shows that the diffuse scattering is strong while the dots are weak. We now present in Fig. 5 the luminescent layer of a sample prepared with H_2O_2 added to the solution, showing reduced substructure (1-2 nm), and all diffuse scattering. The sample has much higher PL efficiency and very broad emission. One may be seeing elongated atomic or molecular sites 0.5 to 0.75 nm long. But there is a some underlying crystalline order. We should note that FTIR measurments and Auger Spectroscopy show that the prepared samples have no oxygen[12].

Figure 3 A region near the middle of the film showing many crystalline islands. The inset is (110) diffraction. The rings are strong while the dots are weak. The directions [001] and [110] are in the plane of the page, vertically up and horizontal.

DISCUSSION

We now address the origin of the diffuse scattering. Crystallinity is found throughout the material, however, there is diffuse scattering that grows with distance from the substrate. The fact that much of the crystallinity in the lower sections is not resolved, coupled with the presence of diffuse scattering, may lead one to assume an amorphous state, in view of the fact that a wide variety of experimental and theoretical work has suggested that amorphous state is more stable than the diamond phase in nanoclusters below ~ 3nm. On the other hand, since crystallinity of most of the nanostructures in the top layer is resolved, one can rule out an amorphous state, at least in this layer. Since intuitively one would expect more crystallinity closer to the substrate, we may conclude that the reduced crystallinity is due to a lack of resolution caused by thickness variations and local nonuniformity, and not due to an amorphous state. It is reasonable to find the upper region more uniform than the lower region since the former is expected to be softer, hence more amenable to ion milling. In fact it has been noted that amorphousity might be confused with purely size effects. For instance, although x-ray diffraction of an amorphous material is distinctly different from that of a crystalline material, there is no sharp division between them. For

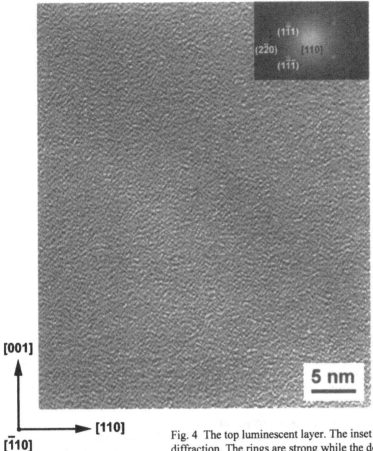

[001]

[110]

[1̄10]

[110]

5 nm

Fig. 4 The top luminescent layer. The inset is (110) diffraction. The rings are strong while the dots are weak.

crystalline powder of smaller and smaller particle size, the pattern lines broaden continuously, and for small enough particles it becomes similar to the amorphous pattern of a liquid or a glass.

We should note that the loss of crystallinity [2-6] has not only been seen in structure but also in optical excitation; it is seen in x-ray and electron diffraction, light scattering, polarization, and reflectance. In all, the loss correlates with size reduction, however, there is no information about its nature. In this regard, we mention a new quantum confinement model which provides structural as well as excitation changes. The model stipulates that, in ultra small nanostructures, restructuring of surface bonds into extended radiative Si-Si dimer bonds (> 0.5 nm) occurs [13-14]. The bonds are stabilized for sizes less than 1.75 nm. In the model the structure and emission of the bonds develop non-crystalline characteristics. We have recent measurement that shows the loss of crystallinity in reflectance is reversible by plating with an ultra thin layer of copper[15].

We acknowledge the US Department of Energy Grant DEFG02-ER45439 and NIH RRO3166.

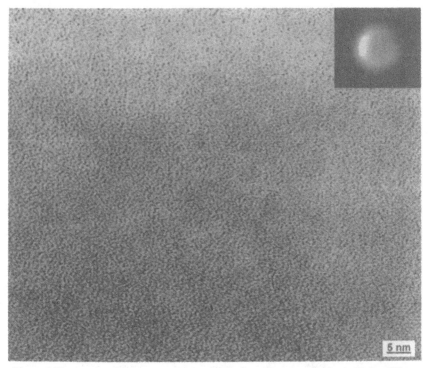

Figure 5 The topmost luminescent layer typical of samples prepared by in H_2O_2 : HF anodization solution that leaves no oxygen. The (110) diffraction shows scattering being completely diffuse with no dots. The directions [001] and [110] are as in previous figures.

REFERENCES

[1] L.T. Canham, Appl. Phys. Lett. **57**, 1046 (1990)

[2] A. G. Cullis, L. T. Canham, and P. Calcott, J. Appl. Phys. **82**, 909 (1997)

[3] R. Vasquez, R. W. Fathauer, T. George, and A. Ksendzov, Appl. Phys. Lett. **60**, 1004 (1992)

[4] J. Perez, J. Villalobos, P. McNeill, J. Prasad, R. Cheek, J. Kelber, J. P. Estrera, P. D. Stevens, and R. Glosser, Appl. Phys. Lett. **61**, 563 (1992)

[5] G. Ambrazevicius, G. Zaicevas, V. Jasutis, and D. Gulbinaite J. Appl. Phys. **76**, 5442 (1994)

[6] R. Fathauer, T. George, A. Ksendzov, and R. Vasquez, Appl. Phys. Lett. **60**, 995 (1992)

[7] A. Cullis, L. Canham, G. M. Williams, P. Smith, and O. Dosser, J. Appl. Phys. **75**, 493 (1994)

[8] R. L. Smith, and S. D. Collins, J. Appl. Phys. **71** (8), R1 (1992)

[9] D. Andsager, J. Hilliard, J. Hetrick, L. Abuhassan, M. Nayfeh, J. Appl. Phys. **74**, 4783 (1993)

[10] J. Hilliard, D. Andsager, L. Abuhassan, H. Nayfeh, M. Nayfeh, J. Appl. Phy. **76**, 2423 (1994)

[11] O. Gurdal, P. Desjardins, J. R. A. Carlsson, N. Taylor, H. H. Radamson, J. E. Sundgren, and J. E. Green, J. Appl. Phys. **83**, 162 (1998)

[12] W. Thompson, Z. Yamani, L. Abuhassan, and M. Nayfeh, Appl. Phys. Lett., **73**, 841 (1998)

[13] G. Allan, C. Delerue, and M. Lannoo, Phys. Rev. Lett. **76**, 2961 (1996)

[14] M. Nayfeh, N. Rigakis, and Z. Yamani, Phys. Rev. B **56**, 2079 (1997)

[15] Z. Yamani, J. Therrien, and M. Nayfeh. (to be published).

ELECTRO-POLYMERIZATION IN POROUS SILICON FILMS

R.K. SONI, L.F. FONSECA, O. RESTO, A. GUADALUPE*, AND S.Z. WEISZ
Department of Physics, University of Puerto Rico, Rio Piedras, San Juan-00931, PUERTO RICO
*Department of Chemistry, University of Puerto Rico, Rio Piedras, San Juan-00931, PUERTO RICO

ABSTRACT

Luminescent porous silicon films were created by electrochemical anodization of n-type substrate under light illumination. Semi-transparent conducting polypyrrole films were deposited by electrochemical polymerization at a low current density. The SEM micrographs showed that the polymer film impregnates into the wide vertical pores of 1-5 μm and grows sideways suggesting strong current distribution on the walls. The AFM images of polymer surface reveal nanometer size polymer aggregates on the porous layer. The impregnation of the polymer film due to sideways growth provides a useful mean to fabricate stable contact for light emitting diodes from porous silicon.

INTRODUCTION

Efficient visible photoluminescence (PL) from porous silicon has resulted in enormous interest in the optoelectronic properties of the material with an eye on silicon based light emitting devices. While stable and highly efficient PL, originating from radiative recombination of carriers in nanocrystalline silicon, is obtained routinely, the electroluminescence (EL) efficiency is rather low and unstable [1]. Early attempts with aqueous electrolytic contact showed high EL efficiency and an ideal rectifying characteristic but poor stability caused by surface oxidation of the top porous layer [2]. Efforts with all solid contacts, such as semitransparent gold, indium tin oxide, have not met with desired efficiency and stability due, partly, to local injection of carriers at the interface between electrode-porous layer [3-4]. For better electrical contacts and uniform carrier injection, it is desirable to completely fill the pores of highly disordered surface of the porous silicon layer with a transparent and conducting polymer. Conducting polymers, such as polypyrrole [5-7] or polyaniline [8], have shown promising results in the formation of rectifying junction with porous silicon film.

Polypyrrole (PPy) is one of the most stable conducting polymer known [9] and can be easily synthesized by electro-chemical oxidation. One of the key advantage of electro-chemical method is that the thickness of the polymer film is proportional to the time integral of the anodic current pulse, thus by controlling either the current or the time of oxidation, polymer film of desired thickness can be grown. Physical properties of PPy are, however, influenced by growth conditions such as monomer concentration, electrolyte concentration, solvent composition, and preparation temperature. Furthermore, during polymerization the PPy is simultaneously oxidized owing to the lower oxidation potential compared to that of monomer and this causes incorporation of counter-anion from the electrolyte during the growth process to maintain electrical neutrality. It is therefore expected that the electrical properties of PPy will be sensitive to the nature of the counter-anion present in the grown film.

We report here the growth of semitransparent conducting polypyrrole film on the n-porous silicon layers with varying surface morphology and formation of a rectifying junction between conducting pyrrole and porous silicon.

EXPERIMENT

Porous silicon layers were created by usual anodic oxidation in a solution of 25% HF in ethanol from n -type silicon wafers of (100) under light illumination of 100 mW/cm^2. The sample was etched with a solution of 20% HF in ethanol for 10 minutes before the anodization and rinsed with ethanol and deionised (DI) water after the completion of the anodization process. Drying was avoided to prevent cracking of the grown porous film. Each sample was checked for PL by exciting with the 442 nm line of a He-Cd laser. The film thickness and pore size was determined from cross-sectional SEM micrographs. Table-I lists the growth parameters of porous films.

Table-I: Growth parameters of porous silicon film

Sample	Substrate	Resistivity ρ (ohm-cm)	Current density J (mA/cm^2)	Anodization time (sec.)	Film thickness μm	Pore size μm
A	n-p junction	4.6	50	600	12	2-6
B	n-type	0.4-1.6	100	120	11	1-2

Polypyrrole film was deposited at low temperature (0 °C) immediately after the porous silicon film was created by applying a controlled current pulse to the cell containing solution of 0.5 M vacuum freshly distilled pyrrole monomer, solvent and 0.5 M HClO$_4$ as the supporting electrolyte. The solvent was either acetonitrile or DI water, we observed that acetonitrile quenches PL from the porous layer, therefore DI water was preferred as solvent as well as for final rinsing

RESULTS

On a bulk silicon substrate the PPy film grows linearly with time at low current density J = 2 mA/cm^2. The grown film appears black in color and inhomogeneous. The extent of oxidation in PPy increases its room temperature conductivity, which is comparable to that of silicon, but also enhances absorption in the visible region of spectrum. In order to study the growth of PPy on highly disordered porous layers, we synthesized porous layers with variety of surface morphology by varying anodization conditions as well as the substrate.

Figure 1 (a) shows cross-sectional SEM micrograph of porous silicon layer (sample A) prepared on an epitaxial layer. A large minority carrier injection in the 20 μm n-type epitaxial layer causes strong anodic reaction and results in a highly disordered comb like structure with 2-5 μm wide vertical pores. The polypyrrole film synthesized on a freshly prepared sample A at current density 2 mA/cm^2 in a solution containing monomer in acetonitrile with 0.5 M HClO$_4$.. The conducting polymer penetrates into the large-scale pore structure of the material and grows on the pore walls, Fig. 1 (b). With increasing pulse duration, the film grows side ways suggesting strong and even distribution of current on the pore walls. The SEM micrograph provides a direct evidence of the extent of PPy incorporation inside the pores. Earlier evidences of polymer incorporation were estimated either by polymer film thickness measurement [6,7] or by micro Raman [5] detection of polymer on porous layer. It appears that the pore filling is mainly through side way growth of polymer rather than from bottom to top [5-7]. When the current is high, voids are observed inside the pore, effective filling of the pores is possible when charge transfer process across the polymer/porous silicon interface is slow.

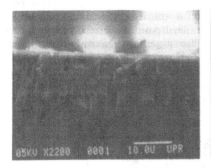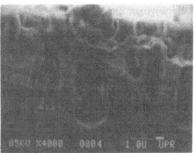

(a)   (b)

Fig. 1 Cross-sectional SEM micrograph of (a) porous silicon layer on an n-type
epitaxial film showing wide pores and (b) impregnation of conducting
polypyrrole in these pores by electro-polymerization at low current density.

The porous film emits strong orange-red PL before the polymerization with a peak at 670
nm, and after the polymerization there is an expected decrease in PL intensity of about 60%
accompanied by a small shift of peak towards lower wavelength, indicating a good transmission
for the visible light. The electrical transport property of the sample was investigated by
depositing semitransparent gold film on the polymerized porous silicon layer. In the forward bias
condition, gold film connected to the positive electrode, the device shows weak conductivity
with rectification ratio of 80 at 20 volts.

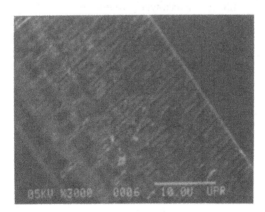

Fig. 2 Cross-sectional SEM micrograph of conducting polypyrrole
film on n-type porous silicon.

Sample B is prepared by anodic reaction with 25% non-ethonic HF solution The porous
layer, consists of a large density of ~1 μm wide vertical pores, emits strong orange-red PL when

excited with 442 nm laser line as well as strong red EL with 5% solution of NaCl. At low current density, the polymer film grows on the surface as well as with in the pore through sidewalls. With increasing current density or time, the polymer film deposits preferentially on the surface so much so that a large number of pores remain unfilled suggesting large charge carrier injection at the surface. Figure 2 shows SEM micrograph of 1 μm pyrrole film deposited on sample B at the current density J = 2 mA/cm^2 for 180 sec. An Atomic Force Microscope (AFM) is utilized to reveal morphology of polymer surface. Figure 3 shows an AFM image of sample B surface. The surface is flat and reveals nanometer size (50-70 nm) polymer aggregates on nanocrystalline silicon surrounding the pore.

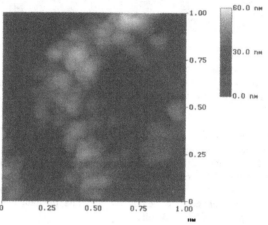

Fig. 3   Atomic force microscopy image of conducting polypyrrole film grown on n-type porous layer.

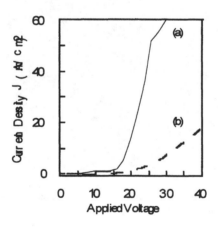

Fig.4   Current-voltage curve of gold/n-porous silicon (a) with and (b) without polymer film.

The current-voltage characteristic of gold/pyrrole/n-porous silicon diode exhibits, Fig. 4, a typical rectifying junction with a marked improvement in the conductivity compared to diode fabricated without pyrrole film. We believe that the large potential drop in the porous layer is a consequence of low carrier mobility across the highly structured interface between polymer aggregates and nanostructured silicon.

CONCLUSIONS

We have shown that it is possible to fill the pores in a porous silicon layer with conducting pyrrole by slow electro-polymerization at very low current density. The polymer grows preferentially on the pore walls, not from the bottom as generally believed, due to large charge injection from the walls. Though SEM micrograph provides a direct evidence of the extent of polymer incorporation inside the pores, further microscopic experiments are required to study the pore filling under various growth conditions. A thin (< 1 μm) and homogeneous layer of the polymer can be grown on the porous silicon surface under suitable growth conditions, AFM images reveal nanometer size polymer aggregates and micro holes at the surface of the polymer. The electrical transport in a diode structure is enhanced by impregnation of conducting polymer, the conductivity is inferior compared to liquid contacts due to highly disorder nature of the polymer surface and polymer/porous silicon interface.

ACKNOWLEDGEMENTS

Authors acknowledge partial support from NASA grant No. NCCW-0088, US ARO grant No. DAAHO4-96-1-0405 and DOE-EPSCoR 046138.

REFERENCES

1. A.G. Cullis, L.T. Canham, and P.D.J. Calcott, J. Appl. Phys. **82**, 909 (1997).
2. M.I.J. Beale, L.T. Canham, and T.J. Cox, in *Mat. Res. Soc. Symp. Proc.* **283**, (1993) p. 377.
3. N. Koshida and H. Koyama, Appl. Phys. Lett. **60**, 347 (1992).
4. F. Namavar, H.P. Muruska, and N.M. Kalkhoran, Appl. Phys. Lett., **60**, 2514 (1992).
5. J.D. Moreno, F. Agullo-Rueda, R. Guerrero-Lemus, R.J Martin Palma, J.M. Martinez-Duart, M.L. Marcos, and J. Gonzalez-Velasco, in *Mat. Res. Soc. Symp Proc.* **452**, (1997) p.479-484.
6. N. Koshida, H. Koyama, Y. Yamamoto, and G.J.Callins, Appl. Phys. Lett. **63**, 2655 (1993).
7. G. Wakefield, P.J. Dobson, Y.Y. Foo, A. Loni, A. Simons, and J.L. Hutchison, Semicond. Sci. Technol. **12**, 1304 (1997).
8. K. Li, D.C. Diaz, Y He, J.C. Campbell, and C Tsai, Appl. Phys. Lett. **64**, 2394 (1994).
9. B.R. Saunders, R.J. Fleming, and K.S. Murray, Chem. Mater, **7**, 1082 (1995).

Part IV

Synthesis and Spectroscopy
of Nanocrystalline Semiconductors

MICROSIZED STRUCTURES FABRICATED WITH NANOPARTICLES AS BUILDING BLOCKS

YONGCHI TIAN*, A. D. DINSMORE, S. B. QADRI, B. R. RATNA
Center for Bio/Molecular Science and Engineering, Naval Research Laboratory, 4555 Overlook Ave., Washington, DC 20375, *Geo Centers, yct@cbmse.nrl.navy.mil

ABSTRACT

Here we report a nanoparticulate route to Y_2O_3 nanofibers (~50 nm in diameter and a few micrometers in length) and for the radial growth of ZnS spheres (200-800 nm diameter). Well-defined higher order structures are developed upon thermostatically aging the dispersions of monomeric nanocrystals. The shapes of the "macromolecules" are correlated to primary monomeric nanocrystallites, the growing time and temperature, and surfactant templating agents. It is anticipated that this approach should inspire fabrication of nanoparticulate structures by using primary nanoparticles as monomers.

INTRODUCTION

Building up chemical entities of controlled shape at the nanometer scale is an important objective of current materials chemistry [1,2]. "Top-down" methods, such as laser ablation or lithography, for nanometer fabrication can reach a minimum size of a few hundred nanometers, below which the process becomes both labor intensive and costly. We suggest herein a "bottom-up" chemical approach which permits the preparation of nanocrystallites 1-10 nm in size as building blocks to construct "macromolecules" of specified shapes and sizes. Understanding the shape of the monomeric units, the primary nanocrystallites, and their physical parameters as a function of generation number, can lead to the development of synthetic route to the construction of submicron structure with complex defined architecture and functionality. Further, surfactant membranes can be used as soft templates to mould the nanoparticles into desired structures. In this paper, we report the growth of Y_2O_3 cubic nanocrystallites into fibers, and of ZnS zincblende nanoparticles into spherical beads. Nanoparticles polymerization is conceptualized by analyzing the growth at different stages.

EXPERIMENT

The primary Y_2O_3 particles were prepared by the hydrolysis of yttrium methoxyethoxide with tetraethylammonium hydroxide (TEAH) at room temperature [3]. Alternatively, an ethanolic dispersion of yttrium ethoxide was employed in the place of methoxyethoxide. Specifically, yttrium(III) acetate was refluxed in the respective alcoholic suspension for 4-8 hours to form an alkoxide subsequent to cooling down to room temperature. A calculated amount of TEAH was introduced in the suspension to complete the hydrolysis. The sol dispersion of the particles was then concentrated by rotary evaporation at 30 °C, yielding a stock solution with concentration of ~0.2 M (yttrium amount). The polymerization of the particles was conducted by one of the two ways: (1) incubating the stock solutions at room temperature for a few days or (2) templating the particles with aqueous bilayer dispersions of dioctadecyldimethyl ammonium bromide (DODAB) or sodium dodecylsulphate (SDS).

The growth of the ZnS nanoparticles was thermally initiated by releasing sulfur anions from thioacetamide (CH_3CSNH_2, TAA) in an aqueous $Zn(NO_3)_2$ system [4,5]. Polymerization followed the primary growth over a prolonged period of non-disturbed aging. The process was terminated by cooling the reaction system down to 0 °C and thoroughly washing away the residual sulfur precursor (unreacted TAA). The morphologies of the products were checked by transmission electron microscopy (TEM) at different stages of the reaction.

RESULTS

Synthesis and Polymerization of Y_2O_3 Nanoparticles

The room temperature sol-gel synthesis led to Y_2O_3 nanocrystallites with average size of 8±2 nm (see Figure 1, left). Observed X-ray and electron diffraction peaks could be indexed to (211), (222), (400), (411), (322) and (431) of a cubic structure with a unit cell size of 596.8 Å³. The ethanolic dispersions of the particles were stable for days (checked with TEM). Gelation occurred after a week of incubation, producing a jelly-like block with slightly yellowish color. Resuspension of the gel in ethanol resulted in a well separated particulate sol, indicating the reversibility of the sol-gel conversion. Polymerization of the nanoparticles was conducted in two steps: (1) forced concentration by driving solvent away and (2) long time incubation. The first step increased the particle density by 5 times while the second step gave the particles opportunity to link via reactive facets.

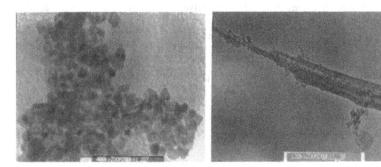

Figure 1. (Left) Morphology of the primary Y_2O_3 nanoparticles freshly prepared in ethanol dispersion before aging. (Right) The fiber block obtained by aging the Y_2O_3 nanoparticles dispersion for two weeks at room temperature. The scale bars represent 25 nm and 83 nm, respectively.

Fiber-like structures were observed in the incubated Y_2O_3 gel two weeks after it was freshly suspended (see Figure 1). A typical single fiber has diameter of 68-100 nm and length of ~2 μm. The electron diffraction of a single fibers could be indexed to the same cubic structure as the nanoparticles. However, the powder diffraction rings, observed for the nanoparticles (See Figure 1 left), becomes spotty for the fibers, suggesting the orientation of the particles to some extent. This result demonstrated that the primary nanoparticles link to form larger aggregates with preferred alignment.

Templated growth was accomplished by dispersing the primary Y_2O_3 nanoparticles in an aqueous bilayer solution of dioctadecyl dimethylammonium bromide (DODAB, 5 x 10^{-3} M) or sodium dodecylsulfate (SDS, 10^{-2} M). TEM [Figure 2 (left)] shows a thin patch of dried foam bubbles flanked by the pools of concentrated Y_2O_3 nanoparticles. These structures were possibly stabilized by encapsulation of the nanoparticles inside the inverted bilayer formed by the adjacent surfactant monolayers within the bubbles, as shown schematically in Scheme 1. This hypothesis is supported by the continuity of the membrane network (dark region of the texture) with the adjacent nanoparticle pools. The presence of Y_2O_3 in the network is confirmed by electron diffraction. As the membrane dries, pulling the surfactant monolayers together, the nanoparticles concentration in this region increases and the particles are forced to align along the bilayer walls.

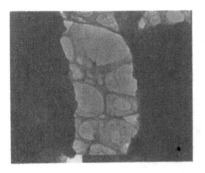

Figure 2. (Left) Y_2O_3 gel confined by DODAB bilayer membranes. The mixed gel of DODAB and Y_2O_3 particles was spread on a carbon coated copper grid followed by drying in vacuum for 30 minutes. The foam bubbles are clearly seen as brighter region in the middle. (Right) Dendritic structures of Y_2O_3 nanoparticles formed in the bilayer confinement. This was obtained after drying the gel sample in vacuum for 48 hours. The scale bars represent 500 nm and 166 nm, respectively.

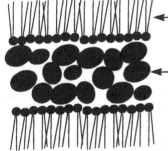 ← **surfactant monolayer**

← **aqueous nanoparticles core**

Scheme 1. Nanoparticles enclosed by two monolayers of surfactants in the bubble membrane. This represents a proposed structure existing in the bubble membranes shown in Figure 2 (Left).

A TEM photograph of a more concentrated surfactant stabilized sample of Y_2O_3, dried in a vacuum desiccator for 48 hours, is shown in Figure 2 (right). This sample shows a fibrous "bird's-nest" morphology. The diameter of the fibers is about 40 nm and uniform. The micrograph clearly shows that each fiber is made of loosely connected Y_2O_3 particles of ~8 nm diameter. This is compatible with a model of membrane-forced organization with poor contact edges between adjacent nanoparticles, rather than through matched crystal facets linked by chemical bonding. The adsorption of the surfactant used in the template on the surface of Y_2O_3 nanoparticles may be responsible for the space between particles in the fiber, preventing the nanoparticles from coming close enough together to form a chemical bond. It is interesting to note that no sheet-like structures were found in the dried sample. A detailed time evolution study of the drying mechanism may provide a clue for the formation of fibers as opposed to sheets.

Polymerization of ZnS Nanoparticles

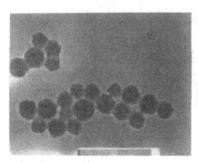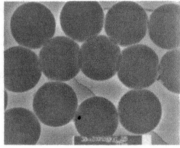

Figure 3. (Left) Radial polymerization of ZnS nanoparticles at early stage. The sample was taken from the reactor a few minutes after initiating the reaction and dropped on a TEM grid. The reaction system contained 2×10^{-2} M $Zn(NO_3)_2$ and 0.1 M TAA bathed at 67 °C for 30 minutes [5]. Radial textures and rough surfaces are evident on each circular feature. (Right) Well-grown spheres developed from ZnS nanoparticles are observed after they grew into matured stage. This sample was taken ~20 minutes after initiating the reaction and aging the system at 70 °C. The reaction system contains 3×10^{-2} M $Zn(NO_3)_2$ and 0.4 M TAA. The scale bars represent 142 nm and 333 nm, respectively.

ZnS, crystallized in cubic lattice (zincblende), is known to possess tetrahedral coordination Using ZnS nanoparticles as the primary building block of the polymerization, we examined the process of forming higher order structures. The nanoparticles were first formed by the controlled introduction of sulfur anion in aqueous $Zn(NO_3)_2$ solution [5]. Purified thioacetamide (TAA) was used as sulfur precursor in the concentration of 0.4 M. Initial reaction of sulfur anions with Zn ions at room temperature led to the ZnS nanoparticles of 3-5 nm in size (checked by TEM). Heating up the reaction system up to 70 °C accelerated the release of the sulfur anion from TAA and, in turn, produced ZnS primary seeds in high density.

Figure 3 shows the growth of ZnS assemblies at different stages. In earlier stage, dendritic features are evident with radial textures extending from the center of the spheres. Electron diffraction pattern shows a typical polycrystalline feature with a set of reflection 3.13, 1.91 and 1.63 X, ascribed to (111), (220) and (311) of zincblende (cubic) lattice structure. Furthermore, spot features can be seen within the submicron particles, indicating that ZnS nanoparticles are the building blocks. Well-grown ZnS spheres at a later stage, shown in Figure 3 (right), exhibit monodisperse size with highly uniform morphology and relatively low surface roughness. Furthermore, the polycrystalline feature was observed in the electron diffraction for micron spheres at both early and later stages. These results lead us to speculate that the growth was initiated on a seed with faces radially directed to all sides and developed by ZnS nanoparticles as building block.

CONCLUSIONS

We have demonstrated that nanoparticles can be used to build up fibrous (Y_2O_3) and spherical (ZnS) micron-sized structures. The formation of organized nanoparticulate assemblies with specified shapes on the microscopic scale have illustrated how nanoparticles might be viewed as active monomers for polymerization and how organic self-organized templates and nanoparticles might be combined for the fabrication of nanoparticulate materials with specific morphology.

ACKNOWLEDGEMENTS

We would like to thank Kristin Buckstad of University of Virginia for help in synthesis, and Dr. R. Price of this Center for help in TEM measurements. This work was financially supported by ONR.

REFERENCES

1. S. A. Davis, S. L. Burkett, Neil H. Mendelson and S. Mann, Nature **385**, p420 (1997).
2. V. Percec, W.-D. Cho, P. E. Moisier, G. Ungar and D. J. P. Yeardley, J. Am. Chem. Soc. **120**, p11061 (1998).
3. R. Rao, J. Electrochem. Soc. **143**, p189 (1996).
4. Y.-D. Kim, T. Yoneyma, S. Nagashima and A. Kato, Nippon Kagaku Kaishi, **9**, p707 (1995).
5. E. Matijevic, Chem. Mater. **5**, p412, (1993).

ELECTRON AND HOLE RELAXATION PATHWAYS IN II-VI SEMICONDUCTOR NANOCRYSTALS

V. I. KLIMOV, Ch. SCHWARZ, X. YANG, and D. W. McBRANCH
Chemical Sciences and Technology Division, CST-6, MS-J585, Los
Alamos National Laboratory, Los Alamos, NM 87545, klimov@lanl.gov

C. A. LEATHERDALE and M. G. BAWENDI
Massachusetts Institute of Technology, 77 Massachusetts Avenue, Cambridge, MA 02139

ABSTRACT

Femtosecond (fs) broad-band transient absorption (TA) is used to study the intra-band relaxation and depopulation dynamics of electron and hole quantized states in CdSe nanocrystals (NC's) with a range of surface properties. Instead of the drastic reduction in the energy relaxation rate expected due to a "phonon bottleneck", we observe a fast sub-picosecond 1P-to-1S relaxation, with the rate enhanced in NC's of smaller radius. We use fs IR TA to probe electron and hole intra-band transitions, which allows us to distinguish between electron and hole pathways leading to the depopulation of NC quantized states. In contrast to electron relaxation, which is controlled by NC surface passivation, depopulation of hole quantized states is extremely fast (sub-ps–to–ps time scales) in all types samples, independent of NC surface treatment (including NC's overcoated with a ZnS layer). Our results indicate that ultrafast hole dynamics are not due to trapping at surface defects, but rather arise from relaxation into intrinsic NC states.

INTRODUCTION

Three-dimensional quantum confinement results in discrete size-dependent energy spectra in semiconductor nanocrystals (NC's) [1, 2]. Structures of inter-band transitions in II-VI NC's have been extensively studied using linear absorption [3], photoluminescence (PL) excitation [4], and transient absorption (TA) [5–8]. Time-resolved TA studies have also been applied to probe energy relaxation and recombination dynamics in NC's [6-8]. In bulk II-VI semiconductors, carrier energy relaxation is dominated by the Fröhlich interaction with longitudinal optical (LO) phonons leading to fast (typically sub-ps) carrier cooling dynamics. In NC's, even in the regime of weak confinement when the level spacing is only a few meV, the carrier relaxation mediated by interactions with phonons is hindered dramatically, because of restrictions imposed by energy and momentum conservation ("phonon bottleneck") [9, 10]. Further reduction in the energy loss-rate is expected in the regime of strong confinement, for which the level spacing can be much greater than LO phonon energies, and hence carrier-phonon scattering can only occur via weak multi-phonon processes [11]. In contrast to predictions of "phonon bottleneck" theories, recent studies [6, 7] indicate extremely fast (sub-ps) energy relaxation in NC's, which is likely due to the opening of new non-phonon relaxation channels, bypassing the "phonon bottleneck".

Energy relaxation leads to establishing quasi-equilibrium populations of electron and hole quantized states. Depopulation of these states can occur via a variety of radiative and non-radiative mechanisms. The radiative electron-hole (e-h) recombination in CdSe NC's is a relatively slow process with ns time constants [12]. The competing non-radiative mechanisms are associated with trapping at defect/surface states [13] and Auger recombination [14], with the latter dominating carrier dynamics in the regime of multiple e-h pair excitations. Carrier dynamics in NC's have been usually studied using fs TA experiments [6, 7]. Since both electrons and holes contribute to inter-band TA signals, inter-band TA spectroscopy does not allow a reliable separation of electron and hole dynamics, which is essential for understanding the mechanisms for carrier trapping and the origin of trapping sites. Additionally, most ultrafast studies of II-VI quantum dots have concentrated on NC/glass samples which have poorly-controlled surface properties. Therefore, it remains unclear whether the fast initial depopulation of quantized states in NC's is entirely due to trapping at surface defects or due to relaxation into intrinsic quantum dot states.

In this paper we present a detailed study of energy relaxation and trapping dynamics in CdSe colloidal NC's with radii (R) from ~4 to ~1 nm, and with a wide range of surface properties.

For all sizes, 1P-to-1S electron relaxation is extremely fast and occurs on the sub-ps time scale. We observe an enhancement in the electron relaxation rate for NC's of smaller sizes, indicating that energy relaxation is dominated by non-phonon energy-loss mechanisms. Comparison of visible and IR TA dynamics enables us to distinguish between electron and hole relaxation pathways leading to the depopulation of NC quantized states. We observe a strong difference in electron and hole relaxation dynamics which allows us to separate conduction- and valence-band contributions to the IR TA. Our data indicate that in contrast to electron dynamics, controlled by NC surface passivation, hole dynamics are extremely fast in all samples with different surface properties, suggesting that they are not due to trapping at surface defects but rather are due to relaxation into intrinsic NC states.

INTRA-BAND ENERGY RELAXATION

To probe carrier dynamics we monitor carrier-induced absorption changes using a fs pump-probe experiment. The samples are pumped at 3.1 eV by frequency doubled 100-fs pulses from an amplified Ti-sapphire laser. In the visible-TA measurements, the transmission of the photoexcited spot is probed by variably-delayed pulses of a fs white-light continuum. TA spectra are detected in the chirp-free configuration [15] with an accuracy up to 10^{-5} in differential transmission. In the near-IR TA measurements, the probe pulses are derived from an IR optical parametric amplifier, tunable in the range 1.1–2.7 μm.

We studied colloidal CdSe NC's synthesized by the organometallic route [16]. NC samples were prepared in two forms: "standard" NC's passivated with large organic molecules [16] and "overcoated" NC's containing a final layer of ZnS [17]. The NC mean radii (R) were from ~1 to ~4 nm, and the NC size dispersion was 4–9%. All measurements were performed at room temperature.

The nonlinear optical response in NC's is dominated by state-filling [6, 18] leading to pronounced bleaching bands at energies of the allowed optical transitions. Due to degeneracy of the valence band and a large difference between electron and hole masses in CdSe ($m_h/m_e \approx$ 6), the bleaching bands at room temperature are dominated by electron populations [19]. TA signals are also affected by Coulomb effects due to two e-h-pair interactions (biexciton effect) [20, 21] and the trapped-carrier-induced dc-Stark effect [22]. While state-filling affects only transitions coupling populated states, the Coulomb interaction influences all transitions in NC's, resulting in transition shifts which are seen as derivative-like features in TA spectra [20-22]. To extract carrier dynamics from TA data, the state-filling-induced portion of the TA must be accurately separated from contributions due to Coulomb effects.

In our study, we concentrate on relaxation between adjacent 1P and 1S electron states. In Fig. 1, we show spectra of pump-induced absorption changes ($\Delta\alpha$) for three "standard" colloidal samples with NC radii 2.23, 2.77, and 4.05 nm detected in the chirp-free mode at $\Delta t = 200$ fs after excitation. To exclude effects of many-particle interactions on carrier dynamics, these data were taken at low pump intensities corresponding to $N_{eh} < 1$ (N_{eh} is the average number of e-h pairs excited per NC). TA spectra in

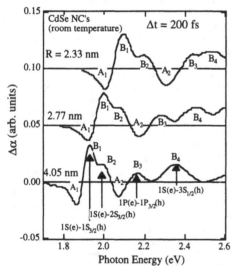

FIG. 1. TA spectra of CdSe NC's of different radii detected at $\Delta t = 200$ fs.

Fig. 1 show several bleaching features (labeled B_1 through B_4) which mark the positions of allowed optical transitions. Using the data from Ref. 4, features B_1, B_2, and B_3 can be assigned to transitions $1S(e)-1S_{3/2}(h)$ (1S transition), $1S(e)-2S_{3/2}(h)$ (2S transition), and $1P(e)-1P_{3/2}$ (h) (1P transition), respectively (for notation of electron and hole states in spherical quantum dots see Ref. 3). The B_4 bleaching can be attributed to transition $1S(e)-3S_{1/2}(h)$ in sample with R = 4.05 nm, and to transition $1S(e)-2S_{1/2}(h)$ in 2.23- and 2.77-nm samples. In addition to state filling, TA is also affected by Coulomb interactions [20-22], leading to transition shifts with associated photoinduced absorption (PA) features ($\Delta\alpha > 0$) A_1 and A_2.

In Fig. 2(a), we display B_1 and B_3 dynamics recorded for the 4.05-nm sample (1P–1S energy separation about 9 LO phonon energies). The decay of B_3 (crosses) is extremely fast (540-fs time constant) which we attribute to the depopulation of the 1P state. The 1S time transient [solid circles in Fig. 2(a)] shows a fast initial rise, followed by a step at ~300 fs, and a slower signal increase (time constant of 530 fs) which is complementary to the B_3 decay. The two different contributions to B_1 can be explained in terms of Coulomb two-pair interactions and state filling [21]. The step in the B1 time transient is a signature of different time offsets for these contributions. The Coulomb interaction leads to an instantaneous red shift of the 1S transition following photoexcitation. This results in a positive TA signal at A_1 and the initial fast onset of B_1, with delayed dynamics almost identical to those of B_3 [Fig. 2(a)]. The delayed growth of B_1 is due to increasing population of the 1S state as a result of carrier relaxation from the 1P state, consistent with the complementary decay of B_3 due to the depopulation of the 1P state.

The measured relaxation constants indicate a very high energy-loss rate of ~0.5 eV ps^{-1}, which is of the same order of magnitude as the estimated rate for the unscreened polar interaction in bulk CdSe with a continuous energy spectrum, and many orders of magnitude higher than the relaxation rate expected for multi-phonon processes [11]. Recent works have suggested that coupling to defects [23] and/or Auger-type interactions [24, 25] can lead to fast energy relaxation not limited by phonon bottleneck. The first of these mechanisms suggests a sequential relaxation involving an electron transition to the defect, defect relaxation, and then transition back to the lower NC level. This scenario is obviously not consistent with measured complementary dynamics of the 1P and 1S states, indicating a direct 1P-to-1S electron transition. Another non-phonon relaxation mechanism [24] involves the Auger-type energy transfer from the electron to high density e-h plasmas (2D plasmas from the adjacent quantum well in Ref. 24). However, this effect cannot be operative under our experimental conditions of low excitation densities (less than 1 e-h pair per NC) and with NC's dispersed in an insulating, optically transparent host. Most likely, the observed fast dynamics can be explained in terms of the Auger mechanism proposed in Ref. 25, which involves *confinement-enhanced energy transfer of the electron excess energy to a hole*, with subsequent fast hole relaxation through its dense spectrum of states.

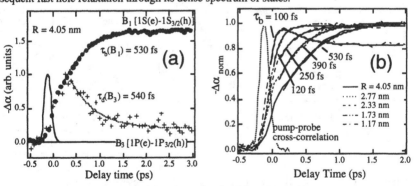

FIG. 2. (a) 1S- and 1P-transition dynamics for CdSe NC's with R = 4.05 nm; (b) 1S-transition build-up dynamics for NC's of different radii.

The important role of confinement in the enhancement of the energy relaxation is evident from a comparison of the 1S state population dynamics in NC's of different radii [see Fig. 2(b)], indicating *a decrease in the 1S build-up time with decreasing NC radius*. This time shortens from 530 fs for R = 4.05 nm to 100 fs for R = 1.17 nm, roughly following a linear dependence.

ELECTRON AND HOLE TRAPPING DYNAMICS

The state-filling-induced bleaching of inter-band optical transitions is proportional to the sum of the occupation numbers of the electron and hole states involved in these transitions. However, due to a high density of valence-band states, room-temperature bleaching signals are dominated by electron populations. This is clearly indicated by fact that for all NC sizes, the 1S and 2S bleaching bands [which involve the same electron (1S) but different hole states] show essentially identical dynamics at $\Delta t > 2$–3 ps, dominated by depopulation of the common 1S electron level. Therefore, these dynamics can be used to monitor electron relaxation behavior. To study hole relaxation dynamics, we use fs IR probe, which allows the spectral separation of electron and hole intra-band signals.

In Fig. 3, we show visible-to-near-IR TA spectra (symbols) for CdSe colloidal NC's with R = 1.73 nm at different pump-probe delay times. Initial electron energy relaxation in this sample occurs within <400 fs, therefore, at $\Delta t \geq 1$ ps the electrons have already reached the lowest 1S state. Population of this state leads to state-filling-induced bleaching of the 1S (2.26 eV) and 2S (2.4 eV) transitions (the visible portion of TA). In the near-IR spectral range, TA is positive indicating *photoinduced absorption due to intra-band transitions*. These transitions couple the lowest electron (hole) state of S symmetry to the higher lying states of P symmetry. For the 1.73-nm sample shown in Fig. 1, in the IR range studied, we can only expect a contribution from the 1S–1P electron transition which has an energy of ~0.45 eV [8]. Assuming a ~0.1 eV transition broadening (estimated for a ~5% size dispersion), the 1S–1P electron absorption does not contribute significantly to the TA signal measured in the 0.55–0.7-eV range, meaning that it is almost entirely due to *intra-valence-band transitions*. Consequently, the dynamics of this signal are indicative of hole relaxation behavior.

In the inset to Fig. 3 and in Fig. 4 (a, b) (circles) we compare time transients recorded for the 1.73-nm sample at visible (1S bleaching) and near-IR (0.69 eV) spectral energies. In contrast to the relatively slow initial sub-100-ps relaxation of the 1S bleaching (electron dynamics), initial relaxation of the near-IR TA (hole dynamics) is extremely fast and occurs on the

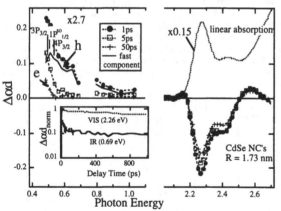

FIG. 3. Visible-near-IR TA spectra of CdSe NC's with R = 1.73 nm.

sub-ps–ps time scale. A pronounced difference in electron and hole relaxation rates allows us to separate electron and hole contributions to IR TA. Since the "hole" signal decays on the ps time scale, the 50-ps IR spectrum in Fig. 3 is characteristic of the electron 1S–1P absorption. To derive the fast hole-related portion of TA (thick solid line in Fig. 3), we subtract the spectrum detected at 5 ps from the 1-ps spectrum. Tentatively, the spectral maximum of the "hole" spectrum (~0.52 eV) can be assigned to three close transitions, involving $1S_{3/2}$ and $3P_{3/2}$, $1P^{so}_{1/2}$, and $4P_{3/2}$ hole states (vertical bars in Fig. 3).

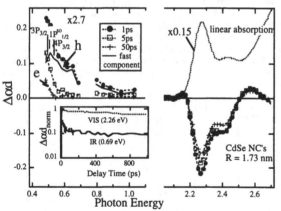

Analysis of the 1S bleaching decay (electron relaxation) shows that it is very sensitive to NC surface properties. In Fig. 4(a) we compare 1S bleaching dynamics for three samples with the same NC radius (1.73 nm), but different degrees of surface passivation. These samples are freshly prepared "standard" NC's passivated with molecules of tri-octylphosphine oxide (TOPO) [16] (circles), the same NC's but eight months after preparation (squares), and freshly prepared NC's "overcoated" with four monolayers of ZnS [17] (crosses). For all samples, 1S bleaching shows a two-component decay: initial sub-100-ps relaxation, followed by slow ns decay. In freshly prepared "standard" NC's, the fast component is about ~15% of the signal amplitude. The fast component is enhanced up to ~50% in the aged sample, but gets reduced (down to ~7%) in the "overcoated" NC's. These comparisons clearly indicate that fast initial electron decay is due to trapping at surface defects. The number of these defects grows with the loss of passivation in aged colloids, resulting in an increased availability of sites for electron relaxation. On the other hand, the number of surface defects is significantly reduced in NC's with epitaxially grown shell of a wide-gap semiconductor, leading to the suppression of the fast component in the electron relaxation. Since in "standard" NC's passivating molecules are coordinated to surface metal ions [16], the electron traps are most likely associated with metal dangling bonds (chalcogenide vacancies) [26].

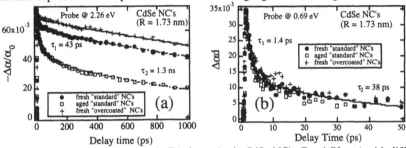

FIG. 4. Visible (a) and IR (b) TA dynamics in CdSe NC's (R = 1.73 nm) with different degrees of surface passivation.

In contrast to the initial electron dynamics, controlled by surface passivation, the hole dynamics are practically unaffected by NC surface properties, as clearly seen from comparison of IR traces [Fig. 4(b)] for the samples described in the previous paragraph. All three samples show nearly identical fast decay with a 1.4-ps time constant, indicating very fast depopulation of hole quantized states, independent of the surface passivation. The fact that hole dynamics are extremely fast in all types of samples, including "overcoated" NC's, indicates that the *sub-ps–ps hole dynamics are not due to trapping at surface defects, but rather are due to relaxation into intrinsic NC states which are likely of surface origin (intrinsic surface states)* [5].

The relaxation data shown above also are relevant to such important technological issue as realization of room-temperature optical gain and lasing in NC media. To explore the possibility of achieving room-temperature gain, we studied the TA pump dependence for "standard" and "overcoated" colloidal CdSe NC's with PL quantum yield up to ~30%. In

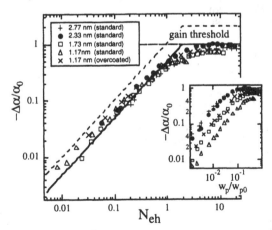

FIG. 5. Pump-intensity dependence of 1S bleaching in NC's of different radii.

215

the inset to Fig. 5 we show normalized 1S-absorption change ($-\Delta\alpha/\alpha_0$, α_0 is the linear absorption) at $\Delta t = 2$ ps vs pump fluence w_p. The gain threshold corresponds to $-\Delta\alpha/\alpha_0 = 1$. *None of the samples shows crossover from absorption to gain even at very high pump densities corresponding to $N_{eh} > 10$.* Interestingly, in the plot of $-\Delta\alpha/\alpha_0$ vs N_{eh} (main frame of Fig. 5), all experimental data points fall along one "universal" curve which shows a linear growth below N_{eh} =1, and saturation at the level $-\Delta\alpha/\alpha_0 \approx 1$ above N_{eh} =1. These data are compared with the dependence expected for state filling in a system for which the lowest optical transition couples a populated electron state to either a populated (dashed line) or unpopulated (solid line) hole state. This comparison strongly suggests that already at 2 ps after excitation all holes are removed from quantized states, consistent with our IR data.

CONCLUSIONS

We have performed fs studies of intra-band energy relaxation in CdSe NC's with radii from ~1 to ~4 nm. We observe a fast sub-ps 1P-to-1S electron relaxation, enhanced in NC's of smaller radius, which directly contradicts predictions for multi-phonon emission and suggests the opening of new confinement-enhanced relaxation channels which likely involve Auger-type electron-hole energy transfer. To distinguish between depopulation dynamics of electron and hole quantized states, we use fs IR TA to probe electron and hole intra-band transitions. We observe extremely fast relaxation of hole intra-band signals, indicating depopulation of hole quantized states on sub-ps-to-ps time scales in all types of samples, independent of NC surface properties. This strongly suggests that hole trapping sites are not defect-related but rather are intrinsic to quantum dots. On the other hand, initial fast electron relaxation is extremely sensitive to the degree of surface passivation, indicating that it is due to trapping at surface defects.

This research was supported by Los Alamos Directed Research and Development funds, under the auspices of the US Department of Energy.

REFERENCES

1. Al. L. Efros and A. Efros, Sov. Phys. Sem. **16**, 772 (1982).
2. L. Brus, Appl. Phys. A **53**, 465 (1991).
3. A. I. Ekimov *et al.*, J. Opt. Soc. Am. B **10**, 100 (1993).
4. D. Norris and M. Bawendi, Phys. Rev. B **53**, 16338 (1996).
5. M. Bawendi *et al.*, Phys. Rev. Lett. **65**, 1623 (1990).
6. U. Woggon *et al.*, Phys. Rev. B **54**, 17681 (1996).
7. V. Klimov and D. McBranch, Phys. Rev. Lett. **80**, 4028 (1998).
8. P. Guyot-Sionnest and M. Hines, Appl. Phys. Lett. **72**, 686 (1998).
9. U. Bockelman and G. Bastard, Phys. Rev. B **42**, 8947 (1990).
10. H. Benisty *et al.*, Phys. Rev. B **44**, 10945 (1991).
11. T. Inoshita and H. Sakaki, Phys. Rev. B **46**, 7260 (1992).
12. Al. L. Efros, Phys. Rev. B **46**, 7448 (1992).
13. V. Klimov, P.H. Bolivar, and H. Kurz, Phys. Rev. B **53**, 1463 (1996)
14. F. de Rougemont et al., Appl. Phys. Lett. **50**, 1619 (1987).
15. V. Klimov and D. McBranch, Opt. Lett. **23**, 277 (1998).
16. C. Murray, D. Norris, and M. Bawendi, J. Am. Chem. Soc. **115**, 8706 (1993).
17. M. A. Hines and P. Guyot-Sionnest, J. Phys. Chem. **100**, 468 (1996).
18. V. Klimov, S. Hunsche, and H. Kurz, Phys. Status Solidi B **188**, 259 (1995).
19. S. Hunsche *et al.*, Appl. Phys. B **62**, 3 (1996).
20. K. I. Kang *et al.*, Phys. Rev. B **48**, 15449 (1993).
21. V. Klimov, S. Hunsche, and H. Kurz, Phys. Rev. B **50**, 8110 (1994).
22. D. J. Norris *et al.*, Phys. Rev. Lett. **72**, 2612 (1994).
23. P. C. Sercel, Phys. Rev. B **51**, 14532 (1995).
24. U. Bockelman and T. Egler, Phys. Rev. B **46**, 15574 (1992).
25. Al. L. Efros, V. A. Kharchenko, and M. Rosen, Solid State Commun. **93**, 281 (1995).
26. N. Chestnoy, T. Harris, R. Hull, and L. Brus, J. Phys. Chem. **90**, 3393 (1986).

STRONG PHOTOLUMINESCENCE IN THE NEAR-INFRARED FROM COLLOIDALLY-PREPARED HgTe NANOCRYSTALS

M.T. HARRISON*, S.V. KERSHAW*, M.G. BURT*
A.L. ROGACH**†, A. KORNOWSKI**, A. EYCHMÜLLER**, H. WELLER**
* BT Laboratories, Martlesham Heath, Ipswich, Suffolk, IP5 3RE, UK
**Institut für Physikalische Chemie, Universität Hamburg, 20146 Hamburg, Germany.
†permanent address: Physico-Chemical Research Institute, Belarussian State University, 220050 Minsk, Belarus

ABSTRACT

We report here the first measurement of strong near-infrared room temperature photoluminescence (PL) from colloidally-prepared HgTe nanocrystals. X-ray diffraction (XRD) and high resolution transmission electron microscopy (HRTEM) measurements indicate that the nanoparticles are in the cubic coloradoite phase, with a diameter of approximately 4 nm. The absorption spectrum shows a pronounced electronic transition in the near-infrared, and the broad PL appears to consist of several overlapping features between 800 and 1400 nm with a peak at 1080 nm, which represent a dramatic shift from bulk HgTe behaviour. The quantum efficiency (QE) of the freshly prepared sample is around 50%, which is among the highest ever reported for a nanocrystalline material. Over a period of several days, the luminescence shifts further into the infrared yielding more dominant longer wavelength features. The observation of this strong infrared luminescence makes this material a promising candidate for application in optical telecommunication systems.

INTRODUCTION

Currently, there is considerable interest in nanometer-sized semiconductor particles, which exhibit unique physical and chemical properties due to the quantum confinement effect and their large surface area-to-volume ratio [1-6]. Recent developments in the physics and chemistry of these materials have broadened the research to real application of nanocrystals in electrical and optoelectronic devices [7-11]. There are now literally dozens of examples of high quality III-V and II-VI nanocrystals prepared either by epitaxial or wet chemical synthetic routes. For the latter, work has tended to concentrate on achieving strong visible luminescence for application in light-emitting diodes and lasers [9-11]. However, the telecommunications industry is far more interested in the near infrared wavelengths of around 1.3 and 1.55 microns, which are the two windows used in optical fibre systems. A material with strong photoluminescence at these wavelengths could be the basis of a broad-band optical amplifier, and by having control of the nanocrystal size distribution, it should also be possible to completely tailor the optical properties of the system. It was decided to attempt a synthesis of HgTe nanocrystals since in the bulk, this material is a semi-metal with an inverted band structure [12], and hence an effectively zero band-gap. The blue shift caused by the quantum confinement effect should therefore yield luminescence in the near infrared.

EXPERIMENT

The preparation of aqueous solutions of colloidal HgTe nanocrystals was simply an extension of the previously published syntheses of cadmium-based systems [13-16]. The stabiliser used was 1-thioglycerol, which has been found previously to be an effective size-regulating capping agent for II-VI nanocrystals [15,16]. A solution of 0.94 g (2.35 mmol) Hg(ClO$_4$)$_2$ and 0.5 mL (5.77 mmol) 1-thioglycerol in 125 mL de-ionized water was placed in a two-necked flask, and was adjusted to pH = 11.2 using 1 M NaOH. The flask was fitted with a septum and valves, and the solution de-oxygenated with nitrogen bubbling for about 30 min. H$_2$Te gas, which was generated by the reaction of 10 mL 0.5 M H$_2$SO$_4$ with 0.08 g Al$_2$Te$_3$, buffered in a slow nitrogen flow was then passed through the stirred solution. Unlike the CdS and CdTe preparations, we did not then apply any heat treatment – refluxing the resulting brown solution appears to quench the photoluminescence rather than grow and stabilise the nanocrystals. A "purified" dry powder was isolated by first concentrating the solution down to about 30 mL on a rotary evaporator and then adding excess propan-2-ol under vigorous stirring to precipitate out the nanocrystals.

The X-ray powder diffraction pattern was recorded on a Philips X'Pert Diffractometer (Cu K$_\alpha$-radiation, variable entrance slit, Bragg-Brentano geometry, secondary monochromator) using finely dispersed HgTe nanocrystals on standard PVC supports. HRTEM images of the nanocrystals were acquired on a Philips transmission electron microscope operating at 300 kV. The TEM samples were prepared by dropping diluted aqueous solutions of the nanocrystals onto 400-mesh carbon-coated copper grids with the excess solvent immediately evaporated. The optical measurements were all obtained using dilute aqueous samples, with the UV-Vis absorption spectra recorded with a Perkin-Elmer Lambda 9 spectrometer. A mechanically chopped Argon-Ion laser running on all-lines (predominantly 488 nm and 514.5 nm) was used as the excitation source for the PL spectra, and an InGaAs photodiode was used as the detector. The PL spectrometer arrangement was calibrated using a dilute sample of Rhodamine 6G organic dye for which the quantum efficiency is know to be ~95%.

RESULTS

By using a combination of powder XRD and HRTEM is was possible to obtain the crystal structure and particle sizes of our HgTe nanocrystals. The diffraction pattern is shown in Fig. 1, where the wide angle peaks clearly indicate a cubic (coloradoite) HgTe phase. The small particle size causes the peaks to be broadened, and by using the Scherrer equation [17] we can estimate a mean diameter of about 3.5 nm from the full width at half maximum of the (111) reflection. The small-angle peak expected from the close-packing of the nanometer-sized crystallites in the powder is only very weakly resolved, which indicates a rather broad size distribution.

Figure 2 shows the HRTEM images of a typical overview of the sample together with a micrograph of an individual HgTe particle and its corresponding Fast-Fourier-Transform (FFT). The overview image confirms the existence of nanometer-sized crystals in our sample, with a large size distribution ranging from 3 to 6 nm. The well-resolved lattice planes indicate that the sample has high crystallinity, and from the images of single particles an inter-planar distance of 3.73 Å can be measured. This value is consistent with the (111) lattice spacing of coloradoite HgTe.

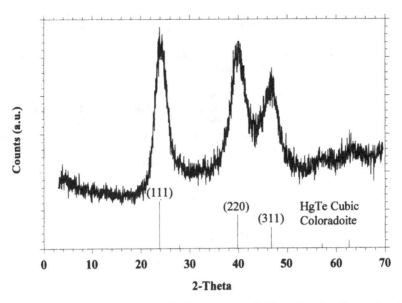

Figure 1 : X-Ray diffraction pattern of HgTe nanocrystals. The reflections from coloradoite HgTe are also shown for comparison.

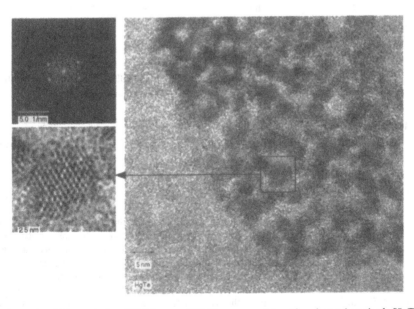

Figure 2 : High-resolution TEM image of HgTe nanocrystals. Also shown is a single HgTe cluster with its corresponding FFT.

The optical absorption spectra, which were recorded for the as-prepared 'fresh' material, following 2 days ageing, and finally after 2 weeks, are shown in fig. 3. The inset is a blow-up of the weak infrared tail which is present at all stages of the nanocrystals' development. We were experimentally restricted to a long wavelength limit of 1350 nm due to the strong water absorption above 1400 nm. Note that the discontinuity at 860 nm in the freshly prepared sample is a spectrometer artefact caused by a detector change. The expected well-developed maxima near the absorption onset, which may be ascribed to the first excitonic transition, are extremely weak and broad but are definitely present. This "excitonic" feature moves to longer wavelengths over time from a not-so-obvious shoulder at 950 nm in the fresh material, through a definite peak after 2 days at 1050 nm, to approximately 1270 nm after 2 weeks. The weakness of these peaks is probably a consequence of the large size distribution seen in the HRTEM images. Also in evidence at 2 days and thereafter is a slight shoulder at around 500 nm.

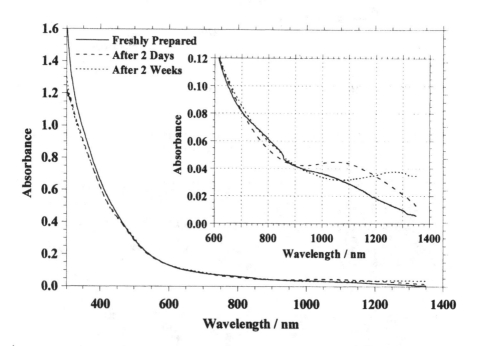

Figure 3 : Optical absorption spectra of HgTe nanocrystals recorded over a period of 2 weeks.

The room temperature PL spectra, again recorded as the material aged in solution over several days, is shown in fig. 4. In the freshly-prepared sample, the intense band-edge luminescence covers the entire spectral region between 800 and 1400 nm, with a peak located at 1080 nm. Numerical integration of this peak yields a value for the quantum efficiency of 48%. The apparent short and long wavelength shoulders at approximately 950 nm and 1250 nm respectively are actually artefacts caused by re-absorption of the PL signal by the aqueous solution. This can be clearly seen from the overlaid transmission spectrum of a 1 mm cell of water (this is approximately the path length of the PL through the solution). As the sample develops over time, two trends are apparent. Firstly, the luminescence shifts to lower energies yielding more dominant long wavelength features (which coincides with the shift seen in the absorption spectrum) giving a peak at 1110 nm after 2 days, and 1270 nm after 2 weeks. Secondly, the quantum efficiency drops to 38% after 2 days, and 1.5% after 2 weeks. However, these values should be treated with caution since the strong water absorption at 1450 nm effectively creates a cut-off and therefore will have a major influence on the appearance of the spectra.

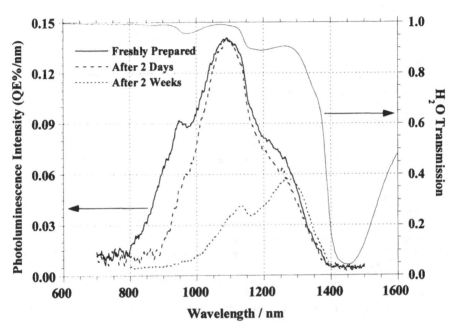

Figure 4 : Room temperature photoluminescence spectra of HgTe nanocrystals recorded over a period of 2 weeks. Also shown is the transmission spectrum of a 1 mm cell of water.

CONCLUSIONS

We present what we believe to be the first wet-chemical synthesis of HgTe nanocrystals on a gram scale and have demonstrated luminescence at infrared wavelengths. In fact, due to the ageing process which occurs over the first few days after synthesis, we already have materials which emit in one of the telecoms windows around 1.3 μm. The clusters are effectively stabilised and capped by 1-thioglycerol, and have the cubic coloradoite crystal structure with sizes ranging from 3 to 6 nm. By using the size-selective precipitation technique [15,16] it should be possible to narrow this broad distribution of sizes, although this may not be significant for broad-band applications. The most exciting feature of this novel material, however, is the extremely high quantum yields of their room temperature infrared luminescence of around 50%.

ACKNOWLEDGEMENTS

We thank J. Ludwig from the Mineralogisch-Petrographisches Institut, Universität Hamburg for the powder XRD measurements. The financial support of the Volkswagen Foundation is gratefully acknowledged.

REFERENCES

1. Al. L. Efros and A. L. Efros, Sov. Phys. – Semicond. **16**, p.772 (1982).
2. R. Rossetti, S. Nakahara, and L. E. Brus, J. Chem. Phys. **79**, p.1,086 (1983).
3. H. Weller, Angew. Chem. Int. Ed. Engl. **32**, p. 41 (1993).
4. H. Weller, Adv. Mater. **5**, p. 88 (1993).
5. A. P. Alivisatos, J. Phys. Chem. **100**, p. 13,226 (1996).
6. J. Z. Zhang, Acc. Chem. Res. **30**, p. 423 (1997).
7. D. L. Klein, R. Roth, A. K. L. Lim, A. P. Alivisatos, and P. L. McEuen, Nature **389**, p. 699 (1997).
8. D. L. Feldheim and C. D. Keating, Chem. Soc. Rev. **28**, p. 1(1998).
9. V. L. Colvin, M. C. Schlamp, and A. P. Alivisatos, Nature, **370**, p. 354 (1994).
10. B. O. Dabbousi, M. G. Bawendi, O. Onitsuka, and M. F. Rubner, Appl. Phys. Lett. **66**, p. 1,316 (1995).
11. M. Gao, B. Richter, S. Kirstein, and H. Möhwald, J. Phys. Chem. B. **102**, p. 4,096 (1998).
12. Landolt-Bornstein, *Numerical Data and Functional Relationships in Science and Technology: New Series. Vol. 17b : Semiconductors*, Springer-Verlag, Berlin, 1982, p. 239 & 465.
13. T. Rajh, O. I. Micic, and A. J. Nozik, J. Phys. Chem. **97**, p. 11,999 (1993).
14. A. Hässelbarth, A. Eychmüller, R. Eichberger, M. Giersig, A. Mews, and H. Weller, J. Phys. Chem. **97**, p. 5,333 (1993).
15. T. Vossmeyer, L. Katsikas, M. Giersig, I. G. Popovic, K. Diesner, A. Chemseddine, A. Eychmüller, and H. Weller, J. Phys. Chem. **98**, p. 7,665 (1994).
16. A. L. Rogach, L. Katsikas, A. Kornowski, D. Su, A. Eychmüller, and H. Weller, Ber. Bunsenges. Phys. Chem. **100**, p. 1,772 (1996).
17. Masao Kakudo and Nobutami Kasai, *X-Ray Diffraction by Polymers*, Kodansha Scientific, Tokyo, 1972, p. 329.

EFFICIENT LUMINESCENCE FROM GaAs NANOCRYSTALS IN SiO₂ MATRICES

Y. KANEMITSU*, H. TANAKA*, S. MIMURA*[1], S. OKAMOTO*[2], T. KUSHIDA*,
K. S. MIN**, and H. A. ATWATER**
*Graduate School of Materials Science, Nara Institute of Science and Technology, Ikoma, Nara
630-0101, JAPAN, sunyu@ms.aist-nara.ac.jp
**Thomas J. Watson Laboratory of Applied Physics, California Institute of Technology,
Pasadena, California 91125, USA

ABSTRACT

We have fabricated zincblende GaAs nanocrystals by means of Ga⁺ and As⁺ co-implantation into SiO₂ matrices. A broad photoluminescence band is observed in the visible spectral region. Under selective excitation at energies within the visible luminescence band, GaAs-related phonon structures are observed at low temperatures. The photoluminescence mechanism in GaAs/SiO₂ nanocomposites is discussed.

INTRODUCTION

There is currently intense interest in optical properties of semiconductor nanocrystals. Recent advances in controlling and characterizing semiconductor nanocrystals have generated considerable interest in exploring new synthesis techniques [1]. Several approaches have been developed for producing nanocrystal materials, including precipitation in solvents (colloidal nanocrystals) [2], self-assembled dot growth (the Stranski-Krastanow growth mode) [3], and electrochemical etching [4], and so on. In addition, high-dose ion-implantation and thermal annealing technique has been shown to provide a versatile technique for creating semiconductor nanocrystals in the surface region of a substrate material, since almost any ions can be implanted into any solid substrates [5,6]. The nanocrystal size can be changed by controlling the ion dose, the implantation energy, and the annealing temperature. In fact, by means of ion implantation and thermal annealing techniques, Si and Ge nanocrystals in SiO₂ matrices have been widely reported [7-10]. However, synthesis of compound semiconductor nanocrystals by ion-implantation has been little reported [11,12].

In this paper, we have fabricated GaAs nanocrystals by means of Ga⁺ and As⁺ co-implantation and have investigated photoluminescence properties of GaAs nanocrystals in SiO₂

---

[1] On leave from Department of Physics and Electronics, Osaka Prefecture University, Sakai, Osaka 599-8531, Japan
[2] On leave from Department of Electrical and Electronic Engineering, Tottori University, Koyama 680-0945, Japan

Mat. Res. Soc. Symp. Proc. Vol. 536 © 1999 Materials Research Society

matrices at low temperatures. The broad photoluminescence band is observed in the visible and near-infrared spectral region. Under site-selective excitation, luminescence fine structures are clearly observed in the visible spectral region. The size-dependent luminescence fine structures are due to the quantum confinement states in GaAs nanocrystals.

SAMPLE PREPARATION AND EXPERIMENT

The nanocrystals were formed in the 100-nm SiO_2 films on (100) crystalline Si substrates. The nanocrystals were synthesized by $3 \times 10^{16}/cm^2$ Ga^+ and $2 \times 10^{16}/cm^2$ As^+ co-implantation at 75 keV into SiO_2. The implanted Ga and As concentration profiles coincide with each other within the oxide film. Subsequent annealing in vacuum resulted in precipitation of GaAs nanocrystals. For a fixed annealing time of 10 min, the annealing temperatures were changed between 800 ℃ and 1100 ℃. For a fixed annealing temperature of 900 ℃, the annealing time was ranged between 0 min and 300 min. After thermal annealing, hydrogen passivation experiments were performed by means of low energy (600 eV) deuterium implantation using a Kauffman ion source and the deuterium dose was determined by elastic recoil spectrometry using a 2.0 MeV $^4He^{++}$ beam. Deuterium was chosen instead of hydrogen in order to facilitate concentration determination. In thermally annealed samples, the presence of nanocrystalline GaAs was verified from lattice images of high-resolution electron transmission micrograph and transmission electron diffraction patterns, as shown in Fig. 1.

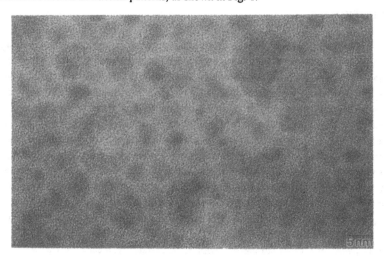

Figure 1: Transmission electron micrograph and transmission electron diffraction pattern of GaAs nanocrystals in SiO_2 matrices.

For photoluminescence measurements, Ar^+, He-Ne or Ti: Al_2O_3 lasers were used as excitation sources. Typical excitation energy density was $\sim$ 30 W/cm². The photoluminescence signals were dispersed by a 50-cm double-grating monochromator and detected by a photomultiplier. The spectral sensitivity was calibrated by using a tungsten standard lamp. The samples were mounted on the cold finger of a temperature-variable closed-cycle He gas cryostat during the measurements. Under 2.707 eV laser excitation at room temperature, we measured the photoluminescence spectra and selected samples showing efficient visible luminescence. The efficient photoluminescence was observed from samples passivated by deuterium implantation at a dose of $3 \times 10^{15}/cm^2$ after annealing at 900 °C for 10 min [12]. Hereafter, luminescence properties of this sample will be reported.

RESULTS AND DISCUSSION

Figure 2 shows photoluminescence spectra in the visible and near-infrared spectral region under 2.707-eV laser excitation at 13 K. The sample shows broad luminescence in the red and near-infrared spectral region near 800 nm, similar to the case of porous GaAs [13]. Other weak luminescence is in the 1-1.4 μ m near-infrared spectral region. The total luminescence spectrum contains contributions from all luminescent centers (nanocrystals, defects, and so on) in the sample and is inhomogeneously broadened. Ion implantation often introduces structural damage into SiO_2 thin films. Even after thermal annealing, there exist defects at the GaAs/SiO_2 interface and in SiO_2 matrices. Therefore, it is believed that defect states contribute to the complicated PL spectrum in GaAs/SiO_2 nanocomposites shown in Fig. 2. The strong luminescence band near 800 nm is close to the band gap energy of GaAs bulk crystal. However, this strong and broad luminescence band can be divided into three Gaussian bands. The peak energies of the PL bands are 1.78 eV (A band), 1.62 eV (B band), and 1.52 eV (C band), as indicated by the arrows. Therefore, we speculate that many different luminescence processes contribute to the strong luminescence band. In order to clarify the origins of these luminescence bands, we applied selective excitation spectroscopy to our GaAs nanocrystal samples.

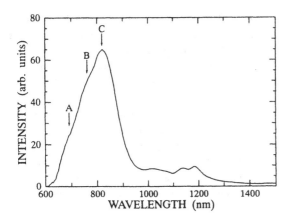

Figure 2: Photoluminescence spectra of GaAs nanocrystals in SiO_2 matrices under 2.707-eV laser excitation at 13 K. A broad luminescence band is observed in the red and near-infrared spectral region.

Figure 3 shows luminescence fine structures under selective excitation at photon energies within the strong luminescence band at 13.3 K. The pronounced fine structures are marked by a solid circle (●). Many different peaks are also observed under selective excitation of the A band. The peaks of fine structures shift to lower energy side with a decrease of excitation photon energies: The peak positions of fine structures depend on the nanocrystal size. When excitation energy decrease and the B and C bands are excited selectively, the fine structures gradually diminish and the sharp peak (●) is only observed.. Selectively excited luminescence spectra suggest that the origin of the A band is different from the defect-like B and C bands.

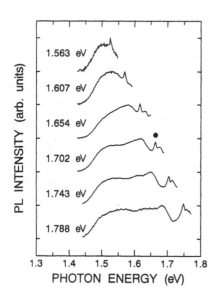

Figure 3: Photoluminescence spectra of GaAs nanocrystals in SiO₂ matrices under various excitation energies at 13.3 K. Fine structures in the PL spectrum are observed under selective excitation at energies within the A band.

The energy difference between the sharp peak energy and the excitation laser energy is ∼ 36 meV, which does not depend on the excitation energy. The observed energy difference of ∼36 meV is in fairly good agreement with LO-phonon energy of bulk GaAs (36.5 meV). On the other hand, the energy differences between the other peaks and excitation energy depend on the excitation laser energy. The energy interval between the weak peaks is also close to the GaAs LO phonon energy.

The energy diagram for efficient luminescence and luminescence fine structures is illustrated in Fig. 4. Since the Stokes shift energy does not depend on the excitation energy, the sharp peak luminescence signal comes from the absorption state like resonant Raman signal. In nanocrystals, the coupling between exciton and phonon increases with a decrease of nanocrystals size [14]. For example, phonon-assisted luminescence is observed in direct-gap CdSe [15] and CdS nanocrystals [16]. Therefore, even in a direct gap semiconductor, the phonon-assisted luminescence is clearly observed. Under selective excitation conditions, the sharp peak is due to the GaAs LO-phonon replica of the band edge emission. The other fine structures located at lower energy are considered to be due to the luminescence from localized states. However, at present, the nature of the localized state is not clear. Another candidate for the origin of phonon replicas in luminescence spectrum is the direct-to-indirect transition in the electronic band structure in GaAs nanocrystals because of the quantum confinement induced Γ-X transition [14,17]. Further experimental studies are needed for the understanding of optical transitions in GaAs/SiO$_2$ nanocomposites. From selectively excited PL spectra, it is concluded that the efficient luminescence band (in particular, the A band) with GaAs LO phonon structures is attributed to quantum confinement states in GaAs nanocrystals. We observe for the first time the intrinsic luminescence of GaAs nanocrystals in SiO$_2$ matrices fabricated by ion-implantation and thermal annealing technique.

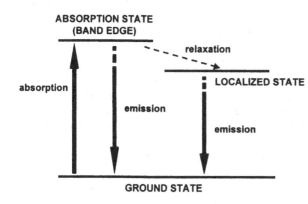

Figure 4: Energy diagram for the efficient red luminescence from GaAs nanocrystals in SiO$_2$ matrices

CONCLUSIONS

In conclusion, we have fabricated GaAs nanocrystals by Ga$^+$ and As$^+$ co-implantation into SiO$_2$ matrices. Efficient and broad luminescence band is observed in the visible and near-infrared spectral region. This efficient luminescence is attributed to both quantum confinement states in GaAs nanocrystals and defects in SiO$_2$. It has been demonstrated that ion implantation and thermal annealing is useful to fabricate compound semiconductor nanocrystals in SiO$_2$ matrices and that selective excitation spectroscopy is a powerful method to determine the origin of the luminescence in complex nanocrystal systems.

ACKNOWLEDGEMENTS

This work was partly supported by a Grant-In-Aid for Scientific Research from the Ministry of Education, Science, Sports and Culture of Japan, The Shimadzu Science Foundation, and The Yamada Science Foundation. S. Okamoto is grateful to the JSPS Research Associate Program (JSPS-RFTF96R12501).

REFERENCES

1. T. Ogawa and Y. Kanemitsu, *Optical Properties of Low-Dimensional Materials* (World Scientific, Singapore, 1995); T. Ogawa and Y. Kanemitsu, *Optical Properties of Low-Dimensional Materials, vol. 2* (World Scientific, Singapore, 1998).
2. C. B. Murray, D. J. Norris, and M. G. Bawendi, J. Am. Chem. Soc. **115**, 8706 (1993); M. Nirmal and L. E. Brus, Mat. Res. Soc. Symp. Proc. **452**, 17 (1997).
3. P. M. Petroff and G. Medeiros-Riberiro, MRS Bulletin **21** (4), 50 (1996).
4. Y. Kanemitsu, Phys. Rep. **263**, 1 (1995); A. Cullis, L. T. Canham, and P. D. J. Calcott, J. Appl. Phys. **82**, 909 (1997).
5. J. D. Budai, C. W. White, S. P. Withrow, R. Z. Zuhr, and J. G. Zhu, Mat. Res. Soc. Symp. Proc. **452**, 89 (1997).
6. H. A. Atwater, K. V. Shcheglov, S. S. Wong, K. J. Vahala, R. C. Flagan, M. L. Brongersma, and A. Polman, Mat. Res. Soc. Symp. Proc. **316**, 409 (1994).
7. T. Shimizu-Iwayama, S. Nakao, and K. Saitoh, Appl. Phys. Lett. **65**, 1814 (1994).
8. T. Komoda, J. Kelly, F. Cristiano, A. Nekim, P. L. F. Hemment, K. P. Homewood, R. Gwilliam, J. E. Mynard, and B. J. Sealy, Nucl. Instrum. Methods Phys. Res. B **96**, 387 (1995).
9. Y. Kanemitsu, N. Shimizu, T. Komoda, P. L. F. Hemment, and B. J. Sealy, Phys. Rev. B **54**, 14329 (1996).
10. K. S. Min, K. V. Scheglov, C. M. Yang, H. A. Atwater, M. L. Brongersma, and A. Polman, Appl. Phys. Lett. **68**, 2511 (1996) and **69**, 2033 (1996).
11. C. W. White, J. D. Budai, J. G. Zhu, S. P. Withrow, R. A. Zuhr, D. M. Hembree, Jr., D. O. Henderson, A. Ueda, Y. S. Tung, R. Mu, and R. H. Magruder, J. Appl. Phys. **79**, 1876 (1996).
12. S. Okamoto, Y. Kanemitsu, K. S. Min, and H. A. Atwater, Appl. Phys. Lett. **73**, 1829 (1998).
13. P. Schmuki, D. J. Lockwood, H. J. Labbe, and J. W. Fraser, Appl. Phys. Lett. **69**, 1620 (1996).
14. Y. Kanemitsu, S. Mimura, and Y. Fukunishi, unpublished.
15. M. Nirmal, C. B. Murry, and M. G. Bawendi, Phys. Rev. B **50**, 2293 (1994).
16. S. Okamoto, Y. Kanemitsu, H. Hosokawa, K. Murakoshi, and S. Yanagida, Solid State Commun. **105**, 7 (1998).
17. A. Franceschetti and A. Zunger, Phys. Rev. B **52**, 14664 (1995).

PLASMONS ON LUMINESCENT POROUS SILICON PREPARED WITH ETHANOL

AND CRITICAL POINT DRYING

O. Resto*, L.F. Fonseca*, S.Z. Weisz*, A. Many**, Y. Goldstein**

*Department of Physics, University of Puerto Rico, Rio Piedras, PR 00931, USA

**Racah Institute of Physics, The Hebrew University, Jerusalem 91904, Israel

ABSTRACT

We investigated the plasmon characteristics on luminescent porous silicon using electron energy loss spectroscopy. The samples were prepared from p-type crystalline silicon, (100) face, using the conventional electrochemical etching technique with the usual solution of HF, ethanol and water, followed by a critical point drying process. The energy of the bulk plasmon was measured both before and after sputter cleaning the sample with argon-ion bombardment. We found that initially the plasmon energy was slightly higher, ~18 eV, than the plasmon energy of crystalline silicon. After sputter cleaning the sample with 5 keV Ar^+ ions, the plasmon energy increased to ~20 eV. Exposure to the electron beam used for the measurements caused a slow upward shift of the plasmon energy as a function of time, toward a saturation energy of 22-23 eV, an energy close to the plasmon energy of SiC. Auger spectroscopy performed in parallel showed an increasing carbon coverage. We prepared also samples without ethanol in the etching solution and/or with no critical point drying. Samples that did not undergo the critical point drying process showed consistently a practically constant plasmon energy, with almost no change upon sputtering and/or exposure to the electron beam. On the other hand, samples that were prepared with or without ethanol but using the critical point drying process, showed an appreciable increase in the plasmon energy upon exposure to the electron beam.

We conclude that traces of CO_2, used in the critical point drying process, are stored in the pores of the porous silicon surface and serve as a source of carbon. Apparently, upon activation by argon bombardment or by the electron beam, the carbon interacts with the porous Si surface forming a carbon-silicon compound, most probably SiC.

INTRODUCTION

Porous silicon [1-4] (PSi), obtained by electrochemical etching procedures applied to crystalline Si surfaces, exhibits high luminescence efficiencies in the visible range. It is quite clear now that the visible luminescence originates from the band-gap enlargement due to quantum confinement [3,4]. At the same time, the reasons for the high-efficiency luminescence are still somewhat under debate [3-5]. It was suggested that it is the amorphous or microcrystalline nature of the porous Si that is responsible for the phenomenon, or that the formation of silicon compounds such as siloxene ($Si_6O_3H_6$) or species of Si-H, Si-O and Si-F bonds are involved in the luminescence [3,4]. The study of the plasmon energies on luminescent PSi is of interest for two main reasons. The nanostructure of PSi was shown [4,5] to have a coral-like structure consisting of a continuous hierarchy of columns and pores. Typical lengths of the columns are a few tens of nanometers, while their average radius is a few nanometers. As such, one may expect that the energy of the surface plasmon will be downshifted with respect to its value on crystalline silicon. This has indeed been found to be the case by Sasaki *et al.* [6]. Another reason is that the plasmon structure is very sensitive to even minute quantities of adsorbates on the surface [7]. Thus a study of the plasmon structure may possibly detect and lead to the identification of surface species that may be involved in the luminescence process in PSi. Berbezier *et al.* [8] concluded from their investigation of the plasmon structure of PSi that the nanocrystalline PSi clusters are surrounded by an amorphous Si surface layer passivated by hydrogen.

In our study of the PSi plasmon structure we too observed the surface plasmon at a reduced energy due to the nanocrystalline structure of PSi [6] and also the plasmon attributed [8] to the collective oscillations of the interface between an amorphous Si-H cap layer and the Si crystallites. However, during our studies we encountered a strange phenomenon, namely we found that the energy of the silicon volume plasmon was shifting during the measurements to higher energies. We decided to investigate this effect and we present here the results and the probable reason for this effect.

RESULTS AND DISCUSSION

The starting material was high-grade p-type silicon of resistivity in the range 0.5 - 1.5 Ω cm. A p+ layer was formed by diffusing metallic Al into the back faces of the silicon wafers to obtain an ohmic contact. The sample was attached to a cylindrical Teflon cell via a Kalrez O-ring, the sample constituting the bottom of the cell, with its front surface, the (100) face, facing upwards. Before anodization, the samples were etched in 20% HF. In order to prepare the porous surface [4], a solution of HF, ethanol and water (1:2:1) was poured into the cell. A platinum electrode was immersed in the solution and a spring contact was attached to the p+ contact. The anodization of the Si surface was carried out with a current density of 100 mA/cm^2. After anodization, the samples were removed from the cell. Some of the samples underwent a critical point drying [9] sequence while others were simply rinsed by de-ionized water and dried with N$_2$ gas.

The photoluminescence of the PSi was excited by a 10 mW He-Cd laser beam ($\lambda = 442$nm). The spectra were measured by an ISA-320 Triax UV-visible spectrometer. Figure 1 shows a typical photoluminescence spectrum measured on one of our PSi samples. The maximum of the spectrum obtains at around 640 nm.

The samples studied were mounted in the vacuum chamber of a Physical Electronics model 560 Auger microprobe. An electron beam of ~100 nA intensity on a ~50x50 μm^2 area was used to excite the plasmons and the scattered electrons were energy analyzed with a cylindrical mirror analyzer. The plasmons were measured mostly using a retarding field applied to the electrons so that they enter the

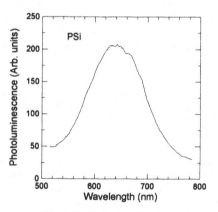

Fig 1. Typical photoluminescence spectrum of our PSi samples.

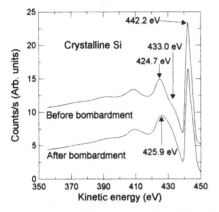

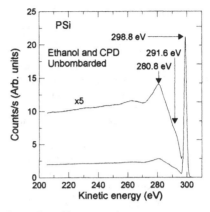

Fig. 2. Electron loss spectra of crystalline Si.

Fig. 3. Electron loss spectrum of unbombarded PSi prepared with ethanol and CPD.

analyzer at a low (constant) energy. The resolution attained in this mode was better than 600 meV. The measurements were done both before and after a bombardment cleaning with Ar ions at 5 keV and at a current intensity of ~200 nA on an area of ~2x2 mm². For comparison purposes we performed measurements on crystalline silicon as well.

Figure 2 shows the plasmon peak (and its satellites) measured on crystalline silicon both before and after argon ion bombardment. The electron beam energy in this measurement was 442.2 eV, as marked in the figure. The plasmon peak energies are marked. We see that the plasmon energy after the bombardment is 16.3 eV, about what is expected for the bulk plasmon of silicon [6,8]. We see also the peaks due to multiple plasmon losses, and their energies scale with the single-plasmon loss. The plasmon energy before the bombardment is somewhat larger, 17.5 eV. Here we see also a shoulder at 9.2 eV due perhaps to some Si-H compounds as suggested in ref. 8. In Fig. 3 we show typical electron loss results from a PSi sample that underwent the critical point drying (CPD) procedure and was not ion bombarded. (The upper curve in the figure is a blow-up by a factor of 5 of the lower curve.) The e-beam energy for this measurement was 298.8 eV, as marked. The energy of the plasmon peak (280.8 eV) is also marked in the figure. We see that the plasmon energy here has already increased to 18.0 eV and similarly the energies of the multiple plasmon losses. There appears also a shoulder at ~7.2 eV, probably due to the surface plasmon, its energy reduced because of the PSi nanostructure [6].

When we ion bombarded the PSi sample, we noticed that the plasmon loss peak shifted to higher energies, to about 20 eV. In addition, we also observed a shift to higher energies just by exposing the PSi samples to the electron beam used to excite the plasmon losses. This is illustrated in Fig. 4 which shows the dependence of the plasmon energy on the exposure time to the electron beam for a PSi sample that had undergone critical point drying, before (triangles) and after (circles) argon ion bombardment. As we can see, before the bombardment the plasmon energy increases with exposure time from a starting value of ~18.3 eV up to ~19.3 eV. After bombardment, the initial plasmon energy is already ~20 eV and with exposure time it continues to increase until it saturates at ~23 eV.

The saturation plasmon energy is close to the plasmon energy of SiC [10]. This prompted the idea that the Ar bombardment and/or electron-beam irradiation promote the formation of SiC on the surface of the PSi. This may be somewhat similar to the phenomenon reported by Heera et al. [11] who found that bombardment by Ge⁻ ions caused a crystallization of amorphous SiC.

We performed Auger surveys to check for possible carbon increase on the PSi surface due to e-beam irradiation. Figure 5 shows two such surveys, the first, curve (a), was taken 17 min. after turning on the e-beam and the second, curve (b), after an additional interval of 7 min. The positions of the Si, C and O peaks are marked in the figure. We see that both the C and O peaks in curve (b) show

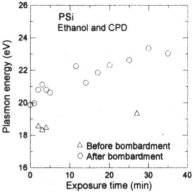

Fig. 4. Plasmon energy as function of exposure time to e-beam for a PSi sample prepared with ethanol and CPD, before (triangles) and after (circles) Ar bombardment.

a marked increase. No similar increase was detected on the crystalline-Si part of the sample.

The preparation of the porous samples involves carbon both in the anodization process, via ethanol, and in the critical point drying process, via liquid CO_2. In order to eliminate these sources of carbon, we prepared a sample without ethanol in the anodizing solution, and cleaned it only by rinsing with de-ionized water and drying with N_2 gas. We measured the plasmon energy for this sample before and after argon-ion bombardment as a function of the exposure time to the electron beam. In both cases the plasmon energy was found to be practically constant, around 17.5 eV, and did not change with exposure time and argon bombardment.

In order to identify which of the forms of carbon, ethanol or CO_2, is responsible for the carbon contamination of the PSi surface, we prepared two types of samples: samples that were anodized without ethanol in the anodizing solution and then went through the critical point drying process, and samples that were anodized with ethanol but did not go through the CPD process. We measured the plasmon energy as a function of exposure time to the e-beam, and the results are shown in Figs. 6 and 7. In Fig. 6 we plot the plasmon energy measured on the PSi sample that was prepared without ethanol but underwent the

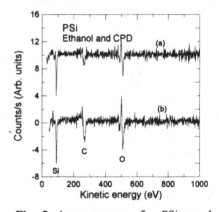

Fig. 5. Auger surveys of a PSi sample prepared with ethanol and CPD. Curve (a) was taken 17 min. after turning on the e-beam and curve (b) after an additional interval of 7 min.

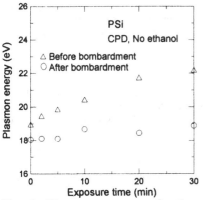

Fig. 6. Plasmon energy as function of exposure time to e-beam for a PSi sample anodized without ethanol and dried by CPD, before and after argon bombardment.

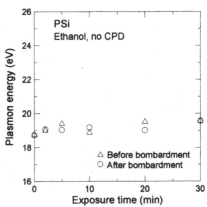

Fig. 7. Plasmon energy as function of exposure time to e-beam for a PSi sample anodized with ethanol but without CPD, before and after argon bombardment.

CPD process. As we can see, the plasmon energy before the bombardment (triangles) starts close to 19 eV and increases with exposure time, even above the plasmon energy for samples prepared with ethanol and CPD (see Fig. 4). The plasmon energy saturates around 22.2 eV. We also measured, in parallel, by Auger electron spectroscopy, the surface carbon concentration and found that it increased by~40% compared to its value before turning on the e-beam. The plasmon energy after the bombardment (circles) starts at a little above 18 eV and changes only slightly and, in parallel, the surface carbon concentration remains practically constant. In Fig. 7 we plot the plasmon energy measured on the other PSi sample, that which was prepared with ethanol but did not undergo the CPD process. Here we see that the plasmon energy is almost unchanged upon exposure to the electron beam and/or upon argon bombardment. The surface carbon concentration increased only by ~10% compared to its value before turning on the e-beam. Apparently, the possible remnants of ethanol in the pores are not an efficient source of surface carbon.

A PSi sample that was prepared without ethanol and without critical point drying, was left to age in room atmosphere. After 4 months we re-measured the plasmon energy and the results are plotted in Fig. 8. We see that after the aging the plasmon energy before sputtering (triangles) starts close to 22 eV and still rises a little upon exposure to the e-beam. This again suggests very strongly that we are looking on

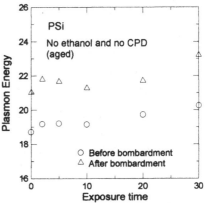

Fig. 8. Plasmon energy as function of exposure time to e-beam for a PSi sample prepared without ethanol and without CPD and aged for 4 months, before (triangles) and after (circles) argon bombardment.

233

plasmons affected by CO_2 and/or CO adsorbed from the atmosphere. After sputtering (circles), the plasmon energy starts at a lower energy (~19 eV) but again increases upon exposure to the e-beam. Our interpretation is that the bombardment removed some of the carbon from the surface but there is still carbon stored in the pores.

CONCLUSIONS

We found that the plasmon energy, measured on porous silicon surfaces prepared with critical point drying, drifts to higher energies upon electron beam irradiation and/or ion bombardment of the surface. The saturation energy of the plasmon is close to the plasmon energy of SiC. Apparently, due to the critical point drying process, CO_2 accumulates in the pores and the electron beam irradiation and/or ion bombardment decompose the CO_2 and promote the formation on the porous surface of a silicon-carbon compound, most probably SiC. Aging the PSi surface in room air causes a similar drift of the plasmon energy, due apparently to adsorbed CO_2.

ACKNOWLEDGMENTS

This work was supported by US ARO grant No. DAAHO4-96-1-0405, by NASA grant No. NCCW-0088, and by the Ministry of Science and Arts, Israel, within the framework of the infrastructure-applied physics support project.

REFERENCES

1. L.T. Canham, Appl. Phys. Lett. **57**, p. 1046 (1990).

2. I. Amato, Science **252**, p. 922 (1991).

3. See, for instance, Z.C. Feng and R. Tsu, Porous Silicon (World Scientific, Singapore, 1994).

4. See review article by A.G. Cullis, L.T. Canham and P.D.J. Calcott, J. Appl. Phys. **82**, p. 909 (1997) and references therein.

5. F. Koch in Silicon Based Optoelectronic Materials, edited by R.T. Collins, M.A. Tischler, G. Abstreiter, and M.L. Thewalt (Mater. Res. Soc. Proc. **298**, Pittsburgh, PA, 1993), p. 319-324.

6. R.M. Sasaki, F. Galembeck, and O. Teschke, Appl. Phys. Lett. **69**, p. 206 (1996).

7. Millo, Y. Goldstein, A. Many, and J.I. Gersten, Phys. Rev. **B 39**, p. 1006 (1989).

8. I. Berbezier, J.M. Martin, C. Bernardi, and J. Derrien, Appl. Surf. Sci., **102**, p. 417 (1996)

9. J. Von Behren, P.M. Fauchet, E.H. Chimowitz, and C.T. Lira in Advances in Microcrystalline and Nanocrystalline Semiconductors, edited by R.W. Collins, P.M. Fauchet, I. Shimizu, J.-C. Vial, T. Shimada, and A.P. Alivisatos (Mater. Res. Soc. Proc. **452**, Pitsburgh, PA 1997), p. 565-570.

10. H. Raether, Excitation of Plasmons and Interband Transitions by Electrons (Springer, Berlin, 1980), Ch. 5.

11. V. Heera, J. Stoemenos, R. Kögler and W. Skorupa, J. Appl. Phys. **77**, p. 2999 (1995).

Part V

Synthesis and Properties of Microcrystalline and Nanocrystalline Semiconductors

NANOPARTICLE PRECURSORS FOR ELECTRONIC MATERIALS

D.S. GINLEY, C.J. CURTIS*, R. RIBELIN, J.L. ALLEMAN, A. MASON,
K.M.JONES, R.J. MATSON, O. KHASELEV*, D.L. SCHULZ
National Center for Photovoltaics and *Basic Energy Sciences Center, National
Renewable Energy Laboratory, 1617 Cole Blvd., Golden, CO 80401-3393

ABSTRACT

The use of nanoparticle precursors for electronic materials including sulfides, selenides, oxides and the elements has potentially wide ranging implications for improving device properties and substantially reducing the deposition costs. To realize this goal the complex interfacial chemistry of these small particles must be controlled. In this paper we present a number of cases demonstrating the complexity of this chemistry. These include $CuInSe_2$ where the kinetics of phase formation dominate the sintering process; CdTe where sintering proceeds with and without the sintering enhancement of $CdCl_2$, but produces materials different electronically than bulk materials; and the use of compound and elemental nanoparticles (Ag, Al, Hg-Cu-Te and Sb-Te) for contacts to elemental and compound semiconductors (Si and CdTe).

INTRODUCTION

The development of large area devices such as photovoltaics and flat panel displays requires a delicate balance between the quality of the materials and the cost of employing them. In order to obtain the best quality, vacuum processing approaches are typically utilized. These include sputtering, evaporation, metal organic chemical vapor deposition and molecular beam epitaxy which can produce very high quality materials, but at the price of high processing costs. Spray deposition and screen printing are low cost large area deposition techniques but don't typically produce as high a quality materials as the vacuum based approaches. This may be a result of the large particles employed in the inks, which sinter to form a porous large grained network. We began a program a few years ago to investigate the possibility of employing nanoparticle inks in the low cost approaches with the hopes of getting vacuum quality films with low cost.

The primary thrust of this work has been in the area of photovoltaics. This is in part because the basic elements of a photovoltaic cell: back contact, absorber, junction former, transparent contact and metallization, are common to a wide variety of other devices and the cell performance is an excellent tool to evaluate the quality of the particle derived layers. Our primary activities have been in the areas of the active absorber layer and in the area of contacts. In this work we report on the development of nanoparticle precursors for $CuInSe_2$ and CdTe absorber layers, Ag and Al for contacts to Si, and Hg-Cu-Te and Sb-Te for contacts to CdTe. In all cases the success of the approach is dominated by the materials science of the interfaces between the particles and between the particles and the substrate. Control of these interfaces can produce high quality and unique materials.

EXPERIMENTAL

CdTe Particles and Films

CdTe nanoparticle colloids were prepared by the reaction of Na_2Te with CdI_2 in methanol as illustrated generically in eq 1 [1]. Two flasks containing the appropriate reagents in dry, degassed methanol cooled to $-78°C$. The Na_2Te solution was added rapidly to the CdI_2 mixture using a large bore cannula. After warming to above 0°C, stirring was stopped and the precipitate was allowed to settle. The remaining red colloidal suspension was transferred to centrifuge tubes and run at 3000 rpm for 10 min. The supernatant was discarded, methanol was added, and the mixture was sonicated for fifteen minutes producing CdTe colloidal suspensions. These colloids were stable for weeks and could be spray deposited directly.

$$aMI_x + bNa_2Q \xrightarrow[-78°C]{MeOH\ -2bNaI} M_aQ_b \qquad (1)$$

$$M = Cd, Cu, Hg, In \qquad Q = Se, Te$$

$$Cu(BF_4)_2 + Na_2Se \xrightarrow{MeOH,\ 2NaBF_4} CuSe \qquad (2)$$

Precursor films were prepared by spraying the purified nanoparticle colloids onto heated CdS coated glass or metal-coated glass. Spray deposition was performed under inert conditions.

Cu-In-Ga-Se Particles and Films

The details of the Cu-In-Ga-Se nanoparticle synthesis are presented elsewhere [2]. In short, nanoparticle colloids were prepared by reacting a mixture of CuI and/or Cu $(BF_4)_2$ and/or InI_3 and/or GaI_3 in pyridine with Na_2Se in methanol at reduced temperature under inert atmosphere as illustrated in eqs 1,2. Colloids with the compositions $CuInSe_{2.5}$, CuSe, In_2Se_3, and $Cu_{1.10}In_{0.68}Ga_{0.23}Se_x$ were prepared in analogy to CdTe (see above).

Precursor films were prepared by spraying purified nanoparticle colloids onto heated molybdenum-coated glass and $CdS/SnO_2/glass$ substrates. Spray deposition of the colloids to form precursor Cu-In-Ga-Se films was performed under inert conditions.

Elemental Contact Particles and Films

Powders of Al and Ag were obtained from Argonide Corp. after preparation by Russian scientists via the electroexplosion (i.e., exploding wire) process. The Al sample was shipped under kerosene, whereas Ag was received as a dry powder. Samples were prepared for TEM by sonication of isopropanol slurries of the dried powders. P-type Si (Wacker, 3.3 Ω•cm) and n-type Si (Wacker, 2.9 Ω•cm) were etched with 5% HF and rinsed with deionized water. The samples for annealing studies were prepared by dropping toluene slurries of the metallic nanoparticles onto the Si substrates using a modified disposable pipette in a He-filled glovebox. The Al on p-type Si samples were annealed at 645°-650°C for 1 h under Ar, whereas the Ag on n-type Si samples were annealed at 882°C for 1 h under Ar. After annealing, the deposits were crumbly and did not provide electrical contact. The residues were removed using an isopropanol-wetted cotton swab, and Ag paint was applied to the alloyed areas and also to non-reacted areas on the Si to provide a control. Electrical testing of the areas showed evidence of contact formation in the areas where the particles had reacted, but none in the control spots. I-V measurements were performed using an Optical Radiation Corporation Solar Simulator 1000 and computer-controlled I-V instrumentation.

Compound Contact Particles and Films

Hg-Cu-Te and Sb-Te materials were prepared by metathesis reactions of metal salts with sodium telluride in methanol at reduced in analogy to CdTe (see above) [3]. TEM, TEM-EDS, and X-ray diffraction (XRD) characterization of as-synthesized Hg-Cu-Te particles showed the samples are micron-sized agglomerates composed of smaller particles (d≤10nm), with the HgTe (Coloradoite) phase (PDF #32-665) and Hg, Cu, and Te present, but no C or O. Characterization of the Sb-Te materials is presently under way. After $CdCl_2$ treatment and NP etch of the CdTe layer, particle contacts were applied. In the case of Hg-Cu-Te, a graphite-based paste was formulated and painted onto the CdTe layer, with subsequent thermal treatment under Ar. Sb-Te particles were applied to CdTe by spray deposition of a methanol ink at 150°C, followed by a thermal treatment under N_2. In both cases, Ag paint was added as the final layer, with devices finished according to standard NREL protocol. CdTe solar cells were characterized by standard light and dark I-V.

RESULTS AND DISCUSSION

Overall the sintering behavior of the nanoparticle precursors is complex depending on the material, the capping agent and the substrate on which the particles were deposited. In what follows we illustrate briefly some of these observations. This in not meant to be a complete iteration of the results to date but rather a snapshot of some of the interesting and not yet understood behavior of these precursors.

CdTe Particles and Films

Figure 1 illustrates the difference in the sintering properties of large grain versus

Figure 1: Illustrates a comparison of the morphologies between large grain precursor (>1 μm grains) and small grain precursor (<100 nm) as deposited in ink form on CdS coated glass and then sintered in the presence of CdCl2 (known to enhance grain growth)

nanoparticle precursors. Note the decreased porosity in the nanoparticle based film and the substantial grain growth which was aided by the presence of $CdCl_2$ introduced by exposing the spray deposited precursor to a saturated $CdCl_2$ solution. These films when fabricated into photovoltaic devices had efficiencies up to 6%. In addition, the presence of CdS (normally a heterojunction partner for the CdTe) also appears to be important. This can be seen in films deposited on Pt coated glass as shown in Figure 2. These films were produced to evaluate the potential of these CdTe materials incorporated into a nanoparticle based photoelectrochemical devices. Based on our work and that of the group at the Weizmann Institute [4] nanocrystalline cells may offer some advantages in terms of light adsorption, electron transfer and control of the solid state properties. Here the film on the left was processed a temperature of 360 °C for 40 min after a 15 min treatment with 75% saturated $CdCl_2$ in MeOH. The one on the right was processed identically except for the sintering temperature was 400°C. In the 400°C sample considerable crystallization is observed, but in both cases the sintering and densification observed on the CdS substrates is not seen. CdTe is known to be very sensitive to the surface chemistry of the redox couple [5-7].

Figure 2: Illustrates the growth of CdTe nanoparticle films on Pt substrates. On the left a film processed at 360°C and on the right at 400°C both for 40 min,

The films on Pt were evaluated in 1M polysulfide solution for their photoelectrochemical response. Figure 3 illustrates the light and dark IV. Interestingly the film shows an n-type photoresponse and a low efficiency. The fact that the normally p-type CdTe has an n-type response is probably due to a reaction with the polysulfide solution to produce an n-type CdS layer and the position of the fermi level in the small grains[DS1].

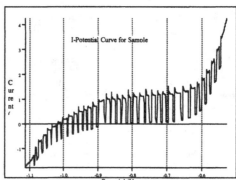

Figure 3: Light induced current as a function of potential for the electrode on the left above. Overall photoresponse is n-type.

Cu-In-Ga-Se Particles and Films

The sensitivity to the nature of the particles and the surface is further illustrated in the case of the $CuInSe_2$ and $Cu(In_{1-x}Ga_x)Se_2$ materials. Figure 4a illustrates a typical

Figure 4: CIS film on the left was synthesized from the binary precursors and deposited on Mo. The film on the right was deposited from a ternary precursor on CdS coated glass.

morphology obtained when spray deposited precursors are selenized with excess Se in vacuum. This film was sprayed from binary (CuSe and In_2Se_3) nanoparticle precursors at 200°C and then processed at 600°C for 20 min[reference us; MRS 1998 (see below)]. The observed morphology indicates very little sintering of the particles is obtained when the precursors are binary selenides, similar results are obtained in the case of ternary precursors.

It appears that in this case the rate of CIS phase formation is faster than the sintering rate and that once the CIS phase is formed then the sintering process is very difficult. Figure 4b shows a similar experiment but on a CdS coated substrate. In this case a ternary $Cu_{1.10}In0_{.68}Ga0_{.23}Se_{1.91}$ precursor sprayed at 225°C and then sintered at 550°C for 10 min. In this case extensive sintering is observed. The CdS acts in some way as a sintering aid. However, in this case clean phase formation is not observed. In both the cases of the CuSe/In2Se3 and $Cu_{1.10}In0_{.68}Ga0_{.23}Se_{1.91}$, there appears to be a large sensitivity to the substrate and to the nature of the grain boundary chemistry. Sintering aids have considerable potential utility in aiding grain growth.

Elemental Contact Particles and Films

Contacts are ubiquitous in the fields of photovoltaics and microelectronics. Nanoparticle precursors to contacts could have the potential advantages of easy deposition, densification and ready ink formulation for spatially resolved contacts [3,8]. In addition, nanoparticles offer the opportunity of formulating compound contacts improving both the ohmic nature of the contact and the thermodynamic stability. To explore this we have investigated metallic contacts to Si and compound contacts to CdTe.

Nanoparticle inks were fabricated from particles produced in Russia by the electroexplosion of Al and Ag wires [reference us; Si workshop and 1998 NCPV (see below)]. Typical morphologies are shown in figure 5. The particles appear to have a bimodal distribution with smaller and a larger component.

241

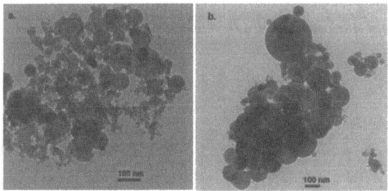

Figure 5: Typical fractions from electroexplosion generated Al particles illustrating a small particle fraction and a large particle fraction.

Contacts were made directly to HF cleaned n- and p-type silicon wafers with Ag and Al inks respectively. In the case of Ag, ohmic contacts were obtained directly after annealing and showed a good linear IV curve. In the case of the Al, initial results showed non-linear IV indicative on non-ohmic contacts as shown in Figure 6a. This was found subsequently to be due to the presence of oxide layers on the as synthesized Al nanoparticles. Two different approaches were employed to alleviate the problem of this surface oxide.

After either the NREL-developed wet-chemical treatment or the improved Al starting material, the Al was observed to contain a much lower amount of O impurity by TEM-EDS. This treated Al was next applied as a contact to p-type Si and annealed as above. I-V characterization of this sample shows a marked improvement in the ohmic character of the treated Al versus untreated Al as shown in Figure 6b.

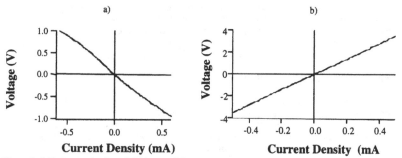

Figure 6: I-V characterization of Al nanoparticle contacts as received (a) and after chemically treatment (b) Al for contact to p-type Si.

Compound Contact Particles and Films

The standard NREL back-contact is a graphite-based Cu-doped HgTe material similar to that described by Britt and Ferekides [9]. The use of solution-synthesized Hg-Cu-Te nanoparticles in graphite-paste contacts was evaluated to determine feasibility versus the micron-sized particles that are normally used. Sb-Te particles were employed in preliminary studies as a contacting material based on the recent report of a Cu-free contact by the Parma group [10].

The amount of Hg-Cu-Te in graphite dag was varied from 18-55 wt.%, and the contact anneal temperature/time was varied from 220°-320°C/20-40 min. Preliminary experiments consisted of two data sets: (1) comparison of 18 vs. 55 wt.% Hg-Cu-Te in graphite dag at three annealing conditions (i.e., 220°C for 20 min, 220°C for 40 min, and 320°C for 20 min); and (2) comparison of 18 vs. 37 wt.% at three annealing conditions (i.e., 220°C for 20 min, 220°C for 40 min, and 270°C for 20 min). In the former, CdTe solar cells contacted with 18 wt.% exhibited better performance in all I-V categories, with optimal performance (i.e., V_{oc} = 834 mV and Eff. = 10.9%) observed for the film annealed at 220°C for 40 min. In the latter, CdTe devices contacted with 37 wt.% exhibited better performance for 220°C anneals, whereas the 18 wt.% seemed better at 270°C (see Table I). The statistical significance of these results is yet to be determined. However, cells contacted with 37 wt% had efficiencies up to 12.6%.

Table I. I-V Data for CdTe Solar Cells Contacted with Nano-Hg-Cu-Te Dag.

	220°C / 20 min		220°C / 40 min		270°C / 20 min	
	18 wt.%	37 wt.%	18 wt.%	37 wt.%	18 wt.%	37 wt.%
V_{oc}	793	811	813	820	812	799
J_{sc}	20.7	22.5	19.9	23.0	21.7	22.5
FF	57.5	63.8	60.8	66.9	67.4	64.2
Eff.	9.4	11.7	9.8	12.6	11.9	11.5

A preliminary study of sprayed Sb-Te contacts on CdTe was also performed. Although V_{oc} > 800 mV was observed for Sb-Te contacted cells annealed at 300°C, efficiencies in these devices were limited to 9.1%, presumably by a large series resistance (>20 Ω) observed in all samples.

CONCLUSIONS

We have demonstrated that a broad range of nanoparticle inks can be fabricated ranging from elemental to quaternary nanoparticles. These inks can be applied by a variety of techniques to produce particle precursors for semiconductors and contacts. Subsequent thermal processing of these precursors produces a wide spectrum of results from simple phase formation to extensive sintering and grain growth. It can be seen that although nanoparticle precursors offer significant potential for improving low-cost, large-area microelectronics this will not be realized until more is understood about the intergrain reactivity of the particles and about the influence of the substrate on grain growth.

ACKNOWLEDGEMENTS

The gift of Al and Ag powder from Fred Tepper (Argonide Corporation) is appreciated. This research was funded by the U.S. Department of Energy, Office of

Energy Research, Chemical Sciences Division and Materials Science Division and the U.S. Department of Energy National PV Program.

REFERENCES

1. Schulz, D. S.; Pehnt, M.; Rose, D. H.; Urgiles. E.; Cahill, A. F.; Niles, D. W.; Jones, K. M.; Ellingson, R. J.; Curtis, C. J.; and Ginley, D. S. *Chemistry of Materials*, **9**, 889(1997).
2. Schulz DL, Curtis CJ, Flitton RA, Ginley DS (1998). Nanoparticulate Film Precursors to CIS Solar Cells: Spray Deposition of Cu-In-Se Colloids. *Materials Research Society Symposia Proceedings* 501:375-380.
3. Schulz DL, Ribelin R, Curtis CJ, Ginley DS. Particulate Contacts to Si and CdTe: Al, Ag, Hg-Cu-Te, and Sb-Te. *1998 National Center for Photovoltaics Annual Review AIP Conference Proceedings* (in press).
4. Hodes, G.; Howell, I. D. J.; Peter, L. M. Dep. Mater. Interfaces, Weizmann Inst. Sci., Rehovot, Israel. J. Electrochem. Soc. 139(11), 3316(1992).
5. Tenne, R.; Lando, D.; Mirovsky, Y.; Mueller, N.; Manassen, J.; Cahen, D.; Hodes, G.. Weizmann Inst. Sci., Rehovot, Israel. J. Electroanal. Chem. Interfacial Electrochem. **143**, 103-12 (1983),.
6. Licht, S.; Tenne, R.; Dagan, G.; Hodes, G.; Manassen, J.; Cahen, D.; Triboulet, R.; Rioux, J.; Levy-Clement, C. Weizmann Inst. Sci., Rehovot, Israel. Comm. Eur. Communities, [Rep.] EUR (1985), (EUR 10025, E.C. Photovoltaic Sol. Energy Conf., 6th, 1985), 59-63.
7. Guo, Yeping; Deng, Xunnan. Dep. Chem., Shanghai Univ. Sci. Technol., Shanghai, Peop. Rep. China. Sol. Energy Mater. Sol. Cells, 29, 115-22(1993),.
8. Schulz DL, Curtis CJ, Ginley DS, Tepper F. Nanosized Al and Ag Particulate Contacts to Silicon: Materials Characterization and Preliminary Electrical Results. Extended Abstracts of the 8th World Conference on Silicon, Copper Mountain, CO
9. Britt, J.S., and Ferekides, C.S. U.S. Patent 5 557 146, 1996.
10. Romeo, N., Bosio, A., Tedeschi, R., and Canevari, V., "High Efficiency and Stable CdTe/CdS Thin Film Solar Cells on Soda Lime Glass," presented at the 2nd World Conference and Exhibition on Photovoltaic Solar Energy Conversion, Vienna, Austria, July 6-10, 1998.

LOW TEMPERATURE ECR –PLASMA ASSISTED MOCVD MICROCRYSTALLINE AND AMORPHOUS GaN DEPOSITION AND CHARACTERIZATION FOR ELECTRONIC DEVICES

Z. HASSAN *, M.E. KORDESCH *, W.M. JADWISIENZAK **, H.J.LOZYKOWSKI**, W. HALVERSON ***and P.C. COLTER ***
*Department of Physics and Astronomy, Ohio University, Athens, OH 45701,
**Department of Electrical Engineering and Computer Science, Ohio University, Athens, OH 45701
***Spire Corporation, Bedford, MA 01730

ABSTRACT

GaN films have been deposited over a range of temperatures from 50 C to 650 C by ECR plasma MOCVD on silicon (111) and (100), sapphire and quartz using triethylgallium and molecular nitrogen or ammonia as reagents. Growth rates of 2 um/hr are achieved on temperature-controlled substrates (total reactor pressure 0.5 mtorr, 250 watts at 2.45 GHz).

Films deposited at 200, 600 and 650 C on sapphire show the GaN(0002) diffraction peak and sharp photoluminescence lines (at 10 K) between 370 and 400 nm and broad emission at 530-550 nm. Broad photoluminescence at 390 nm is observed from GaN/Si(111). Films deposited at 50 and 100 C show no evidence of a crystalline phase or GaN(0002) diffraction peak. The films are smooth and optically transparent. A broad photoluminescence peak at 520 nm, with a fwhm of about 150 nm is also observed (at 10K). The optical bandgap is measured to be about 2.6-2.7 eV. All of these films show a GaN LO phonon mode at 736 cm^{-1}. IR spectra indicate some hydrocarbon impurities in the low temperature films.

Prototype devices (Schottky barrier diodes) have been made from MOCVD GaN and amorphous GaN.

INTRODUCTION

Group III nitride semiconductors have received increasing attention in recent years due to advances in the epitaxial growth of high quality films as well as potential application for these materials as blue/UV-emitters/detectors and high power or high temperature electronic devices [1]. Recent calculations by Stumm and Drabold suggest that amorphous GaN may have utility as an electronic material. They predict a band gap of about 2.8 eV and no mid-gap states [2]. The special attribute of amorphous materials is the ability to deposit the material inexpensively over large areas. Thus, whereas crystalline GaN (c-GaN) is the material of choice at present for short wavelength emitters and detectors, the development of amorphous GaN (a-GaN) or microcrystalline GaN (μc-GaN) could open up opportunities for large-area devices such as UV detector arrays, high temperature transistors and large area LED arrays at low cost.

Growth of the GaN films was carried out using ECR (electron cyclotron resonance) plasma-enhanced MOCVD (metalorganic chemical vapor deposition). To date, there have only been reports on growth of a-GaN and μc-GaN using reactive sputtering [3,4]. The application of these films to transparent thin film transistor (TFT) have been demonstrated [5]. In this paper, the properties of a-GaN and μc-GaN and the fabrication of Schottky barrier diodes using μc-GaN are reported.

245

Mat. Res. Soc. Symp. Proc. Vol. 536 ° 1999 Materials Research Society

EXPERIMENTAL

Spire's ECR-heated plasma reactor, shown schematically in Figure 1, is an aluminum vacuum chamber 20 cm in diameter, 80 cm long, and evacuated by a 10" cryopump and a turbomolecular pump. A pair of solenoid coils form a magnetic "mirror" and an octupole permanent magnet array between the coils provide partial magnetic confinement for the plasma and electron cyclotron resonance conditions for the 2.45 GHz microwaves that ionize and heat the plasma.

The "hybrid polarizer" changes the linearly polarized TE_{01} waveguide mode of the microwaves to a circularly polarized TE_{01} mode. The right-handed circular polarization (RHCP) can propagate as "whistler" waves at plasma densities higher than classical cut-off (7.4×10^{16} m^{-3} for 2.45 GHz) and is strongly absorbed by parametric decay and electron cyclotron resonance. RHCP microwaves couple more efficiently with the plasma than linearly polarized waves; the polarizer also significantly reduced power reflected back into the magnetron tube by shunting left-handed circularly polarized (LHCP) waves from the plasma chamber into the "dummy" load that is coupled to the "left-hand" port of the hybrid polarizer.

The microwave source is a 2.45 GHz magnetron (AsTex Model AX2107) that provides up to 700 W of power. Forward and reflected power is sampled by a 60 dB directional coupler and diode detectors; the directional coupler is followed by a three-stub tuner to maximize forward power transmission, and, because reflected power is so low, no isolator is necessary to protect the magnetron. The microwaves are coupled to the "right-hand" arm of a hybrid polarizer (Atlantic Microwave Corp. Model WR340) with a transition from square to 9.09 cm diameter circular waveguide at the output end. The flange of the circular waveguide mates with a fused quartz window of 9.09 cm clear aperture and 6.35 mm thickness; the quartz window is the vacuum barrier between the waveguide and end-flange of the reactor chamber. An air-cooled load is coupled to the "left-hand" arm of the hybrid polarizer; this load absorbs LHCP waves that are reflected back from the plasma, quartz window, or the interior of the vacuum chamber. A directional coupler and diode measures LHCP power absorbed by the dummy load.

Nitrogen gas is admitted to the reactor through a variable flow valve that is feedback-controlled by the ion gauge. To provide a source of atomic N, a feedback-controlled ammonia source was installed on the system. The ammonia is metered by a commercial (Unit Instruments Model 1660) mass-flow controller with a maximum flow rate of 20 sccm; valves and piping for by-passing and purging the flow controller were provided to assure cleanliness and safe operation. The ammonia is injected through a central port in the vacuum chamber, between the solenoid magnet coils shown in Figure 1.

In addition to standard vacuum instrumentation, the principal plasma diagnostics on the ECR PECVD reactor were a Langmuir probe and an optical emission spectrometer. The Langmuir probe measures the local plasma density and floating potential, average electron energy (temperature), and energy distribution function in the region sampled by the probe. Optical emission spectroscopy (OES) provides information on the atomic and molecular species in the plasma as a function of time, position, and extrinsic parameters such as gas flow rates, pressure, microwave heating power, and magnetic field strength. The OES system can also make rough measurements of plasma density and electron temperature by comparing the intensity ratios of selected spectral lines.

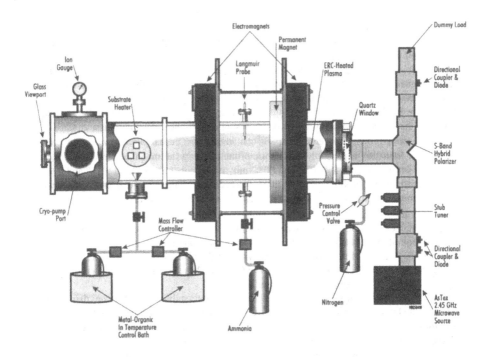

Figure 1: ECR Plasma enhanced CVD system.

GaN films were deposited over a range of temperature from 50°C to 650°C on silicon (100), and (111), sapphire and quartz substrates. The surface morphology of the films was observed using scanning electron microscopy (SEM). X-ray diffraction was used to assess the structure of the deposited films. These data were collected as θ-2θ scans on a Rigaku diffractometer with a wide angle automated goniometer and computer based data acquisition and analysis system. Photoluminescence (PL) measurements were performed at low temperatures (10K-12K) with excitation by light at 325 nm from a He-Cd laser. Optical transmission and absorption measurements at low and room temperature were done on GaN films deposited on quartz and sapphire substrates. The dark counts, transmission of the substrate and transmission of the film were collected through a lens, monochromator spectrometer and detected with a CCD camera. A 100W tungsten halogen lamp was used as the light source. Schottky barriers were formed by sputtering of Ni or Cr through a mask to form 0.25 mm diameter contacts (about 100 nm thick). Ti was used as the ohmic contact (about 150 nm thick). Film thickness and IR spectra were measured with an IR microscope.

RESULTS AND DISCUSSION

All films have very smooth surface morphology, regardless of deposition temperature, No microstructure was resolved in scanning electron microcopy to 0.1μm.

X-ray diffraction measurements of these samples reveal no crystalline structure for GaN films grown at 100°C and below. The (0002) peak for GaN at 34.6° is present for films grown at 200°C and above, which is indicative of a crystalline or microcrystalline phase due to the low intensity of this peak. The x-ray scans for samples that were grown after a modification of the microwave system in order to improve the microwave coupling with the plasma differ from the earlier results. This is probably due to the increased deposition rate as a result of more efficient plasma heating. For these samples, the x-ray diffraction spectra for films grown at 200°C and below do not exhibit any peaks corresponding to the crystalline phase of GaN. However a weak broad peak centered at ~35°can be seen for the sample that was grown at 600°C. This structure is probably a signature of the microcrystalline phase for GaN. The x-ray spectra for samples grown at 200°C and 80°C only reveal substrate peaks at 33.4° and 69.6° which correspond to Si(200) and (400) planes respectively. The (400) peak for Si is several hundred times more intense than the (200) peak. X-ray diffraction peaks for GaN should be observed at around 32.3°, 34.6° and 36.8° which correspond to ($10\bar{1}0$), (0002) and ($10\bar{1}1$) planes of the hexagonal c-GaN. The absence of the sharp crystalline peaks and a broadening centered around the crystalline region could be an indication that these films are a mixed phase of crystalline and amorphous structure. The width of the Si substrate peaks in our measurements average at about 0.1 degree, the GaN(0002) peaks measured for the samples deposited above 200 C are about 0.4 degrees fwhm.

The PL spectra of selected amorphous and microcrystalline GaN film are displayed in Figure 3. It should be noted here that the x-ray diffraction spectrum shows a weak (0002) GaN peak for the film that was deposited at 200°C which is a signature of a crystalline or microcrystalline structure. On the other hand, the film deposited at 100°C was amorphous based on the x-ray scans for samples that were deposited at this temperature using the same growth conditions.

A sharp peak at 369 nm near the band edge was observed for samples grown on sapphire at 650°C and 200°C and was absent for the sample that was grown at 100°C. The intensity of the near band edge peak relative to the "yellow" emission was greatest for GaN grown at 650°C. However in the case of PL spectra for amorphous GaN grown at 100°C, a broad deep level related emission centered at 619 nm (2.0 eV) was observed with very weak emission at 374 nm (3.31 eV) and 395.3 nm (3.13 eV). Similar PL spectra were also observed for GaN grown on quartz at 200°C and 100°C. The luminescence of the films is dominated by a broad deep level related emission centered at ~600 nm. A weak band-edge transition peak at 371 nm (3.34 eV) and a donor acceptor recombination peak at 391 nm (3.17 eV) were also observed. These two samples are amorphous based on the XRD scans of samples that were grown under the same conditions at this temperature which explains the remarkable similarity of the spectra even though the films are deposited at different temperatures. PL spectra of films grown on Si (111) at 650°C exhibited a broad "blue" emission band centered at 391 nm (3.17 eV). In contrast, the sample that was grown at 100°C which is amorphous exhibited a broad "green" emission centered at ~510 nm (2.40 eV). A similar PL spectra were obtained for a-GaN deposited at 100°C and 50°C on Si(100). Both samples exhibited a broad "green" emission centered at ~2.4 eV.

From the transmission data, the plot of the square of the absorption coefficient versus photon energy showed an optical band gap of 2.6-2.7 eV (at room temperature) and ~2.8 eV (at 22K).

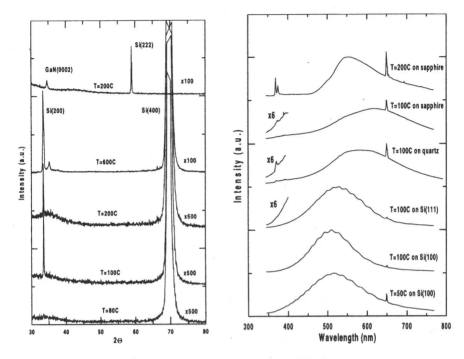

Figure 2: X-ray diffraction spectra
of amorphous and microcrystalline GaN films.

Figure 3 Photoluminescence spectra of
amorphous and microcrystalline GaN films

The I-V characteristic for Ni/GaN Schottky barrier diodes that were fabricated from samples grown on Si (111) at 200°C is shown in Figure 4. Prior to diode fabrication, x-ray diffraction spectra of this sample exhibited a weak (0002) peak for GaN at 34.6° which is indicative of a crystalline or microcrystalline phase due to the low intensity of this peak. Also RBS measurement performed on this sample shows a Ga to N ratio of 1:1. No oxygen is detected in this film. It should be noted here that an annealing treatment at ~400°C in a nitrogen ambient was performed after the deposition of Ti ohmic contacts. There was no annealing treatment conducted for the Schottky contacts.

Finally, cathodoluminescence images were obtained in the SEM. Consistent with the PL data, uniform luminescence was observed from all of the films. No microcrystalline grains or other structures could be resolved in the images. The IR spectra taken of the low temperature samples (50C) indicate that at this temperature some of the precursors, most probably the triethylgallium, may not completely react and some hydrocarbon impurities remain in the film. Unintentional doping by carbon can not be ruled out for these samples. The low temperature deposition samples were also highly strained, shredding spontaneously over time or upon insertion into the SEM or the contact deposition system. No diodes were fabricated from the 50 C samples for this reason.

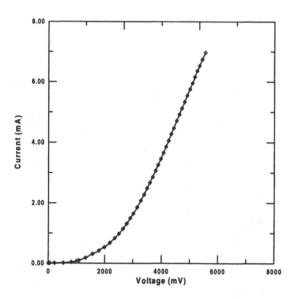

Figure 4 I-V characteristic of Ni/GaN Schottky diode.

CONCLUSIONS

The deposition of GaN films using ECR plasma enhanced CVD has been demonstrated. Under the experimental conditions used for this study, micro- or polycrystalline GaN is deposited above 200 C, and amorphous GaN is deposited at lower temperatures. The films are smooth, and can be deposited over large areas (typically 4 inch wafers, 2 µm/hr), with optical bandgaps measured to be in the range 2.6-2.8eV. Schottky barrier diodes were successfully fabricated from the microcrystalline GaN. PL data suggest that the films deposited at temperatures below 600 C do not show "blue" light emission, and may not be competitive with MOCVD GaN in those applications. The presence of hydrocarbon contaminants in films deposited at 50C may enhance luminescence at longer wavelengths due to unintentional doping, but may also cause mechanical stress in the films.

Acknowledgements:
This work was supported by the Ballistic Missile Defense Organization grant N00014-96-1-0782 and SBIR Phase II Contract 33615-96-C-2622 from the Air Force Research Laboratory.

1. S. Strite, H. Morkoc, J. Vac. Sci. Technol. **B10**, 1237 (1992).
2. P. Stumm, D. A. Drabold, Physical Review Letters 79 (4), 677 (1997).
3. S. Nonomura, S. Kobayashi, T. Gotoh, S. Hirata, T. Ohmori, T. Itoh, S. Nitta, K. Morigaki, Journal of Non-Crystalline Solids **198-200**, 174 (1996).
4. T. Hariu, T. Usuba, H. Adachi, Y. Shitaba, Appl. Phys. Lett. 32 (4), 52 (1978).
5. S. Kobayashi, S. Nonomura, T. Ohmori, K. abe, S. Hirata, T. Uno, T. Gotoh, S. Nitta, Applied surface Science **113/114**, 480 (1997).

NANOSTRUCTURED ARRAYS FORMED
BY FINELY FOCUSED ION BEAMS

R. A. Zuhr*, J. D. Budai*, P. G. Datskos*, A. Meldrum*, K. A. Thomas*, R. J. Warmack, C. W. White*, L. C. Feldman*·**, M. Strobel***, and K.-H. Heinig***
* Oak Ridge National Laboratory, Oak Ridge, TN 37831
** Vanderbilt University, Nashville, TN 37235
*** Research Center Rossendorf, D-01314 Dresden, Germany

ABSTRACT

Amorphous, polycrystalline, and single crystal nanometer dimension particles can be formed in a variety of substrates by ion implantation and subsequent annealing. Such composite colloidal materials exhibit unique optical properties that could be useful in optical devices, switches, and waveguides. However colloids formed by blanket implantation are not uniform in size due to the nonuniform density of the implant, resulting in diminution of the size dependent optical properties. The object of the present work is to form more uniform size particles arranged in a 2-dimensional lattice by using a finely focused ion beam to implant identical ion doses only into nanometer size regions located at each point of a rectangular lattice. Initial work is being done with a 30 keV Ga beam implanted into Si. Results of particle formation as a function of implant conditions as analyzed by Rutherford backscattering, x-ray analysis, atomic force microscopy, and both scanning and transmission electron microscopy will be presented and discussed.

INTRODUCTION

It has been shown that amorphous, polycrystalline, and single crystal nanoparticles can be formed in a variety of substrates by ion implantation under prescribed conditions combined with subsequent thermal annealing [1,2,3]. Typical sizes of these particles range from a few nanometers to several hundred nanometers in diameter. They may be metals or semiconductors and can be made of either single or multiple elements, including compound semiconductors, by using sequential implantation techniques. These composite materials frequently exhibit unique optical properties and may find applications in optical devices. One problem of composite materials formed in this way is that the particle size is generally not uniform. In addition, the particles are randomly distributed throughout the implanted region. The purpose of this work is to make uniformly spaced lattices of colloidal particles of more uniform size by using a finely focused ion beam (FEI, FIB200) to implant ions only into a microscopic region (typically 60 nm diameter) at each point of a two-dimensional array. Simulations using a kinetic lattice Monte Carlo code indicate that such localized implants should form a localized distribution of colloids that, under proper conditions, may Ostwald ripen into a single large colloid at each lattice site [4]. Under these ideal circumstances the colloids formed would be of nearly uniform size because of the identical particle dose implanted at each spot, and in addition they would be uniformly arranged on a two-dimensional lattice. Such a composite would exhibit greatly enhanced optical characteristics.

EXPERIMENT

Bulk implants of Ga and As into Si(100) at energies of 35 to 160 keV were used to study ion retention vs. dose and GaAs formation. They were made on an Extrion 200-1000 ion implantation accelerator at different doses and substrate temperatures in order to understand the behavior of Ga and As during the low energy implantation which is typical of focused ion beam (FIB) formation of patterned samples. The FIB samples themselves were implanted at 30 keV and room temperature for a variety of different doses, dose rates, beam diameters, and spot patterns using an FEI FIB-2000 focused ion beam system. Substrate temperatures were not controlled during FIB processing.

A number of techniques were employed to study the completed FIB patterned samples. Rutherford backscattering (RBS) using He-4 at 2.3MeV and a scattering angle of 160 degrees was used to determine the absolute amounts of Ga retained in the samples after processing. Scanning electron microscopy (SEM) on a JEOL 840 operating at 10 keV was used to observe the implanted pattern as well as the size and shape of the individual dots. Atomic Force Microscopy (AFM) with a digital Instruments Nanoscope III was effective in determining the geometry of the individual implanted dots, including the depth of holes and the height of raised dots, as well as the uniformity of the over all pattern. Low resolution TEM images were taken on a Philips EM400 operating at 100 keV, while high resolution images and EDS spectra were made on a Philips CM200 at 200 keV.

RESULTS AND DISCUSSION

Saturation Concentration

Since the FIB was limited to 30 keV, it was necessary to determine the sputtering coefficient for 30 keV Ga and, as a result, the saturation concentration of Ga that could be implanted at this energy. Book values for the sputter coefficient ranged around 2 [5]. We measured the retained saturation amount of Ga directly as a function of incident fluence for Ga on Si. These results are shown in Fig. 1, which includes values at different substrate temperatures, as well as values for As at the same energy, and a retention curve calculated with the analytical code Profile [6]. In order to fit the experimental Ga data at liquid nitrogen temperatures, a sputtering coefficient of 2.8 was required in Profile. This is close to the tabulated value assumed by Profile (3.1) and is the value that will be used in our discussions. Note that above room temperature, very little of the incident Ga is retained, due to evaporation from the surface, while at room temperature an interesting phenomenon occurs. It appears that significantly more Ga can be retained at room temperature than at LN$_2$. More careful

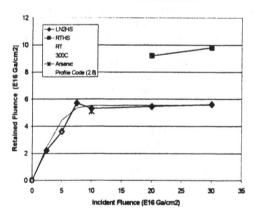

Fig. 1 Retained vs. incident fluence for 35 keV Ga and As on Si measured by RBS at liquid nitrogen, room, and 300C temperatures. Also shown is a Profile Code calculation for a sputter coefficient of 2.8.

investigation of the RBS spectra reveals that this Ga is concentrated on the surface as evidenced by the shift of the Si edge to lower energy. It appears that for the room temperature samples, the Ga is sufficiently mobile to form a surface layer of near liquid (Ga melts at 29C) Ga globules that is less effectively sputtered by the incident Ga beam. These globules can be observed optically and are sufficiently "liquid" that they can be blown from the surface with air.

Low Energy Formation of GaAs by Blanket Implants

The next step was to determine whether GaAs could be formed in Si by ion implantation at energies as low as 30 keV as required by the FIB system. We had previously shown that crystalline GaAs could be formed during ion implantation of the individual elements at both 500 keV and 150 keV with subsequent thermal annealing,* but there were indications that proximity to the surface might result in loss of material rather than compound formation. To study this, Si(100) samples were implanted sequentially at 35 keV with 5×10^{16} each of Ga and As/cm^2 at 500C, both with As first, and with Ga first. As seen in the RBS spectra of Fig. 2a, most of the Ga is lost from samples in which the Ga is implanted first, so studies were concentrated on

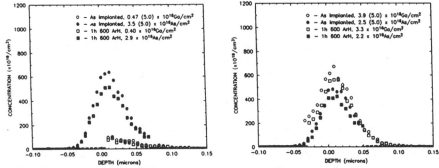

| Fig. 2a RBS spectra of 35 keV Ga + As, 5×10^{16} each in Si, as implanted, and annealed at 600C. | Fig. 2b RBS spectra of 35 keV As + Ga, 5×10^{16} each in Si, as implanted, and annealed at 600C. |

samples implanted first with As and subsequently with Ga. Fig. 2b indicates that most of the Ga and As are retained when implanted in this order. Although similar samples implanted at 150 keV show clear x-ray evidence for the formation of crystalline GaAs, the samples implanted at FIB energies (35 keV) do not. Since formation of GaAs in blanket implanted samples had not been demonstrated, this work on FIB formation of colloids was directed initially toward the formation of elemental Ga particles.

Theoretical Simulation of FIB Implantation

Kinetic 3D lattice Monte-Carlo simulations are a suitable tool for modeling all basic physical processes of ion beam synthesis of nanocrystals.* For the simulations presented here, a kinetic Ising model is used to incorporate effective Ga-Ga interactions within a neutral matrix (Ga-Si interactions can thus be ignored). For a symmetric initial distribution of implanted ions, the code in general predicts aggregation into multiple colloids, with possible Ostwald ripening into single particles at each lattice site. Discussing the present physical parameters, it is obvious that no spherical single precipitates are to be expected. For a nominal beam diameter of 120 nm (FWHM) and a projected range of about 25 nm, at best a single disk should be formed because of the large lateral to vertical ratio of the effective deposition volume. For simplicity the

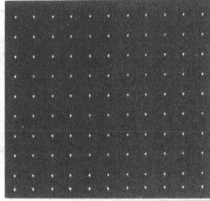

Fig. 3a Optical micrograph of 30 eV Ga 64x64 row pattern on Si. Doses vary from $2x10^{10}$ on the left to $1x10^6$ on the right. Square is 1.1x1.1 mm.	Fig. 3b Optical micrograph of Ga dots on Si at 30 keV and $2x10^{10}$ Ga/dot (left side of Fig. 2a) Spacing between dots is 20 microns.

evolution of a single spot was studied starting with an idealized initial configuration matching the assumed Gaussian-like deposited Ga density used in this work. According to the preliminary MC simulation for this set of parameters, formation of a central disk-like nanocrystal is not clear.

Array Formation Using a 30 keV Ga Focused Ion Beam

The 30 keV Ga focused ion beam from the FIB-2000 was used to produce arrays of implanted spots at a variety of different doses. Nominal beam diameters of 60 and 120 nm and currents of 170 and 1000 pA, respectively, were used to form 1.1 mm square patterns of implanted spots in 2-dimensional rectangular lattices on Si substrates. In general, in spite of careful iterative focusing of the ion beam, the observed implant spots are far larger (200 x in area) and less symmetric than the nominal beam size. Optical micrographs of Fig. 3 show a grid of spots implanted with the 30 keV 120 nm beam at spacings of 20 microns between spots. Doses on this sample were increased by a factor of 4 after every 8 rows from a minimum of $1.25x10^6$ to $2.05x10^{10}$ Ga/spot over the 64 row grid, so that a wide range of incident fluences

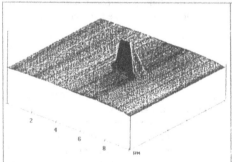

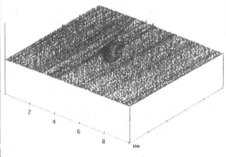

Fig. 4a AFM inverted micrograph of deep hole produced by $2x10^{10}$ Ga/spot at 30 keV (Region 1).	Fig. 4b AFM micrograph of raised dimple produced by $2x10^7$ Ga/spot at 30 keV (Region 6).

Table I: Doses Based on AFM Measured Spot Areas

Region	Ions/Spot	Area cm2	Dose Ions/cm2	Topography
1	2.05E+10	2.85E-08	7.18E+17	Deep Hole (2 micron)
2	5.12E+09			Deep Hole
3	1.28E+09	1.24E-08	1.03E+17	Hole
4	3.20E+08			Hole
5	8.00E+07	1.12E-08	7.14E+15	Dimple with Hole
6	2.00E+07			Dimple
7	5.00E+06			Smeared
8	1.25E+06			No Spots

would be present. The purpose of this sample was to determine, in light of the disparity between nominal and observed spot sizes, the most effective dose per area for colloid formation. The results in terms of ions/cm^2 of observed spot area and surface topography as measured by AFM are shown in table I. As shown in Fig. 4, the higher doses produce deep holes, the lightest doses leave the substrate unaffected, and intermediate doses produce a slight rise in the implanted region. The raised areas at moderate doses are presumed to be due to incorporation of Ga along with the conversion of the area from single crystal Si to the less dense amorphous phase. The high aspect ratio holes at high doses clearly represent a poor geometry from which to form the desired single uniform size colloid at each lattice site, while the lowest doses provide insufficient Ga to produce a colloid. Therefore, the moderate dose regions, such as region 5 in the table, with a dose of 7E15 ions /cm^2, should be most interesting.

The presence of Ga in the implanted dots has been observed by several techniques. Confirmation of the retention of Ga has been shown by RBS to be 1.1×10^{14} Ga/cm^2 over the entire pattern, which is in agreement with the amount expected to be retained after sputtering. The qualitative presence of Ga in each individual lattice spot has been demonstrated by x-ray analysis by scanning the pattern with a 1.5 micron white x-ray beam and observing the resulting Ga line, which is observed only when the beam is on a spot. Fig. 5 shows a TEM image of a particle of Ga formed near the edge of an implanted spot taken from region 4 in the table. The particle is about 18 nm in diameter and can be positively identified as Ga from the EDS spectrum in the figure. Therefore small Ga colloids are formed at the lattice points, but uniform single colloids have not yet been observed.

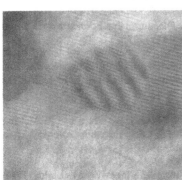
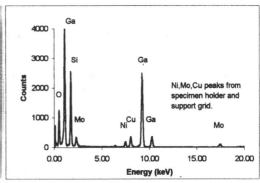

| Fig. 5a TEM micrograph showing Moire fringes of Ga particle formed by 3.2×10^8 30 keV Ga/spot (Region 4). | Fig. 5b EDS spectrum from particle in 5a indicating that the particle is Ga. |

255

CONCLUSION

It has been demonstrated that a 2-dimensional lattice of Ga dots can be written on a Si substrate using a programmed finely focused ion beam at 30 keV. Aggregation of the roughly Gaussian implanted profile into a single particle at each of the lattice sites has not been demonstrated, but it is clear that Ga is retained in the substrate and does form small crystalline colloids. The dose and energy, and thus the depth, of the implanted ions appears to be critical in the formation of larger single particles, since the Ga ions must be implanted into a relatively undisturbed region of the substrate in order to form a single colloid. High doses drill holes into the substrate surface (See Fig. 5), leaving Ga spread around the wall and bottom of the hole, which produces an unsuitable geometry from which to diffuse the Ga into a single particle. Therefore the fluence should be limited to a value which leaves the implanted regions of the surface relatively flat. Based on the current work, such doses are on the order of 1×10^{16}ions/cm^2. In addition, for the formation of compound semiconductors, the individual elements, Ga and As, must buried deep enough beneath the surface so that they do not escape from the surface during annealing or even during the implant itself. To date, using a blanket implant of 35 keV As followed by localized 30 keV FIB implants of Ga, we have not been able to produce GaAs quantum dots, although we have demonstrated that GaAs nanoparticles can be formed by blanket sequential implantation of the individual elements at 150 keV. Several techniques are proposed to produce such compound semiconductor quantum dots by the FIB process in the future, including higher energy (150 keV) FIB, and capping of the pattern with SiO$_2$ prior to high temperature annealing to form the GaAs. These ideas are currently undergoing further study.

ACKNOWLEDGMENT

The authors are indebted to Chuck Egert, who recently passed away, for arranging this collaboration. Oak Ridge National Laboratory is managed by Lockheed Martin Energy Research Corp. for the U. S. Department of Energy under contract number DE-AC05-96OR22464.

REFERENCES

1. G. W. Arnold, J. A. Borders, J. Appl. Phys. **489**, p. 1488 (1977).

2. C. W. White, et al., "Encapsulated Nanocrystals and Quantum Dots Formed by Ion Beam Synthesis," Nucl. Instrum. Meth. Phys. Res. B **127/128**, p. 545 (1997).

3. R. A. Zuhr and R. H. Magruder, III, "Optical Properties of Multi-Component Metallic Nanoclusters Formed in Silica by Sequential Ion Implantation," invited paper, Journal of the Surface Science Society of Japan **18, #5**, p. 269 (1997).

4. M. Strobel, K.-H. Heinig, W. Moller, A. Meldrum, D. S. Zhou, C. W. White, and R. A. Zuhr, "Ion Beam Synthesis of Gold Nanoclusters in SiO2: Computer Simulations Versus Experiments," Nucl. Instrum. Meth. Phys. Res. B, to be published.

5. *Sputtering by Particle Bombardment*, ed. by R. Behrisch, Springer-Verlag, Berlin, 1981, p.169.

6. The Profile Code is a product of Implant Sciences Corporation, Wakefield, MA.

OPTICALLY DETECTED MAGNETIC RESONANCE AND DOUBLE BEAM PHOTOLUMINESCENCE OF CdS/HgS/CdS NANOPARTICLES

H. PORTEANU*, A. GLOZMAN*, E. LIFSHITZ*, A. EYCHMÜLLER**, H. WELLER**
*Department of Chemistry and Solid State Institute, Technion, Haifa 32000, Israel,
e-mail: ssefrat@tx.technion.ac.il
**Department of Chemistry, Hamburg University, Hamburg, Germany,
e-mail: weller@chemie.uni-hamburg.de

ABSTRACT

CdS/HgS/CdS nanoparticles consist of a CdS core, epitaxially covered by one or two monolayers of HgS and additional cladding layers of CdS. The present paper describes our efforts to identify the influence of CdS/HgS/CdS interfaces on the localization of the photogenerated carriers deduced from the magneto-optical properties of the materials. These were investigated by the utilization of optically detected magnetic resonance (ODMR) and double-beam photoluminescence spectroscopy. A photoluminescence (PL) spectrum of the studied material, consists of a dominant exciton located at the HgS layer, and additional non-excitonic band, presumably corresponding to the recombination of trapped carriers at the interface. The latter band can be attenuated using an additional red excitation. The ODMR measurements show the existence of two kinds of electron-hole recombination. These electron-hole pairs maybe trapped either at a twin packing of a CdS/HgS interface, or at an edge dislocation of an epitaxial HgS or a CdS cladding layer.

INTRODUCTION

Several recent reports have described the development of hetero-nanoparticles [1-7]. The latter materials are composed of a semiconductor core that is coated with several shells of different semiconductors. Examples are the CdS core particles coated with $Cd(OH)_2$, the CdSe core coated with ZnS or CdS, and the HgS core coated with CdS. The most advanced form of hetero-nanoparticles are the onion-like particles, built up by a core of CdS covered by an interior shell of HgS and additional outer cladding layers of CdS. The CdS/HgS/CdS nanoparticles can be synthesized by a wet chemistry method, leading to the formation of or tetrahedral [7] shapes. The wide band gap (2.5 eV) of CdS and the narrow band-gap of HgS (0.5 eV), cause a localization of some of the electron and hole states within the HgS internal layer. Therefore, the last can be viewed as a quantum well, and the entire CdS/HgS/CdS structure, as a quantum-dot quantum-well (QDQW). Then, the electron-hole (e-h) excitation and recombination energy in these structures is determined by the global confinement of the entire nanoparticle, local confinement in the internal quantum well, and the electron-hole pair interaction.

While the excitonic transitions in QDQW samples had been investigated in the last few years, the understanding of the interface influence on the optical properties has been neglected. Indeed, for particles in such a small regime, a large percentage of the atoms is on/near the internal interfaces or the external surfaces. These sites may act as electron and hole traps. Thus, the present work describes our attempts to clarify the optical properties associated with trapped electrons and holes, utilizing optically detected magnetic resonance (ODMR) and double-beam photoluminescence spectroscopy.

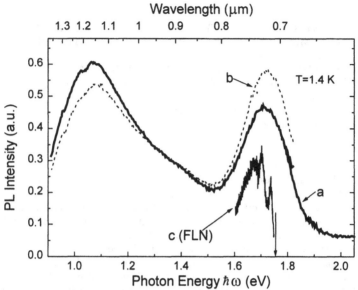

Figure 1: Photoluminescence spectra of the 1.3 CdS/HgS/CdS nanoparticles excited with (a) 2.7 eV, (b) 2.7eV and 2.0 eV, and (c) 1.75eV (FLN, shown by arrow in the figure).

EXPERIMENT

The synthesis of the CdS/HgS/CdS nanoparticles was carried out in a colloidal solution, by the procedure discussed in reference [8]. The present study included the investigation of CdS core (5.4 nm in diameter) covered by one or two monolayers of HgS, and one up to three cladding layers of CdS. The prepared samples are labeled hereon as X.Y (X = number of HgS monolayers, Y = number of CdS cladding monolayers). The particles were covered with organic surfactant (tetra-butyl-ammonium hydroxide), in order to protect the external surface and enable to re-dissolve the particles in polymer solution of polyvinyl butyral covinyl alcohol. The resulted polymer films, embedded with CdS/HgS/CdS nanoparticles, were utilized for the optical measurements discussed below. The PL-spectra were recorded using an Acton Spectrometer and Si and Ge detectors. For ODMR we used a 10 GHz microwave setup and an optical cryostat equipped with a 3 T superconducting magnet [9].

RESULTS

The PL spectra of CdS/HgS/CdS samples were recorded at 1.4 K excited under different conditions. Representative spectra of the 1.3 sample are shown in figure 1 (a-c). Spectrum (a) was excited above the CdS band-gap. It consists of an excitonic and an additional low energy band. Spectrum (b) was excited with two laser beams, above the CdS band-gap (modulated) and just above the excitonic line (CW). Addition of the red excitation quenched partially the low energy luminescence band and enhanced the excitonic band. Spectrum (c) was resonantly excited with an energy within the excitonic band, leading to a fluorescence line narrowing (FLN) of the exciton spectra and to a resolution of LO phonon bands. The ODMR spectra were obtained by recording the change in the luminescence intensity at energies<1.46 eV versus strength of the

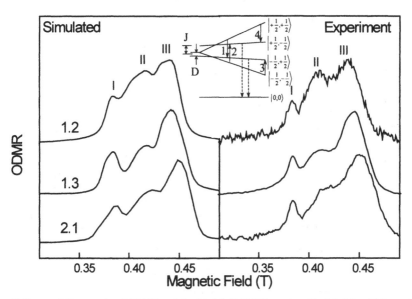

Figure 2: Representative experimental (right) and simulated (left) ODMR spectra of the 1.2, 1.3 and 2.1 samples.

external magnetic field, in the presence of a modulated microwave field. Representative ODMR spectra of the samples 1.2, 1.3 and 2.1 are shown on the right column of figure 2. The left column corresponds to simulated line-shapes, while the inset corresponds to the spin manifold at the excited state (vide infra). All the ODMR spectra of the various samples consist of three magnetic resonance bands (labeled I, II and III in the figure 2), ranging between 0.35-0.50 Tesla, with corresponding g-factors as discussed in conclusion.

The ODMR spectra of the mentioned samples were examined at different experimental conditions. The spectra of the 1.3 sample, recorded at different microwave modulation frequencies are shown in figure 3, as a representative example. Careful observation of the last figure suggests that resonance bands III and I alter simultaneously, while resonance II has different characteristic. Similar behavior was observed upon a change in the laser excitation power (not shown). This suggests that the ODMR spectra consist of two overlapping magnetic resonance events, and more likely, resonance III and I correspond to a common incident. Examination of the microwave modulation frequency and laser power dependence of the ODMR spectra of samples 1.2 and 2.1 shows a similar behavior to that described for the 1.3 sample.

DISCUSSION AND CONCLUSIONS

The PL spectra, shown in figure 1, contain an excitonic band that is red-shifted with respect to the CdS energy band, but is blue-shifted with respect to the HgS bulk energy gap. Furthermore, the energy of this exciton is predominantly influenced by the thickness of the HgS shell (not shown), while this energy is only slightly adjusted by the thickness of the CdS cladding layer. Thus, it was previous suggested [7] that the exciton is associated with optical transitions between the lowest quantized energy levels of the HgS quantum well. The PL spectra also suggest the existence of trapped electron-hole recombination. The mechanism of this

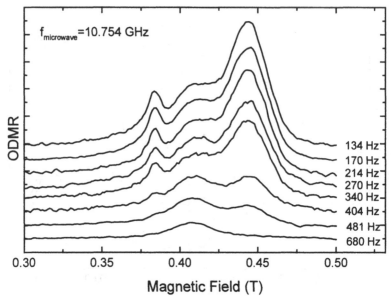

Magnetic Field (T)

Figure 3: The ODMR spectra of the 1.3 sample, recorded at different microwave modulation frequencies.

recombination and the identification of the recombining sites were investigated by the analysis of the ODMR measurements.

The ODMR spectra of the various QDQW samples consist essentially of the same resonance signals, as shown in figure 2. Although resonance I appears slightly narrower than the others, the dependence on the experimental parameters suggests that resonance III and I are associated with a similar incident while resonance II has an independent behavior. Thus, the ODMR spectra consist of two overlapping electron-hole recombination events. Each one of them can be simulated by the following phenomenological spin Hamiltonian:

$$Hs = \beta S_e g_e H_0 + \beta S_h g_h H_0 + S_e D S_h + J S_e S_h \tag{1}$$

The first two terms in equation (1) correspond to the effective Zeeman interaction of an electron and hole, the third term corresponds to the zero field splitting (may be overlapped by an anisotropic exchange interaction) while the last term, to the isotropic exchange interactions. The excited state can simply be described as a spin manifold of the electrons (e) and holes (h), each with an effective spin quantum number of S=1/2, and corresponding spin projections on the external magnetic field direction of m_S (e)=±1/2 and hole m_S (h)=±1/2. This spin manifold is shown in the inset of figure 2, where the numbers 1 - 4 indicate the spin resonance transitions, while the dashed lines designate the optical transitions. J corresponds to an isotropic e-h exchange interaction while D to the zero-field splitting.

Resonance signals III and I resemble a case with a relatively weak exchange interaction. Thus, in the first stage we simulated a situation with a relatively small J and D and isotropic g-factor. This led to a narrow and symmetric resonance signal, in contradiction with the experimental spectra. Incorporation of anisotropy in the g-factor showed an asymmetric broadening or singularities in the resonance signals. The last broadening did not predict the sharpness of resonance I and the relative intensities between resonance I and III. However, the latter discrepancies were overcome by the consideration of thermalization of population between levels. In order to make a best-fit with the experimental spectra, additional broadening associated with the distribution of J values was essential. Thus, consideration of anisotropic g, distribution of J values and thermalization factor leads to a best fit of resonance I and III. Observation of resonance II suggests that it is associated with electron-hole pair recombination with relatively strong isotropic exchange interaction (J»gβH₀) and D=0.02 meV«J. The small D and the extremely large J value leads to coalescence of the resonance into a single band with an average g-factor $g_{avr}=(g_e+g_h)/2$ [10]. Thus, the summation of the simulation of resonance I & III and II, leads to the best-fit of the experimental spectrum, as shown by the left column of figure 2.

The g values of the different resonance signals range between 1.65-1.96 with about 10% anisotropy. These values deviate substantially from those of the valence and conduction band [11] and thereupon, suggest that the ODMR phenomenon corresponds to a recombination between states with wavefunctions that are not necessarily associated with CdS or HgS band edges. The J value of resonance I & III is about 0.04 meV, while that of resonance II is J»gβH₀. Both values deviate from a typical exciton exchange interaction in II-VI nanoparticles [12] (0.03-0.3 eV), supporting the non-excitonic character of the studied recombination emission.

The anisotropy of the trapping sites, associated with resonances I and III excludes the possibility of their location at substitutional or interstitial sites within a cubic zinc blende nanoparticle. Instead, it suggests localization at an interface or surface. It is presumed that resonance II is associated with different chemical imperfections. Several suggestions regarding the chemical nature of the trapping sites can be raised. Mews *et al.* [7] showed the crystallization of CdS/HgS/CdS nanoparticles have tetrahedral shapes with (111) facets. They also indicated the possibility of twin grain boundaries between the HgS and the CdS cladding layer. Such a boundary causes the appearance of vacancies at every second atom, as shown schematically in inset (a) of figure 4.

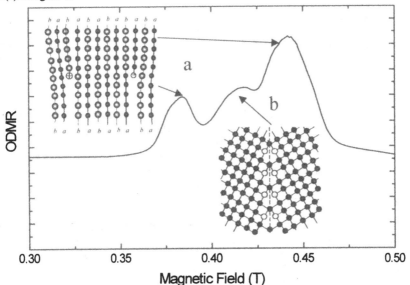

Magnetic Field (T)

Figure 4: Plausible trapping sites: (a) twin grain boundary in the (112) interface in a cubic crystal, (b) edge dislocations in *AB* - compounds.

Furthermore, in the zinc blende crystallographic unit, with three atomic layers (ABC), these vacancies can be either in the metal (Cd, Hg) or in the S atomic layer. Cd^{+2} vacancies and S^0 vacancies act as double acceptors. However, trapping of a single hole alone leads to the creation of paramagnetic centers that can be detected by the ODMR method. Cd^0 and S^{-2} vacancies act as double donors, and in a similar manner, the trapping of one electron will be detected by the ODMR method. In other words, photo-generated electrons and holes can be trapped at vacancy defects that are created at the twin grain boundaries in QDQW structures. The unique principal axis, (the spectroscopic g-factor) is considered to be normal to the boundary while the other axes are tangential to it. Accordingly, the distribution of exchange interaction values can be due to the existence of an ensemble of e-h pair distances, more likely, around the periphery of an interface. The twin grain boundaries can very well be described by resonance signal II in the ODMR spectra shown in this paper. The occurrence of a vacancy on any second atom at the twin boundary, and the appearance of the latter in 50% of the QDQW particles [7], both lead to a condensed distribution of trapped electrons and holes that consequently causes a strong exchange interaction.

Also, it was shown that the growth of a monolayer is not perfect [7] and it can be terminated by an edge dislocation, as shown schematically in inset (b) of figure 4. In a similar manner to the described vacancies, such terminations have unsaturated chemical bonds, creating electron and hole trapping sites. Likewise, these edges have asymmetric local chemical bonds, reflected in an asymmetric g-factor. Since these edge dislocations are distributed randomly, more likely, they will be spaced apart from more than two-chemical bonds and thereupon, the exchange interaction of carriers trapped in those sites will be relatively small. Hence, the edge dislocations can be associated with trapped e-h pairs that produce the resonance signals III and I in the studied ODMR spectra.

ACKNOWLEDGMENTS

The author express their deep gratitude for the support of the German-Israel Foundation, grant no. 414.021.10. HP express his thank for the Israeli Council for Higher Education Postdoctoral Fellowship.

REFERENCES
1. M. G. Bawendi, P. J. Carrol, W. L. Wilson and L. E. Brus, J. Chem. Phys. **96**, 946 (1992).
2. A. Eychmüller, A. Hasselbarth, L. Katsikas and H. Weller, Ber. Bunsenges. Phys. Chem. **95** (1), 79 (1991).
3. L. Spanhel, M. Haase, H. Weller and A. Henglein, J. Am. Chem. Soc. **109**, 5649 (1987).
4. U. Woggon, W. Petri, A. Dinger, S. Petillon, M. Hetterich, M. Grun, K. P. O'Donnell, H. Kalt, and C. Klingshirn, Phys. Rev. B **55** (3), 1364 (1997).
5. Yongchi Tian, J. Phys. Chem. **100**, 8927 (1996).
6. H. S. Zhou, I. Honma, J. W. Haus, H. Sasabe, H. Komiyama, J. of Lumin. **70**, 2 (1996).
7. A. Mews, A. V. Kadavanich, U. Banin, A. P. Alivisatos, Phys. Rev. B **53** (20), 13242 (1996).
8. A. Eychmüller, A. Mews, H. Weller, Chem. Phys. Letters. **208**, (1) 59 (1993).
9. E. Lifshitz, I. Dag, I. D. Litvin, G. Hodes, J. Phys. Chem. (1998) in press.
10. E. Lifshitz, I. D. Litvin, H. Porteanu, Chem. Phys. Lett. 11591, (1998) in press.
11. J. J. Hopfield and D. J. Tomas, Phys. Rev. B **122**, 35 (1961).
12. M. Nirmal, D. J. Norris, M. Kuno, M. G. Bawendi, Al. L. Efros and M. Rosen, Phys. Rev. Lett. **75**, 3728 (1995).

OPTICAL AND ELECTRICAL PROPERTIES OF CdTe NANOCRYSTAL QUANTUM DOTS PASSIVATED IN AMORPHOUS TiO$_2$ THIN FILM MATRIX

A. C. RASTOGI*, S. N. SHARMA AND SANDEEP KOHLI
Division of Electronic Materials, National Physical Laboratory, K.S.Krishnan Road,
New Delhi 110 012, INDIA, *E Mail : alok@csnpl.ren.nic.in

ABSTRACT

CdTe nanocrystal quantum dots sequestered in TiO$_2$ thin film matrix have been synthesized by r.f. sputtering from a composite CdTe/TiO$_2$ target. CdTe nanocrystal formation is nucleation controlled as their size (11-25 nm), dispersion and volume fraction (0.065-0.2) increases with film thickness, substrate temperature (100° C) and thermal treatment. The optical band gap derived from the onset of absorption coefficient showed blue shifts concurrent with the CdTe nanocrystal size reduction due to quantum size effects. These shifts, not consistent with theoretical models based on strong or weak confinement regimes, are explained on the basis of anisotropic growth and formation of CdTe nanocrystal clusters. TiO$_2$, in addition to being an ideal passivator and providing a barrier for carrier confinement to observe quantum effects, shows O$_2$ vacancy dependent conductivity modulation. Electrical conductivity variation with CdTe nanocrystal size and density is attributed to electrical coupling and tunneling behavior of carriers between CdTe nanocrystallites.

INTRODUCTION

Composite thin films of semiconductor nanocrysta ls sequestered in a dielectric matrix represent a simple zero-dimensional structure which exhibit quantum size effects relevant to the size of the nanoparticles. Novel properties [1,2], such as strong optical nonlinearity, increased optical band gap, size dependent electroluminescence emission, exhibited by these films arise due to strong 3-dimensional quantum confinement. In these nanocrystals, often referred to as quantum dots, charge carrier confinement causes band like energy level to become discrete and oscillator strength gets concentrated due to sharp electron-hole transitions. Consequently, profound changes observed in optical properties of nanocrystal quantum dots, especially in direct gap II-VI semiconductors have been a subject of intensive study due to possible photonic applications [3-6]. CdTe nanocrystals embedded in SiO$_2$ and glass matrix have been widely studied in this context [5,6]. We have prepared CdTe nanocrystal quantum dot structures by surface passivation in TiO$_2$ thin film matrix by r.f. sputtering technique. The choice of TiO$_2$ is unique, as due to large band gap (3.5 eV), it not only provides high barrier for carrier confinement in CdTe nanocrystals, its electrical conductivity can be changed by several orders depending on oxygen vacancies present. This offers immense possibility to electrically couple CdTe nanocrystals in search for novel electrical properties. This paper presents early results on synthesis, optical and electrical properties of CdTe nanocrystal quantum dots passivated in TiO$_2$.

EXPERIMENT

TiO$_2$ thin films with CdTe nanocrystal dispersions were prepared by r.f. magnetron sputtering from a composite target consisting of thin 10 mm dia CdTe pellets placed over 50 mm dia sintered TiO$_2$ disc. Sputter deposition was done in Ar ambient at pressure of 0.02 mbar and r.f. power of 240-270 Watts. Thin films were deposited over polished quartz substrates for optical measurements, while Corning glass and conducting oxide coated glass substrates for electrical and cleaved rocksalt crystals for electron microscopy studies. Two sets of films, one having CdTe volume fraction < 0.2 and the other ranging from 0.2 to 0.6 were prepared. In the former case the deposition rate was 0.2-0.3 Å/s and variation in CdTe nanocrystal dispersion density was obtained by increasing the deposition temperature to 100°C, changing the film thickness and annealing of films at 400°C in H_2+N_2 ambient for 15-30 min. Increase in CdTe nanocrystal density over 0.2 volume fraction was achieved by increasing the number of CdTe pellets covering the TiO_2 target as well as depositing at elevated temperature. These films were of constant thickness of 0.5 μm deposited at rates 0.4-0.8 Å/s. The optical studies were carried out by measuring reflectance (R_f) and transmittance (T_f) and the data were corrected for interface and substrate attenuations using standard procedure. The absorbance (A_f) was evaluated using the relation $A_f=1-(R+T_f)$. Microstructure was observed in Jeol electron microscope JEM 200CX and crystallographic structure by X-ray diffraction (XRD) using CuKα radiation.

RESULTS AND DISCUSSION

Structure of CdTe Nanocrystals

Fig. 1 shows transmission electron micrograph of a 1500 Å thick CdTe TiO_2 composite film. CdTe nanocrystals have a nearly spherical shape and are of an average size ~ 12.5 nm. These are homogeneously dispersed in an amorphous TiO_2 thin film matrix which essentially has no structural features. The electron diffraction pattern in the inset shows diffused ring-like pattern. Due to low (~0.065) dispersion density of CdTe nanocrystals in the composite film, the diffraction pattern is superimposed with broad halos from amorphous TiO_2 matrix preventing any crystallographic analysis. Crystal structure was thus studied by XRD analysis in thicker, 0.5 μm composite films having fraction of CdTe dispersion >0.2. XRD spectra of these film in Fig. 2 show a well centered peak at 2θ=25° which becomes sharper as CdTe fraction in TiO_2 matrix increases. It corresponds to (111) diffraction plane from cubic (zinc blende) structure. Due to nano size of crystals, diffraction gives fewer peaks and peak spread decreases with increase in crystal size and dispersaion density.

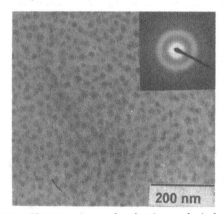

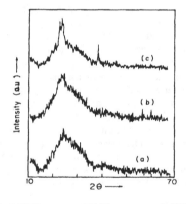

Fig.1 Electron micrograph showing a spherical 12.5nm CdTe nanocrystal dispersion in TiO_2 film

Fig.2 XRD spectra of CdTe nanocrystal: TiO_2 film with increasing fraction curves (a) to (c)

Optical Properties

TiO_2 films were sputter deposited under conditions identical to formation of CdTe nanocrystal. Initailly, TiO_2 composite films were studied. These did not show any absorption features in the spectral region of interest for CdTe. TiO_2 films showed high >85% transmission in the wavelength range 350-1200 nm and no sub-band absorption from defect states due to oxygen vacancies. Optical absorption behaviour of nanocrystal CdTe:TiO2 composite films was strongly modulated depending on the size and concentration of CdTe nanocrystallites. These attributes of CdTe crystallites are controlled by thickness and the substrate temperature of film growth. Detailed microstructure studies [7] have shown that in composite films, increase in thickness by continued sputtering results in nucleation of additional crystallites rather than increase in size of those already nucleated. The TiO_2 co-depositing at a higher rates rapidly passivates the CdTe nanosize particles, preventing their further growth by direct condensation. On the other hand, film depositions carried out at 100°C show an increase in size and dispersion density due to coalescence. Optical absorption spectra of a 2500

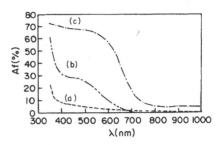

Fig 3 Absorption spectra (a)RT (b)100°C deposited 2500Å thick nanocrystal CdTe: TiO_2 film (c) polycrystalline CdTe film.

Å film deposited at RT and 100°C is shown in Fig.3. Generally, a shift in absorption edge of sputter deposited polycrystalline CdTe film (curve c) towards higher energy side (blue shift) along with a reduction in absorbance for CdTe nanocrystal:TiO_2 thin films is observed. The absorbance of composite film deposited at 100°C having an average CdTe nanocrystal size ~25 nm is characterized by curve b. Room temperature deposited films have much smaller crystals typically, 12.5 nm and show a further shift towards higher energies. The change in slope of absorption curve below 400 nm is characteristic of direct band transition belonging to TiO_2 thin film matrix.

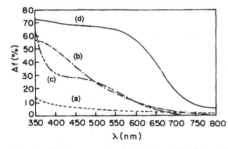

Fig.4 Thickness dependence of absorption spectra of nanocrystal CdTe:TiO_2 film deposited at 100°C (a) 500 (b)1500 (c) 2500Å (d) polycrystalline CdTe film

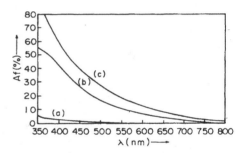

Fig.5 Change in absorption values for after thermal treatment in H_2 at 400°C (a) 500 (b)1500 (c)2500Å nanocrystal CdTe:TiO_2 film

A systematic effect of film thickness on the optical absorption spectra for composite films deposited at 100°C in as-deposited state and after thermal treatment at 400°C is shown in Figs. 4 and 5, respectively. A direct band to band transition energy of 1.55 eV for polycrystalline CdTe film was evaluated from the absorption curve (d) in Fig.4 by fitting the data to the Tauc relation $(\alpha h\nu)^2 = h\nu - E_g$, applicable to direct gap semiconductors. Similar analysis was applied to CdTe nanocrystals. The results are shown in Fig. 6 and 7. The optical gap for nanocrystal CdTe:TiO_2 films increases from a polycrystalline value of 1.55 eV to 1.75, 1.87 and 2.07 eV as the thickness of composite film decreases to 2500, 1500 and 500Å, respectively. These optical gap values show a red shift on thermal treatment as optical gap decreases to 1.65, 1.7 and 1.95 eV for corresponding films (Fig.7)

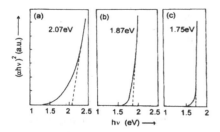

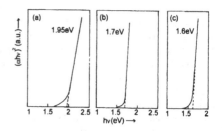

Fig.6 $(\alpha h\nu)^2$ vs photon energy plots of absorption data in Fig.4 for determining band gap

Fig.7 $(\alpha h\nu)^2$ vs photon energy plots of absorption data in Fig.5 for determining band gap

In as-deposited films as thickness increases, a gradual shift in absorption edge to longer wavelength side is attributed to change in CdTe nanocrystal size which is ˜15 nm for a 500Å thick film and increases to 25 and 40 nm for 1500 and 2500 Å thick films. The shift in absorption edge is accompanied by increase in absorption coefficient due to increased volume fraction of CdTe nanocrystals in TiO_2 matrix. On thermal treatment, the red shifts in absorption edge for corresponding films is attributed to fast coalescence of CdTe nanocrystals providing increase in their size. The absorption coefficient increases more appreciably for thicker films which strongly suggest coalescence process, since in thinner films, widely spaced CdTe nanocrystllites would show a minimal change as is indeed observed.

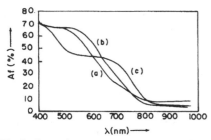

Fig.8 Absorption spectra of nanocrystal CdTe:TiO_2 (.5μm) films of high fraction (>0.2) of CdTe nanocrystal dispersions

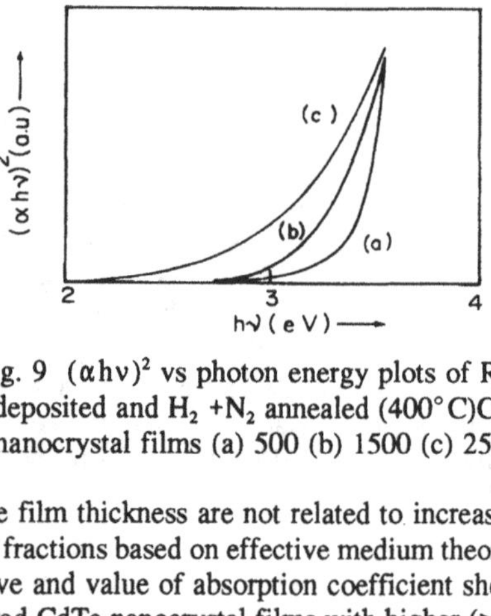

Fig. 9 $(\alpha h\nu)^2$ vs photon energy plots of RT deposited and H_2 +N_2 annealed (400°C)CdTe nanocrystal films (a) 500 (b) 1500 (c) 2500Å

The red shift associated with increase in the film thickness are not related to increase in relative concentration of CdTe nanocrystal and TiO_2 fractions based on effective medium theories. In fact in this case only the slope of absorption curve and value of absorption coefficient should increase and no shifts should occur. We have prepared CdTe nanocrystal films with higher (>0.2) volume fraction by increasing the area of CdTe sputter target and keeping the film thickness constant. Fig. 8 shows optical absorption spectra in which curves a, b and are due to films deposited by increasing the surface area of CdTe chips covering 4,8 and 12% of TiO_2 sputter target. In these curves, slope and magnitude of absorption coefficient increases without any corresponding red shifts in optical edge.

Fig. 9 shows $(\alpha h\nu)^2$ vs photon energy plots for RT deposited nanocrystal CdTe:TiO_2 films after thermal treatment in H_2 at 400°C. Thickness and other sputtering conditions of these were identical to those of Fig.5. In as-deposited state 500 and 1500 Å thick films for which we have the microscopy data, average nanocrystal sizes were 11 and 12.5 nm and dispersion fraction of .02 and .065, respectively. Band edge due to direct transition between highest valence band and conduction band does not obey familiar linear Tauc relation for crystalline semiconductors in this size regime. Instead the absorption values show gradual change characteristic of smearing of band edges. Low absorption from CdTe nanocrystals is due to extremely low density. In higher absorption region > 3 eV, the absorption spectra is heavily modulated by features of TiO_2 band to band transition.

Our results clearly show a reasonable correlation between the change in size of CdTe nanocrystals and the shifts in optical band gap. Large blue shifts in the absorption edge from a polycrystalline value towards CdTe in the nanocrystal size regime are ascribed to quantum confinement effects pertaining to the nature of Coulomb interaction between the electron-hole pair within CdTe nanocrystals. Various theoretical models [8,9] based on effective mass approximation used to describe optical behavior and band structure recognize two cases of confinement. (i) When nanocrystal size R_{NC} is greater than Wannier exciton Bohr diameter R_B, R_{NC}/R_B ~4, strong coupling between electron and holes due to Coulomb interaction leads to translational motion confinement of exciton. (ii) Single particle confinement of decoupled electrons and holes occur when nanocrystal size is lower than $2R_B$ The band edge shifts are given by $\hbar \pi^2 /(2m_{eff} R^2_{NC})$. The effective mass m_{eff} for case (i) is translational exciton mass ($m_e + m_h$) and for case (ii) is reduced mass of electron hole pair, $1/m_{eff} = (1/m_e + 1/m_h)$. For CdTe, R_B is 15 nm. In composite films under inverstigation, the ratio R_{NC}/R_B ranges between 0.8-2.6 which fall within the range of quantum size effects. Quantum size effects in CdTe nanocrystal-glass films have been observed earlier [5] in the size range 4.6-15.8 nm. The band gap shifts observed here show linear relation with inverse square of nanocrystal size in accordance with theory but show a much larger shift than predicted by it. This discrepancy is attributed to strong passivation provided by co-deposited TiO_2 to the CdTe nuclei in the formative stages. A typical CdTe nano size particle may in fact comprise of several passivated CdTe nanocrystallites of much smaller sizes. Cluster containing passivated CdTe nanocrystals presents a situation where differences that exist between surface atoms and those within the nanocrystals is large in favor of surface atoms. Vast interface between CdTe nanocrystals and TiO_2 passivating medium will have a pronounced effect on optical and electrical properties [10]. This cluster of CdTe nanocrystals could result in over estimation of their size by a factor of 2.5 to 3 times. Taking this into account the band edge shifts observed here agree well with the strong confinement regime for single particle confinement of decoupled electrons and holes.

Electrical Conductivity of CdTe Nanocrystals in TiO_2 Matrix

Fig. 10 and 11 summarize the in-plane electrical conductivity of composite films of different thicknesses deposited at RT and 100°C, respectively. Conduction in all films in the temperature range 300-500 K follows an activated process characterized by $\sigma = \sigma_0 \exp (- E_a /kT)$. The σ-1/T plots give single activation energy, E_a. In polycrystalline CdTe films, $E_a = 0.19$ eV corresponds to ionized Cd vacancy [V_{Cd}^{2-}]. High conductivity ~10^{-2} S/cm of these films decreases by 4-5 orders in CdTe nanocrystal composite films. E_a changes from 0.83 eV for a 500 Å film having average crystal size of 11 nm to 0.74 eV for a 2500Å thick film having CdTe nanocrystals of average size of 40 nm. These values represent intrinsic conduction process. There are no native defect states present in these crystallites. RT conductivities in these films decrease for thicker films which are characterized by increased density and size of CdTe nanocrystals. This is ascribed to the effect of CdTe in localizing O_2 - vacancies in TiO_2 film matrix. These O_2 - vacancies are known to act as donors and increase conductivity. The thermal traetment in H_2 also has the effect of increasing the O_2 vacancies and increase the conductivity(Fig11). Films having a high fraction

267

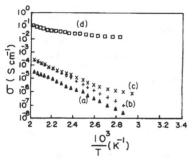

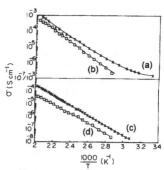

Fig. 10 Activation energy plot of (a)2500 (b)1500 (c) 2500 Å CdTe nanocrystal TiO_2 films(d)polycrystalline CdTe

Fig.11 Activation energy plots of films of c and b after thermal treatment in H_2: 400° C

conductivity. Thermal treatment in H_2 also has the effect of increasing the O_2 -vacancies and increasing the conductivity (Fig.11).Films having a high fraction of CdTe nanocrystals dispersions exhibit a systematic dependence of conductivity of composite film as shown in Fig.12. RT conductivity of these films scales with CdTe volume fraction and usually shows a higher value for high temperature deposited films in the range 3.1×10^{-6} to 1.7×10^{-4} S/cm. High conductivities are attributed to thermally assisted tunneling process The CdTe nanocrystals are thus electrically coupled to each other through TiO_2.

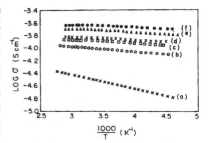

Fig.12 σ vs 1/T : CdTe films of >0.2 fraction

CONCLUSIONS

Nanocrystalline CdTe dispersions in TiO_2 films show quantum size effects as described by shifts in optical edge which correlate with crystal size. Large shifts are due to CdTe nano-clusters comprised of finite number of passivated CdTe crystals. Electrical conduction of coupled nanocrystals is facilitated by O_2 deficient state of TiO_2 interface and shows an increase with CdTe volume fraction.

REFERENCES

1. A.Vanhaudenarke, M.Trespidi and R. Frey, J. Opt. Soc. Am., **B6**, 818 (1994)
2. B.O Dabbousi,M.G.Bawendi,O.Onitsuka and P.Rubner, Appl. Phys. Lett., **66**, 1316 (1995)
3. D.Nesheva and Z. Levi, Semicond. Sci. Technol., **12**, 1318 (1997)
4. I.Tanahashi, A.Tsujimura,T.Mitsuyu and A.Nishino, Jpn. J. Appl. Phys., **29**, 2111 (1990)
5. B.G. Potter Jr. and J.H.Simmons, J. Appl. Phys., **68**, 1218 (1990)
6. H.Nasu, K.Tsunetomo, Y.Tokumitsu and Y.Osaka, Jpn. J. Appl. Phys., **28**, L862 (1989)
7. A.C.Rastogi, S.N.Sharma and Sandeep Kohli, (To be published, 1998)
8. Al.L. Efros, A.L. Efros, Sov. Phys. Semicond., **16**, 772 (1982)
9. N.F.Borrelli, D.W. Hall, H.J.Holland and D.W.Smith, J. Appl. Phys., **61**, 5399 (1987)
10. G.M.Whitesides, J.P.Mathias and C.T.Seto, Science, **254**, 1312 (1991)

OPTICAL PROPERTIES OF SELF-ORGANIZED InGaAs/GaAs QUANTUM DOTS IN FIELD-EFFECT STRUCTURES

A.BABIŃSKI*, T. TOMASZEWICZ*, A. WYSMOŁEK*, J. M. BARANOWSKI* ,
C. LOBO**, R. LEON***, C. JAGADISH**
*Institute of Experimental Physics, Warsaw University, ul. Hoża 69, 00-681 Warszawa, Poland
**EME Dept, RSPhysSE, ANU, Canberra ACT 0200, Australia
***Jet Propulsion Laboratory, California Institute of Technology, 4800 Oak Grove Drive,
Pasadena, CA 91109-8099

ABSTRACT

The results of photoluminescence (PL) and electroreflectance (ER) measurements on InGaAs/GaAs self-organized quantum dots (QDs) in field-effect structure are presented. It has been found that the QDs PL can be completely quenched in reversely biased structure both at room temperature and at T=4.2K. A non-monotonic dependence of QDs PL peak energy with applied bias is observed at low temperature, which is attributed to the band-gap re-normalization due to QDs charging and size distribution effects. The electric field dependence of the QDs ER feature at room temperature has been analysed. A red shift of that feature with increasing electric field has been observed.

INTRODUCTION

One of the approaches to the QDs investigation is the use of field-effect structures. The QD's electron occupancy can be controlled in such structures by applied bias and electrical and optical properties of QDs charged with a few electrons can be investigated [1]. The InAs/GaAs self-organized QDs in field-effect structures were intensely measured both electrically and optically. Confinement energies of carriers in QDs and Coulomb energies of QDs were established from capacitance-voltage (CV) measurements [2]. The effect of QDs charging on absorption [3] and photoluminescence (PL) [4,5] has been recently investigated. In the present communication we investigate the effect of electric field (and QDs electron occupancy) on optical properties of self-organized InGaAs/GaAs QDs in a field - effect structure. Results of PL at RT and at T=4.2K and electroreflectance (ER) at RT are presented and discussed.

EXPERIMENTAL PROCEDURE

Samples investigated in this work were grown using a low pressure Metal Organic Vapor Phase Epitaxy. Investigated samples were grown on the [100] - oriented semi-insulating GaAs substrate and consisted of 200 nm of Si doped ($n = 2 \times 10^{18}$ cm^{-3}) n$^+$GaAs layer acting as a back contact, followed by 20 nm of undoped GaAs tunnelling barrier, the In$_{0.6}$Ga$_{0.4}$As layer with wetting layer (WL) and QDs grown in the Stranski-Krastanow mode (the 0QD sample) or In$_{0.6}$Ga$_{0.4}$As layer with WL and no QDs (the QW sample), the 20 nm undoped GaAs spacer, 31 nm Al$_{0.3}$Ga$_{0.7}$As barrier, and 25 nm GaAs cap. The Al$_{0.3}$Ga$_{0.7}$As barrier was grown in order to improve the Schottky structure. A In$_{0.6}$Ga$_{0.4}$As layer was deposited at the growth temperature of 550° C, GaAs was grown at 650° C and Al$_{0.3}$Ga$_{0.7}$As at 750° C. Surface topography of QDs was examined using an Atomic Force Microscopy on a sample with uncapped QDs. It was found that the lens shaped islands of average basal size 890 nm^2 ± 22% with a surface density 7.1×10^9 cm^{-2} were formed in the 0QD sample [6].

The PL measurements were performed with the sample cooled in a continuous flow Oxford CF-1204 cryostat. A laser illumination ($\lambda=780$ nm) was used for a non-resonant PL excitation. PL spectra were dispersed by a SPEX spectrometer and detected with a liquid nitrogen-cooled Ge p-i-n diode. For the bias dependent PL and the ER measurements a circular (1.5 mm diameter) semitransparent (30 nm thick) Ni/Cr Schottky contact was evaporated through a shadow mask on the top of the structure. The back gate was formed by alloying an In contact at the top of n^+ GaAs layer after part of the sample had been electro-chemically etched off. A negative bias denotes in our paper a reverse polarization of the Schottky structure. Bias dependent PL measurements were taken with a laser illumination through a Schottky gate. The ER measurements were performed at room temperature using modulated bias voltage applied between two contacts of the sample.

RESULTS AND DISCUSSION

Photoluminescence

The PL spectra of investigated samples measured at RT are presented in Fig.1. A strong peak at 1.283 eV dominates the PL spectrum of the sample consisting QDs (0QD). We relate this PL peak to the electron-hole recombination within the QDs. This peak is not present in the sample without QDs (QW). Moreover the band-to-band GaAs PL can be observed in both structures. The bias dependence of the PL in the 0QD sample with the Schottky gate measured at RT is shown in Fig. 2. The intensity of the dominant QDs PL peak strongly depends on the electrical polarization of the structure. The QDs PL peak can be completely quenched by negative bias. As it can be seen, no changes of the PL peak position with bias can be observed at RT. In addition to the dominant QDs PL peak, two other features can be seen in the PL spectrum from the Schottky structure. The bias independent feature at 1.37 is probably due to the recombination in the WL. The broad bias-dependent PL band below 1.2 eV is attributed to the large InGaAs islands present in our structure with a low surface density ($\sim 10^8$ cm^{-3}). It has to be noted that no GaAs band-to-band PL can be observed from Schottky structure. This is probably due to large electric field in the investigated structure ($\sim 10^5$ V/cm).

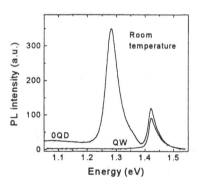
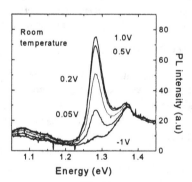

Fig. 1. Photoluminescence in the samples with InGaAs/GaAs QDs (0QD) and InGaAs/GaAs wetting layer (QW)

Fig. 2 Bias dependence of the photoluminescence in the 0QD sample at room temperature.

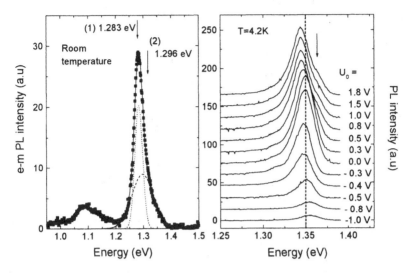

Energy (eV)   Energy (eV)

Fig. 3. Electrically modulated photoluminescence in InGaAs/GaAs QDs at room
temperature. Theoretical fit (solid) with two
Gaussians with E_1 = 1.283 eV (dotted) and
E_2 = 1.296 eV (dashed) is also presented.

Fig. 4. The bias dependence of the
photoluminescence in InGaAs/GaAs QDs at
T = 4.2K. Vertical offset added for more
clarity.

A strong bias dependence of the QDs PL can be used for electrically modulated PL (e-m PL)
measurement. A structure is continuously illuminated with a laser beam, and it is biased with a
sum of DC bias U_0 and AC modulation signal ΔU in such an experiment. The latter AC voltage (f
= 18Hz, ΔU= 0.2V) is used also as a reference for a lock-in amplifier. The e-m PL signal depends
on the radiative recombination processes sensitive to the applied bias and it can be regarded as a
derivative of the PL intensity on the bias. The e-m PL spectrum measured at RT at U_0 = 0V is
presented in Fig. 3. The dominant peak in the e-m PL spectrum was fitted with two Gaussians
with energy $E_{1(2)}$ = 1.283 eV (1.296 eV). The PL at E_2 is very likely due to a recombination from
an excited state within the QDs. The E_2 - E_1 energy separation (13 meV) is lower than observed
previously in similar QDs (~40 meV) [7]. Reduction of an intersublevel energy spacing results
from an In intermixing, which took place during the AlGaAs layer growth at elevated temperature
(750°C). The In intermixing can be also responsible for relatively high energy of the QDs PL in
the 0QD sample [8-9]. The PL band at 1.08 eV seen in Fig. 3. is probably due to partially relaxed
larger InGaAs islands observed in our samples.

The bias dependence of the QDs PL at T = 4.2K is shown in Fig. 4. Spectra shown in Fig. 4
were obtained after subtracting the „background" PL measured at U_0 = -1.5V, which is bias-
independent. The PL intensity continuously increases at bias up to U_0 = 0.3 V. At bias $U_0 > 0.3$V
an additional feature (marked with an arrow) becomes apparent on high energy tail of the QDs

PL. The PL intensity can be related to the electron occupancy of QDs, deduced from the CV characteristic of the investigated sample (see Fig. 5). At bias $U_0 < -0.5$ V the QD's ground state is above the Fermi level and there is no steady electron occupation of the QDs. The capacitance of the structure reflects geometrical distance between the back contact and the top Schottky gate. The treshold of capacitance at $U_0 = -0.2$ V is related to the QDs charging. Formation of the two-dimensional electron gas (2DEG) in the WL is responsible for the feature observed in CV at $U_0 = 0.7$ V. An increase of capacitance at $U_0 = 0.8$ V results from the 2DEG formation at the AlGaAs/GaAs interface. A weak PL

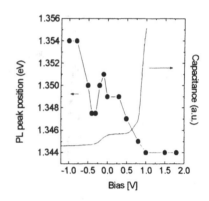

Fig. 5. The bias dependence of the PL peak at T = 4.2 K and the CV characteristic of the 0QD sample at T = 6K.

signal (compare Fig. 4) can be seen at $U_0 < -0.5$ V , when the QDs are empty. Charging of QDs with electrons from the back contact results in the PL signal recovery. We relate the PL quenching in high electric field observed at RT and T = 4.2K to the dissociation of an exciton in high electric field. Both photo-generated electron and hole must be trapped onto the QD in order to recombine radiatively. Under high electric field they are removed from the dots before thermalization to the ground state. As a result a radiative recombination within the QDs is reduced. Charging of QDs increases the recombination probability. Similar effect has been recently observed at low temperature in InAs QDs [4]. Another explanation to the bias induced quenching of the QD's PL signal would be the spatial separation of electron and hole wave-functions due to the Quantum Confined Stark Effect (QCSE) [10]. However no PL energy shift as a function of bias can be seen at RT (see Fig. 2), which should accompany the QCSE.

As it can be seen in Fig. 5 the PL peak energy at T = 4.2K depends non-monotonically on the bias. The main trend of a PL bias dependence is a red shift with the QDs occupancy increase ($U_0 > -0.1$V). This effect can be attributed to the band gap re-normalization in QDs [11]. Such an effect was reported previously in InAs/GaAs QDs [4]. The PL peak bias dependence at $U_0 < -0.1$V is not fully understood at the moment. The PL blue shift with decreasing electric field observed in the bias region of QDs charging with first electron (-0.5V $> U_0 > -0.1$V) was tentatively attributed to the effect of QDs size distribution [5], however more investigation is necessary to support this hypothesis.

Electroreflectance

The ER spectra measured on the 0QD sample at room temperature are presented in Fig. 6. Lines above 1.4 eV are related to GaAs. The ER features due to QDs (1.25 eV - 1.4 eV) can be seen in the whole bias range. An electric field in GaAs layers, as found from an analysis of Franz-Keldysh oscillations in the investigated structure, ranged from 60 kV/cm to 260 kV/cm at bias voltage from $U_0 = 1$V to $U_0 = -2$V respectively. It has to be noted that even at zero bias there is a strong built-in electric field in the structure, which is due to surface states. A monotonic decrease

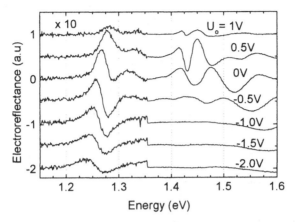

Fig. 6. The bias dependence of the electrorefletance in InGaAs/GaAs QDs at room temperature. Offset is added for more clarity.

of the QDs ER feature energy an electric field can be observed in the whole investigated bias range. The ER results can be compared with the PL data at room temperature. In contrast to the PL peak, the E_0 energy of QD ER feature decreases with increasing electric field. This difference is due to different processes involved in both experiments. An excitonic recombination from the ground state of QDs is observed in PL. In ER transitions to excited states of QDs can be observed as well. In our opinion the red shift of the ER feature with increasing electric field results from changes in QDs electron occupation (Burstein-Moss effect). Similar effect was observed in optical absorption from QDs [12]. A possible role of the QCSE in the red shift of this ER feature should be also investigated. We believe that measurements at lower temperatures should clarify the QCSE influence on the QDs ER spectrum.

CONCLUSIONS

In conclusion the PL and ER from the self-organized InGaAs/GaAs QDs in field-effect structures have been studied. It has been found that the QDs PL can be quenched in reversely biased structure. The PL recovery can be correlated to the steady electron occupation of QDs. The PL peak energy does not depend on bias at RT whereas a non-monotonic dependence of that energy can be observed at T = 4.2K. The systematic red shift of the QDs ER feature with increasing electric field observed at RT was explained in terms of Burstein-Moss effect.

ACKNOWLEDGMENTS

This work was supported in part by the Polish Committee for Scientific Research Grant no. PBZ-28.11. A.B. gratefully acknowledges financial support from The Stefan Batory Foundation. The work at ANU is supported in part by the Australian Agency for International Development through IDP Education Australia.

REFERENCES

1. H. Drexler, D. Leonard, W. Hansen, Phys. Rev. Lett. **73**, 2252 (1994)

2. B. T. Miller, W. Hansen, S. Manus, R. J. Luyken, A. Lorke, J. P. Kotthaus, S. Huant, G. Medeiros - Ribeiro, P. M. Petroff, Phys. Rev. B **56**, 6764 (1997); G. Medeiros-Ribeiro, D. Leonard, P. M. Petroff, Appl. Phys. Lett. **66,** 1767 (1995); G. Medeiros-Ribeiro, F. G. Pikus, P. M. Petroff, A. L. Efros, Phys. Rev. B **55**, 1568 (1997)

3. R. J. Warburton, C. S. Dürr, K. Karrai, J. P. Kotthaus, G. Medeiros-Ribeiro, P. M. Petroff, Phys. Rev. Lett. **79**, 5282 (1997)

4. K. H. Schmidt, G. Medeiros-Ribeiro, P. M. Petroff, Phys. Rev. B **58**, 3597 (1998)

5. A. Babinski, A. Wysmolek, T. Tomaszewicz, J. M. Baranowski, R. Leon, C. Lobo, C. Jagadish, Appl. Phys. Lett. **73**, 2811 (1998)

6. C. Lobo and R. Leon, J. Appl. Phys. **83**, 4186 (1998)

7. R. J. Warburton, C. S. Dürr, K. Karrai, J. P. Kotthaus, G. Medeiros-Ribeiro, P. M. Petroff, Phys. Rev. Lett. **79**, 5282 (1997)

8. R. Leon, Yong Kim, C. Jagadish, M. Gal, J. Zou, D. J. H. Cockayne, Appl. Phys. Lett. **69**, 1888 (1996)

9. C. Lobo, R. Leon, S.Fafard, P. G. Piva, Appl. Phys. Lett. **72**, 2850 (1998)

10. G. Bastard, E. E. Mendez, L. L. Chang, L. Esaki, Phys. Rev. B **28**, 3241 (1983)

11. A. Wojs, P. Hawrylak, Phys. Rev. B **55**, 13066 (1997)

12. V. Ya. Aleshkin, B. N. Zvonkov, I. G. Malkina, E. R. Lin'kova, I. A. Karpovich, D. O. Filatov, in *Proc. 23rd Int. Conf. Phys. Semicond.*, edited by M. Scheffler and R. Zimmermann, (World Scientific, Singapore 1996), p.1397

GIANT ANISOTROPY OF CONDUCTIVITY IN HYDROGENATED NANOCRYSTALLINE SILICON THIN FILMS

A.B. PEVTSOV, N.A. FEOKTISTOV, V.G. GOLUBEV
A.F. Ioffe Physico-Technical Institute, 194021 St. Petersburg, Russia, alex@pevtsov.spb.su

ABSTRACT

Thin (<1000 Å) hydrogenated nanocrystalline silicon films are widely used in solar cells, light emitting diodes, and spatial light modulators. In this work the conductivity of doped and undoped amorphous-nanocrystalline silicon thin films is studied as a function of film thickness: a giant anisotropy of conductivity is established. The longitudinal conductivity decreases dramatically (by a factor of 10^9-10^{10}) as the layer thickness is reduced from 1500 Å to 200 Å, while the transverse conductivity remains close to that of a doped a-Si:H. The data obtained are interpreted in terms of the percolation theory.

INTRODUCTION

Some devices based on hydrogenated amorphous silicon (a-Si:H) and nanocrystalline silicon (nc-Si:H) require use of doped layers with low conductivity along the film surface. This is the case for, e.g., vidicons and photosensitive cells operating under low-illumination conditions [1] and spatial light modulators [2].

In the present paper we demonstrate that thin (200-250 Å) nc-Si:H films can be used as doped layers having very low surface leakage current and high transverse conductivity.

In fact the synthesized films are a system of crystalline quantum dots introduced into an amorphous matrix. The interface nanocrystallite/amorphous matrix is a heterojunction with a tunnel-transparent dielectric [3]. The size of crystallites may be in the range from 20 to 100 Å, and their volume fraction varies within 0-50% [4-6]. On the other hand, it was shown in [6,7] that abrupt conductivity changes observed in nc-Si on increasing the volume fraction of crystallites can be explained in terms of phenomenological concepts of the percolation theory, without invoking quantum phenomena. For example, when the volume fraction of nanocrystallites achieves a critical value of about 16%, a percolation cluster of nanocrystallites is formed and the conductivity of a film increases dramatically. When the volume fraction is lower than the critical value, corresponding to the percolation threshold, no percolation cluster is formed and the conductivity is determined by the properties of the high-resistance amorphous phase.

The percolation theory approach, taking into account the fact that the presence of film boundaries hinders formation of an infinite cluster, has been used [8] to explain the exponential dependence of the longitudinal conductivity on the film thickness. The available experimental data (see references in [8]) refer to low temperatures where the Mott law is obeyed for the electrical conductivity and the reduction of the longitudinal conductivity does not exceed four orders of magnitude at about 100 K. At a sufficiently high (>16%) volume fraction of nanocrystallites in amorphous-nanocrystalline silicon (a/nc-Si:H) films the conduction between the nanocrystallites is governed by the conventional tunneling of electrons through an amorphous spacer of size L. In this case the conductivity $\sigma \sim \exp(-2L/a)$ and depends on temperature only slightly (a is the radius of localized state). The occurrence of a natural dispersion with respect to amorphous spacer size allows considering the objects of this work as a network of resistors with an exponentially wide scatter of their values. From this approach follows that the longitudinal conductivity must exponentially decrease at room temperature with decreasing thickness of a/nc-Si:H.

In the case of transverse conductivity we assumed, following [9], that at small thickness of the nc-Si film the infinite cluster will be shunted by isolated chains of resistors connecting metallic sandwich electrodes. These chains are formed by nanocrystallites located anomalously close to one another. Even though the probability of formation of such chains is exponentially low, their conductivity is exponentially higher than that of the infinite 3D cluster, so that just these chains will determine the transverse conductivity of the film.

SAMPLES

Nanocrystalline silicon was manufactured by plasma enhanced chemical vapor deposition (PE CVD), using silane strongly diluted with hydrogen. The parameters of the technological process were the following: silane concentration in hydrogen 2-3%, working mixture pressure 0.2-0.4 Torr, working mixture flow rate 10-20 sccm, substrate temperature 200-300°C, specific rf power 0.3-1 W/cm^2, frequency 17 MHz.

The deposition was performed using a conventional PE CVD diode scheme in which the rf voltage was applied to the rf electrode, and substrates were placed on the heated second electrode. The heated electrode was grounded for the ˙rf component, and there was a possibility of applying to it a dc negative bias in the range 0-300 V. The growth rate and the optical parameters were monitored *in situ* by laser interferometry, with the laser beam falling onto the substrate at the Brewster angle. To obtain doped nc-Si:H films of n-type, phosphine was added to the gas mixture: $PH_3/(SiH_4+PH_3) \sim 1\%$.

As substrates were used quartz and crystalline silicon. The growth rate of the films was 0.3-1.0 Å/s. The volume fraction and size of crystallites were determined by numerical processing of Raman spectra, with account taken of the spatial confinement of optical phonons [10], to be ~30% and ~40 Å, respectively. The film thickness was varied from 1500 to 200 Å.

To compare how the conductivity of nanocrystalline and amorphous films depends on their thickness, doped a-Si:H films were deposited at a silane concentration in the gas mixture exceeding 10% and a reduced rf power (<0.1 W/cm^2).

RESULTS AND DISCUSSION

Figure 1 shows as a function of film thickness the longitudinal conductivity of films various composition: nanocrystalline, both undoped and phosphorus-doped, and weakly doped amorphous ($PH_3/SiH_4 \approx 0.01\%$).

It can be seen that nanocrystalline films show a very strong dependence of conductivity on film thickness. The longitudinal conductivity decreases by more than 9 orders of magnitude on reducing the thickness from 1500 to 200 Å. At the same time the conductivity of amorphous films changes by less than an order of magnitude on varying their thickness within the same limits.

276

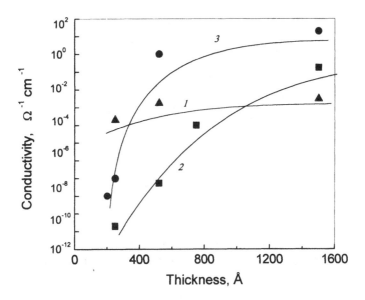

Fig. 1: Dependencies of conductivity vs thickness of nanocrystalline (2,3) and amorphous (1)
silicon films. Phosphorus doped (1,3) and undoped (2) films.

Let us discuss the results obtained. From the percolation theory follows that the
structure of the percolation cluster formed from nanocrystallites must depend on the film
thickness d when d is shorter than the correlation radius of infinite cluster, but longer than the
average distance between the crystallites.   A   consequence   of   this fact is   the   linear
dependence [8]:

$$\lg\lg\frac{\sigma(\infty)}{\sigma(d)} \propto -\frac{1}{\nu}\lg d \ , \qquad (1)$$

where $\sigma(\infty)$ is equal to the conductivity of the thick film, ν is the 3D index of the correlation
radius. We replotted the data of Fig. 1 in the $\lg\lg\sigma[(\infty)/\sigma(d)] - \lg d$ coordinates and found
that the experimental   points for nc-Si films (curves 2 and 3 in Fig.2) fall (to a good
approximation) on a straight line. However, the ν values obtained from the slopes of these
straight lines lie within the interval 0.4-0.6, markedly differing from the known theoretical
value v=0.9. Thus, we may conclude that the percolation theory is insufficient for quantitative
interpretation of the obtained data, and there is empirical reason to believe that the
microscopic properties of the material may vary when its thickness is reduced.

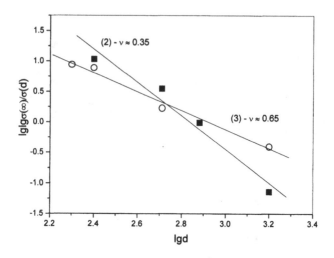

Fig. 2: Experimental data on conductivity approximated by the expression (1): 2-undoped film, 3-doped film.

We now consider the behavior of the transverse conduction. It should be noted in the first place that in our experiments, with the area of the upper metal contact of $\sim 10^{-3}$ cm^2, the measured transverse resistances of ~ 1000-Å -thick films must exceed significantly the resistance of the bottom electrode (~ 10 Ω) only at conductivities lower than 10^{-4} Ω^{-1}cm^{-1}. It is this value that limited the observable maximum transverse conductivity. Therefore, to measure the transverse conductivity (Fig. 3), we fabricated special, sufficiently high-resistance nc-Si samples with a volume fraction of crystallites lower than the critical value (16%).

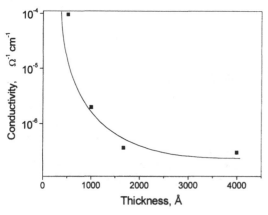

Fig. 3: Dependence of transverse conductivity vs thickness.

A tendency is clearly seen for the conductivity to increase dramatically with decreasing thickness. This, in our opinion, is due to the decisive contribution to the transverse conductivity from the optimal chains of nanocrystallites. The actual value of transverse conductivity for thin films with high content of crystallites is presumably not lower than the bulk conductivity of thick (>1500 Å) nc-Si films.

Figure 4 shows a set of current-voltage characteristics of 500 Å -thick nc-Si:H films sandwiched between titanium electrodes. At the beginning the current is determined by the most short chain of nanocrystallites (curve 1). As the voltage is increased the current channel is destroyed by the Joule heating. Afterwards, the current will be determined by longer nanocrystalline chains with higher resistance (curves 2, 3). Thus the observed switching from the curve 1 to the curve 2 and then to the curve 3 can also be understood as a successive destruction of optimal nanocrystallite chains with gradually increasing resistance.

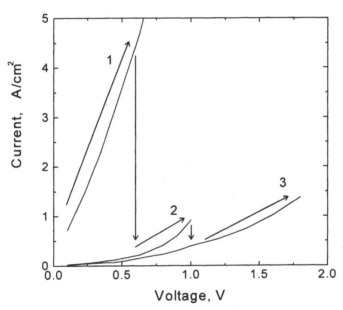

Fig. 4: Set of the current-voltage characteristics of 500 Å thick nc-Si:H film.

CONCLUSION

Thus, the dependence of the conductivity of nc-Si:H films on their thickness has been studied. A gigantic anisotropy of conductivity (difference between longitudinal and transverse σ values) has been discovered. The conductivity along the layer decreases by 8-10 orders of magnitude, when the thickness is reduced from 1500 to 200 Å, to become as low as

$10^{-11}\Omega^{-1}$ cm^{-1}. The observed dependence of the longitudinal conductivity can be quantitatively understood in terms of the percolation theory. The transverse conductivity is determined by isolated high-conductivity optimal chains formed from silicon nanocrystallites. Its value $(10^{-1}\text{-}1)$ Ω^{-1} cm^{-1} is comparable with that of thick nanocrystalline films. The obtained thin (200-250 Å) nc-Si:H films can be used as doped layers in p-i-n structures for diminishing the effect of leakage current over the layer surface with rather high transverse conductivity retained. In particular, p-i-n structures with layers of this kind have been used for controlling liquid-crystal spatial light modulator and increased severalfold the contrast of read-out image.

ACKNOWLEDGMENTS

The work was supported by the Russian Foundation of Basic Research (grant No. 98-02-17350) and INFO-COPERNICUS (grant No.PL97-8104).

REFERENCES

1. *Amorphous Semiconductor Technologies and Devices*, edited by Y.Hamakawa (Elsevier Science Publishers, New York, 1985), p.495.
2. K.Akiyama, A.Takimoto, A.Ogivara, A.Ogava, Jpn. Appl.Phys. **33**, part 1, 590 (1993).
3. G.Y. Hu, R.F.O'Connel, Y.L.He, M.B.Yu, J.Appl.Phys. **78**, 3945 (1995).
4. X. Liu, S.Tong, L.Wang, X.Rao, J. Appl.Phys. **78**, 6143 (1995).
5. A.B.Pevtsov V.Yu.Davydov, N.A.Feoktistov, V.G.Karpov, Phys.Rev. **B52**, 955 (1995).
6. V.G.Golubev, V.Yu.Davydov, A.V.Medvedev, A.B.Pevtsov, N.A.Feoktistov, Fiz.Tverd. Tela, **39**, 1348 (1997) [Phys. Solid State **39**, 1197 (1997)].
7. R.Tsu, J.Gonzales-Hernandes, S.S.Chao, S.C.Lee, K.Tanaka, Appl.Phys.Lett. **40**, 534 (1982).
8. B.I.Shklovskii, phys.stat.sol.(b) **83**, K11 (1977); A.L.Efros and B.I.Shklovskii. *Electronic Properties of Doped Semiconductors* (Springer, Berlin, 1984), p.420.
9. A.V.Tartakovskii, M.V.Fistul', M.E.Raikh, I.M.Ruzin, Fiz. Tekh. Poluprovodnn. **21**, 603 (1987) [Sov. Phys. Semicond. **21**, 312 (1985)].
10. I.H.Cambell, P.M.Fauchet, Solid State Commun. **58**, 739 (1986).

THE INVERTED MEYER-NELDEL RULE IN THE CONDUCTANCE OF NANOSTRUCTURED SILICON FIELD-EFFECT DEVICES

R.E.I. SCHROPP and H. MEILING
Debye Institute, Utrecht University,
P.O. Box 80 000, 3508 TA Utrecht, The Netherlands

ABSTRACT

Thin film transistors (TFTs) offer the possibility to study the electronic transport properties of an intrinsic semiconductor as a function of the Fermi level position without the introduction of dopants and/or doping related defects. Recently, we reported on the first TFTs incorporating nanostructured silicon deposited with the Hot-Wire Chemical Vapor Deposition technique. These structures offer significant advantages over conventional plasma-deposited amorphous silicon TFTs. First of all, the HW deposited nanocrystalline silicon (nc-Si:H) TFTs do not show any threshold voltage shift upon prolonged gate voltage stress. Therefore, it is now possible to study the transport characteristics at a relatively large gate voltage in a controlled fashion, unhampered by any drift of the characteristics due to the creation of metastable electronic defect states and/or charge trapping. Second, the result of the field effect is that the Fermi energy moves into the conduction band of the virtually defect-free nanocrystalline domains in the channel region of the TFT. As the effective mobility gap of the surrounding amorphous phase is higher than that of the silicon crystallites, the Fermi energy is driven deep into the band-tail distribution of the amorphous phase, a situation that could never be achieved in purely amorphous silicon TFTs nor by heavily doping an amorphous semiconductor. Thus, the nanostructured nature of the silicon thin film near the gate insulator allows to shift the Fermi level far into the tail states region of the amorphous phase. This situation reveals for the first time the inverted Meyer-Neldel relationship in an *intrinsic* semiconductor.

INTRODUCTION

In many disordered materials various physical quantities exhibit thermally activated behaviour. For instance, the electrical conductivity σ of hydrogenated amorphous silicon, a-Si:H, obeys $\sigma = \sigma_0 \exp(-E_a/kT)$, with E_a the activation energy, k Boltzmann's constant, and T the absolute temperature. An exponential relation is found between the activation energy and the pre-exponential factor σ_0, a feature that is generally known as the Meyer-Neldel rule (MNR) [1]: $\sigma_0 = \sigma_{00} \exp(E_a/E_{MN})$, with E_{MN} the characteristic Meyer-Neldel energy ($E_{MN} > 0$) and σ_{00} is a constant. Between various properties and materials, E_{MN} is fairly similar and varies between 0.03 and 0.1 eV [2]. The Meyer-Neldel behaviour is related to the disordered or inhomogeneous nature of the material. The microscopical background, however, remains a subject of debate. Jackson [2] described the connection between multiple trapping transport, as in hydrogen diffusion, and the MNR. On the other hand, various authors argue that the statistical shift of the Fermi level E_F with temperature is the origin of Meyer-Neldel behaviour in electrical conductivity measurements [3-5]. A positive value of E_{MN} arises from a movement of the Fermi level up to, but not into, the conduction band-tail density-of-states (DOS). $E_{MN} > 0$ has been observed for doped and undoped a-Si:H and for lightly doped and undoped μc-Si:H. A negative value of $E_{MN} > $ is also possible, and arises from a Fermi level that is moved deeply into the bandtail. The latter phenomenon is called the inverse MNR. Both were theoretically predicted for doped a-Si:H by Overhof and Beyer [3]. It was recently experimentally observed in heavily doped microcrystalline silicon [6-8]. In heavily doped amorphous silicon the inverse MNR has not been observed so far, because it is fundamentally difficult to dope a-Si:H heavily enough to move E_F deeply into the tail-DOS, due to disorder-induced broadening of the tail state distribution. In field-effect structures such as thin-film transistors, TFTs, the position of the Fermi level can be altered by applying a voltage to the gate electrode rather than by doping the semiconductor. The TFT thus offers the possibility to study the electronic transport properties of an intrinsic semiconductor as a function of the Fermi level displacement without the introduction of doping-related electronic defects. Various authors have used this, see, e.g., Ref. [9], where the conventional MNR is readily observed. In addition, Kondo et al. recently observed, albeit weakly, the inverse MNR in intrinsic plasma-CVD a-Si:H, by incorporating the material in a TFT using Mg source-drain contacts instead of conventional n+/metal contacts [10]. Recently [11], we reported on the first TFTs incorporating nanostructured silicon, which was deposited with the hot-wire chemical vapor deposition (HWCVD) technique [12], also referred to as catalytic CVD [13]. Cross-sectional transmission electron microscopy images show that this nanostructured material consists of crystallites which are embedded in an amorphous phase [14,15]. These heterogeneous structures offer significant advantages over conventional plasma-enhanced CVD-based a-Si:H TFTs. First, the HWCVD nanostructured silicon TFTs do not show any shift of the threshold voltage upon prolonged gate voltage stress, in contrast to the conventional a-Si:H TFTs. Therefore, it is now possible to study the transistor characteristics at a relatively large gate voltage in a controlled fashion, without any drift of the characteristics due to the creation of metastable electronic defect states and/or charge trapping in the insulator [16]. Second, the nanostructured nature of the channel material, i.e, the thin layer of silicon near to the gate insulator, in principle allows for a shift of the Fermi level deeper into the tail state region of the amorphous phase than in purely amorphous films.

EXPERIMENTAL

Table 1: Summary of the hot-wire CVD process settings for deposition of the semiconductor material and the main TFT results. SiH_4/H_2 is the silane fraction in the gas phase, T_w is the wire temperature, r_d is the silicon deposition rate, μ_s is the saturation mobility, V_{th} is the threshold voltage, and S is the subthreshold swing.

	Process settings			Semiconductor layer Nanocryst. inclusions?		TFT results		
Sample	SiH_4/H_2	T_w ($°C$)	r_d ($\text{Å}/s$)	Raman	HR-TEM	μ_s ($cm^2 \cdot V^{-1}s^{-1}$)	V_{th} (V)	S (V/ decade)
HW-A	100 %	1900	27	no	—	0.8	8.2	0.8
HW-B	10 %	1900	9	yes	yes	1.1	5.7	0.5
HW-C	10 %	1850	9	yes	yes	1.5	5.9	0.5

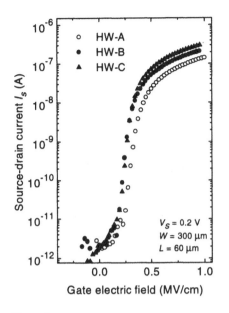

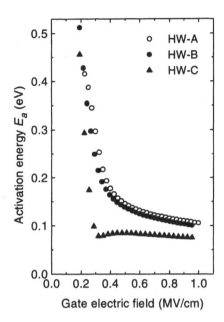

Figure 1:
Transfer characteristics of three TFTs including hot-wire CVD silicon

Figure 2:
Gate-voltage dependence of the source-drain current activation energy E_a.

Inverted-staggered TFTs were deposited with hot-wire CVD on highly doped thermally oxidized single-crystal silicon substrates. A summary of the deposition conditions and main TFT results is given in Table 1 including the device labeling. The last two sub-columns of the Semiconductor layer column of the table refer to Raman spectroscopy measurements of single films and high resolution transmission electron microscopy (HR-TEM) measurements of TFT devices. It is observed that the silicon layers made using 100 % SiH_4 do not show any c-Si contribution in the Raman signal. The TFTs made with H_2 dilution show a substantial c-Si contribution in the bulk material, both in the Raman signal and in the HR-TEM images, although a highly ordered thin amorphous silicon incubation layer is present at the interface with the insulator, in which nanocrystalline domains are embedded [14,15]. In Fig. 1 the transfer characteristics (source-drain current I_s vs. gate electric field) are plotted for the three different TFTs. The three HWCVD transistors behave essentially the same and show characteristics that are typical for state-of-the-art PECVD a-Si:H TFTs: the saturation mobility $\mu_s = 0.8$-1.5 cm^{2}V^{-1}s^{-1}, the threshold voltage $V_{th}= 6$-8 V, and the subthreshold swing S $= 0.5$-0.8 V/decade. The current switching ratio is more that 5 orders of magnitude at a source-drain voltage V_s of only 0.2 V, for transistors with a channel length L of 60 μm and a width W of 300 μm. The similarity of the three curves confirms that the electronic transport in the channel of all three transistors is dominated by the amorphous phase. Details of the fabrication and stability performance of these TFTs can be found elsewhere [17]. Under electron accumulation the Fermi level E_F moves towards the conduction band E_C, when the gate voltage is increased

283

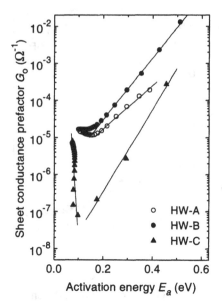

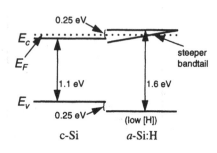

Figure 3:
Sheet conductance prefactor G_0 versus the activation energy E_a. Solid lines are linear fits to the straight parts of the data.

Figure 4:
Schematic representation of the band alignment upon gate bias application at the interface of the crystallites and the amorphous silicon tissue around it, showing the Fermi level positioned deeply into the conduction band tail-DOS of the a-Si:H.

above the threshold voltage V_{th}. The rate at which E_F moves towards E_C depends on the density of electronic defect states in the bandgap and on the width of the conduction band tail-state distribution, assuming a trap-free oxide insulator. The sheet conductance G ($= I_s L V_s^{-1} W^{-1}$) is proportional to $\exp(-E_a/kT)$, with a gate-voltage dependent (and temperature dependent) activation energy E_a. In Fig. 2 we plot the activation energy at a range of gate voltages, where the open symbols reflect the properties of the TFT with the silicon deposited with 100 % SiH$_4$ (HW-A), and the closed symbols represent TFTs with the channel deposited using H$_2$ dilution of SiH$_4$ (HW-B and HW-C). The activation energy saturates around 100 meV at a high gate electric field for all TFTs. However, also some distinct differences can be observed. First, at the maximum field applied of 1 MV/cm the highest-mobility TFT (HW-C) has the lowest sheet conductance activation energy: 75 ± 5 meV. Second, in the highest-mobility TFT the Fermi level moves the fastest through the bandgap when a small gate voltage above V_{th} is applied, indicating that its defect density in the bandgap is the lowest of the three TFTs. Third, a significantly different trace of the decrease of E_a is followed for the HW-C transistor, as compared to the HW-B or the a-Si:H (HW-A) TFT.

The activation-energy dependence of the sheet conductance prefactor G_0 is shown in

Fig. 3. The a-Si:H TFT shows the conventional Meyer-Neldel behaviour above $E_a = 0.2$ eV, with a characteristic slope of 0.074 ± 0.003 eV. The HW-B TFT shows the conventional MNR in the same range, with a slope of 0.051 ± 0.003 eV. These values are in good agreement with previous work on bulk a-Si:H and TFTs. Further, the HW-A and HW-B TFTs fail to show the inverse MNR as noted earlier in a-Si:H TFTs with conventional n+/Al contacts [10]. The HW-C TFT also shows conventional Meyer-Neldel behaviour, with a characteristic slope of 0.039 ± 0.003 eV, but in addition, when $E_a < 0.1$ eV a very pronounced inverse MNR is observed, with a slope approaching negative infinity.

DISCUSSION

As described above, some of the HWCVD deposited materials that are used here have shown to include crystalline domains with a size of several nanometer, which are embedded in an amorphous matrix at the interface with the insulator. The amount of crystallites strongly depends on the process conditions during deposition, and is especially sensitive to the wire temperature. Also, the hydrogen concentration in this material was shown to be low, at 2 at.-% [15], which indicates that the mobility gap will be smaller than that of conventional a-Si:H. With this in mind, we put forward the following explanation for the occurrence of the inverse MNR in our intrinsic nanostructured silicon. The explanation is based on the band alignment of the constituent phases of nanocrystalline domains embedded in an amorphous matrix, and is consistent with the model given by Lucovsky, that described and explained the inverse MNR in heavily doped microcrystalline silicon [6]. The low hydrogen content of this HWCVD material results in a-Si:H tissue with a bandgap of approximately 1.6 eV [12]. At the interface with the insulator this a-Si:H accounts for the largest fraction of material, although nanocrystalline inclusions are present. The Fermi level in these nanocrystalline inclusions is around midgap, that is, at 0.55 eV below the conduction band. In the a-Si:H phase the Fermi level is 0.8 eV below E_C, while the band offset between E_C of the nanocrystalline inclusions and E_C of the surrounding a-Si:H amounts to 0.25 eV. The width of the a-Si:H tail-state density of states amounts to 0.3 eV [6]. The situation is schematically drawn in Fig. 4. The shift of E_F upon gate voltage increment in the nanocrystalline inclusions is very rapid, due to the absence of defect states within the bandgap. In the crystalline phase, E_F moves above E_C and now it is apparent that E_F can move deeply into the tail states of the surrounding a-Si:H. At high positive gate voltage, the transport is still governed by the amorphous tissue, while E_F is at a level that cannot be achieved by doping the a-Si:H with phosphorus, nor by using the field effect in a purely amorphous TFT. In the latter case, a faint inverse MNR is observable only if the TFT is supplied with special Mg source and drain contacts [10]. Now, due to the heterogeneous nature of the channel material a far more pronounced inverse MNR can be observed even with conventional n+/Al contacts.

CONCLUSION

In summary, we have shown that it is possible to experimentally observe the inverse Meyer-Neldel rule in the conductivity of intrinsic nanostructured silicon, by incorporating hot-wire CVD silicon made under specific conditions into conventional inverted-staggered thin-film transistors. This material has superior structural order as compared to conventional PECVD a-Si:H, which enables one to shift the Fermi level deeply ($E_a < 0.08$ eV) into the conduction band tail state distribution by utilizing the field effect.

ACKNOWLEDGEMENT

J. Holleman (MESA Research Institute, University of Twente) is gratefully acknowledged for the photolithography and etching of the TFT structures.

REFERENCES

1. W. Meyer and H. Neldel, Z. Tech. Phys. **18**, 588 (1937).
2. W. B. Jackson, Phys. Rev. B **38**, 3595 (1988).
3. H. Overhof and W. Beyer, Philos. Mag. B **47**, 377 (1983).
4. B.-G. Yoon and C. Lee, Appl. Phys. Lett. **51**, 1248 (1987).
5. X. Wang, Y. Bar-Yam, D. Adler, and J. D. Joannopoulos, Phys. Rev. B **38**, 1601 (1988).
6. G. Lucovsky, and H. Overhof, J. Non-Cryst. Solids **164-166**, 973 (1993).
7. A. Rubino, M. L. Addonizio, G. Conte, G. Nobile, E. Terzini, and A. Madan, in *Amorphous Silicon Technology - 1993*, edited by E. A. Schiff, M. J. Thompson, A. Madan, K. Tanaka, and P. G. LeComber (Mater. Res. Soc. Proc. **297**, Pittsburgh, PA, 1993), p. 509.
8. R. Brüggemann, M. Rojahn, and M. Rösch, Phys. Stat. Sol. **166**, R11 (1998).
9. R. Schumacher, P. Thomas, K. Weber, W. Fuhs, F. Djamdji, P. G. Le Comber, and R. E. I. Schropp, Philos. Mag. B **58**, 389 (1988).
10. M. Kondo, Y. Chida, and A. Matsuda, J. Non-Cryst. Solids **198-200**, 178 (1996).
11. H. Meiling and R. E. I. Schropp, Appl. Phys. Lett. **70**, 2681 (1997).
12. A. H. Mahan, J. Carapella, B. P. Nelson, R. S. Crandall, and I. Balberg, J. Appl. Phys. **69**, 6728 (1991).
13. H. Matsumura, Appl. Phys. Lett. **51**, 804 (1987).
14. J. K. Rath, F. D. Tichelaar, H. Meiling, and R. E. I. Schropp, in *Amorphous and Microcrystalline Silicon Technology - 1998*, edited by R. Schropp, H. Branz, S. Wagner, M. Hack, and I. Shimizu (Mater. Res. Soc. Proc. **507**, Pittsburgh, to be published).
15. A. M. Brockhoff, E. H. C. Ullersma, H. Meiling, F. H. P. M. Habraken, and W. F. van der Weg (unpublished).
16. C. van Berkel, in *Amorphous and Microcrystalline Semiconductor Devices Vol. II: Materials and Device Physics*, edited by J. Kanicki (Artech House, Boston, London, 1992), p. 397; and references therein.
17. H. Meiling, A. M. Brockhoff, J. K. Rath, and R. E. I. Schropp, in Amorphous and Microcrystalline Silicon Technology 1998, edited by R. Schropp, H. Branz, S. Wagner, M. Hack, and I. Shimizu (Materials Research Society, Pittsburgh, to be published).

IN SITU STUDIES OF THE VIBRATIONAL AND ELECTRONIC PROPERTIES OF Si NANOPARTICLES

J. R. FOX*, I. A. AKIMOV*, X. X. XI*, AND A. A. SIRENKO*
* Department of Physics, Penn State University, University Park, Pennsylvania 16802

ABSTRACT

We report on *in situ* studies of the vibrational properties of ultra-thin Si layers grown by dc magnetron sputtering in ultrahigh vacuum on amorphous MgO and Ag buffer layers. The average thickness of the Si layers ranged from monolayer coverage up to 200 Å. The interference enhanced Raman scattering technique has been used to study changes in the phonon spectra of Si nanoparticles during the crystallization process. Marked size-dependencies in the phonon density of states of the Si quantum dots and the relaxation of the **k**-vector conservation condition with decrease in size of the Si nanoparticles have been detected. Electron energy loss spectra have been collected for amorphous and crystallized Si nanoparticles on SiO_2 buffer layers and the difference in the onset of the electronic transitions have been found.

INTRODUCTION

As the size of condensed matter systems is reduced in one or more dimensions to nanoscale levels, the electron and phonon states are influenced by confinement and surface effects. The discovery of visible luminescence at room temperature from porous Si increased interest in the modification of silicon's electronic and vibrational properties with size [1,2]. Nanometer-sized Si-based structures prepared by a wide variety of methods have been intensively investigated [2-5]. Recent experimental studies, focused on the modification of the electronic states of nanocrystalline Si (nc-Si), report substantial changes in the luminescence properties and optical gap which are attributed to quantum confinement of the electronic states and a breakdown of **k**-vector conservation [6-8].

Theoretical [9,10] and experimental [11] studies of semiconductor nanocrystallites (or quantum dots) demonstrate the modification of the phonon spectra due to the confinement of the optical and acoustic vibrations. With decreasing particle size, Raman spectra change due to relaxation of **k**-vector conservation allowing observation of the scattering from entire branches of the acoustic and optical phonons [12]. At the same time, the fraction of the surface to interior atoms increases in smaller particles leading to further modification of the phonon density of states [12,13]. Confinement of the electronic states strongly affect the electron-phonon interaction, which should be also taken into account for analysis of the Raman spectra [14,15].

The phonon states of *isolated* Si nanoparticles have been relatively unexplored experimentally -- publications on *in situ* measurements of nc-Si are rare [16]. Previous Raman studies have generally focused on samples with Si particles within a matrix or with ligands [2,4,5,17,18]. In these systems, the influence of a matrix or surface contamination on the phonon spectra cannot be neglected. Thus, in the analysis of the observed shifts and broadening of the optical phonon lines, it was difficult to distinguish between effects due to phonon confinement, surface state modification, and matrix-induced stress.

Mat. Res. Soc. Symp. Proc. Vol. 536 © 1999 Materials Research Society

In this paper, we report on *in situ* Raman studies of ultra-small Si particles grown in ultrahigh vacuum. We have studied the phonon states of the Si nanoparticles free of chemisorbed species as well as changes in the phonon states induced by the crystallization process. The main experimental problem was a weak signal from the system with extremely small scattering volume. The situation has been drastically improved by using interference enhanced Raman scattering (IERS) [19] combined with the multichannel detection [20]. Raman signal from the ultrathin films was enhanced by a factor of 15-20 due to the bilayer structure, which makes IERS sensitive even to submonolayer coverage. Recently this technique provided useful information about the vibrations of surface and near-surface atoms of carbon nanocrystallite clusters and ultrathin films [13] and isolated Ge nanoparticles [21].

Reflection electron energy loss spectroscopy (EELS) involves the scattering of a monochromatic beam of low energy electrons from a sample and analyzing the energy distribution in the reflected beam. For losses arising from the dipole scattering mechanism, the observed spectrum is related to the dielectric function of the material. EELS has been shown to be a valuable tool in the measure of surface optical constants [22] and the study of the absorption edge and subgap absorption in thin films of amorphous semiconductors [23] and isolated nanoparticles [24]. In this paper, we report preliminary *in situ* EELS measurements of the optical gap of Si nanoparticles.

EXPERIMENT

Ultrathin layers of Si were grown on amorphous MgO or Ag buffer layers at room temperature by dc magnetron sputtering in an ultrahigh vacuum (UHV) chamber. A power of 6 W/cm^2 was applied to the Si sputter target. The background pressure of Ar was kept around 4 mTorr during the Si growth process. Before deposition of Si, the substrates were annealed at 500°C. The growth rate of Si, calibrated by both *ex situ* elipsometry and profilometry, was about 0.5 Å/sec. The average thickness of the Si layers d ranged from monolayer coverage up to 200 Å. In the following, the Si samples will be identified by their value of d. To crystallize the Si ultrathin layers, the samples were heated up to about 500°C by electron emission and thermoradiation. The temperature of the sample was estimated by means of both a thermocouple attached to the sample holder and a pyrometer.

Silicon forms nanometer-size clusters of a hemispherical shape on the substrate. The average size and concentration of the Si clusters varied with d. For example, in a $d = 30$ Å film on Ag, the Si clusters had a $20 - 50$ Å base width and were $15 - 30$ Å high, as measured by high resolution electron microscopy. Our preliminary results with *in situ* core-level x-ray photoemission (XPS) show that at $d \leq 50$ Å, Si forms nanoscale islands on MgO substrates. For thicker layers of Si, the characteristic MgO signal in the XPS spectra quickly disappears, corresponding to a complete coverage of the buffer-layer surface with Si atoms. It is similar to the growth regime of Ge nanoparticles investigated previously with the same apparatus [21]. It was shown that the Ge clusters have a hemispherical shape and each of them contains about $(10 - 20)d$ [Å] atoms. Electron microscopy confirms that this relation can be also used in our case for estimating the Si particle size.

The Raman measurements were performed *in situ* at room temperature and at a pressure of 2 x 10^{-10} torr in the UHV system. Interference enhancement of the laser electric field close to the surface (the position of the Si nanoparticles) was achieved by utilizing a bilayer structure, which consists of a transparent dielectric film (MgO) deposited on a metallic layer (Ag) [19,25]. The entire structure was grown on a Si substrate. The thickness of the amorphous Ag, which works in the bilayer structure as a light reflector, was about 2000 Å. This is thick enough to completely screen any possible Raman signal from the Si substrate. The 5145 Å line (2.4 eV) of an Ar^+-ion laser was used for excitation of the Raman spectra. The MgO layer was grown with a thickness of 500 Å, as calculated to give optimal interference enhancement at the exciting wavelength. The choice for the dielectric buffer layer (MgO) was determined by the two following reasons. First, the Raman signal of MgO is weak at the green light excitation and, second, the analysis of the photoemission spectra is easier than, e.g. in case of a SiO_2 buffer, which also contains Si atoms.

Raman spectra were measured with a SPEX Triplemate spectrometer equipped with a charge-coupled device (CCD) detector cooled with liquid nitrogen. The resolution was about 1.5 cm^{-1}. The intensity of the Raman signal was analyzed in the conventional backscattering configuration with the laser beam and scattered light perpendicular to the plane of the sample.

Electron energy loss spectra were measured with an LK2000 HREELS spectrometer with a rotatable analyzer operated with 5 eV pass energies yielding a resolution of approximately 30 meV. Ultrathin Si was deposited on SiO_2 buffer layers (prepared by dc magnetron sputtering, about 240 Å thick) on Si substrates. Incident electrons were at an energy of 50 eV.

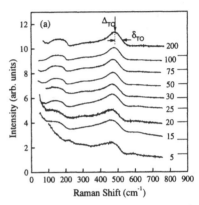

RESULTS AND DISCUSSION

Raman spectra for samples with a different average thickness of the Si layer grown on MgO are shown in Fig. 1. They are similar to that of amorphous Si (a-Si) and consist of the TA, 2TA and TO strong bands centered around 150, 300, and 480 cm^{-1}, respectively, and the weak LA shoulder at 370 cm^{-1} [26-28]. As the film thickness decreases, the high frequency half width at half maximum (HWHM) of the TO band, δ_{TO} defined as shown in Fig. 1, changes from 35 to 50 cm^{-1}. At the same time the position of the maximum Δ_{TO} moves towards the low Raman shifts. The dependencies of Δ_{TO} and δ_{TO} on d are presented in Fig. 2. Similar features of the phonon spectra were reported for a-Si thick films grown at different conditions and, e.g., it has been shown Δ_{TO} reflects the distribution of the sp-bonds in a-Si [12].

We explain the broadening of our spectra by the modification of the phonon density of states due to changes in the

Figure 1. Normalized Raman spectra of amorphous Si layers measured with excitation at 2.4 eV. The respective average thicknesses are given next to the spectra in Å. The spectra were vertically shifted for clarity. The horizontal solid lines indicate the zero-signal levels.

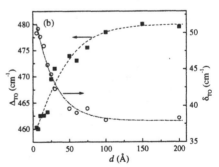

Figure 2. The HWHM δ_{TO} and the position Δ_{TO} of the TO Raman peak as functions of the average thickness of the Si layers. Dashed lines guide the eye.

angular distribution of the *sp*-bonds in the near-surface atomic layers and dangling bonds on the surface. Indeed, in ultra-thin Si films, the fraction of the surface atoms to interior atoms increases making the surface effects more pronounced. Our experiments are important to distinguish between size-dependent modifications of the phonon density of states from that due to contamination of the Si surface with hydrogen or oxygen. The latter might be neglected for our samples, but could be very strong in, e.g., porous Si. Note that the significant variation in the phonon spectra occurs for the samples with $d < 50$ Å. Recall that our preliminary XPS results demonstrate that near this value of d, Si forms nanoscale islands on the substrate.

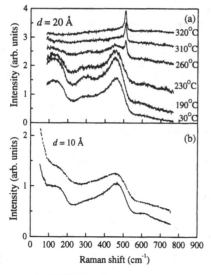

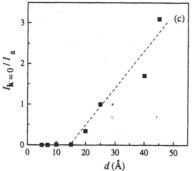

Figure 3. (a) Normalized Raman spectra of the 20 Å sample measured at different temperatures during crystallization process. The spectra were vertically shifted and the constant background was subtracted for clarity. (b) Raman spectra of the 10Å sample before (solid line) and after annealing (dashed line). (c) The ratio of the k=0 peak intensity $I_{k=0}$ to that of the amorphous background I_a as a function of the average thickness of the Si layers. Dashed line guides the eye.

After heating the samples, a remarkable transformation in the Raman spectra was observed. Fig. 3(a) shows several spectra for the 20 Å sample taken at different temperatures. The sharp crystalline peak arises in the Raman spectra when the temperature increases to more than T ≈ 260°C. At the same time, the intensity of the amorphous-like broad features decreases disappearing at T > 320°C. This indicates a partial recovering of the **k**-vector conservation condition with crystallization of the Si layers allowing only scattering by optical phonons at the zone center (**k**=0), where the TO and LO phonons in bulk crystalline Si are degenerated [29]. At room temperature, the position of the crystalline peak in our samples is about 519 ± 1 cm^{-1}, which is close to that in bulk [30]. Its full width at half maximum is about 7 cm^{-1}, which is about two times wider than that in Si single crystal. Although one expects a red shift in the position of the optical phonon line in nanoparticles due to confinement, no strong systematic shift of this peak was observed with our experimental accuracy (1.5 cm^{-1}). In our system, it can be explained by the partial compensation of the confinement effect by compressive strain, which works in the opposite direction. High-resolution Raman measurements are required for detailed studies of these effects in nc-Si. However, these experiments would be limited by the extremely small scattering volume of Si nanoparticles

As the average thickness of the Si layer decreases, the relative intensity of the k=0 peak to the intensity of the amorphous-like broad peak at 480 cm^{-1} decreases and the k=0 peak disappears for Si films with $d \leq 10$ Å. The ratio of the k=0 peak intensity to the amorphous-like background intensity is shown in Fig.3(c). In the case of the relatively thick films, with d greater than 50 Å, the complete k-vector conservation could be achieved after annealing, which manifests itself by the disappearance of the amorphous-like features in the Raman spectra. For Si films with d less than 10 Å, the Raman spectra do not change with annealing and practically no modifications have been observed even after heating up to 500°C. Two corresponding spectra of the 10 Å sample grown on a Ag buffer layer and measured at room temperature before and after annealing are shown in Fig. 3(b).

This result is important for the determination of the minimum size of a crystalline system, which can satisfy k-vector conservation conditions. According to our estimations, the transition between crystalline- and amorphous-like behavior takes place in the samples with an average number of Si atoms in one hemispherical particle equal to about three hundred. The role of surface and near-surface layers requires additional theoretical studies. The comparative study of the Raman scattering spectra and the high resolution electron microscopy images will be published elsewhere [31].

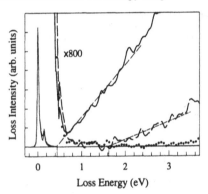

Figure 4. EELS spectra for SiO₂ substrate (dotted) and ultrathin (d=5Å) Si: as-deposited (solid) and annealed (dashed). Light dashed lines guide the eye to the onset of electronic transitions.

Electron energy loss spectra for a sample with an average thickness $d = 5$ Å of Si grown on SiO₂ are shown in Fig. 4. The three spectra were normalized to the elastic peak intensity. An intense peak at 145 meV, appearing in all three spectra, is due to a dipole-active phonon in the SiO₂ substrate [24]. On the same plot, the lower intensity electronic transitions are shown with magnification to illustrate the onset of electronic transitions in the SiO₂ substrate, the as-deposited Si layer, the annealed (T > 260°C) Si layer. The SiO₂ substrate has no transitions evident below 3.8 eV. The onset of electronic states in the pristine sample occurs at lower energy than that seen in the annealed sample. Qualitatively, it can be estimated that the onset of electronic transitions is at about 0.4 eV and 1.6 eV before and after annealing, respectively. A similar, quantitative result can be extracted from the EELS data by computing the absorption coefficient $\alpha(\omega)$ [23] and using analogous techniques as those employed to estimate the optical gap in amorphous semiconductors [32]. We have thus demonstrated the ability to measure optical gap differences in nanoparticles of silicon.

CONCLUSIONS

In conclusion, we studied the phonon states of Si nanoparticles grown and measured in ultrahigh vacuum. Marked transformation in the phonon density of states and the relaxation of the k-vector conservation condition with decrease in size of the Si nanoparticles were detected.

ACKNOWLEDGMENTS

The authors are grateful to S. Rouvimov for high resolution electron microscopy measurements and to V. I. Merkulov for useful discussions.
This work was supported by USDOE Grant DE-FG02-84ER45095 and NSF Grant DMR-96-23315.

REFERENCES

1.   A. G. Cullis and L. T. Canham, Nature **353** (6342), 335 (1991).
2.   A. G. Cullis, L. T. Canham, and P. D. J. Calcott, J. Appl. Phys. **82** (3), 909 (1997).
3.   K. Eberl, K. Brunner, and W. Winter, Thin Solid Films **294** (1), 98 (1997).
4.   E. W. Forsythe, E. A. Whittaker, F. H. Pollak et al., in Microcrystalline and Nanocrystalline Semiconductors, edited by R. W. Collins (MRS, Pittsburgh, 1994), Vol. 358.
5.   X. S. Zhao, Y. R. Ge, J. Schroeder et al., Appl. Phys. Lett. **65** (16), 2033 (1994).
6.   W. L. Wilson, P. F. Szajowski, and L. E. Brus, Science **262** (5137), 1242 (1993).
7.   S. Schuppler, Y. J. Chabal, F. M. Ross et al., Phys. Rev. Lett. **72** (16), 2648 (1994).
8.   D. Kovalev, H. Heckler, B. Averboukh et al., Phys. Rev. Lett. , accepted for publication (1998).
9.   R. Alben, D. Weaire, Jr. J. E. Smith et al., Phys. Rev. B **11**, 2271 (1975).
10.   T. Takagahara, J. Lumin. **70** (1), 129 (1996).
11.   A. Ekimov, J. Lumin. **70** (1), 1 (1996).
12.   J.S. Lannin, "Raman Scattering of Amorphous Si, Ge, and their Alloys," in Semiconductors and Semimetals, edited by J. I. Pankove, v. 21B (Academic Press, Orlando, 1984), p. 159.
13.   J.S. Lannin, V.I. Merkulov, and J.M. Cowley, in Advances in Microcrystalline and Nanocrystalline Semiconductors : Materials Research Society Symposia Proceedings No. 452, edited by R. W. Collins, et al., (MRS, Pittsburgh, 1996), p. 225.
14.   C. Trallero-Giner, A. Debernardi, M. Cardona et al., Phys. Rev. B **57** (8), 4664 (1998).
15.   A. A. Sirenko, V. I. Belitsky, T. Ruf et al., Phys. Rev. B **58** (4), 2077 (1998).
16.   S. Hayashi and H. Abe, Jpn. J. Appl. Phys. **23**, L824 (1984).
17.   Z. Igbal, S. Veprek, A.P. Webb et al., Solid State Comm. **37**, 993 (1981).
18.   Y. Gao and T. López-Ríos, Solid State Commun. **60**, 55 (1986).
19.   G.A.N. Connel, R. J. Nemanich, and C.C. Tsai, Appl. Phys. Lett. **36**, 31 (1980).
20.   J.C. Tsang, in Light Scattering in Solids, edited by M. Cardona and G. Güntherodt, Topics in Applied Physics, v. V (Springer-Verlag, Berlin, 1989), p. 233.
21.   J. Fortner and J. S. Lannin, Surf. Sci. **254** (1), 251 (1991).
22.   H. Froitzheim, H. Ibach, and D.L. Mills, Phys. Rev. B. **11** (12), 4980 (1975).
23.   G.P. Lopinski and J.S. Lannin, Appl. Phys. Lett. **69** (16), 2400 (1996).
24.   Gregory P. Lopinski, Vladimir I. Merkulov, and Jeffrey S. Lannin, Phys. Rev. Lett. **80** (19), 4241 (1998).
25.   W. S. Bacsa and J. S. Lannin, Appl. Phys. Lett. **61** (1), 19 (1992).
26.   J. E. Smith, M. H. Brodsky, B. L. Crowder et al., Phys. Rev. Lett. **26**, 642 (1971).
27.   J.S. Lannin and P.J. Carrol, Philos. Mag. **45**, 155 (1982).
28.   G. Nilsen and G. Nelin, Phys. Rev. B **6**, 3777 (1972).
29.   Peter Y. Yu and Manuel Cardona, Fundamentals of Semiconductors: Physics and Materials Properties (Springer-Verlag, Berlin, 1995) , p 103.
30.   D. Bimberg, et al., in Numerical Data and Functional Relationships in Science and Technology, edited by O. Madelung, Landolt-Börnstein, New Series, Group III, v. 17a (Springer, Berlin, 1982), p. 72.
31.   A. Sirenko, J.R. Fox, S. Rouvimov et al., "To be published."
32.   G.D. Cody, "The Optical Absorption Edge of a-Si:H," in Semiconductors and Semimetals, edited by J. I. Pankove, v. 21B (Academic Press, Orlando, 1984), p. 11.

X-RAY REFLECTIVITY STUDY OF POROUS SILICON FORMATION

V. CHAMARD*, G. DOLINO*, J. EYMERY**
*Laboratoire de Spectrométrie Physique, UMR 5588 CNRS, Université J. Fourier, BP 87, 38402 Saint Martin d'Hères, France, virginie.chamard@ujf-grenoble.fr
**Département de Recherche Fondamentale sur la Matière Condensée (SP2M/PSC), 17 avenue des Martyrs, CEA-Grenoble, 38054 Grenoble Cedex 9, France

ABSTRACT

X-ray reflectometry is used to study the first stages of formation of thin n-type porous silicon layers. Results on classical n⁻-type porous silicon prepared under illumination are first reported. Then, the effect of the illumination during the formation is observed by comparing $n^{+/-}$-type samples prepared in darkness or under illumination. X-ray specular reflectivity measurements allow to observe an increase of the surface porosity even for the short formation times and a macroporous layer under the nanoporous layer is also identified for illuminated samples. The presence of a crater at the top of the layer is observed by profilometer measurements, especially in the case of illuminated samples. Specular and diffuse x-ray scattering results show important effects of light during the porous silicon formation.

INTRODUCTION

Since the discovery of room temperature luminescence of porous silicon (PS) in 1990, this material has raised a strong interest [1]. However, the formation mechanisms [2] and the structure of this nanocrystalline material are not well understood, in particular for the formation of n-type PS under illumination. In this case, a two-layer structure, consisting of a macroporous under a nanoporous layer often associated to a crater formation, has been observed, but not clearly explained [3]. In addition to the usual anodisation parameters (current intensity (j), HF concentration ([HF]) and anodisation time (t_f)), the light features (wavelength, intensity) must be considered.

We present an x-ray reflectivity (XRR) study of thin layers of n-type PS to get information about the formation process. XRR is a well adapted technique for measurements of thin PS layer [4,5]. In the specular XRR geometry, only the direction perpendicular to the sample surface is investigated. Moreover, in the off-specular geometry, the measurements are also sensitive to the in-plane and interfaces correlations [6].

A preliminary characterisation of n⁺-type and n⁻-type PS has already been obtained by specular XRR techniques [7]. In this article, we first present experimental results obtained with specular XRR and diffuse scattering measurements on classical n⁻-type PS sample prepared under illumination. Then, we study intermediate doped ($n^{+/-}$-type) PS, prepared in darkness and under illumination. In addition to XRR studies, the samples characterisation was done by profilometer measurements.

EXPERIMENTAL RESULTS

Experimental conditions

The PS samples are produced by an electrochemical etching of n-type (001) silicon substrate of 5 Ω.cm resistivity for n$^-$-type and of 10^{-2} Ω.cm resistivity for n$^{+/-}$-type samples. The porous layer is formed by anodisation of silicon in a 15 % HF solution (HF:H$_2$O:C$_2$H$_5$OH (3:7:10)) under a constant current density of 5 mA/cm^{-2} for formation times (t$_f$) between 1 and 600 s, and then rinsed for 5 min with deionized water. For samples prepared under illumination, an halogen lamp is used with a filter giving an illumination of 40 mW.cm^{-2} with wavelength larger than 660 nm. A Tencor profilometer is used to measure the crater depth and the nanoporous layer thickness after an NaOH etching[3].

Most of the specular XRR measurements are performed with a high resolution Philips apparatus, using an x-ray tube and a four-reflection Ge monochromator [5]. Because of the low intensity of the diffuse scattering signal, the off-specular measurements are performed with a 12 kW rotating anode set-up using a graphite monochromator and slits to define the divergence. In both instruments, the radiation wavelength is the CuK$_{\alpha 1}$ radiation (0.154 nm). The incidence and detector angles are named respectively ω and 2θ.

n$^-$-type sample

In a previous work [7], we started to study the evolution of the porous layer with specular XRR for the typical cases of n$^-$-type PS prepared under illumination and n$^+$-type PS prepared in darkness. In Fig. 1, the first measurements of x-ray diffuse scattering (ω scans) performed with an homogeneous porous layer on a silicon substrate are presented. They correspond to a n$^-$-type PS sample prepared under illumination with t$_f$ = 15 s.

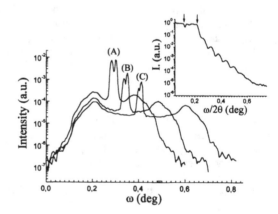

Figure 1: ω-scans x-ray diffuse scattering for n$^-$-type PS for three positions of the detector : (A): 0.6°; (B): 0.7°; (C): 0.82°. The insert shows the specular XRR measurements; the arrows indicate the positions of the critical angles of PS and bulk silicon.

The ω-scans show two symmetric peaks (Yoneda wings) on both sides of the coherent reflection (specular peak) [6]. The simulation of the diffuse intensity, not presented in this paper, gives mainly information about in-plane correlation lengths and interfaces correlations. The bimodal shape of the specular peak, clearly observed in curves (B) and (C), comes from a slight non homogeneous curvature of the sample. The specular XRR shown in the inset is obtained by recording the larger peak of curves (B) and (C). Two critical angles (ω_{PS} for PS and ω_{Si} for bulk silicon) and some interference fringes are observed in this insert. Several parameters of the PS layers can be deduced from these measurements:
- the porosity (P) is related to ω_{PS} by:

$$1 - P = (\omega_{PS}/\omega_{Si})^2, \qquad (1)$$

- the layer thickness is deduced from the period of the interference fringes,
- the surface and interface roughnesses are related respectively to the decrease of the intensity at large angles and to the damping of the interference fringes.

The aim of this paper is to point the general differences resulting from the preparation conditions of the samples, so that the quantitative treatment of the data will not be developed.

$n^{+/-}$-type PS prepared in darkness

The evolutions of the nanoporous layer thickness and of the crater depth as a function of the formation time are presented respectively on Fig. 2(a) and Fig. 2(b). An important feature of the $n^{+/-}$-type PS samples prepared in darkness is that at the beginning ($t_f < 60$ s) the growth of the porous layer is quite fast, then the growth rate reaches a slower regime for larger formation time. Moreover, no crater has been observed for $t_f < 225$ s; for $t_f = 600$ s, the crater depth is only of 20 nm.

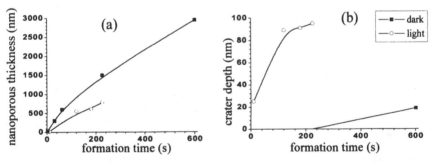

Figure 2: *Profilometer measurements of the nanoporous layer thickness (after NaOH etching) (a) and of the crater depth (b) of $n^{+/-}$-type PS samples prepared in dark and under light. Lines are just a guide for the eye.*

The results of specular XRR measurements for samples of different formation times are presented on Fig. 3(a). The critical angle of the porous layer decreases with increasing formation time, showing the highly surface sensitivity of the technique. Although the thickness of the PS layer is small for short formation times ($t_f < 30$ s), no interference fringes are observed. This comes from the important roughness of the PS/bulk silicon interface which completely destroys constructive interference resulting from an homogeneous layer. Actually, the p-type PS/bulk silicon interface is known to be rougher than the external surface [4]. For longer formation time (t_f = 600 s), the decrease of the intensity is much faster, due to an increase of the surface roughness, probably related to the crater formation. The increase of the surface porosity from 5 % to 70 % as a function of the formation time is presented in the insert of Fig. 3(a). For long formation times, this is due to the chemical dissolution of the top of the porous layer whereas for short formation times, the strong increase of the surface porosity from 5 % to about 40 % in only 60 s may be related to pores nucleation [8].

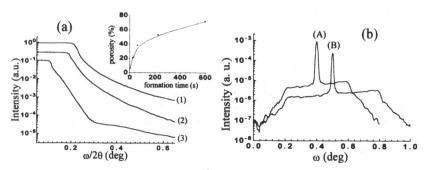

Figure 3(a): *Specular XRR for $n^{+/-}$-type PS samples prepared without illumination, for several formation times: (1): 1s; (2): 30 s; (3): 600 s. The insert presents the evolution of the porosity with the formation time. (b): Diffuse scattering curves measured on sample (2) for two positions of the detector: (A): 0.4°, (B): 0.5°.*

Fig. 3(b) shows ω-scan diffuse scattering curves for different detector positions for the 30 s $n^{+/-}$-type PS sample. The diffuse scattering is not clearly structured compared to Fig. 1. curves; the Yoneda wings are not observed and a very flat plateau is measured. This is related to the large roughness of the PS/bulk silicon interface.

$n^{+/-}$-type PS prepared under illumination

The porous layer thickness and crater depth obtained for $n^{+/-}$-type samples prepared under illumination with formation time between 5 and 225 s are presented on Fig. 2. As observed for samples prepared without illumination, the growth of the porous layer is faster for short formation time. With these formation conditions, a crater is measured even for the short formation time. For t_f = 5 s, the crater depth equals the nanoporous layer thickness (about 25 nm). For larger thickness, it is well known that a macroporous structure appears below the nanoporous layer [3]. When the nanoporous layer is removed by NaOH etching, the sample surface has a grey and mat appearance, typical of a macroporous layer.

The specular XRR results, presented on Fig. 4(a), are very different from those of Fig. 3(a). For short formation times (curve (1), t_f = 5 s), careful measurements allow to detect the PS critical angle and, for larger angles, a weak interference fringe resulting from the PS layer appears. For a 30 s formation time (curve (2)), interference fringes between the critical angles of porous and bulk silicon are observed. For longer formation time (t_f = 225 s), a supplementary critical angle is detected at about 0.1° (arrow on curve (3)). This angle is related to the porosity of the macroporous layer under the nanoporous layer. Actually, after NaOH etching of the sample, the specular XRR of the porous layer shows the same critical angle.

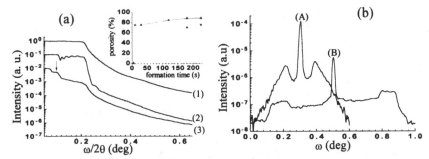

Figure 4(a): *Specular XRR for $n^{+/-}$-type PS samples prepared under illumination, for several formation times: (1): 5s; (2): 30 s; (3): 225 s. The arrow indicates the position of the macroporous critical angle. The insert presents the evolution of the porosity of the nanoporous (▲) and of the macroporous (▼) layers with the formation time.* **Figure 4(b):** *Diffuse scattering curves measured on sample (2) for two positions of the detector: (A): 0,3°; (B): 0.5°.*

The evolution of the porosity as a function of the formation time is presented in the insert of Fig. 4-a. The porosity of the surface layer is about 80 % for formation times between 10 and 225 s. The nucleation stage is much shorter than for samples prepared without illumination (a few seconds). The porosity of the macroporous layer, which is first observed at $t_f = 180$ s, for a nanoporous thickness of about 500 nm, is around 75 %.

X-ray diffuse scattering measurements presented on Fig. 4(b) have been performed on sample (2). Yoneda wings are observed on both sides of the specular peak, their shapes are broadened due to the existence of the two critical angles (porous layer and Si substrate).

Complementary measurements have been performed by X-ray diffraction [9]. Diffraction Bragg peaks have already been observed on n^+-type samples prepared in darkness but not on n^--type samples formed under light [7]. On the present $n^{+/-}$-type samples formed without light, the porous layer Bragg peak has been clearly observed for formation time around tf = 60 s. However for longer t_f, the Bragg peak is reduced to a broad shoulder. For $n^{+/-}$-type samples formed under light only a weak shoulder is observed for all t_f values.

CONCLUSION

The aim of this work was to study thin PS layers on n-type samples. Both profilometer and x-ray reflectivity measurements allow to see some differences between samples prepared with and without illumination.

With profilometer measurements, a crater is observed, even for short formation times, for $n^{+/-}$-type samples formed under light, while for similar samples formed in darkness a crater appears only for formation time longer than 600 s. The growth of the nanoporous layer shows a fast initial regime followed by a slower growth rate.

The XRR is a sensitive non destructive method to study thin PS layers, giving different results for n-type samples formed with or without light: the most spectacular difference is related to the very different roughness of the layer-substrate interface. The samples formed under light have a sizeable roughness which produce only a damping of the thickness fringes but does not prevent their observations. In this case, the diffuse scattering is clearly observed with well formed Yoneda wings. On the other hand, the samples formed in darkness have a large roughness which prevents the observation of thickness fringes and produces an intense diffuse scattering without Yoneda wings. More quantitative results for these porous structures can be obtained from XRR simulations.

A great interest of the XRR method is the possibility to perform measurements of the surface porosity. An important result obtained in the present work is the large initial variation of the surface porosity as a function of time, which is probably related to pore nucleation at the sample surface. Indeed, Brumhead et al. [8] have proposed that all the pores are not initiated instantaneously but nucleate progressively during a finite period of time, producing a rapid initial increase of the surface porosity. Then at longer time, the porosity is slowly increased by the usual chemical dissolution of silicon in HF. A possible explanation is that the top of the very porous layer disintegrate in drying, producing a rough crater.

In this work, we have presented some results illustrating the power of the XRR to study PS layers. This technique is well suited for measurements of thin layers, classically under a 0.4 μm thickness. For PS, it is shown that even with a strong interface roughness, this laboratory technique gives the possibility to study the initial stage of pore nucleation.

REFERENCES

[1] *Properties of Porous Silicon*, edited by L. T. Canham, EMIS, Datareviews Series No. 18, (INSPEC, Institute of Electrical Engineers, London, 1997).

[2] Y. Kang and J. Jorné, J. Electrochem. Soc. **140**, 2258 (1993); A. Valance, Phys. Rev. B **55**, 9706 (1997).

[3] L. C. Lévy-Clément in *Porous Silicon Science and Technology*, edited by J. C. Vial and J. Derrien (Springer, Berlin, 1994), p. 329.

[4] T. R. Guilinger, M. J. Kelly, E. H. Chason, T. J. Headley and A. J. Howard, J. Electrochem. Soc **142**, 1634 (1995).

[5] D. Buttard, G. Dolino, D. Bellet, T. Baumbach and F. Rieutord, Solid State Comm., to be published (1999).

[6] D. Bahr, W. Press, R. Jebasinski and S. Mantl, Phys. Rev. B **47**, 4385 (1993).

[7] V. Chamard, G. Dolino, G. Lérondel, S. Setzu, Physica B **248**, 101 (1998).

[8] D. Brumhead, L. T. Canham, D. M. Seekings and P. J. Tufton, Electrochim. Acta **38**, 191 (1993).

[9] D. Buttard, D. Bellet, G. Dolino and T. Baumbach, J. Appl. Phys. **83**, 5814 (1998)

LIGHT INDUCED ESR MEASUREMENTS ON MICROCRYSTALLINE SILICON WITH DIFFERENT CRYSTALLINE VOLUME FRACTIONS

J. MÜLLER, F. FINGER, P. HAPKE, H. WAGNER
Forschungszentrum Jülich, ISI-PV, D-52425 Jülich

ABSTRACT

Microcrystalline silicon with various crystalline volume fractions was prepared by plasma enhanced chemical vapour deposition. The material was studied by steady state and transient electron spin resonance in the dark and under light illumination. The observed resonances at g-values of 2.01, 2.0052, 2.0043, 1.998 can be attributed to the amorphous and microcrystalline constituents, and their respective intensities change as the ratio of amorphous to crystalline volume is varied. The origin of a fifth resonance at g = 1.995 remains unclear. Smaller crystalline volume fractions lead to lower spin densities and affect the recombination behaviour of photo-generated charge carriers. The recombination behaviour in highly crystalline material is also influenced by moderate Fermi level shifts, where differences show up between n-type (or undoped) and p-type samples. The differences are attributed to trapping of photo-generated holes in deep states within the disordered regions.

INTRODUCTION

The successful application of microcrystalline silicon as low-bandgap absorber layer in solar cell devices [1], has initiated a growing interest in the technology and the physical properties of this material. When incorporated into solar cells the device quality depends on cell design, interfaces and the material parameters of the intrinsic absorber layer. An interesting question is for example the relationship between the respective crystalline and amorphous volume fractions in the i-layer material and the cell performance [2]. Some insight can be gained by studying the excitation and recombination behaviour of light induced charge carriers as a function of the crystalline volume fraction. This was done in the present work using electron spin resonance (ESR), which allows to distinguish between different electronic states within the material. In the past, microcrystalline material (intrinsic, p- and n-type) with crystalline volume fractions ≥ 90% has been studied by ESR in great detail (see [3] and references therein). In the present study we systematically vary the ratio of amorphous to crystalline volume over a wide range. Slightly n- and p-doped samples are included to examine the effects of Fermi level shifts. The material is characterized by ESR in the dark and under different light illumination conditions (white and infrared light). Transient experiments monitor the signal intensity of the conduction electron (CE) resonance [3,4,5] upon various dark and illumination sequences to study excitation and recombination processes. The light induced ESR (LESR) signals of samples with lower crystallinity consist of a superposition of both the amorphous and the crystalline constituents. Remarkable differences depending on the respective volume fractions and the type of doping show up in the transient spectra.

EXPERIMENT

Samples were prepared by plasma enhanced chemical vapour deposition at excitation frequencies between 13 and 115 MHz and a substrate temperature of 200 °C. Deposition rates r (between 0.4 and 4.4 Å/s) and crystallinity of the material were varied by adjusting the silane to hydrogen flow ratios ([SiH4] / [SiH4]+[H2] = 3 - 8%) and by changing gas pressure (300 - 900 mTorr) and

Mat. Res. Soc. Symp. Proc. Vol. 536 © 1999 Materials Research Society

discharge power (5 - 50 W). As a measure for the crystalline volume fraction of the material we use the intensity ratios $R_C = I_C / I_C + I_a$ of the Raman peaks at approximately 520 cm^{-1} and 500 cm^{-1} ($I_C = I_{520} + I_{500}$, attributed to the crystalline phase) to the overall Raman intensity $I_C + I_a$ with $I_a = I_{480}$ attributed to the disordered phase [6,7]. We keep in mind that the intensity ratio R_C can only serve as an estimate of the crystalline volume fraction and usually underestimates the real values (see [7] and references therein). For doping phosphine or diborane were added. ESR was measured on powdered samples with an X-band spectrometer. Further details can be found in [3]. For LESR we used a 100 W halogen lamp with either a heat-filter (transmission for light energies hv > 1.75 eV, which will be referred to as "white light" in the following), or a germanium filter (transmission for hv < 0.67 eV, "IR-light"). The light source can be switched on and off with a mechanical shutter, and the subsequent transient behaviour of the LESR signal is measured at constant magnetic field at the position of the respective resonances with a modulation amplitude of 10 G.

RESULTS AND DISCUSSION

Steady State Measurements
Fig. 1 shows ESR and LESR spectra of an amorphous (a-Si:H) and two microcrystalline samples with Raman intensity ratios R_C of 35 % (r = 3.8 Å/s) and 85 % (r = 2.4 Å/s) at a temperature of 20 K under white light illumination. The dark spectra contain the resonances of defects at g-values of 2.0055 in a-Si:H and at both 2.0052 and 2.0043 in μc-Si:H [3,8,9]. In the sample with R_C = 85 % the conduction electron (CE) line at g = 1.998 appears. For all samples the signal intensity is strongly enhanced upon illumination and further resonances - depending on R_c - can be identified. In the amorphous sample the LESR spectrum consists of the Si dangling bond resonance at g = 2.0055 and two lines at g = 2.01 and g = 2.0043 due to holes in valence band (VB) tail and electrons in conduction band (CB) tail states. Going to the mixed-phase material (R_C = 35 %) two additional resonances show up. One (g = 1.998) can be identified as the CE resonance, while the origin of the second line at g = 1.995 is unclear. Indications for such an additional resonance have already been found in slightly n-type high-R_C material [3]. In the sample with R_C = 35% the resonance at g = 2.0043 can be attributed to both electrons in the CB tails of the amorphous phase and electrons in - probably oxygen related - defect states typical for microcrystalline material [9]. Finally, in the highly crystalline sample (R_C = 85 %) the typical μc-Si:H LESR spectrum is found with no indication of tail state resonances from a-Si:H. Note that already for the sample with R_C = 55% (not shown), which should still contain a considerable

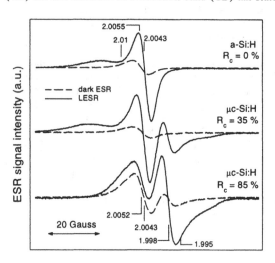

Fig. 1: ESR and LESR spectra of a-Si:H and μc-Si:H samples with different crystallinity.

amount of amorphous phase, the resonance of holes in VB tail states of a-Si:H has almost disappeared. This observation is indicative of a reduced lifetime of photoexcited holes in VB tail states as the fraction of amorphous phase decreases.

Fig. 2 shows the total ESR spin densities of the two defect resonances at g-values of 2.0043 and 2.0052 as a function of the intensity ratio R_C for undoped samples. The respective growth rates are also indicated. The dashed vertical line separates samples where a characteristic fast transient ("spike") is observed in time dependent measurements (see below, Figs. 3 - 5) from samples where this feature is not visible. With decreasing crystallinity the spin density decreases. This indicates either a reduction in the number of defects or changes in their charge state ($D^0 \rightarrow D^-/D^+$), as only neutral (singly

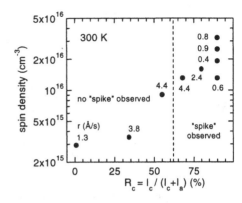

Fig. 2: Spin density of defects at 2.0052 and 2.0043 in µc-Si:H as a function of the Raman intensity ratio R_C. The numbers denote the respective growth rates (Å/s) and "spike" means the fast transient shown in Figs. 3 and 5.

occupied, D^0) defects are paramagnetic and hence visible in ESR experiments. At the lowest value of R_C (about 1%) the spin density is $3 \cdot 10^{15}$ cm^{-3}. On the other hand, there is no straightforward relation between growth rate and spin density or between growth rate and crystalline volume fraction.

Transient Measurements

Light induced ESR measurements presented so far were performed with light energies larger than 1.75 eV. Additional information on excitation and recombination mechanisms is obtained using infrared light with hv < 0.67 eV, a value smaller than both the amorphous and the crystalline silicon bandgap, and *combinations* of white and infrared light (Figs. 3, 4 & 5). IR light excitation also leads to an LESR response for samples with $R_C \geq 35$ %. The signal intensity depends on R_C. For the a-Si:H sample and for the sample with only small traces of crystalline content ($R_C \approx 1$ %) no IR-LESR signal was observed. The transient behaviour of the conduction electron signal upon white light and IR light illumination is shown in Figs. 3 (a) and (b) for two samples with high and intermediate R_C (R_C = 90 % and 55 %, respectively). The signal heights of the CE line were monitored as a function of time (here: for a total of 600 s) upon the following dark and illumination sequence: At times t < 0 the signal level corresponds to the signal intensity of the CE line in the dark. At time t = 0 white (upper curves) or infrared light (lower curves) is turned on. In case of white light illumination the light source is switched off after 300 s. After another 120 s the samples are re-illuminated by infrared light for the rest of the time. Within the time frame of the experiments (up to 2700 s) the signal intensity upon IR illumination remains always below the level obtained after the white light / dark / IR sequence. This was found for all but one undoped sample prepared at high depositon rate and also for n-type samples, while for p-type material the same signal levels are reached with both illumination sequences (Fig. 5).

As another important feature, an additional fast transient signal enhancement (at the point denoted by "IR on") appears in samples with high R_C. The dependence of this spike on various illumination sequences and IR light energies has already been studied in the past [3].

301

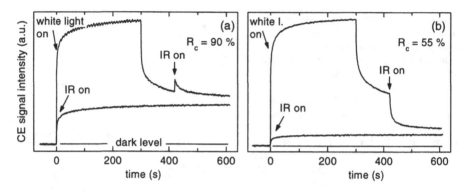

Fig. 3: CE signal intensity as a function of time upon IR illumination (lower curves) and for a white light / dark / IR illumination sequence (upper curves) for samples with different crystalline volume fractions, i.e. (a) $R_C = 90\%$ and (b) $R_C = 55\%$.

In Fig. 4 the signal enhancements (light induced minus dark signal, i.e. the signal levels monitored in Fig. 3 relative to the dark level) of the conduction electron (CE) resonance achieved by IR and white light illumination are shown as a function of R_C. The ratio of IR light signal enhancement to white light signal enhancement clearly increases with increasing crystalline volume fraction. Like in purely amorphous material, no infrared response at all is found for the sample with $R_C \approx 1\%$.

To study the influence of the Fermi level position we repeated the same transient measurements in p- and n-type samples (Fig. 5 a,b) with high crystallinity ($R_C = 90\%$). Doping levels of 2 ppm (n-type) and 3 ppm (p-type) lead to changes in the dark conductivities of four (n-type) and two (p-type) orders of magnitude as compared to intrinsic material. From this a shift of the Fermi level of 0.2 eV can be estimated [3,10]. Early studies have shown that the spin density of defect states remains almost constant over a wide doping range [9]. Instead, the recombination behaviour of charge carriers is already affected by small amounts of doping: The spike found for undoped high-R_C material also appears in doped samples. Like in undoped material (Fig. 3), for the n-type samples the signal level upon IR illumination remains below the level obtained after white light / dark / IR light. For the p-type sample instead both signal levels are the same. In addition, the initial decay after turning off the white light as well as the IR LESR rise-time are faster in p-type samples as compared to n-type and undoped material.

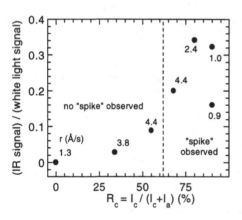

Fig. 4: Ratio of the signal enhancements achieved by IR and white light as a function of the Raman intensity ratio R_C. The numbers denote the respective growth rates (Å/s).

The above results show that the mechanisms of carrier excitation and recombination are very sensitive even to small Fermi level shifts and to changes in the crystalline volume fraction. A detailed interpretation

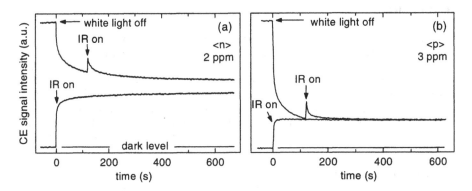

Fig. 5: Variation of CE signal intensity upon the same illumination sequences as in Fig. 4 for an (a) n-type and (b) p-type sample with Raman intensity ratios R_C of 90 %.

would require knowledge of the excitation processes for infrared light, the subsequent recombination mechanisms that determine the carrier lifetime and the influence of the structural changes when R_C is varied. As detailed information is not available so far, only a qualitative discussion is possible at this point.

What can be said is that infrared light applied to pure a-Si:H does not lead to any signal enhancement in contrast to what is found for μc-Si:H, where the infrared response is strongest in samples which are almost totally crystalline. One can expect that a larger volume fraction of amorphous regions will reduce the IR-induced signal as found in Fig. 4.

The existence of the distinct "spike", which always appears in samples with high R_C, but which is absent for $R_C \le 55$ %, has been explained assuming the following mechanism (for details see [3]): Electrons and holes excited by white light do not only stay within the crystalline grains (where electrons occupy CB states and CB tail states, visible as CE resonance in ESR), but also have enough excess energy to cross potential barriers between crystalline and the remaining disordered regions (for which the amorphous silicon bandgap is assumed) where they get trapped. The threshold for this trap-filling process was found to be at light energies hv $\ge$ 0.75 eV [3,10]. Subsequent IR light (hv < 0.67 eV) will then re-emit these carriers from their traps into the crystallites, where the electrons lead to a transient increase of the CE line ("spike") before they recombine.

Finally we comment on the differences in the signal levels reached under IR light illumination with or without preceding white light, which depend on the type of doping (Figs. 3 and 5). Apparently in p-type μc-Si:H the electrons and holes, which are trapped in states that have been filled by white light, can be re-emitted by IR illumination and subsequently recombine. On the other hand, in n-type or undoped material a certain fraction of the photoexcited electrons survives IR-quenching. This could be caused by the following effect: In n-type μc-Si:H, with a large number of negatively charged defects, some of the holes excited into disordered regions will fall into deep traps making the transition $D^{-/0}$ + hole $\rightarrow D^{0/+}$. They get lost for re-emission and hence increase the electron lifetime, leading to a residual CE signal which is larger than in case of pure IR light. On the other hand, p-type material contains much less D^- or D^0-states, so that holes will stay in shallow traps before the IR light is applied. As a result, all of the white-light induced charge carriers can recombine with each other after their re-emission by IR light, and the resulting signal level is identical to the one reached with pure IR illumination.

CONCLUSIONS

The ratio of amorphous and crystalline volume fraction in μc-Si:H strongly influences the spin density and the excitation and recombination behaviour of light induced charge carriers. The transition from amorphous to microcrystalline material is nicely reflected in light-induced ESR measurements, which show a superposition of microcrystalline and amorphous constituents. Even for small amounts of crystalline phase the signals typical for μc-Si:H can be observed in ESR and LESR. Doping-induced Fermi level shifts do not change the spin density of defect states, whereas they have a major effect on the lifetimes of optically excited carriers. In particular, the lifetime of photo-generated holes seems to be very sensitive to changes in crystalline volume fraction and/or Fermi level position.

ACKNOWLEDGEMENTS

We thank R. Carius for helpful discussions and J. Wolff, A. Lambertz and D. Steinbacher for technical assistance. This work was supported by the Bundesministerium für Bildung, Wissenschaft, Forschung und Technologie.

REFERENCES

1. J. Meier, R. Flückiger, H. Keppner, and A. Shah, Appl. Phys. Lett. **65**, 860 (1994)

2. O. Vetterl, P. Hapke, O. Kluth, A. Lambertz, S. Wieder, B. Rech, F. Finger, and H. Wagner, to be published in: Solid State Phenomena, *Polycrystalline Semiconductors V - Bulk Materials, Thin Films and Devices*, edited by J. H. Werner, H. P. Strunk, and H. W. Schock (Scitech Publ., Uettikon am See, 1999)

3. F. Finger, J. Müller, C. Malten, and H. Wagner, Phil. Mag. B **77**, 805 (1998)

4. S. Hasegawa, S. Narikawa, and Y. Kurata, Phil. Mag. B **48**, 431 (1983)

5. J. Müller, F. Finger, R. Carius, and H. Wagner, Mat. Res. Soc. Symp. Proc. **507**, 1998

6. P. Hapke and F. Finger, J. Non-Cryst. Solids **227-230**, 861 (1998)

7. L. Houben, M. Luysberg, P. Hapke, R. Carius, F. Finger, and H. Wagner, Phil. Mag. A **77**, 1447 (1998)

8. J. Müller, C. Malten, F. Finger, and H. Wagner in *The Physics of Semiconductors*, edited by M. Scheffler and R. Zimmermann (World Scientific, Singapore, 1997), pp. 2697-2700

9. J. Müller, F. Finger, C. Malten, and H. Wagner, J. Non-Cryst. Solids **227-230**, 1026 (1998)

10. J. Müller, PhD thesis, Rheinisch-Westfälische Technische Hochschule Aachen, 1998

VLS Growth of Si nanowhiskers on a H-terminated Si{111} surface

N. Ozaki, Y.Ohno, S.Takeda, and M.Hirata.

Department of Physics, Graduate School of Science, Osaka University
1-16 Machikane-yama, Toyonaka, Osaka 560-0043, Japan

ABSTRACT

We have grown Si nanowhiskers on a Si{111} surface via the vapor-liquid-solid (VLS) mechanism. The minimum diameter of the crystalline is 3nm and is close to the critical value for the effect of quantum confinement. We have found that many whiskers grow epitaxially or non-epitaxially on the substrate along the <112> direction as well as the <1̄1̄1̄> direction.

In our growth procedure, we first deposited gold on a H-terminated Si{111} surface and prepared the molten catalysts of Au and Si at 500℃. Under the flow of high pressure silane gas, we have succeeded in producing the nanowhiskers without any extended defects. We present the details of the growth condition and discuss the growth mechanism of the nanowhiskers extending along the <112> direction.

INTRODUCTION

It is well known that the quantum confinement comes out in nanodots and nanowires of compound semiconductor. In silicon, it is predicted that wires or dots smaller than 2~3nm in diameter may exhibit visible photoluminescence due to the quantum effects. Hence, various efforts have been made to fabricate silicon-based nanostructures[1-8]. It was recently reported that silicon whiskers of several tens of nanometers in diameter are grown from molten Au-Si catalysts on a silicon substrate[4,5] by the VLS growth technique. More recently, silicon nanowhiskers have been grown from freestanding catalysts via the laser ablation technique[6-8]. These studies claimed that when the whiskers are grown on a substrate, there is a lower limit for their diameters, i.e., 10nm. Using a hydrogen terminated Si{111} surface as a substrate, we have succeeded in growing silicon nanowhiskers by the VLS mechanism. Si nanowhiskers (3.0nm in minimum silicon-core diameter and about 2 μ m in maximum length) can indeed be grown on the substrate by the VLS technique. We show the experimental conditions for the catalyst formation as well as the whisker growth.

GROWTH PROCEDURES AND CHARACTERIZATIONS

The procedure for whiskers growth is described schematically in Fig.1. The whiskers were observed by means of transmission electron microscopy (TEM) and diffraction (TED). The whiskers were also examined by Raman spectroscopy, in which an Ar⁺-ion laser of 514.5nm line was used for excitation. The laser power on the whiskers was estimated to be about 2mW. The spectral resolution of the detector was about 5cm^{-1}.

RESULTS AND DISCUSSION

Structures of Si nanowhiskers grown on a H-terminated surface.

Fig.2(a) shows Si nanowhiskers grown on a H-terminated Si{111} surface. The whiskers' outside diameter ranges from 6.0 to about 30 nm. A Au-silicide particle on the top of a whisker

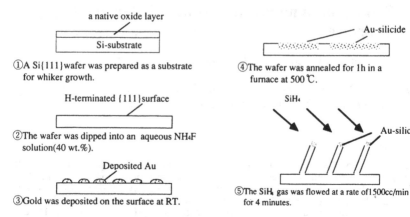

a native oxide layer

| Si-substrate |

①A Si{111}wafer was prepared as a substrate for whiker growth.

H-terminated {111}surface

②The wafer was dipped into an aqueous NH₄F solution(40 wt.%).

Deposited Au

③Gold was deposited on the surface at RT.

Au-silicide

④The wafer was annealed for 1h in a furnace at 500 ℃.

SiH₄

Au-silicide

⑤The SiH₄ gas was flowed at a rate of 1500cc/min for 4 minutes.

Fig.1 Growth procedure for Si nanowhiskers on a H-terminated Si{111} surface.

(as indicated by arrows in Fig. 2(a)) is the evidence of VLS growth. Fig.2(b) shows selected area TED spots from the whiskers which were located in the circle indicated in Fig.2(a). The Debye-Scherrer rings from the whiskers prove that the whiskers are of crystalline silicon.

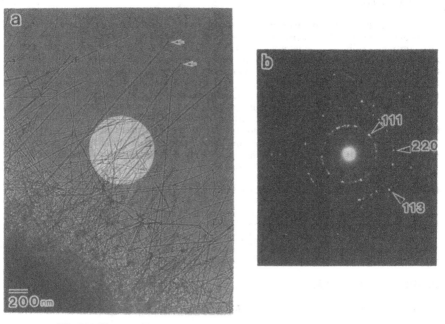

Fig.2(a) Si nanowhiskers grown on a H-terminated Si {111} surface.
(b) Debye-Scherrer ring obtained from the Si nanowhiskers.

As shown in Fig.3, the diffraction spots from nanowhiskers are accompanied with diffuse streaks. The directions of the diffuse streaks in Fig.3(b) are normal to the anisotropic direction of the nanowhiskers shown in Fig.3(a). So,we found that the whiskers indicated by · single and double arrows in Fig.3(a) grow along the <112> directions. The growth direction differs from that in the conventional VLS grown Si whiskers[4,6,9], i.e., the <111> direction. This peculiar direction is also observed in free-standing Si nanowires[8].

Observing the nanowhiskers by high-resolution TEM, we confirmed that many of them grew along the <112> direction (e.g., Fig.4,5). It was thought[10] that the VLS growth whiskers grew epitaxially along <111> even on the {112} substrate. In contrast, the recent studies of whiskers grown from free-standing catalysts (without a substrate) imply that the growth along <111> originates from the stable interface between molten catalysts and silicon crystal, i.e., Si {111}. We think that the solid-liquid interface becomes unstable, when the whiskers are thinner, or the interface is smaller. Thus, the interface may be lying on not only Si {111}, but also the other planes whenever the interface energies are sufficiently low such as the {112} plane. Furthermore, whiskers grown along <112> can be bounded by surfaces of low energy, i.e., {111}, {113} and {110}[11]. This argument is supported by the experimental result that the nanowhiskers along <112> do not always exhibit the epitaxial relationships with the substrate as seen in Fig.3: In the TED pattern (Fig.3(b)), the diffraction spots indicated by the single arrows correspond to whiskers grown epitaxially on the substrate, while the double arrows correspond to whiskers grown nonepitaxially on the substrate. Knowing that there are various low-energy grain boundaries in Si, we believe that grain boundaries of silicon crystals are probably left at the roots of the whiskers.

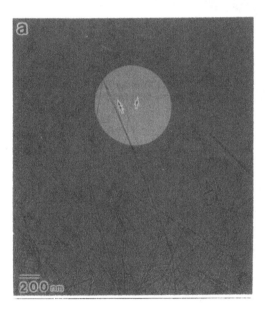
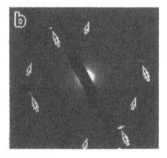

Fig.3
Transmission electron diffraction from nanowhiskers.
The incident electron beam is parallel to the <110> directions of whiskers shown in (a). In(b),the {111}type diffraction spots from the whiskers, indicated by single and double arrows in (a), are accompanied with the diffuse streaks normal to the growth directions, i. e., the <112> directions.

HRTEM observations have also shown that the diameters of the silicon core of whiskers range from 3.0nm to 25nm, and the core is covered with a thin SiO_2 amorphous layer whose thickness is estimated to be less than 2.0 nm. In the Fourier filtered image in Fig.5(b), the {111} lattice planes in the core of the whisker are clearly seen. The minimum outside diameter is much smaller than that of whiskers[4,5] grown on an unprepared oxide surface, i.e., 15 to 20 nm. In fact, we confirmed that under our growth conditions, only a few whiskers grew on unprepared surfaces. Wagner et al.[9,10] have already showed that the diameter decreases when the growth temperature is lowered. According to published studies[4], a higher partial pressure of SiH_4 tends to grow thinner whiskers. However, whiskers grown at lower temperatures (from 320 to 440 ℃) and under a partial SiH_4 pressure of 0.01-1 torr, have a number of structural defects such as kinks, even though their diameters reach 15 to 20nm[4]. Therefore, it was thought that the conventional VLS technique can not yield straight nanowhiskers[4]. In our study using the H-terminated surface, we have grown nanowhiskers at lower temperatures and under much higher SiH_4 partial pressures. Therefore, we have found that the growth of silicon nanowhiskers depends on the preparation of the surface.

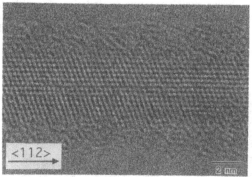

<112>

Fig.4
HRTEM image of a Si nanowhisker whose diameter is 4nm.

Fig.5 (a) HRTEM image of the thinnest whisker in our work (3nm in diameter).
(b) Fourier filtered image of (a).
Clearly, the two kinds of {111} type lattice planes are seen in the core of the whisker in the Fourier filtered image.

In situ TEM observation of catalysts formation on a H-terminated surface.

We studied the formation process of catalysts on a silicon surface by preparing two kinds of thin silicon specimens: one with hydrogen-terminated surfaces and the other with untreated surfaces naturally covered with an oxide layer. After gold was deposited on one surface of the two kinds of thin specimens in the same way as described above, the specimens were observed by means of *in situ* plan-view TEM. The vacuum in the electron microscope was estimated to be 2.6×10^{-7} Torr. Just after deposition, we could not find any difference in morphology between the two kinds of surfaces. The grains of crystalline fcc gold form on both treated (Fig.6(a)) and untreated (Fig.6(b)) surfaces. However, observing *in situ* the gold islands on the hydrogen-terminated surface at 500℃, which is equal to the temperature of preannealing and VLS growth, we have found that many molten Au-Si islands form (Fig.6(c)). The corresponding electron diffraction pattern (the inset of Fig.6(c)) exhibits the halo rings (indicated by the arrow) due to molten islands of Au and Si. The sizes of the islands range from about 4nm to 40nm, probably corresponding to the diameters of the nanowhiskers. It is known that gold and silicon mutually diffuse regardless of the existence of hydrogen at a surface[12]. Therefore, the Au-Si interdiffusion takes place even in a conventional high vacuum environment ($\sim 10^{-7}$Torr) whenever the preannealing temperature exceeds the eutectic temperature, i.e., 363 ℃. The molten Au-Si catalysts are formed as easily as in ultrahigh vacuum environment. On the other hand, the grains of crystalline gold coarsened at 500℃ on the untreated surface (Fig. 6(d)), since the native oxide layer prevents gold from diffusing into a silicon substrate vice versa even above the eutectic temperature[13]. Therefore, on the oxidized surface, preannealing at a higher temperature is needed, since crystalline gold islands must absorb silicon from SiH_4 in order to form molten Au-silicide catalysts for VLS growth. Oxygen is presumably involved in this process, probably causing various growth defects such as kinks as observed in the previous work[4].

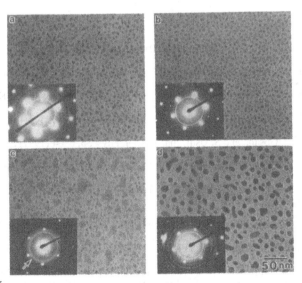

Fig. 6
(a) Deposited Au on a H-terminated Si{111} surface observed at RT.
(b) Deposited Au on a Si{111} surface with a native oxide layer observed at RT.
(c) An *in situ* image of (a) after annealing for 1h at 500℃
(d) An *in situ* image of (b) after annealing for 1h at 500℃

Raman spectroscopy

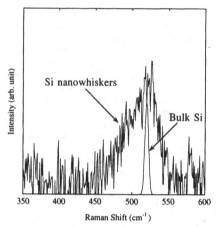

Fig.7 Raman spectrum from whiskers.

Figure 7 depicts the Raman spectrum from the whiskers shown in Fig.2(a). We observe an asymmetric peak at 520cm^{-1} with a linewidth of 50cm^{-1}. It differs considerably from the symmetric Raman peak of bulk silicon whose linewidth is 5.5cm^{-1}. The asymmetry and the larger linewidth of the Raman peak also appear in porous silicon[14] and free-standing nanowire[7] and they are attributed to the quantum confinement effect.

SUMMARY

We have succeeded in growing silicon nanowhiskers on a hydrogen-terminated silicon {111} surface by the VLS mechanism. The one-dimensional nanostructures exhibit a Raman spectrum consistent with quantum confinement effects.

ACKNOWLEDGMENTS

This work is in part supported by the Grant-in-Aid for Scientific Research from the Ministry of Education Science, Sports and Culture, Japan.

REFERENCES

[1] H. I. Liu, N. I. Maluf, R. F. W. Pease, D. K. Biegelsen, N. M. Johnson, and F. A. Ponce, J. Vac. Sci. Technol. B **10**, 2846 (1992).

[2] H. I. Liu, D. K. Biegelsen, F. A. Ponce, N. M. Johnson, and R. F. W. Pease, Appl. Phys. Lett. **64**, 1383 (1994).

[3] Y. Wada, T. Kure, T. Yoshimura, Y. Sudo, T. Kobayashi, Y. Goto, and S. Kondo, Jpn. J. Appl. Phys. **33**, 905 (1994).

[4] J. Westwater, D. P. Gosain, S. Tomiya, S. Usui, and H. Ruda, J. Vac. Sci. Technol. B **15**, 554 (1997).

[5] J. Westwater, D. P. Gosain, and S. Usui, Jpn. J. Appl. Phys. **36**, 6204 (1997).

[6] A. M. Morales and C. M. Lieber, Science **279**, 208 (1998).

[7] Y. F. Zhang, Y. H. Tang, N. Wang, D. P. Yu, C. S. Lee, I. Bello, and S. T. Lee, Appl. Phys. Lett. **72**, 1835 (1998).

[8] D. P. Yu, C. S. Lee, I. Bello, X. S. Sun, Y. H. Tang, G. W. Zhou, Z. G. Bai, Z. Zhang, and S. Q. Feng, Solid State Commun. **105**, 403 (1998).

[9] R.S.Wagner and W.C.Ellis, Appl.Phys.Lett. **4**, 89 (1964).

[10] R.S.Wagner, in Whisker Technology, A.P.Levitt, Ed .(Willey-Interscience, N.Y., 1970) pp. 47-119.

[11] D. J. Eaglesham, A. E. White, L. C. Feldman, N. Moriya, D. C. Jacobson, Phys. Rev. Lett. **70**, 1643 (1993).

[12] M. Iwami, M. Nishikuni, K. Okuno and A. Hiraki, Solid State Comm. **51**, 561 (1984).

[13] Y. Golan, L. Margulis, S. Matlis, and I. Rubinstein, J. Electrochem Soc. **142**, 1629 (1995).

[14] Z. Sui, P. P. Leong, I. P. Herman, G. S. Higashi, and H. Temkin, Appl. Phys. Lett. **60**, 2086 (1992).

THERMODYNAMICS AND KINETICS OF MELTING AND GROWTH OF CRYSTALLINE SILICON CLUSTERS

P. KEBLINSKI
Materials Science Division , Argonne National Laboratory, Argonne, IL, 60439, USA
and Forschungszentrum Karlsruhe, 76021 Karlsruhe, Germany, keblinski@anl.gov

ABSTRACT
Molecular-dynamics (MD) simulations and the Stillinger-Weber three-body potential are used to
study the growth and stability of silicon clusters of diameters from 2 to 5 nm embedded in the
melt. Our simulations show that the melting temperature of such nano-clusters is lower than the
bulk melting temperature by an amount proportional to the inverse of the cluster size. We also
show that the nature of the kinetics of such small Si clusters is essentially the same as that of the
homoepitaxial growth. In particular, we show that the mobility of the highly-curved crystal-
liquid interface is controlled by diffusion in the adjacent melt, and is characterized by the same
activation energy.

INTRODUCTION

Significant experimental and theoretical effort to understand and control nucleation, growth
and subsequent microstructural evolution of polycrystalline silicon is motivated by the need to
obtain high-quality material for electronic and optical applications. However, very little is
known about the initial stages of growth of small (nanometer-sized) crystal clusters from the melt
or amorphous matrix. This important issue, largely determining the resulting microstructure of a
polycrystal, is extremely difficult to study experimentally due to the small size of the clusters
and small time scale involved.

By contrast, the kinetics and thermodynamics of the homoepitaxial melting and
solidification of silicon has been studied extensively by experiment [1-3] and MD simulation [4-
5]. Both the experimental data and the results of the MD simulations on the homoepitaxial
growth and melting of Si are well understood in terms of the transition-state theory of crystal
growth [6]. According to transition-state theory the driving force, F, for the movement of the
liquid-crystal interface is the difference between *bulk* liquid and crystal free energies. This
difference is approximately proportional to the magnitude of the undercooling, T_m-T, where T_m
is the melting temperature. The velocity of the moving interface, V, is proportional to the
driving force, V=kF, where k is the mobility of the liquid-crystal interface. This interfacial
mobility is determined by the movement of the atoms in the liquid phase since the atoms residing
in the crystalline phase are far less mobile. Therefore, usually it is supposed that this mobility is
proportional to the thermally activated atomic diffusion in the liquid phase [6]. Indeed, in case of
silicon, both experiment [1-3] and MD simulation [4-5] show that the mobility of the liquid-
crystal interface is thermally activated and that the activation energy is equal to that for the
diffusion in the liquid phase.

As is well established by theory and experiment, the melting point of finite-size clusters is
reduced from the bulk value due to the contribution of the free-energy of the liquid-crystal
interface. However, the kinetics of the highly-curved, liquid-crystal interface are not clear. The
first MD study of the small Si cluster kinetics were done by Uttomark *et al*. [7]. This study was
essentially limited to the simulations of crystal dissolution and did not investigate systematically
effects of the cluster size. More recent MD simulations suggested that the kinetics of the
nanosized clusters are fundamentally different from those of epitaxial growth [8]. This
difference manifested itself in much lower values of the liquid-crystal interfacial mobility with
its activation energy about three times higher than that for diffusion in the liquid [8]. This result
was indeed surprising considering the fact that the atomic mobility in the liquid surrounding the
nanoclusters should be the same as in the liquid adjacent to the planar interface. It suggested that
there is a fundamentally new mechanism controlling the growth of the nano-sized grains. Since
in the limit of the large cluster size the kinetics of the homoepitaxial growth must be recovered,

311

low interfacial mobility of the small clusters compared with those of large clusters would lead to the pattern of more rapid growth of the larger clusters, despite their lower curvature. In this paper we use MD simulations and the Stillinger-Weber (SW) empirical potential for Si [9] to study the growth and dissolution of the Si nano-clusters systematically as a function of the cluster size and temperature. We show that there are actually no significant differences between the growth of nanoclusters and planar interfaces, in contrast to the previous studies [8]. We find that whereas the melting temperature of a cluster is lowered with the respect to the bulk value by an amount proportional to the inverse of the cluster size (i.e. the surface to volume ratio, in accord with simple thermodynamics) the activation energy for the mobility of the liquid-crystal interface is essentially the same as that for liquid diffusion. The discrepancy between the results of our studies and the previous MD simulations [8] is due to the fact that we simulate the growth and melting of nanoclusters at constant (zero) pressure condition whereas the previous simulations were performed at constant volume with, as we show, a very large negative pressure leading to an incorrect thermodynamical path for the study of growth and melting of Si clusters. (We note that the studies described in ref. [7] were also performed at constant volume, however, greater care was taken to ensure that the pressure was approximately zero during the simulations.)

THERMODYNAMICS AND KINETICS OF Si NANOCLUSTERS

Kluge and Ray [4] studied homo-epitaxial growth on the (100) surface of silicon using the SW potential. They found that growth of the undercooled melt in contact with the crystal occurs between $T \cong 1680$ K and $T \cong 1000$ K, the upper limit corresponding to the melting point, T_m, and the lower limit corresponding to the melting temperature of the model amorphous Si, T_{a-l} [10] (experimentally T_{a-l} is around 1400K [11]). Below T_{a-l} the undercooled liquid Si freezes into the amorphous solid with very slow solid-like diffusion not observable on the MD time scale. The maximum growth velocity for the (100) substrate orientation, $V_{(100)} \cong 20$ m/s, was obtained for $T = 1350$ K; both the position and the maximum growth rate were found to be in good agreement with experiment on the (100) epitaxial solidification of crystalline silicon [2]. By contrast, melting was observed above $T_m=1680$ K with the velocity of the liquid-crystal interface rapidly increasing with the magnitude of the overheating. By fitting the simulated data, the mobility of the homoepitaxial crystal-liquid surface was found to be thermally activated with the activation energy equal to that for diffusion in the liquid [4].

In our studies of growth and melting of Si we monitor the crystal front velocity using the fact that the SW potential consists of additive two-body and three-body energy terms [7]. The three-body term is zero for the perfect-crystal structure at $T = 0$K, but even at high temperature assumes relatively low values in the crystalline phase (e.g. the three-body energy is about 0.1 eV/atom at $T \sim 1200$ K). By contrast, the liquid phase is characterized by much larger three-body energy (~ 1 eV/atom). Using this large difference we calculate the amount of crystal and liquid phase present in the simulation cell simply by monitoring the total three-body energy and using as reference the corresponding values for the bulk liquid and bulk crystal at the same temperature. To validate further our method, in our preliminary study of homoepitaxial growth we calculate the growth velocity independently by monitoring the plane-by-plane structure factor, $S(k)$, calculated for a reciprocal lattice vector, k, of the planar crystal structure [12]. This structure factor assumes large values for crystalline planes and is essentially zero in liquid. Our simulations show that the front velocity measured by the three-body energy is the same (within the statistical error) as obtained by the planar structure factor calculations.

Since our primary goal is to study growth and melting of three-dimensional clusters, which are delimited by various crystallographic planes, so as to establish a reference we first study homoepitaxial growth of several crystallographically different interfaces. First we reproduced the result of Kluge and Ray [4] for the growth of the (100) crystal-liquid interface with the maximum growth velocity of $V_{(100)}=19.8 \pm 2.0$ m/s at $T=1350$ K. For the (110) and (111) crystal-liquid interfaces, the growth velocities at $T=1350$ K was determined to be $V_{(110)}= 16.5 \pm 2$ m/s and $V_{(111)}=11.5 \pm 2$ m/s (in agreement with [5]). .

In the studies of homoepitaxial growth described above flat, infinitely extended, planar interfaces, i.e., with no curvature were studied. By contrast, 3d-clusters embedded into melt are

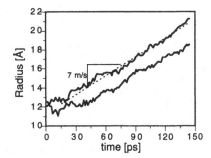

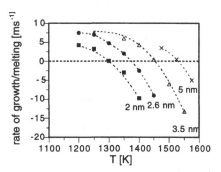

Figure 1: Radius vs. time of a growing single crystalline cluster with an initial size of 2.6 nm at T=1250K for two independent runs. The slope corresponds to the growth velocity of the liquid-crystal interface of ~ 7m/s.

Figure 2: Growth (positive) or melting (negative) rate of silicon clusters vs. temperature for different clusters sizes: 2 nm, solid squares, 2.6 nm, solid circles, 3.5 nm, triangles, 5.0 nm, crosses. The dotted lines represent fits of the data according to Eq. 4.

delimited not only by more or less well-defined, relatively small planar facets but also by edges and corners where these facets meet; these can be expected to slow down the crystal-growth kinetics as well as to lower the melting point due the non-negligible surface contribution to the free-energy of the crystalline cluster embedded in the melt.

In order to assess the influence of these finite size effects on the growth kinetics we first simulate the growth of a single, spherical seed from its melt at 1250 K. The initial configuration consists of a 5x5x5 cubic simulation cell containing approximately 6000 atoms on the perfect diamond lattice. The system is melted at 3000 K at constant volume while the 250 atoms forming the spherical seed (corresponding to a seed diameter, d, of about 2.2 nm) are kept fixed. The melt is then cooled down to 1250 K and the whole system (the melt plus the seed) evolves freely with no further constraints imposed on the seed atoms. A constant-pressure Andersen algorithm [13] is used to assure zero external pressure during the growth process. The typical MD involved 20000 - 50000 MD time steps (we used a MD time step of 2.75×10^{-15}s, for several test runs we obtained statistically the same results with twice smaller time step).

The seed was observed to grow more or less isotropically, thus retaining its roughly spherical shape; however, in agreement with previous simulations [8] we found that the fastest growing (100) interfaces facet into (111) segments, presumably because it can lower its energy by lying on the densest rather than third-densest plane in the cubic diamond lattice. Once faceted, the growth velocity of the cluster is controlled by the growth of the easy identifiable (111) facets. During growth no defects in the crystalline structure were observed, except for numerous twins created on the (111) crystal planes, which for the short range SW potential have zero energy. The average growth velocity obtained by monitoring the radius of the crystalline cluster vs. time was found to be ~7m/s (see Fig. 1). This value is somewhat less than that for the (111) epitaxial growth at the same temperature. As we discuss below, it is due to the reduction of the melting temperature of a finite-size cluster resulting in a lower driving force for crystallization. We also note that thermal equilibration of the seeds, that were kept frozen during the melting, results in a transient behavior leading to the initial differences in the growth rate as seen in Fig. 1.

The dependence of the melting temperature, $T_m(d)$, of the cluster size, d, is investigated by simulating the growth behavior of clusters with various initial sizes as a function of temperature. Using the above described procedure we simulated four cluster sizes of 2.0, 2.6, 3.5, and 5.0 nm in diameter, respectively. To eliminate the initial transient in the growth rate all seeds are thermally equilibrated at temperatures for which the growth rate was essentially zero. Fig. 2 shows the velocity of the growth (and melting for $T>T_m(d)$) as a function of temperature

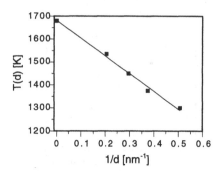

Figure 3: The melting temperature of the clusters, $T_m(d)$, vs. the inverse of the cluster size. The solid line is a linear fit to the data

(averaged over five independent runs for each cluster size and temperature). All clusters melt well below $T_m = 1680$ K, with their melting temperature decreasing with decreasing cluster size. ($T_m(d)$ was determined as the temperature at which the fit to the data crosses zero.) Fig. 3 shows the relationship between $T_m(d)$ (obtained from the fits in Fig. 2) and the inverse of the cluster diameter, $1/d$, which is a natural measure for the studied system corresponding to the cluster surface-to-volume ratio. According to Fig. 3 the reduction of $T_m(d)$ below T_m is proportional to the inverse of the cluster size.

This behavior can be understood within classical nucleation and growth theory. The free energy of the cluster of size d, can be approximated by a surface contribution, U_s, proportional to the surface area times the liquid-crystal free energy, γ_{ls}, such that $U_s = A\gamma_{ls}d^2$ where A is a geometrical constant of order one (for a spherical cluster A= π), and a bulk contribution, U_b, proportional to the volume of the cluster and the difference between crystal and liquid free energy densities, Δu, such that $U_b = B\Delta u d^3$ where B is another geometrical constant of order one (for a spherical cluster B= $\pi/6$). The difference between crystal and liquid free energy densities at the vicinity of the melting point is proportional to the magnitude of undercooling (or overheating), $\Delta u = u_0(T-T_m)$, where u_0 is a constant (note that Δu correctly vanishes at the melting point). For a given temperature, the critical cluster size corresponds to the maximum of the free energy $U = U_s + U_b = A\gamma_{ls}d^2 + Bu_0(T-T_m)d^3$. By differentiation of the free energy with respect to the cluster size, d, one finds the maximum at $T = T_m - C\gamma_{ls}/d$, where C is a constant depending A, B and u_0. The linear dependence of the decrease of the melting point as a function of the inverse of the crystalline size (see Fig. 3) implies that the interfacial energy, γ_{ls}, does not change significantly with temperature.

The value of γ_{ls} can be estimated assuming that the geometry of the cluster is spherical (A=π and B=$\pi/6$) and using the values of the free energies of crystal and liquid silicon calculated for the SW potential by Broughton and Li [14]. We find that the thus estimated interfacial free energy is 300 ± 30 erg/cm^2 (consistent with an estimate obtained in ref. [7]). Notably, our simulation results are consistent with the experimental nucleation data on undercooling of liquid Si [15]. From experiment the value of crystal-liquid interfacial free energy was estimated to be ~ 400 erg/cm^2 at T=1400K [15].

Finally we turn to the analysis of the growth rate vs. temperature dependence in terms of undercooling (driving force) and thermally activated interfacial mobility. In order to understand this dependence we recall classical nucleation theory [16]. Assuming that growth take place on an atom by atom basis, the average rates of crystallization and dissolution are

$$v_+ = v \cdot \exp \frac{\Delta u - (A_{n+1} - A_n)\gamma_{ls}}{2k_B T} \quad \text{and} \quad v_- = v \cdot \exp \frac{-\Delta u + (A_{n+1} - A_n)\gamma_{ls}}{2k_B T} \quad (1)$$

respectively, where Δu is per atom free energy difference between the crystalline and liquid phases, $A_{n+1} - A_n$ is an increase in the interfacial area due to the attachment of an atom to the crystal and v is the jump frequency of the interfacial atoms. The jump frequency is usually assumed to be thermally activated, $v = v_0 \exp(-E/k_B T)$. Then the cluster growth velocity resulting from the difference between v_+ and v_- can be written as:

$$V_c \sim \exp(-E/k_B T) \sinh \frac{\Delta u - (A_{n+1} - A_n)\gamma_{ls}}{2k_B T}. \quad (2)$$

314

The argument of the hyperbolic sine is small at the vicinity of the melting temperature (it is exactly zero at the melting point of the cluster) therefore Eq. 2 can be approximated by:

$$V_c \sim \exp(-E/k_BT) \frac{\Delta u - (A_{n+1}-A_n)\gamma_{ls}}{2k_BT}. \qquad (3)$$

Eq. 3 shows that the rate of the growth/melting is driven by the lowering of the free energy , Δu - $(A_{n+1}-A_n)\gamma_{ls}$, while the interfacial mobility is determined by the activation energy, E, for diffusional jumps of the interfacial atoms. Noticing that $A_{n+1}-A_n$ is proportional to $1/d$ and $\Delta u=u_0(T-T_m)$, we will recover the already discussed results: T_m - $T_m(d) \sim \gamma_{ls}/d$. ($T_m(d)$ is the temperature at which $V_c=0$.) For planar growth the interfacial contribution to the free energy disappears; thus V_c is zero exactly at the bulk melting point (at $\Delta u=0$).
For a given cluster size the free energy term can be expanded around its melting point such that:

$$V_c \sim \exp(-E/k_BT) \frac{T(d)-T}{T} \qquad (4)$$

By fitting our data to Eq. 4 (see Fig. 2) we verified that, indeed, the nucleation-growth theory described above describes well the kinetics of nano-cluster growth and dissolution. The activation energies, E, obtained for the best fits to the data are 0.75 ± 0.05 eV for 2.0 and 2.6 nm clusters and 0.85 ± 0.05 eV for 3.5 nm clusters, respectively. Due to the small number of data points, the results for the 5.0 nm cluster were fitted only to find its melting point with no attempt made to fit the activation energy.
The activation energy for the interfacial mobility of ~ 0.7-0.8 eV should be compared with the activation energy for diffusion in supercooled liquid. An Arrhenius plot for the diffusion constant in the supercooled melt for temperatures within the 1100-1600K range (not shown) demonstrates that diffusion is thermally activated with the activation energy of 0.7 ± 0.1 eV. (We note that this activation energy is higher than the activation energy obtained for the temperatures range above the melting point.) The above result shows that the mobility of the liquid-crystal interface is determined by the diffusion in the adjacent bulk liquid, exactly as in the case of homoepitaxial growth. Our findings are in disagreement with the previous MD simulations [8], where a much higher activation energy for the interfacial mobility (~1.7 eV) was found. The previous MD simulations, however, were performed at constant volume at the density equal to about 97% of the perfect crystal density at T=0.
For comparison with our constant-pressure simulations, we performed constant volume simulations at the same density as used in [8] at T=1400K for d=2 nanoclusters. As illustrated in Fig. 4 for the constant-volume run the cluster *grows* with a velocity of about 2m/s (in

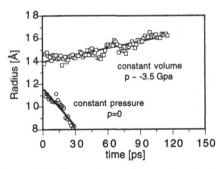

Figure. 4: Radius vs. time of a growing single crystalline cluster with the initial size of 2.0 nm at T=1400K for the constant (zero) pressure simulation and constant volume simulations.

agreement with [8]). By contrast, in the constant (zero)-pressure run the same cluster *melts* with a velocity of about 10 m/s. This result show that there is a qualitative difference in the behavior of the same clusters subjected to the two different conditions. To understand the origin of this difference we monitored the pressure for the constant volume run. The pressure increased with time; however, it always had large negative values of about -3.5 GPa. This large negative pressure originates from the fact that liquid silicon (both experimentally and as modeled by the SW potential) has a higher density than crystalline silicon. Melting, without allowing for the decrease of the simulation cell volume leads to the large negative pressure. At this large negative pressure the bulk melting point increases significantly leading to the corresponding increase in the melting temperature of the cluster. Due to the above reason, a cluster at a given temperature can melt or grow depending on the pressure conditions.

DISCUSSION AND CONCLUSIONS

The main purpose of our studies was to investigate the kinetics and thermodynamics of the growth and dissolution of single crystalline nano-clusters. Our main conclusions are as follows: (1) Small cluster sizes (thus high interfacial curvatures) lead to a significant lowering of the melting point. (2) The lowering of the melting temperature is proportional to the inverse of the cluster size i.e. the surface to volume ratio, implying that the free energy of the liquid-crystal interface does not vary strongly with temperature. (3) The kinetics of the growth process of nano-sized clusters are essentially the same as those of homoepitaxial growth. (4) The activation energy for the mobility of the liquid-crystal interface is equal to that for diffusion in the undercooled liquid, as in the case of homoepitaxial growth [4, 5]. We also explained why the results of the previous MD studies [8] exhibit a much higher activation energy for the interfacial mobility of small Si clusters. We have shown that the source of the discrepancy of our results and those obtained by Marques et. al. studies [8] lies in the fact the previous simulations were performed under the constant volume condition, leading to high negative pressures arising from the significantly higher density of the liquid phase over that of the crystal phase (~8% higher at T_m). Most importantly our study demonstrates that in the description of the experimental results on the growth of the small Si clusters one can use experimentally accessible parameters obtained in the studies of homoepitaxial growth. These parameters combined with the knowledge of the liquid-crystal interfacial free energy allows the thermodynamics and kinetics of the growth and melting of small Si clusters to be fully described.

ACKNOWLEDGMENTS

We have greatly benefited from the discussions with our colleagues S. R. Phillpot, D. Wolf and H. Gleiter. We also gratefully acknowledge support from the Alexander von Humboldt Foundation.

REFERENCES

1. M. O. Thompson et. al., Phys. Rev. Lett. **52**, 2360 (1984).
2. G. J. Galvin, J. W. Mayer, and P. S. Peercy, Appl. Phys. Lett. **46**, 644 (1985).
3. B. C. Larcen, J. Z. Tischler, and D. M. Mills, J. Mater. Res. **1**, 144 (1986).
4. M. D. Kluge, and J. R. Ray, Phys. Rev. B **39**, 1738 (1989).
5. M. H. Grabow, G. H. Gilmer, and A. F. Bakker, Mater. Res. Soc. Symp. Proc. **141**, 349 (1989).
6. see for example: K. A. Jackson, in Crystal Growth and Characterization, edited by R. Ueda and J. B. Mullin (North-Holland, Amsterdam, 1985).
7   M. J. Uttomark, M. O. Thompson, and P. Clancy. Phys. Rev. **B 47**, 15717 (1993).
8. L. A. Marques, M-J. Carturla, T. D. de la Rubia, and G. H. Gilmer, J. Appl. Phys. **80**, 6160 (1996).
9. F. H. Stillinger and T. A. Weber, Phys. Rev. B **31**, 5262 (1985).
10. W. D. Leudtke and U. Landman, Phys. Rev. B **40**, 1164 (1989).
11. E. P. Donovan et al., J. Appl. Phys. **57**, 1795 (1995).
12. P. Keblinski, S. R. Phillpot, D. Wolf and H. Gleiter, J. Am. Ceram. Soc. **3**, 717 (1997).
13. H. C. Andersen, J. Chem. Phys. **72**, 2384 (1980).
14. J. Q. Broughton and X. P. Li, Phys. Rev. **B 35**, 9120 (1987).
15. F. Spaepen and Y. Shao, Mater. Res. Soc. Symp. Proc. **398**, 39 (1996).
16. D. Turnbull and J. C. Fisher, J. Chem. Phys. **17**, 71 (1949).

MICROSTRUCTURE AND SIZE DISTRIBUTION OF COMPOUND SEMICONDUCTOR NANOCRYSTALS SYNTHESIZED BY ION IMPLANTATION

A. Meldrum[1], S.P. Withrow[1], R.A. Zuhr[1], C.W. White[1], L.A. Boatner[1], J.D. Budai[1], I.M. Anderson[2], D.O. Henderson[3], M. Wu[3], A. Ueda[3], and R. Mu[3]

[1] Oak Ridge National Laboratory, Solid State Division, Oak Ridge, TN 37831
[2] Oak Ridge National Laboratory, Metals and Ceramics Division, Oak Ridge, TN 37831
[3] Fisk University, Dept. of Physics, Nashville, TN 37208

Abstract

Ion implantation is a versatile technique by which compound semiconductor nanocrystals may be synthesized in a wide variety of host materials. The component elements that form the compound of interest are implanted sequentially into the host, and nanocrystalline precipitates then form during thermal annealing. Using this technique, we have synthesized compound semiconductor nanocrystal precipitates of ZnS, CdS, PbS, and CdSe in a fused silica matrix. The resulting microstructures and size distributions were investigated by cross-sectional transmission electron microscopy. Several unusual microstructures were observed, including a band of relatively large nanocrystals at the end of the implant profile for ZnS and CdSe, polycrystalline agglomerates of a new phase such as γ-Zn_2SiO_4, and the formation of central voids inside CdS nanocrystals. While each of these microstructures is of fundamental interest, such structures are generally not desirable for potential device applications for which a uniform, monodispersed array of nanocrystals is required. Methods were investigated by which these unusual microstructures could be eliminated.

I. Introduction

Nanocrystalline semiconductors may show unique opto-electronic properties as a result of quantum confinement in three dimensions. Below a material-specific critical radius, confinement of the exciton wave function results in a bandgap shift that is particle-size dependent. The optical absorption and photoluminescence properties of quantum-confined semiconductor nanocrystals are, therefore, strongly dependent on the particle size distribution. The electronic properties of nanophase semiconductors are potentially useful for optical switching and optical memory devices. The main device requirements for semiconductor nanocrystals are that a uniform, high volume-filling fraction of monodispersed precipitates be embedded in an optically transparent host material. Ion implantation is a flexible technique whereby semiconductor nanocrystals can be formed in a variety of crystalline or glassy hosts [1,2,3,4,5]. However, the resulting particle size distributions are such that, as yet, no fine structure has been observed that could be attributed to quantum confinement effects.

In the present work, the size distributions and microstructure of ZnS, CdS, PbS, and CdSe compound semiconductor nanocrystals were investigated. Nanocrystals were formed by ion implantation into fused silica substrates followed by thermal processing. X-ray diffraction and transmission electron microscopy (TEM) were used to characterize the resulting precipitates, and optical absorption measurements were performed on selected samples in order to determine whether the particles showed quantum confinement effects. Our main objective was to investigate means by which simple microstructures and narrow size distributions might be obtained.

II. Experimental

Ion implantation of the component elements followed by thermal annealing was used to form compound semiconductor nanocrystals in fused silica hosts. Ion energies were selected using the TRIM-96 code so as to give overlapping concentration profiles of the implanted elements, and equal doses were used to give the proper stoichiometry. Specimens were annealed for 1 hour at 1000°C in a flowing Ar+4%H$_2$ atmosphere unless otherwise specified. Rutherford Backscattering Spectrometry (RBS) using 2.3 MeV He$^+$ ions and a detector angle of 160° was used to measure the concentration profiles of the as-implanted and annealed samples. Optical absorption measurements from 185 to 3500 nm were performed using a dual-beam Hitachi 3501 spectrometer. Cross-sectional specimens were prepared for TEM as described elsewhere [2]. Standard bright- and dark-field TEM analysis was performed using a Philips EM400 electron microscope operated at 100 keV, and high-resolution images were obtained using a Philips CM200 FEG STEM.

III. Results and discussion

A. ZnS nanocrystals

A fused silica specimen was implanted at room temperature to a dose 1 x 10^{17} ions/cm^2 of zinc followed by an equal dose of sulfur and was subsequently annealed at 1000°C in Ar only. The specimen was not heat sunk during implantation. RBS analysis showed that the implanted zinc and sulfur remained in the near-surface region after annealing. The concentration profiles for the implanted elements were slightly sharpened (i.e., lower concentration in the tails of the profile compared to the center), and a peak appeared in the concentration profile at a depth of 280 nm. X-ray diffraction analysis confirmed the presence of randomly-oriented wurtzite-structure ZnS in the fused silica. A relatively small amount of zincblende-structure material was also detected. The optical absorption measurements (Fig. 1) show an absorption onset at 338 nm corresponding to the bandgap of bulk ZnS, indicating that the particles are probably not quantum confined. The Bohr exciton radius for ZnS is only 2 nm, suggesting a minimum size for the particles produced in fused silica.

Cross-sectional TEM analysis demonstrated that ZnS nanocrystals were indeed present in the implanted region (Fig. 2), but two unexpected microstructural developments were observed. First,

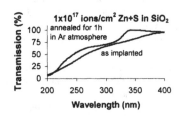

Fig. 1. Optical transmission spectra for a fused silica specimen implanted with zinc and sulfur. The dose is given in ions/cm^2.

Zn(320 keV,1x10^{17} ions/cm^2) S(180 keV, 1x10^{17} ions/cm^2) annealed 1000°C/1h/Ar

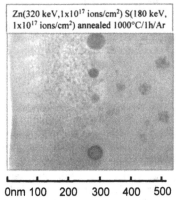

0nm 100   200   300   400   500

Fig. 2. Cross-sectional TEM image showing ZnS nanocrystals in SiO$_2$.

the size distribution of precipitates is strongly bimodal, with a band of large nanocrystals located at a depth corresponding to the maximum depth of the irradiation damaged silica. Second, at depths well beyond the end-of-range of the implanted ions, irregular polycrystalline spherulites were observed (Fig. 2). The average size of the ZnS nanocrystals is 8.2 nm and the standard deviation is 6.0 nm – a large value that represents the bimodal size distribution. These observed particle sizes are consistent with the observed lack of quantum confinement effects in the optical absorption spectra (Fig. 1).

The bimodal size distribution of ZnS precipitates is attributed to the combined effects of rapid diffusion of Zn in fused silica, the effects of radiation damage, and the spatial distribution of the implanted species [6]. Thus, thermal annealing appears to generally result in a bimodal distribution of ZnS precipitates, even for annealing at lower temperatures or for shorter times [2].

Fig. 3. Fused silica specimen implanted with Zn only, showing the formation of the zinc silicate phase beyond the calculated end-of-range of implanted the Zn ions.

The polycrystalline spherulites occurring at depths greater than 300 nm were sufficiently large to obtain selected-area electron-diffraction patterns, which show that they are actually a polycrystalline assemblage of γ-Zn_2SiO_4. Thus, some of the Zn diffuses beyond the implanted region and reacts with the host to form a new silicate phase. This was confirmed by implanting a specimen with Zn only and checking for the formation of γ-Zn_2SiO_4 after annealing (Fig. 3). Both of these problems must be eliminated if device-quality specimens are to be obtained.

The present and previous [2] experiments suggest that ZnS nanocrystal formation by methods other than thermal annealing is desirable in order to obtain simple microstructures. Accordingly, we have attempted to form ZnS nanocrystals by electron irradiation techniques. A fused silica specimen was implanted at liquid nitrogen temperature (LN2) in order to prevent nanoparticle formation during the implantation process. Electron irradiation was then performed on a cross-sectional sample *in-situ* in the TEM. The beam current was 1 nA and temperature rise was calculated by Fisher's model [7] to be less than 50°C. The results illustrated in Fig. 4 show that ZnS nanocrystals can, in fact, be formed as a result of electron irradiation of implanted specimens. The size distribution was narrow (Fig. 4), and the formation of γ-Zn_2SiO_4 was entirely suppressed. Other experiments at lower electron energies suggest that the precipitation of ZnS is mainly caused by ionization processes, but since the specimen heating by the electron beam cannot be directly measured, its effect cannot be entirely ruled out. Experiments are currently underway to see if this process will be similarly effective in bulk specimens.

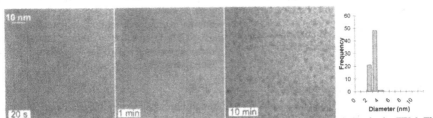

Fig. 4. ZnS nanocrystals formed by ion implantation followed by electron irradiation in the TEM. The beam current was 1 nA (current density = 1.4 A/cm²). The size distribution histogram corresponding to 10 minutes of irradiation is given at the far right.

B. CdS nanocrystals

To form monodispersed CdS nanocrystals, specimens were implanted with several ion energies so as to obtain a relatively flat concentration profile from the surface to a depth of 200 nm. Three specimens were prepared, with implanted concentrations of 0.8, 2.0, and 5.3 x 10^{21} ions/cm^3, respectively. The specimens were not heat sunk during implantation. X-ray diffraction analysis showed that the particles were randomly oriented and, as for ZnS, have the hexagonal wurtzite structure. The optical absorption spectra, normalized to an equivalent concentration (*i.e.*, absorption coefficient), show a shift in the bandgap to higher energies as the implanted concentration is decreased (Fig. 5). This result suggests that a significant fraction of the particles in the low-concentration specimen show quantum confinement effects and are, therefore, smaller than the Bohr exciton radius of 2.5 nm.

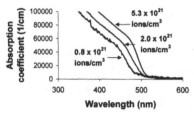

Fig. 5. Absorption coefficient for three fused silica specimens implanted with different concentrations of Cd + S.

TEM analysis of the three Cd+S – implanted fused silica specimens was reported elsewhere [2]. While the average diameter of the particles was 4.9, 6.5, and 9.8 nm in the low, medium, and high concentration samples, respectively, a significant fraction of CdS particles was, in fact, below the confinement radius, particularly in the low-dose specimen (0.8 x 10^{21} ions/cm^3). At high implanted concentrations of Cd+S, another unusual microstructure was observed. Most of the resulting CdS precipitates, especially near the middle of the implanted concentration profile, were hollow (Fig. 6). In subsequent experiments, we have found that hollow particles may form when at least one of the implanted elements has a sufficiently high vapor pressure [8] – either at the implantation temperature or during subsequent thermal processing.

According to these observations, the vapor pressure of the sulfur should be lowered during implantation in order to prevent the formation of large voids. This can be accomplished by performing the implants at lower temperature. An additional sample was prepared that was heat sunk during implantation. Fig. 7 shows that, in fact, the formation of internal voids was entirely suppressed in this case. These results, therefore, provide one method by which simpler microstructures for CdS nanocrystals may be obtained.

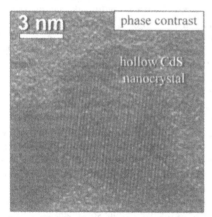

Fig. 6. High resolution image showing a CdS precipitate from the fused silica specimen implanted to a concentration of 5.3 x 10^{21} ions/cm^3. EELS analysis confirmed that the central light-contrast feature is, in fact, a void.

C. PbS and CdSe nanocrystals

PbS nanocrystals could also be formed by ion implantation into fused silica (Fig. 8). The PbS precipitates had, as expected, the cubic NaCl structure. In the case of PbS, the particles were closer to the specimen surface because of the shorter range of the heavy Pb ions. In general, no peculiar microstructures were observed for the PbS nanocrystals, although at higher doses, the

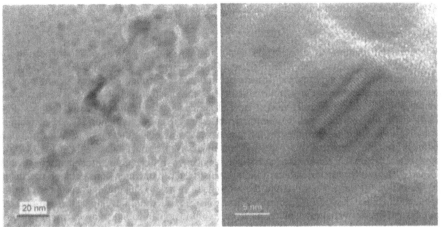

Fig. 7. Cross-sectional and high-resolution images showing CdS nanocrystals in a fused silica specimen that was heat sunk during implantation. The ion dose was 7.5×10^{16} ions/cm^2, and hollow particles were not observed. A second sample was implanted to the same dose at room temperature and hollow particles similar to that shown in Fig. 6 were observed, indicating the importance of the substrate temperature in determining the microstructure of the resulting nanocrystals.

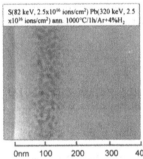

S(82 keV, 2.5×10^{16} ions/cm^2) Pb(320 keV, 2.5 $\times 10^{16}$ ions/cm^2) ann. 1000°C/1h/Ar+4%H$_2$

Onm   100   200   300   400

Fig. 8. PbS nanocrystals in SiO$_2$.

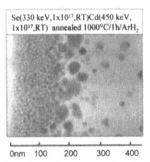

Se(330 keV,1×10^{17},RT)Cd(450 keV, 1×10^{17},RT) annealed 1000°C/1h/ArH$_2$

Onm   100   200   300   400

Fig. 9. CdSe nanocrystals in SiO$_2$. Note the large precipitates at depths below 150 nm.

size distribution became bimodal with large precipitates located near the peak of the implanted ion concentration. The Bohr exciton radius for PbS is 180 nm but the optical absorption spectra showed no obvious structure that could be correlated to a bandgap – possibly indicating that the size distribution was too broad for such structure to be observed.

The CdSe precipitates formed in fused silica had the wurtzite structure, as for the cases of ZnS and CdS. The utility of the implantation technique for forming compound semiconductor nanoparticles is well-illustrated by the success at forming all four compositions attempted in the present experiments. The microstructure of the Cd+Se – implanted specimen did, however, demonstrate a peculiar characteristic (Fig. 10). The particles located at depths between 0 and 150 nm were well formed and showed a monomodal size distribution. However, at depths greater than the calculated end-of-range of the implanted ions, a new feature was observed that was not previously documented in an identical specimen [9] – essentially because the cross-sectional image in the earlier work was cut off at a depth of ~150 nm. At depths between 150 and 400 nm, a band of randomly dispersed larger particles was observed (Fig. 9). High-resolution imaging showed that, in fact, areas of light or dark contrast in these particle actually moved during TEM observation – as if they possibly contained a liquid phase. Research is currently ongoing to determine the nature and origin of this unusual microstructure.

IV. Conclusions

Ion implantation is a versatile technique for forming compound semiconductor nanocrystals in fused silica. ZnS, CdS, PbS, and CdSe nanocrystals were produced by sequential ion implantation, and the crystal structures and size distributions were determined by transmission electron microscopy. In the case of ZnS, the size distribution of the precipitates was strongly bimodal and the formation of a new phase occurred. Preliminary results suggest that both of these phenomena could be eliminated by using electron irradiation instead of thermal processing to nucleate the particles. In fact, the narrowest size distribution of particles from ion implanted samples reported thus far was obtained by this method. In the case of CdS nanocrystals, the optical absorption spectra show evidence of quantum confinement effects, but TEM analysis revealed the formation of hollow nanocrystals in the high-concentration specimen. The formation of large voids could be prevented by heat sinking the specimen during implantation. Initial results for CdSe precipitates also revealed the development of an unusual feature at depths below the calculated end-of-range of the implanted ions. Large, possibly liquid-filled "particles" were observed whose internal structure changed during observation in the TEM.

Acknowledgments

A.M. acknowledges support from NSERC Canada. Electron microscopy was performed at the ShaRE Facility at ORNL. Oak Ridge National Laboratory is managed by Lockheed Martin Energy Research Corp. for the U.S. Department of Energy under contract number DE-AC05-96OR22464.

References

[1] C.W. White, J.D. Budai, S.P. Withrow, J.G. Zhu, , E. Sonder, R.A. Zuhr, A. Meldrum, D.J. Jr. Hembree, D.O. Henderson, and S. Prawer. Nucl. Instr. Meth. Phys. Res. **B141**, 228 (1998).

[2] A. Meldrum, C.W. White, L.A. Boatner, I.M. Anderson, R.A. Zuhr, E. Sonder, and J.D. Budai, and D.O. Henderson, Nucl. Instr. Meth. Phys. Res. (in press).

[3] C.W. White, A. Meldrum, J.D. Budai, S.P. Withrow, E. Sonder, R.A. Zuhr, D.M. Hembree, M. Wu, and D.O. Nucl. Instr. Meth. Phys. Res. (in press).

[4] C.W. White, J.D. Budai, S.P. Withrow, J.G. Zhu, S.J. Pennycook, R.A. Zuhr, D.M. Hembree, D.O. Henderson, R.H. Magruder, M.J. Yacaman, G. Mondragon, and S. Prawer, Nucl. Instr. Meth. Phys. Res. **B127/128**, 545 (1997).

[5] C. Bonafos, B. Garrido, M. Lopez, A. Romano-Rodriguez, O. Gonzáles-Varona, A. Pérez-Rodriguez, and J.R. Morante, Appl. Phys. Lett. **72**, 3488 (1998).

[6] A. Meldrum, R.A. Zuhr, I.M. Anderson, C.W. White, J.D. Budai, L.A. Boatner, and D.O. Henderson, J. Mater. Res. (in prep.).

[7] S.B. Fisher, Radiat. Eff. **5**, 239 (1970).

[8] A. Meldrum, R.A. Zuhr, I.M. Anderson, C.W. White, J.D. Budai, L.A. Boatner, and D.O. Henderson, J. Appl. Phys. (submitted).

[9] C.W. White, J.D. Budai, J.G. Zhu, S.P. Withrow, D.M. Hembree, D.O. Henderson, A. Ueda, Y.S. Tung, and R. Mu, Mater. Res. Soc. Symp. Proc. **396**, 377 (1996).

SYNTHESIS OF BORON CARBIDE NANOWIRES AND NANOCRYSTAL ARRAYS BY PLASMA ENHANCED CHEMICAL VAPOR DEPOSITION

Daqing Zhang*, B. G. Kempton*, and D. N. McIlroy*, Yongjun Geng** and M. Grant Norton**
*Department of Physics, Engineering and Physics Building, University of Idaho, Moscow, Idaho 83844-0903.
**School of Mechanical and Materials Engineering, Washington State University, Pullman, Washington 99164-2920

ABSTRACT

A plasma enhanced chemical vapor deposition technique has been developed to grow single crystal boron carbide nanowires and nanonecklace arrays using the single precursor $closo$-1,2-dicarbadodecaborane ($C_2B_{10}H_{12}$). Nanowire and nanonecklace growth is expedited by Fe seeding of the substrate. Using the compound Fe-$(C_5H_5)_2$ as an Fe source, it has been demonstrated that the density of nanowires, as well as the types of nanostructures that grow, can be tailored by controlling the concentration of Fe deposited onto the substrate surface prior to boron carbide deposition.

INTRODUCTION

Single crystal boron carbide is a high temperature refractory material with a melting temperature in excess of 2400 °C. The crystal structure of boron carbide is rhombohedral (hR15; S.G. R3m) and consists of 12-atom icosahedral units located at the corners of a rhombohedral unit cell connected by C-B-B or C-B-C chains lying along the cell diagonal [1]. The equilibrium compound in the B-C system is B_4C [2]. Other compounds within this system are regarded as solutions of boron in B_4C or solid solutions with carbon. Single crystal boron carbide has potential application in thermoelectric power conversion [3,4]. A polaronic transport model has been developed to explain the thermoelectric conversion properties of single crystal boron carbide, where polaron hopping between icosahedra is phonon-assisted i.e., thermally activated [1,3-6]. As a consequence of this phonon-assisted transport mechanism and the unique geometry of the boron carbide nanowires, we hypothesize that their thermopower properties will differ from that of bulk boron carbide due to the introduction of new phonon modes. In addition to finite size effects, we anticipate that quantum confinement will have a significant effect on the transport properties of boron carbide nanowires, especially at low temperatures.

In this paper we present the results of our recent efforts to grow boron carbide nanowires (quantum wires) and arrays of nanoparticles (quantum

Mat. Res. Soc. Symp. Proc. Vol. 536 © 1999 Materials Research Society

dots), from here on referred to as nanonecklaces, by plasma enhanced chemical vapor deposition (PECVD), where nanowire growth is explained in terms of the vapor-liquid-solid (VLS) growth mechanism [7,8]. Vapor-liquid-solid growth occurs as a consequence of the formation of a metallic eutectic, in this case Fe-4wt%, which absorbs boron and carbon molecular fragments from the vapor which in turn are secreted out the base of the Fe eutectic once the droplet becomes saturated. The secreted material is in the form of a continuous wire of single crystal boron carbide. By controlling the amount of Fe deposited onto the substrate prior to nanowire growth the density, as well as the types of nanostructures that grow, can be controlled.

EXPERIMENT

The nanowires were grown on (100)-oriented silicon substrates. Prior to insertion into the chamber the Si substrates were cleaned for five minutes in a 5% hydroflouric/deionized water solution followed by a rinse in deionized water. The boron carbide nanowires were grown in a custom parallel plate 13.56 MHz PECVD chamber, which has been discussed in more detail elsewhere [9]. The substrates were located on the grounded electrode during deposition. The deposition temperature was typically 400 °C for Fe deposition and in the range 1100 - 1200 °C for boron carbide growth, respectively, with a plasma power of 50 W. The source compounds were Fe-$(C_5H_5)_2$, or ferrocene, and closo-1,2-dicarbadodecaborane ($C_2B_{10}H_{12}$), which will be referred to from here on as orthocarborane. Argon was used as the carrier gas and was allowed to flow through the source bottles that were held at 50 °C during deposition. Fe deposition ranged from 1-5 min. at 80 mTorr of Ar (10 sccm) and 8 mTorr (0.9 sccm) ferrocene/Ar. Boron carbide films were grown for 20 min. with a gas mixture of 80 mTorr of Ar (10 sccm) and 8 mTorr of orthocarborane/Ar (9 sccm) (conditions 1), or 2 hrs. with a gas mixture of 60 mTorr of Ar (8 sccm) and 25 mTorr of orthocarborane/Ar (3.5 sccm) (conditions 2).

RESULTS

At deposition temperatures in the range of 1100 - 1200 °C single crystal boron carbide forms [8], as compared to amorphous-like growth below 1000 °C [10]. The B_4C crystal structure of the boron carbide nanostructures have been verified with selected area electron diffraction [8]. In Fig. 1 we present an SEM image of an extremely low density film of boron carbide nanowires and nanocrystal arrays grown under condition 1 with 1 min. of Fe seeding of the substrate. These nanocrystal arrays, or nanonecklaces, consist of 100 nm nanoparticles connected at their apex by 20-30 nm diameter nanowires [8]. Consistent with the VLS growth mechanism, the nanowires and nanonecklaces grow vertically from the surface for the first few micrometers, after which they begin to bend either back in a direction towards the substrate or parallel to the surface [8]. This suggests that the nanowires and nanonecklaces are very resilient. The nanocrystals exhibit the symmetry of the rhombohedral unit cell

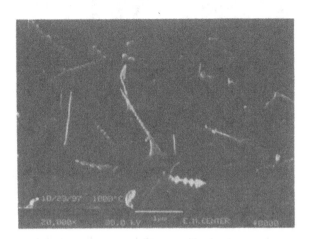

Figure 1. An SEM image of individual nanowires and nanonecklaces.

of boron carbide [8], which suggests that nanocrystal formation is a consequence of a variation in the contact angle of the liquid droplet or an instability in the nanowire, where the structure of the nanocrystal corresponds to a local minimum of the free energy. Nanowire growth is typically more common than nanonecklace growth which suggests that the wire geometry is the lower free energy state of the two types of growth.

A variety of nanoscale morphologies can be achieved under growth conditions 2 and depending on the amount of Fe seeding of the Si substrate prior to boron carbide deposition. For 2 min. of Fe seeding, which is twice the seeding time used to grow the sample in Fig. 1, in combination with 2 hrs. boron carbide deposition a mat of boron carbide nanowires grows (Fig. 2). Consistent with previous observations [8], there are three distinct types of nanostructure growth: cylindrical nanowires with smooth surfaces and an average diameter of 20 nm (Fig. 2), cylindrical with rough, faceted surfaces and an average diameter of 50 nm [8], and linear arrays of approximately equally spaced rhomboidal nanostructures (Fig. 1).

Under similar conditions as the sample shown in Fig. 2, but an Fe seeding time of 5 min., nanowire growth is no longer favorable and dendritic growth now becomes the preferred means of growth. A sample of this type of growth is displayed in Fig. 3. As before, this growth is promoted by Fe seeding, yet in this case the excess amounts of Fe results in too many nanowire

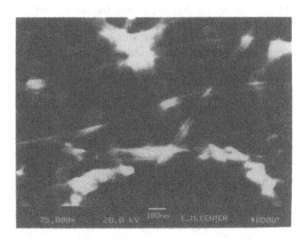

Figure 2. An SEM image of a thick mat of boron carbide nanowires.

nucleation sites. As a consequence, nanowires either collide with nanowires from adjacent nucleation sites early in their growth and or secondary growth

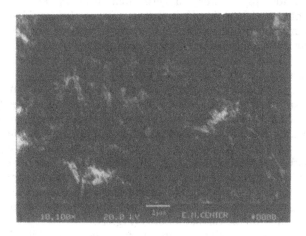

Figure 3. An SEM image of dendritic growth of boron carbide.

from the sides of nanowires occurs. In the presence of too much Fe seeding the morphology can vary dramatically [11]. By reducing the Fe seeding to 1 min.

and growth under the same conditions as the sample in Fig. 1, low nanowire density films with nanowires in excess of 15 μm can be obtained [8].

Summary

Through the manipulation of the vapor-solid-liquid growth mechanism of boron carbide it is possible control the types of nanostructures which form using the technique of PECVD. Nanowire growth of boron carbide can be promoted over dendritic growth by limiting the amount of Fe on the surface prior to deposition. Although it was not discussed in detail, nanowire growth is also dependent on the substrate temperature and the vapor density of the boundary layer in the vicinity of the substrate surface. Future work will focus on controlling the size and distribution of Fe seeds on the substrate surface in order to develop a technique for selectively depositing boron carbide nanowires in an ordered array.

Acknowledgments.

D.N.M. would like to acknowledge the support of the NSF-Idaho EPSCoR project under NSF Cooperative Agreement number EPS-9350539, a seed grant from the University of Idaho Research Council (KDY817), the Petroleum Research Foundation (PRF# 32584-G5), and NASA ISGC.

References

1. O. Chauvet, D. Emin, L. Forro, T. L. Aselage, and L. Zuppiroli, Phys. Rev. B 53, 14450 (1996).
2. ASM Handbook, Volume 3, Alloy Phase Diagrams, (edited by H. Baker), ASM International, Metals Park, Ohio (1992).
3. C. Wood and D. Emin, Phys. Rev. B 29, 4582 (1984).
4. C. Wood, D. Emin, and P. E. Gray Phys. Rev. B 31, 6811 (1985).
5. I. A. Howard, C. L. Beckel, and D. Emin, Phys. Rev. B 35, 2929 (1987).
6. I. A. Howard, C. L. Beckel, and D. Emin, Phys. Rev. B 35, 9265 (1987).
7. I. D. R. Mackinnon and K. L. Smith, MRS Proc. 97, 127 (1987).
8. Daqing Zhang, D. N. McIlroy, Yongjun Geng and M. Grant Norton, J. Mat. Sci. Lett., in press.
9. S. Lee, J. Mazurowski, G. Ramseyer, and P. A. Dowben, J. Appl. Phys. 72, 4925 (1992).
10. Daqing Zhang, D. N. McIlroy, W. L. O'Brien, Gelsomina De Stasio, J. of Mat. Sci., in press.
11. D. N. McIlroy et al., to be published.

Germanium Nanostructures Fabricated By PLD

K.M. Hassan, A.K. Sharma, J. Narayan, J.F. Muth[1] and C.W. Teng[2]
Department of Materials Science and Engineering, NSF Center for Advanced Materials
Processing and Smart Structures, North Carolina State University, Raleigh, NC 27695.
[1,2] Department of Electrical and Computer Engineering, North Carolina State University,
Raleigh, NC 27695

Quantum confined nanostructures of semiconductors such as Ge and Si are being actively studied due to their interesting optical and electronic transport properties. We fabricated Ge nanostructures buried in the matrix of polycrystalline-AlN grown on Si(111) by pulsed laser deposition at lower substrate temperatures than that used in previous studies. The characterization of these structures was performed using high resolution transmission electron microscopy (HRTEM), photoluminescence and Raman spectroscopy. HRTEM observations show that the Ge islands are single crystal with a pyramidal shape. The average size of Ge islands was determined to be 15 nm, considerably smaller than that produced by other techniques. The Raman spectrum reveals a peak downward shift, upto 295 cm[-1], of the Ge-Ge mode caused by quantum confinement in the Ge-dots. Photoluminescence (PL) was observed both with a single layer of Ge nanodots embedded in the AlN matrix and from ten layers of dots interspersed with AlN. The PL of the dots was blue shifted by ~0.266 eV from the bulk Ge value of 0.73 eV at 77 K, resulting in a distinct peak at ~1.0 eV. The full width at half maximum (FWHM) of the peak was 13 meV, for the single layer and 8 meV for the ten layered sample, indicating that the Ge nanodots are fairly uniform in size, which was found to be consistent with our HRTEM results. The importance of pulsed laser deposition (PLD) in fabricating novel nanostructures is discussed.

INTRODUCTION

Semiconductor nanostructures are being investigated extensively to study the effect of the confinement of their carrier motions on the optical properties such as photoluminescence. Indirect bandgap semiconductors such as Si and Ge have very poor luminescent efficiency as their lowest energy electronic transition is optically forbidden. However, by making the crystallite size of these semiconductors smaller than the exciton Bohr radius, one would expect a sharp increase in the oscillator strength due to quantum confinement effect [1]. Since, photoluminescence depends primarily on the energy gap and the oscillator strength, one should expect to see the nanocrystals of these semiconductors luminescent in the infrared and visible regions. A large body of literature already exists on the visible photoluminescence of nanocrystalline Si [2], also called porous silicon. However, nanostructures of Ge have received very little attention. It must be mentioned that the excitonic Bohr radius of bulk Ge is larger (17.7 nm) than that of Si (4.9 nm) [3] and, therefore, the quantum size effects will be more prominent in Ge nanocrystals even at larger sizes of their crystallites.

The optical properties of Ge nanoparticles embedded in dielectric materials, such as SiO_2, have been investigated recently [4]. Aluminum nitride is a wide bandgap material (E_G~6.2 eV) with dc dielectric constant ~9.14 [5]. Higher energy confinement and dielectric confinement would be expected if Ge nanocrystals were embedded is this oxygen-free matrix. We have investigated the structure, bonding characteristics and optical properties of Ge nanocrystals embedded in AlN matrix. In the present work, we have employed pulsed laser ablation of AlN and Ge targets to deposit several alternate multilayers of Ge and AlN on p-type Si(111) substrate. Pulsed laser ablation technique is known for growing films at lower substrate temperatures with better microstructural quality.

EXPERIMENT

We have used pulsed excimer laser system ($\lambda\sim$248 nm, t_p=25 ns) with multiple target holder to ablate Ge and AlN targets in high vacuum ($\sim5\times10^{-7}$ torr) environment. The laser energy density used in this experiment was $\sim$2.5 J/cm^2 and the substrate temperature was $\sim$ 500 ^{0}C . The p-type Si(111) wafers were cleaned in acetone and methanol ultrasonically prior to surface oxide removal by 10% HF solution. We deposited Ge dots either in vacuum or in the presence of 300 mtorr partial pressure of nitrogen. The AlN layers are much thicker than those of Ge. We have grown ten multilayers of Ge interspersed with AlN layers.

Topcon 002B transmission electron microscope (resolution $\sim$0.18 nm) was used to analyze these nanostructures. The size distribution of the Ge nanodots, crystallinity of the AlN matrix and the thickness of the different multilayers were investigated by cross-sectional HRTEM. Raman spectroscopy was performed on the samples using Ar$^+$ ion laser ($\lambda\sim$514.5 nm) in the backscattering geometry. For photoluminescence studies the samples were mounted on oxygen-free heat conductive copper plug and placed on the cold finger of a liquid nitrogen dewar. The luminescence was excited by an argon ion laser ($\lambda\sim$514.5 nm) at 77 K. Photoluminescence from the sample was chopped at 1.3 kHz and collected using a 0.5-m spectrometer and liquid nitrogen cooled germanium photovoltaic detector combined with transimpedance preamplifier. The lock-in technique was used to get a high signal to noise ratio.

RESULTS AND DISCUSSIONS

Figure 1 shows the cross-sectional TEM image of the Ge layers embedded in AlN matrix.

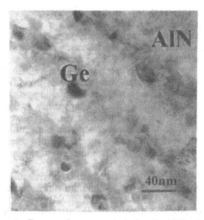

Figure 1. Ge nanodots appear darker in AlN matrix.

The selected area diffraction (figure 2) reveals that AlN is polycrystalline and Ge nanodots have diamond cubic structure. The high resolution image (figure 3) reveals that

330

the Ge nanodots are pyramidal shaped with height ~15 nm and base dimension ~20 nm. The average size of the nanodots is ~15 nm, which is smaller than the excitonic radius of bulk Ge [3]. The size distribution is fairly uniform. HRTEM images reveal that the islands are strained and dislocation free. The pyramidal shaped nanodots have a domed top, possibly to decrease the surface energy.

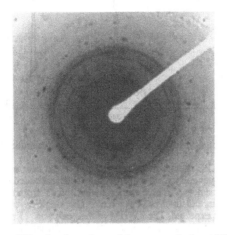

Figure 2. Selected area diffraction from the multilayer reveals that AlN is polycrystalline and Ge nanodots have diamond cubic structure.

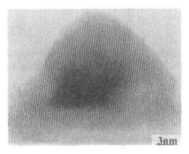

Figure 3. HRTEM image of a pyramidal shaped Ge nanodot. Height ~15nm. Base ~20nm.

Figure 4 is the Raman spectrum of the sample measured under the backscattering geometry. The Ge-Ge peak from single layer and multilayers of Ge nanodots is ~295 cm^{-1} and ~290 cm^{-1}, respectively. The large downward energy shift is expected to be due to phonon confinement in the Ge nanostructures [6]. It is to be noted that strain effects induce an upward shift of the peak. The Ge-Ge peak from bulk Ge target was recorded to be ~305 cm^{-1}.

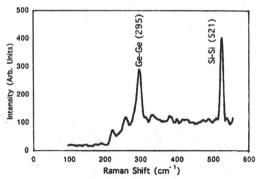

Figure 4. Raman spectra from a single layer of Ge nanodots embedded in AlN matrix.

Figure 5 shows PL spectra measured at 77 K in the near infra-red (IR) range for samples with a single layer and ten layers of Ge nanodots embedded in the AlN matrix. The photoluminescence of the dots was blue shifted by ~0.266 eV from the bulk value of 0.73 eV at 77 K, resulting in a distinct peak at ~1 eV. The single-layer sample has a weaker luminescent intensity as expected. The FWHM of the peak was 13 eV, for the single layer and 8 eV for the ten layered sample, indicating that the Ge nanodots are fairly uniform in size

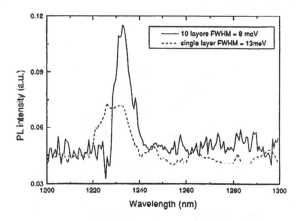

Figure 5. Photoluminescence spectra measured at 77K

Three major causes can lead to the observed blue shift of the PL peak. Firstly, strain due to dot-matrix interaction which has a minor contribution to the blue shift. The second is carrier recombination of the interface states. This mechanism is responsible for visible photoluminescence from oxidized Si or Ge nanoparticles embedded in oxide matrix. Lastly, the bandgap widening is considered to be due to the quantum confinement effect of electrons, holes, and excitons [7]. While ignoring the first two effects, a spherical potential treatment suggested by Brus [8] was used to make a rough estimate. The confinement energy could be expressed as

$$\Delta E = \frac{\Pi^2 \hbar^2}{2 \mu R^2}$$

where μ is the reduced effective mass of electron-hole pairs, $\hbar$ is the Plancks constant and R is the dot radius. Using $\mu = 0.082 m_0$ for Ge, an energy blue shift 0f 0.266eV corresponds to an average dot size ~14 nm, which is not far from ~15 nm dot size measured from HRTEM analysis in our samples.

The intensity of the photoluminescence indicated that the radiative recombination process was very efficient. To confirm that AlN capping layer does not act as the source of recombination states contributing to IR photoluminescence, bare AlN films (without Ge nanodots) were deposited by PLD under similar conditions. These bare samples did not reveal any signal in the IR range when they were excited with 514.5 nm Ar^+ ion laser.

CONCLUSIONS

We have successfully fabricated Ge nanostructures embedded in AlN matrix with fairly uniform size distribution. Raman signatures suggest phonon confinement due to downward shift of the Ge-Ge peak. High resolution microscopy revealed dislocation free pyramidal shaped Ge nanocrystals with average size ~15 nm. Quantum confinement effect was confirmed by the distinct blue shifted photoluminescence peak observed at ~1.0 eV. The sharp FWHM of the PL peak suggests a uniform size distribution of the nanodots. Thus PLD is an attractive technique for fabricating nanostructures, at low substrate temperature, with a narrow size distribution. The PLD method and the use of AlN matrix also demonstrates the ability to build multilayers of nanodots in a controlled manner and place them in an oxygen free environment. This will permit the investigation of interface and strain effects by embedding similar sized Ge nanodots in different matrices.

ACKNOWLEDGEMENT

This work was supported by NSF Center for Advanced Materials Processing and Smart Structures, North Carolina State University, Raleigh, NC 27695.

References

1. B. Delley and E.F. Steigmeier, Phys. Rev. B 47, 1397 (1993).
2. L.T. Canham, Appl. Phys. Lett. 57, 1046 (1990).
3. A.G. Cullis, L.T. Canham, and P.D.J. Calcott, J. Appl. Phys. 82, 909 (1997).
4. Shinji Takeoka, Phys. Rev. B 58, 7921 (1998).
5. A.T. Collins et.al., Phys. Rev. B 158, 833 (1967).
6. Xun Wang, Zui-min Jiang, Hai-jun Zhu, Fang Lu, Daiming Huang, and Xiaohan Liu, Appl. Phys. Lett. 71, 3543 (1997).
7. V.Ranjan and Vijay A. Singh, Phys. Rev. B 58, 1158 (1998).
8. L.E. Brus, IEEE J. Quantum Electron. 22, 1909 (1986).

Part VI
Oxide and Chalcogenide Semiconductors

IN-SITU X-RAY DIFFRACTION STUDY OF LITHIUM INTERCALATION IN NANOSTRUCTURED ANATASE TITANIUM DIOXIDE

R. VAN DE KROL*, E. A. MEULENKAMP**, A. GOOSSENS*, AND J. SCHOONMAN*

* Delft Interfaculty Research Center: Renewable Energy, Delft University of Technology
Laboratory for Inorganic Chemistry, Julianalaan 136, 2628 BL Delft, The Netherlands
** Philips Research Laboratories, Prof. Holstlaan 4, 5656 AA Eindhoven, The Netherlands

ABSTRACT

Electrochemical lithium intercalation in nanostructured anatase TiO_2 is investigated with *in-situ* X-ray diffraction. A complete and reversible phase transformation from tetragonal anatase TiO_2 to orthorhombic anatase $Li_{0.5}TiO_2$ is observed. The difference of the XRD spectra before and after insertion can be fitted with the lattice parameters of the two phases as fit parameters. The maximum amount of lithium that can be dissolved in anatase TiO_2 before the phase transformation occurs is found to be very small.

INTRODUCTION

Recent studies have shown that anatase TiO_2 is a promising candidate for devices based on electrochemical intercalation of lithium ions. Charge storage in the form of lithium ions is the basis for its use as an electrode material in lithium ion batteries [1], while the optical absorption of the charge-compensating electrons can be employed in electrochromic windows [2]. A porous nanostructured morphology greatly enhances the response times and the efficiencies of these devices.

Usually, optical and electrical characterization techniques are employed to study the intercalation process. Detailed (*ex-situ*) structural characterization is seriously hampered by the formation of lithium compounds and de-lithiation upon exposure to air. Here, structural changes during lithium insertion in nanoporous anatase TiO_2 are studied with *in-situ* X-ray diffraction.

EXPERIMENTAL

Kapton foil with a thickness of 125 μm was used as a substrate. The foil was sputter-coated with a 100 nm gold film to serve as a current collector. Nanostructured porous anatase TiO_2 was prepared with a slightly modified method as described by O'Regan *et al.* [3]. The sol was deposited on the Kapton/gold substrates by doctor-blading, using three layers of Scotch-tape (3M) to determine the film thickness. After drying at room temperature, the samples were fired for 12 hours at 400°C in air.

The samples were mounted in a specially designed air-tight electrochemical cell for *in-situ* X-ray diffraction [4]. The active geometric surface area of the samples was 4.1 cm². Lithium foil was used for the counter and reference electrodes. The electrolyte was 1.0 M $LiClO_4$ in propylenecarbonate (Merck). The potential was controlled by a model 263A (EG&G) potentiostat. The cell was incorporated in a Philips PW1835 X-ray diffractometer, employing Cu-K_α radiation. All measurements were made while the system was in equilibrium, *i.e.* without significant charge flow during the measurement.

RESULTS AND DISCUSSION

Figure 1 shows the difference between two XRD spectra, recorded before and after insertion of the maximum possible amount of lithium. The fact that some peaks disappear while others appear at different angles clearly indicates a structural phase transformation. The negative peaks result from the disappearing anatase TiO_2 phase, while the positive peaks represent the

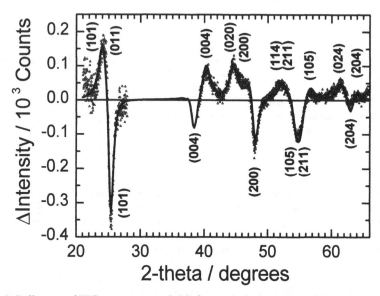

Figure 1. Difference of XRD spectra recorded before and after insertion of the maximum amount of lithium ions (1.3 C per geometric cm^2). Positive peaks indicate a newly formed phase, while negative peaks represent the disappearing anatase TiO_2 phase. The corresponding (hkl) indices of both phases are indicated. The solid line is a fit of the measured data calculated from the unit cell parameters.

phase into which the anatase TiO_2 is transformed. This new phase is found to correspond to anatase $Li_{0.5}TiO_2$, which is reported by Cava et al. [5] in a neutron diffraction study on the reaction product of anatase TiO_2 and n-butyllithium, i.e. chemical lithiation. The phase transformation is accompanied by a change in symmetry from tetragonal (TiO_2, space group $I4_1/amd$) to orthorhombic ($Li_{0.5}TiO_2$, space group Imma). The occurrence of a phase transformation is supported by frequent observations of constant potential plateaus during a constant current discharge of lithiated anatase TiO_2 [1,6]. XRD spectra before insertion and after extraction of lithium are identical, which shows that the phase transformation is completely reversible.

The measured data are corrected for substrate absorption, angle-dependence of the diffraction volume, and height errors caused by substrate bending [4]. Lorentz-shaped peaks are found to best fit the data. The $Li_{0.5}TiO_2$ unit cell constants, the peak widths, and the integrated peak intensities are used as fit parameters. It proved necessary to include the c-axis of the TiO_2 unit cell as an additional fit parameter, the a- and b-axes were fixed to previously reported values [5]. The particle size of ~15 nm in diameter induces an angle-dependent peak width (Scherrer equation), which has been included in the fit procedure. Since the anatase TiO_2 peaks disappear completely, 100% of the anatase TiO_2 is transformed into $Li_{0.5}TiO_2$. The final fit parameters are presented in Table 1.

Table 1. Fit parameters for a least-squares fit of the data in Figure 1. The space groups and unit cell dimensions (in Å) from literature are listed between parentheses [5]. The estimated accuracy is 0.02 Å.

Parameter	anatase TiO_2 ($I4_1/amd$)		$Li_{0.5}TiO_2$ (Imma)	
a	3.78	(3.784)	3.85	(3.808)
b	3.78	(3.784)	4.06	(4.077)
c	9.40	(9.515)	8.95	(9.053)
V (Å^3)	134.6	(136.2)	139.9	(140.5)
Σ peak areas (a.u.)	1.0		1.1	

Considering the quality of the measured data, a good agreement is found with values reported in the literature. The fitted values for the c-axes seem rather small for both phases. Front-side XRD measurements on other batches of nanostructured TiO_2 in air, *i.e.* without liquid

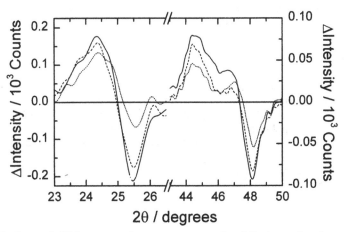

2θ / degrees

Figure 2. Differential XRD spectra during extraction of a fully intercalated sample with a 0.5 Ccm^{-2} equivalence of Li atoms (solid line), after subsequent extraction of 25% (dashed line) and 50% of the inserted lithium ions (dotted line). The reference spectrum is measured at 3.4 V vs Li. The data are recorded with a total integration time of 20 sec. per 0.02° 2θ, and Fourier-smoothed for clarity.

electrolyte, show similar small values for the c-axis. The origin of this effect is currently unclear. The sum of the integrated peak areas is approximately equal for both phases, which indicates that the anatase TiO_2 lattice is not greatly distorted upon transformation to $Li_{0.5}TiO_2$ [5]. The phase transformation is accompanied by a 4% increase in volume.

To further investigate the phase transformation, X-ray diffraction spectra are measured at various stages of extraction of a fully intercalated sample. The results are shown in Figure 2 for 0%, 25%, and 50% extraction of the total inserted charge. No significant peak shifts are observed during extraction, which implies a two-phase equilibrium instead of a solid solution. This agrees with the often observed constant potential plateaus during slow cyclic voltammetry and slow galvanostatic extraction of lithium from intercalated anatase TiO_2 [1,6]. However, a small amount of solid solution formation cannot be excluded on the basis of the present data. Zachau-Christiansen et al. reported that lithium dissolves up to x = 0.05 before the lithium-rich phase nucleates [6].

CONCLUSIONS

In-situ X-ray diffraction is a powerful tool to study structural changes during electrochemical insertion of lithium ions into anatase TiO_2. The use of an *in-situ* technique avoids complications due to formation of lithium compounds at the surface and related de-lithiation of the sample due to air exposure. In addition, very small changes of the signal in the order of a few percent can still be measured accurately.

A phase transformation to anatase $Li_{0.5}TiO_2$ is observed, in agreement with earlier reported observations on chemical lithiation of anatase TiO_2. The unit cell parameters obtained from a fit of the measured diffraction data agree well with previously reported values. During extraction of lithium, no significant shift in the peak positions is observed, which demonstrates that the transformation occurs via a two-phase equilibrium with very little dissolution of lithium in the TiO_2 phase.

LITERATURE

1. S. Y. Huang, L. Kavan, I. Exnar, and M. Grätzel, J. Electrochem. Soc., **142**, L142 (1995).
2. A. Hagfeldt, N. Vlachopoulos, and M. Grätzel, J. Electrochem. Soc., **141**, L82 (1994).
3. B. O'Regan, J. Moser, M. Anderson, and M. Grätzel, J. Phys. Chem., **94**, 8720 (1990).
4. E. A. Meulenkamp, J. Electrochem. Soc., **145**, 2759 (1998).
5. R. J. Cava, D. W. Murphy, S. Zahurak, A. Santoro, and R. S. Roth, J. Solid State Chem., **53**, 64 (1984).
6. B. Zachau-Christiansen, K. West, T. Jacobsen, and S. Atlung, Solid State Ionics, **28-30**, 1176 (1988).

NOVEL ELECTRONIC CONDUCTANCE CO_2 SENSORS BASED ON NANOCRYSTALLINE SEMICONDUCTORS

M.-I. BARATON, L. MERHARI *, P. KELLER **, K. ZWEIACKER **, J.-U. MEYER **
SPCTS UMR 6638 CNRS, Faculty of Sciences, F-87060 Limoges, France, baraton@unilim.fr
*CERAMEC R&D, F-87000, Limoges, France
**Fraunhofer Institute for Biomedical Engineering, Sensorsystems/ Microsystems Department, Sankt Ingbert, Germany

ABSTRACT

We have recently demonstrated that screen-printed sensors using a 20 nm- instead of microsized $BaTiO_3$-CuO-additives powder exhibit up to one order of magnitude higher sensitivity to CO_2. In this paper, we focus on both the surface chemistry of the nano-$BaTiO_3$-CuO-additives powder (mix-$BaTiO_3$) and electrical changes during O_2 and CO_2 adsorptions. We show by Fourier transform infrared (FTIR) spectrometry, thus without using electrodes, that the mix-$BaTiO_3$ system behaves like a p-type semiconductor at the operating temperature. The variations of the electrical conductivity versus CO_2 concentrations are followed *in situ* by FTIR spectrometry and prove to be dependent on the surrounding oxygen. These IR results are then correlated to the electrical measurements performed on the sensor. Preliminary electrical response modelling shows a good agreement with the surface barrier layer theory.

INTRODUCTION

Gas concentration cells, solid-electrolyte type sensors, infrared sensors and oxide-based sensors are all capable of detecting CO_2 concentrations in air [1]. However, most of these devices suffer from poor selectivity, poor long-term stability, and significant cross-sensitivity to relative humidity (RH). Rugged low-cost portable sensors having a low power consumption are currently the object of a soaring demand for detecting and controlling CO_2 concentrations in areas as diverse as biotechnologies, agriculture, indoor air quality monitoring, air conditioning, and medical services. Compared to the sensitive and reliable infrared sensors, oxide-based sensors have the major advantage of a low cost. Ishihara *et al.* [2] in Japan have investigated a promising material based on a $BaTiO_3$-CuO mixed oxide for capacitive-type CO_2 sensors. Häusler and Meyer [3] in Germany designed a conductance sensor based on $BaTiO_3$-CuO mixed oxide and additives offering a high selectivity to CO_2 and negligible cross-sensitivity to RH. A major improvement of the CO_2 sensitivity has been the use of nanosized powders [4]. Indeed, their high specific surface area and enhanced reactivities are useful to design superior materials [5]. As the mechanism of gas detection by oxide-based sensors rests on the fundamental *gas-surface interactions* leading to both chemical and electrical reversible changes [6], it is important to rely on an investigation tool monitoring all these changes *in situ* during gas adsorptions. In this paper, we mainly used FTIR spectrometry to gain an insight into the complex detection mechanism of CO_2 by mix-$BaTiO_3$-based conductance sensors.

EXPERIMENTAL

$BaTiO_3$-CuO-additives nanosized powder synthesis and sensor prototype fabrication

It has been proven that the surface chemical composition of the nanosized powders must be characterized and controlled to ensure their optimum use in high-added-value applications [5].

This obviously calls for versatile clean synthesis methods. The present 20 nm mix-BaTiO$_3$ powder was synthesized at the University of Clausthal, Germany, by laser ablation of a pressed microsized powder blend (BaTiO$_3$:CuO=1:1, and additives including La$_2$O$_3$ and CaCO$_3$), followed by condensation of the vapour in a controlled atmosphere. The yield of this surface-controlled nanosized powder exceeds 30g/hour with the current production apparatus built around a 1200W Nd:YAG laser [7]. XRD spectra (not shown) of the mix-BaTiO$_3$ powder essentially reveal two separate phases corresponding to BaTiO$_3$ and CuO. Additional smaller peaks assigned to BaCO$_3$, TiO$_2$ and Cu$_2$O also appear.

The prototype sensors were fabricated using the conventional screen-printing method. The mix-BaTiO$_3$ powder was dispersed in an organic solution so as to obtain a homogeneous paste which was then applied on the interdigitated electrodes printed on one side of a thin alumina substrate. A heater element printed on the back of the substrate makes it possible to operate the sensor up to 600°C. Thermal treatments of the sensitive nanocrystalline layer include drying at 80°C and firing up to 750°C. In the above process, the most challenging problems to overcome are powder agglomeration in the dispersion, control of the layer surface tension during drying and grain growth during thermal treatment.

Characterization techniques

The electrical characterization of the sensors consists in measuring the variation of their impedance when subjected to various concentrations of CO$_2$ in synthetic air. The sensors were characterized in a flow through chamber connected to a computer-controlled gas blender. All the measurements were performed at 1 kHz using a computerized LCR meter. In the following, we define the sensor sensitivity toward CO$_2$ as S = R$_{CO2}$/R$_{air}$ where R$_{CO2}$ and R$_{air}$ are the resistances measured in air containing known CO$_2$ concentrations and in clean synthetic air respectively. For sake of convenience, throughout the paper, the sensor response to CO$_2$ will be labelled 'positive' if S > 1, and 'negative' if S < 1.

The FTIR analysis was either directly performed on the sensor prototypes by means of a diffuse reflectance setup (DRIFTS), or on plain pellets made of pressed mix-BaTiO$_3$ powder simulating the sensor. These plain pellets were inserted in a specially designed heatable vacuum cell [8] placed inside the spectrometer sample compartment and were analyzed in the transmission mode. All the spectra were recorded by means of a Perkin-Elmer Spectrum 2000 FT-IR spectrometer equipped with an MCT cryodetector. The analyzed spectral range extended from 500 to 6500 cm^{-1} with a 4 cm^{-1} resolution. Whatever the IR setup, controlled pressures of gases were adjusted through a precision valve system.

INVESTIGATION OF THE CO$_2$ SENSING PROPERTIES OF BaTiO$_3$-CuO SYSTEMS

Electronic properties variations of the BaTiO$_3$-CuO sensors versus CO$_2$ and O$_2$ adsorptions

The sensitivity to 0.01-10 vol. % CO$_2$ concentrations in synthetic air for nano- and micro-sized BaTiO$_3$-CuO based sensors operated at 550°C is plotted in Fig. 1. The striking result is that the use of nanoparticles increases the sensitivity to CO$_2$ by a factor ranging from 3 to 10. Moreover, the detection limit improves from 0.1 to 0.01 vol. %. It must be noted that the response is always 'positive'. Detection of CO$_2$ in an oxygen-free environment (pure N$_2$) was attempted and showed a negative response (Fig. 2). Again, the use of nanoparticles significantly increases the sensitivity to CO$_2$, while a response drift is observed. These results show that oxygen must play an important role in the detection mechanism. Moreover, as suggested by Ishihara [2], BaTiO$_3$-CuO semiconducting junctions are involved in the above

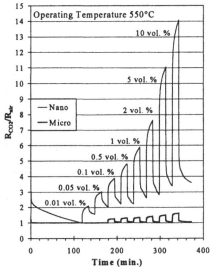

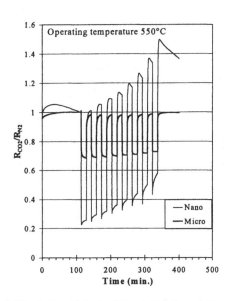

Fig. 1. Sensitivity to CO_2 concentrations in synthetic air vs particle size of $BaTiO_3$-CuO-based sensors.

Fig. 2. Sensitivity to 5% vol. CO_2 in pure N_2 vs particle size of $BaTiO_3$-CuO-based sensors.

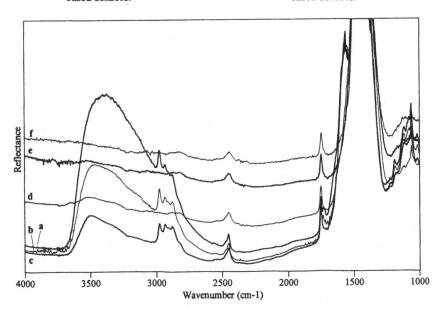

Fig. 3. DRIFT spectra of nano-$BaTiO_3$-CuO-based sensors vs temperature in air: a) room temperature; b) 100°C; c) 200°C; d) 300°C; e) 400°C; f) 500°C.

mechanism, which is in agreement with our results showing that increasing the number of junctions per surface unit (case of nanosized powders) leads to a higher CO_2 sensitivity.

The FTIR spectrum of the sensor was recorded in the DRIFT mode *in situ* under air at increasing temperatures from room temperature to 500°C (Fig. 3). The very broad band centered at 3000 cm^{-1} is due to the $v(OH)$ stretching vibration of water molecules adsorbed through hydrogen bonds on surface hydroxyl groups. This broad band steadily disappears when the temperature increases (Fig. 3d-f). The group of bands in the 3000-2800 cm^{-1} region are assigned to the $v(CH)$ stretching vibrations of the organic binder eliminated at 300°C. The other bands mainly originate from $BaTiO_3$ and $BaCO_3$. Particularly, the very intense band centered at 1500 cm^{-1} proves the presence of a significant amount of $BaCO_3$. It is worth noting that when the temperature increases, the baseline of the spectrum moves toward higher reflectance values. This indicates an increase of the average absorption of the infrared incident beam by the sensor, thus revealing an increase of the electrical conductivity. This conductivity increase under oxydizing conditions definitely proves that the mix-$BaTiO_3$ material behaves like a p-type semiconductor. Indeed, it is known that the modulation of the IR energy by oxidizing or reducing gaseous environments leads to a modulation of the free carriers population right at the surface of the semiconductors [9].

Transition from a 'negative' to a 'positive' sensor response to CO_2 versus O_2 concentrations

The observation of 'positive' and 'negative' responses depending on the surrounding O_2 content during CO_2 detection suggests a CO_2/O_2 critical ratio at which the transition would occur. This is proven for mix-$BaTiO_3$ based sensors operated at 550°C in Fig. 4 where the sensitivity to 0.1-10 vol. % CO_2 concentrations in a nitrogen environment containing 1% vol. O_2 is plotted. The transition from a negative to a positive response occurs between 0.5 and 1 vol. % CO_2. It is worth noting that introducing 1 vol. % O_2 in pure N_2 (at t=30 min.) decreases the sensor resistance, thus confirming the p-type semiconductivity of the material.

Similar results are obtained in Fig. 5 where the IR energy transmitted through a mix-$BatiO_3$ pellet simulating the sensor is recorded versus O_2 and CO_2 introductions. Reliable measurements obtained once the device stabilizes show an increase of conductivity upon O_2 introduction, and a transition from a 'negative' to a 'positive' response versus CO_2 concentrations. Clearly, FTIR spectrometry can be used as a rapid and cost-effective method for quality control of the nanosized powder batches before the actual fabrication of the sensors.

TENTATIVE MECHANISM OF CO_2 DETECTION

To determine the surface reactions taking place during CO_2 addition at 450°C, transmission infrared studies were performed *in situ* on the mix-$BaTiO_3$ powder pressed into a thin pellet simulating the sensor. Indeed, in this particular case, the DRIFT method is not sensitive enough to allow the direct study of the evolution of the surface species on the real sensor under weak CO_2 pressures. Nevertheless, it has been checked that similar results are obtained at high CO_2 pressures with the sensor in the DRIFT mode and with the nanopowder pellet in the transmission mode. Figure 6 shows difference spectra, that is the differences between the spectra recorded before and after CO_2 addition to the nanopowder pellet at 450°C. It must be noted that the difference around 1500 cm^{-1} has no real meaning because of the very strong absorption of the $BaCO_3$ mode (cf. Fig. 3). The first spectrum (Fig. 6a) corresponds to CO_2 adsorption in absence of oxygen. We note an increase of the $BaCO_3$ content and the formation of surface carbonate species. These carbonates, similar to those observed on $BaTiO_3$ (containing no CuO) and on TiO_2 during CO_2 adsorption, indicate that CO_2 reacts with basic

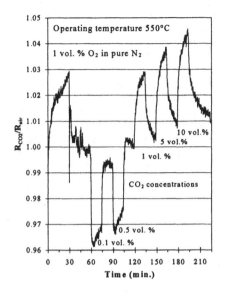

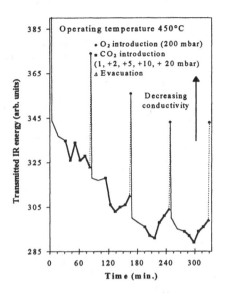

Fig. 4. Sensitivity to CO_2 concentrations of $BaTiO_3$-CuO-based sensors in a blend of pure nitrogen and 1 vol. % oxygen.

Fig. 5. Variations of the IR energy transmitted through nano-$BaTiO_3$-CuO based pellets vs O_2 and CO_2 adsorptions.

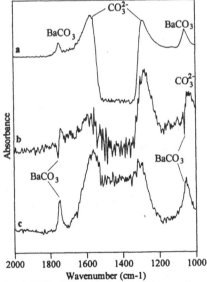

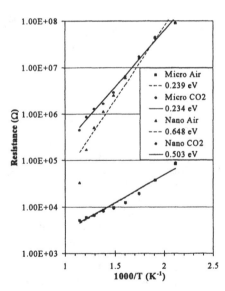

Fig. 6. FTIR difference spectra of the surface species formed during CO_2 adsorption: a) no O_2; b) negative response; c) positive response.

Fig. 7. Effect of air, and 7.5 vol. % CO_2 in air on the potential barrier height at the $BaTiO_3$-CuO grain boundaries vs powder particle size.

345

O^{2-} surface sites. When CO_2 in added at 450°C in presence of oxygen, the same new species appear (Fig. 6c) only in case of a positive response. Since oxygen is present as ionosorbed O^- (eventually O_2^-) species, CO_2 preferentially reacts with O^- to form carbonate species. The carbonate formation leads to a weakening of the electron withdrawing power of ionosorbed oxygen, and thus to a decrease of the electrical conductivity. In case of a negative response, carbonates are also formed but in a lower amount (the spectrum in Fig. 6b has been enlarged about 5 times). Moreover, the $BaCO_3$ content appears to slightly decrease.

In an attempt to relate the resistance (R) changes of the sensor during CO_2 detection with any variation of the energy barrier at the CuO-$BaTiO_3$ junctions, ln (R) in air with and without 7.5 vol. % CO_2 is plotted versus $1/T$ in Fig. 7. Straight lines reasonably fit the data points in the 200°C-600°C range. Moreover, CO_2 appears to systematically decrease the barrier height. This effect is clearly more pronounced in the case of nanoparticles. Finally, the behavior of this p-type mix-$BaTiO_3$ system subjected to CO_2 appears to obey the barrier layer theory.

CONCLUSION

FTIR spectrometry proves to be a valuable technique for fundamental studies of conductance gas sensors based on nanosized oxides as it makes it possible to investigate the chemical reactions occuring right at the real surface of the material and, simultaneously, access without contact perturbation to the electronic phenomena. Our preliminary results on the p-type $BaTiO_3$-CuO system show that the surrounding O_2 concentration plays a central role in the CO_2 detection mechanism. Reversible carbonate formation on adsorbed oxygen species is believed to modulate their electron withdrawing power and thus modulate the conductance.

ACKNOWLEDGMENTS

This work has been performed in the framework of a BRITE-EURAM III project (contract number BRPR-CT95-0002) funded by the European Commission. The authors acknowledge Prof. Werner Riehemann, Dr Hans Ferkel and Mr Jürgen Naser of the Technical University of Clausthal, Germany for the synthesis of various surface-controlled nanosized powders.

REFERENCES

1. T. Seiyama (Editor), *Chemical Sensor Technology*, Vol. 1, (Kodansha, Tokyo/Elsevier, Amsterdam, 1988).
2. T. Ishihara, K. Kometani, Y. Nishi and Y. Takita, Sensors &Actuators B **28**, 49-54 (1995).
3. A. Häusler and J.-U. Meyer, Sensors & Actuators B **34**, 388 (1996).
4. P. Keller, H. Ferkel, K. Zweiacker, J. Naser, J.-U. Meyer, and W. Riehemann, Proceedings of Eurosensors XII, Southampton, UK, 1998.
5. K.E. Gonsalves, M.-I. Baraton, R. Singh, H. Hofmann, J.X. Chen, and J.A. Akkara (Editors), *Surface-Controlled Nanoscale Materials for High-Added-Value Applications*, (Mater. Res. Soc. Proc. **501**, Warrendale, PA, 1998).
6. S.R. Morrison, *The Chemical Physics of Surfaces*, 2nd ed., Plenum Press, London, 1990.
7. W. Riehemann in *Surface-Controlled Nanoscale Materials for High-Added-Value Applications*, edited by K.E. Gonsalves, M.-I. Baraton, R. Singh, H. Hofmann, J.X. Chen, and J.A. Akkara (Mater. Res. Soc. Proc. **501**, Warrendale, PA, 1998) pp. 3-14.
8. M.-I. Baraton, Sensors & Actuators B **31** (1-2), 33-38 (1996).
9. N.J. Harrick in *Internal Reflection Spectroscopy*, Interscience, Wiley, New York, 1967. Second printing by Harrick Scientific Corporation, Ossining, N.Y. 1979.

GROWTH KINETICS OF QUANTUM SIZE ZnO PARTICLES

E. M. WONG*, J. E. BONEVICH**, P. C. SEARSON*
*MSE Department, The Johns Hopkins University, Baltimore, MD 21218
**Metallurgy Division, NIST, Gaithersburg, MD 20889

ABSTRACT

Colloidal chemistry techniques were used to synthesize ZnO particles in the nanometer size regime. The particle aging kinetics were determined by monitoring the optical band edge absorption and using the effective mass model to approximate the particle size as a function of time. We show that the growth kinetics of the ZnO particles follow the Lifshitz, Slyozov, Wagner theory for Ostwald ripening. In this model, the higher curvature and hence chemical potential of smaller particles provides a driving force for dissolution. The larger particles continue to grow by diffusion limited transport of species dissolved in solution.

Thin films were fabricated by constant current electrophoretic deposition (EPD) of the ZnO quantum particles from these colloidal suspensions. All the films exhibited a blue shift relative to the characteristic green emission associated with bulk ZnO. The optical characteristics of the particles in the colloidal suspensions were found to translate to the films.

INTRODUCTION

Zinc oxide is of interest for use as a phosphor material in display applications. The recent demand for portable display devices has created interest in materials that can operate efficiently at low voltages. While efficiencies of 15 to 20 % can be obtained above 10 keV for many phosphors, zinc doped zinc oxide (ZnO:Zn) is one of the few systems that has demonstrated reasonable (1 to 5 %) efficiencies at low voltages (10 to 50 eV) [1]. Furthermore, contrary to the case of high-voltage phosphors, where larger particle size is needed for higher efficiency, it has been suggested recently that the cathodoluminescence of small sized phosphors is superior to large sized phosphors under low-voltage excitation [2,3]. Thin films of ZnO have been deposited by a number of techniques, including rf sputtering [4], chemical vapor deposition [5], ionized cluster beam deposition [6], electron cyclotron resonance sputtering [7], pulsed laser deposition [8], and electrodeposition [9]. Most of these techniques require elaborate preparation and the use of costly equipment, which can be contrasted to EPD, where quantum particles are directly deposited from the suspension thereby providing an inexpensive and simple approach that can be performed in ambient conditions.

EXPERIMENT

The ZnO colloidal suspensions were prepared, following the method of Bahnemann and Hoffmann [10]. For all experiments reported here, the colloids were aged for 2 hours and subsequently stored at 5°C to minimize room temperature aging effects.

Particle sizes were determined using a JEOL3010 HRTEM [11]. Samples were prepared by placing a drop of the colloidal suspension on a holey carbon coated 2.3 mm copper grid for 30 seconds. The specimens were allowed to dry by evaporation in air.

Absorption spectra were recorded in ambient conditions using a Shimadzu UV-2101PC scanning spectrophotometer. A tin oxide coated glass slide (Libby Owens Ford) was used as the reference sample for the films and a blank solution of 2-propanol was used for the colloidal suspensions.

Suspensions for EPD were prepared by increasing the particle concentration via rotary evaporation of the colloidal suspension at 25°C to a concentration of 0.004 M. The suspension was then stirred for 30 minutes and maintained in an ice bath during deposition. A tin oxide coated glass slide (2.5cm × 7.5 cm × 1.3 mm) with a deposition area of 3.75 cm^2 was used as the anode and a stainless steel plate (2.5 cm × 10 cm × 0.1 mm) was used as the cathode. The electrodes were placed parallel to each other separated by a distance of 2 cm and immersed into

347

the ZnO suspension. A constant current of 10 mA was applied for a period of up to 2 hours using an HP 6209B power supply. The films were then removed from the solution and allowed to dry at ambient conditions.

RESULTS

Particle Growth Kinetics

In order to study the kinetics of the growth process it is necessary to determine the particle size as a function of time for different aging temperatures. For convenience we use the particle size inferred from the absorption edge. Although this approach has certain limitations, the particle size obtained from the absorption edge was consistently slightly larger than the most probable value obtained from analysis of the HRTEM images and hence represents a reasonable approximation for kinetic analysis. Single optical absorption spectra from four aging temperatures after 2 hours of aging are shown in Figure 1. A red shift in the absorption onset is seen for increasing aging temperatures but equivalent aging times. An HRTEM image of particles

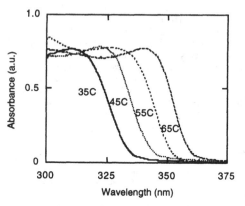

Figure 1. Optical absorbance spectra from four aging temperatures after 2 hours.

from a colloid aged at 35°C for 2 hours is shown in Figure 2 where the particle shown in the inset demonstrates strong facetting with a surface step consisting of a single atomic layer.

The average particle size as a function of time was determined from the absorption spectra using the effective mass model derived by Brus [12]. In the strong-confinement regime, the confinement energy of the first excited electronic state can be approximated by:

$$E^* \cong E_g^{bulk} + \frac{\hbar^2\pi^2}{2r^2}\left(\frac{1}{m_e^* m_o} + \frac{1}{m_h^* m_o}\right) - \frac{1.8e^2}{4\pi\varepsilon\varepsilon_o r} - \frac{0.124e^4}{\hbar^2(4\pi\varepsilon\varepsilon_o)^2}\left(\frac{1}{m_e^* m_o} + \frac{1}{m_h^* m_o}\right)^{-1} \quad (1)$$

where E_g^{bulk} is the bulk band gap, $\hbar$ is Planck's constant, r is the particle radius, m_e^* is the effective mass of electrons, m_h^* is the effective mass of holes, m_o is the free electron mass, e is the charge on an electron, ε_o is the permittivity of free space, and ε is the relative permittivity. The particle size was obtained from the band gap inferred from the optical absorption spectra taking $E_g^{bulk} = 3.4eV$, $m_e^* = 0.24$, $m_h^* = 0.45$, and $\varepsilon = 3.7$ [12].

Coarsening effects due to capillary forces at phase boundaries are generally termed Ostwald ripening [13-15]. Lifshitz and Slyozov [14] and Wagner [15] developed a rigorous mathematical approach to Ostwald ripening, referred to as the LSW theory, in which either mass transport or reaction at the interface is the rate limiting step.

For a species present at a solid/liquid interface, the local equilibrium concentration of the species in the liquid phase is dependent on the local curvature of the solid phase. Differences in the local equilibrium concentrations, due to variations in curvature, set up concentration gradients that lead to transport of species from the regions of high concentration (high curvature) to regions of low concentration (low curvature). These capillary forces provide the driving force for the growth of larger particles at the expense of smaller particles.

Figure 2. HRTEM image of quantum ZnO particles aged 35°C for 2 hours.

The concentration of a species in a liquid phase in equilibrium with a spherical solid particle is given by the Gibbs-Thomson equation:

$$C_r = C_\infty \exp\left(\frac{2\gamma V_m}{RT}\frac{1}{r}\right) \qquad (2)$$

where C_r is the equilibrium concentration for a particle of radius r, C_∞ is the equilibrium concentration at a flat surface, γ is the interfacial energy, V_m is the molar volume of the solid phase, R is the gas constant, and T is the temperature.

For diffusion controlled growth, the flux of a species to a growing particle can be obtained by considering Fick's first law in spherical geometry. For the case where the diffusion length is much greater than the particle radius and considering only first order terms (i.e. small deviations from the bulk concentration), we can write:

$$J = -D\left(4\pi r^2\right)\left(\frac{C_b - C_r}{r}\right) = -D4\pi r\left(C_b - C_\infty - \frac{2\gamma V_m C_\infty}{RT}\frac{1}{r}\right) \qquad (3)$$

where D is the diffusion coefficient and C_b is the bulk concentration at sufficiently large distances from the particle.

The flux of species to a growing particle must obey the conservation of mass such that:

$$J = \frac{1}{V_m}\frac{dV}{dt} = \frac{4\pi r^2}{V_m}\frac{dr}{dt} \qquad (4)$$

The growth law can be obtained by combining equations (3) and (4) and integrating under the appropriate conditions. For a system of highly dispersed particles where the growth is controlled by diffusion, the rate law is given by:

$$\bar{r}^3 - \bar{r}_o^3 = \left(\frac{8\gamma D V_m^2 C_\infty}{9RT}\right)\cdot t = Kt \qquad (5)$$

where $\bar{r}$ is the mean particle radius and $\bar{r}_o$ is the initial particle radius. Figure 3 shows the cube of the particle radius plotted versus aging time for the different aging temperatures. From this figure it can be seen that the particle growth follows the Ostwald ripening growth law.

Further confirmation of the Ostwald ripening model can be obtained by consideration of the constant K. Since γ and V_m can be taken from the literature, and C_r can be determined experimentally, values for the diffusion coefficient, D, can be acquired from the slopes (obtained by linear regression) in Figure 3 and compared to the Stokes - Einstein model. A value for the interfacial energy of 0.24 mJ/m^2 was assumed [16]. The concentration of Zn^{2+} in solution, C_r, for a colloids aged at 65°C for 2 hours was found to be 1.63×10^{-7} mol/L from atomic absorption spectroscopy. Since the concentration difference is relatively small ($2\gamma V_m/RTr \ll 1$), we make the approximation that $C_\infty \approx C_r$. Figure 4 shows the diffusion coefficient obtained from equation (5) versus reciprocal temperature. From this figure it is seen that the diffusion coefficient obtained from the value of K is on the order of 10^{-5} cm^2/sec, consistent with typical values for ions in solution at room temperature [17,18].

The diffusion coefficient for Zn^{2+} obtained from the Ostwald ripening model can be compared to the Stokes-Einstein model for ionic diffusion:

$$D = \frac{kT}{6\pi\eta a} \quad (6)$$

where η is the viscosity of the solvent, a is the hydrodynamic radius of the solute, and kT has the usual meaning. The viscosity for 2-propanol as a function of temperature is known [17] and the hydrodynamic radius for Zn^{2+} in methanol is 0.51 nm [19]. The temperature dependence of the diffusion coefficient obtained from equation (6) is shown in Figure 4. Comparison of the diffusion coefficients obtained from the experimental data using the LSW model and the Stokes-Einstein equation

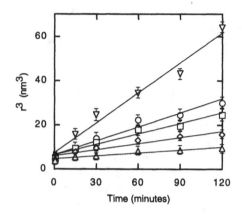

Figure 3. Cube of the particle radius versus aging time for (Δ) 35°C, ($\Diamond$) 45°C, ($\Box$) 55°C, (O)65°C, and (∇)75°C.

shows good agreement, providing further confirmation of the Ostwald ripening model for the growth kinetics of the particles.

The LSW model assumes that ion transport between particles is diffusion limited, however, for colloidal suspensions both diffusion and convection can contribute to ion motion. The relative influence of convective and diffusive transport is expressed by the Peclet number, Pe [20]:

$$Pe = \frac{U_0 L}{D} \quad (7)$$

where U_0 is the characteristic flow velocity, L is the characteristic length along which the major change in concentration takes place, and D is the diffusion coefficient. From equation (7), it can be seen that diffusion is dominant when Pe < 1; conversely, when the Peclet number is large, the concentration distribution is determined by convective transport, and diffusion can be neglected. For our system, the characteristic length corresponds to the interparticle spacing which can be estimated from the particle concentration. For example, at 35°C, taking a particle radius of 2.0 nm, the particle concentration is determined to be ~10^8 particles/cm^3, resulting in an average particle spacing of ~10^{-5} cm. The diffusion coefficient is on the order of 2×10^{-5} cm^2/s and from equation (7) we obtain an upper limit on the flow velocity of 2 cm/s to satisfy the condition that diffusion is dominant

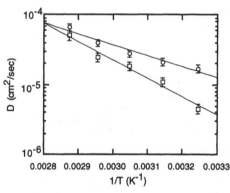

Figure 4. Arrhenius plot of the diffusion coefficient for Zn^{2+} obtained from the LSW model ($\Box$) and the Stokes-Einstein equation (O).

(Pe < 1). Under ambient conditions, the convective velocity is expected to be much less than this value such that molecular diffusion is expected to be the predominant mechanism of mass transport [20].

Film Fabrication and Properties

Electrophoretic deposition involves the motion of charged particles in a suspension under the influence of an applied electric field followed by deposition at the oppositely charged electrode. In constant current EPD, the electric field is maintained constant by increasing the total potential drop between the electrodes. Constant current EPD thus avoids the limited deposition and deposition rate problems of constant voltage EPD.

The films deposited in this study were prepared directly from the concentrated colloidal suspension (without additives) such that deposition occurred at the tin oxide cathode. Constant voltage experiments were found to be limiting in the respect that high quality films could not be obtained and very little mass was deposited even with deposition times approaching several hours. Constant current experiments performed at 10 mA yielded significantly more uniform films with reduced deposition times (~ 1 hour). The films appeared to have uniform coverage over the deposition area and were translucent and white in color. Longer time deposition yielded increasingly opaque films. Optical absorbance spectra of the thin films prepared by EPD from the colloidal suspensions are shown in Figure 5. Here we note that the films exhibit the

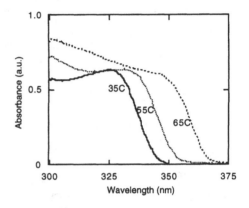

Figure 5. Absorbance spectra of thin films prepared by EPD from colloidal suspensions of quantum particle ZnO.

expected size dependent properties. The slight red-shift in the onset of absorbance as compared to the colloid (seen in Figure 1) is attributed to aging effects occurring during deposition. The optical properties are maintained from the colloid to the films suggesting that EPD does not significantly alter the optical properties of the quantum particles.

CONCLUSIONS

We have shown that the growth law for nanometer size ZnO particles is of the form $\bar{r} = (Kt)^{1/3}$, consistent with Ostwald ripening kinetics. Comparison of the diffusion coefficient obtained from the constant K shows good agreement with the value predicted by the Stokes-Einstein equation. The distribution of particle sizes predicted by the LSW theory was in reasonable agreement with the distributions obtained from analysis of HRTEM images.

Thin phosphor films of quantum particles with diameters from ~40-60Å were prepared by electrophoretic deposition directly from colloidal suspensions of ZnO. The properties of the particles in solution were found to translate to the deposited thin films with a small red shift due to aging effects occurring during deposition.

ACKNOWLEDGMENTS

This work was supported by the US Army Microelectronics Research Collaboration Program. Additional support was provided by the NSF Materials Research Science and Engineering Center on Nanostructured Materials at JHU. EMW acknowledges support from the JHU Materials Initiative Fellowship.

REFERENCES

1. A. Vecht, D.W. Smith, S.S. Chadha, C.S. Gibbons, J. Koh, D. Morton, J. Vac. Sci. Technol. B **12** (2), 781 (1994).
2. J.S. Yoo and J.D. Lee, J. Appl. Phys. **81** (6), 2810 (1997).
3. E.T. Goldburt, B. Kulkarni, R.N. Bhargawa, J. Taylor, M. Libera, Mat. Res. Symp. Proc. **424**, 441 (1997).
4. T. Yamamoto, T. Shiosaki, A. Kawabata, J. Appl. Phys. **51** (6), 3113 (1980).
5. T. Shiosaki, S. Ohnishi, A. Kawabata, J. Appl. Phys. **50** (5), 3113 (1979).
6. K. Matsubara, I. Yamada, N. Nagao, K. Tominaga, T. Tagaki, Surf. Sci. **86**, 290 (1979).
7. M. Kadota, T. Kasanami, M. Minakata, Jpn. J. Appl. Phys. **31**, 3013 (1992).
8. S. Hayamizu, H. Tabata, H. Tanaka, T. Kawai, J. Appl. Phys. **80** (2), 787 (1996).
9. M. Izaki, T. Omi, Appl. Phys. Lett. **68** (17), 2439 (1996).
10. D.W. Bahnemann, C. Kormann, M.R. Hoffmann, J. Phys. Chem. **91**, 3789 (1987).
11. **The use of brand or trade names does not imply endorsement by NIST.**
12. L.E. Brus, J. Chem. Phys. **80** (9), 4403 (1984).
13. G.W.Greenwood, Acta Metall. **4**, 243 (1956).
14. I.M. Lifshitz and V.V. Slyozov, J. Phys. Chem. Solids **19**, 35 (1961).
15. C. Z. Wagner, Z. Elecktrochem. **65**, 581 (1961).
16. D.W. Marr, A.P. Gast, Langmuir **10** (5), 1348 (1994).
17. B. Tremillon, D. Inman, *Reactions in Solution: An Applied Analytical Approach* (John Wiley & Sons, Chichester, England, 1997) p. 471.
18. S.F. Patil, A.V. Borhade, M.J. Nath, Chem. Eng. Data **38**, 574 (1993).
19. R. Lovas, G. Macri, S. Petrucci, J. Am. Chem. Soc. **92** (22), 6502 (1970).
20. V.G. Levich, *Physicochemical Hydrodynamics* (Prentice-Hall, Inc., New Jersey, 1962) pp. 49-52.

SYNTHESIS AND CHARACTERIZATION OF Mn DOPED ZnS QUANTUM DOTS FROM A SINGLE SOURCE PRECURSOR

M. AZAD MALIK,[a] PAUL O'BRIEN,[a*] and N. REVAPRASADU[ab]
a. Department of Chemistry, Imperial College of Science, Technology and Medicine, Exhibition Road, London, SW7 2AZ, UK. b. Department of Chemistry, University of Zululand, Private Bag X1001, Kwadlangezwa, 3886. SA. Email: p.obrien@ic.ac.uk; m.malik@ic.ac.uk; n.revaprasadu @ic.ac.uk

ABSTRACT

Nanoparticles of ZnS and Mn-doped ZnS capped with TOPO (tri-n-octylphosphine oxide) and close to mono-dispersed have been prepared by a single source route using bis(diethyl-dithiocarbamato)zinc(II) as a precursor. The nanoparticles obtained show quantum size effects in their optical spectra and ZnS nanoparticles exhibit near band-edge luminescence. Clear difference in photoluminescence results between ZnS and ZnS:Mn samples. Changes in Mn-doping levels are shown by the changes in photoluminescence intensity. The most intense photoluminescence was observed for 1% and the least intense for 5% doping level. ESR spectra and ICP results confirm the presence of Mn in ZnS quantum dots and also correspond to the amount of Mn in each ZnS:Mn sample.

The Selected Area Electron Diffraction (SAED), X-ray diffraction (XRD) pattern and Transmission Electron Microscopy (TEM) show the material to be of the hexagonal phase. The crystallinity of the material was also evident from High Resolution Transmission Electron Microscopy (HRTEM) which gave well-defined images of nanosize particles with clear lattice fringes.

INTRODUCTION

Semiconductor nanoparticles have attracted considerable attention because of their unique optical properties which are due to quantum size effects. Mn doped ZnS quantum dots have demonstrated an increase in photoluminescent quantum efficiency and enhanced transition rates.[1-4] This is due to the hybridisation of the sp-electron states of ZnS with the d-electron states of Mn., allowing the otherwise spin forbidden, $^4T_1$-$^6A_1$ electronic transition of Mn. The location of the Mn in the dots is important as it effects the optical properties.[5] Manganese doped samples displays orange emission (585 nm) whereas the ZnS dots with Mn on the surface emit in the ultraviolet (435 nm).[5]

Mn doped ZnS have been synthesised via aqueous solution methods,[5,6] in structured media[1,2] and in non structured media where the addition of an organic additive eg. methacrylic acid prevents aggregation of the particles.[4,7] The local structure of these particles have been studied by transmission electron microscopy (TEM), X-ray diffraction (XRD), X-ray absorption fine structure (EXAFS) and electron paramagnetic resonance (EPR).

We report the synthesis and complete characterization of TOPO capped Mn doped ZnS nanoparticles using a novel, stable single source precursor, bis(diethyldithiocarbamato)zinc(II) [Zn(S$_2$CNEt$_2$)$_2$] and manganese dichloride.

EXPERIMENTAL

Chemicals

Bis(diethyldithiocarbamato)zinc(II), tri-*n*-octylphosphine oxide (TOPO), tri-*n*-octylphosphine (TOP, 90%) and manganese(II) chloride were purchased from Aldrich Chemical Company Ltd and methanol, toluene were from BDH. The *bis*(diethyldithiocarbamato)zinc(II) was further purified by recrystallization from toluene and then dried under vacuum. TOPO was purified by vacuum distillation at *ca.* 250 °C (0.1 Torr). The solvents used for air sensitive chemistry were distilled, deoxygenated under a nitrogen flow and stored over molecular sieves (type 4 Å, BDH) before use.

Instrumentation

UV/VIS Absorption Spectroscopy: A Philips PU 8710 spectrophotometer was used to carry out the optical measurements of the semiconductor nanoparticles. The samples were placed in silica cuvettes (1 cm path length). Photoluminescence Spectroscopy: A Spex FluoroMax instrument with a xenon lamp (150 W) and a 152 P photomultiplier tube as a detector was used to measure the photoluminescence of the particles. Good spectral data was recorded with the slits set at 2 nm and an integration time of 1 second. The samples were quantitatively prepared by dissolving 25 mg in 10 ml toluene. The samples were placed in quartz cuvettes (1 cm path length) used as blank for all measurements. The wavelength of excitation was set at a lower value than onset of absorption of a particular sample. X-Ray Diffraction (XRD): X- Ray diffraction patterns were measured using a Philips PW 1700 series automated powder diffractometer using Cu-K (radiation at 40 kV/40 mA) with a secondary graphite crystal monochromator. Samples were prepared on glass slides (5 cm). A concentrated toluene solution was slowly evaporated at room temperature on a glass slide to obtain a sample for analysis. Electron microscopy: A Joel 2000 FX MK 1 operated at 200 kV electron microscope with an Oxford Instrument AN 10000 EDS Analyser was used for the conventional TEM (transmission electron microscopy) images. Selected area electron diffraction (SAED) patterns were obtained using a Jeol 2000FX MK 2 electron microscope operated at 200 kV. The samples for TEM and SAED were prepared by placing a drop of a dilute solution of sample in toluene on a copper grid (400 mesh, agar). The excess solvent was wicked away with a paper tip and completely dried at room temperature. Electron spin resonance (ESR): Measurements were made on a Bruker 200D X-band spectrometer employing 100 kHz modulation, magnetic field markers from an NMR gaussmeter and an external microwave frequency counter.

Synthesis

All experiments were carried out by using dry solvents and standard Schlenk techniques. Glassware was dried in the oven before use and TOPO (tri-*n*-octylphosphine oxide) was used after purification by vacuum distillation

Preparation of ZnS quantum dots

The method used was essentially as described by Trindade and O' Brien.[8,9] [Zn(S$_2$CNEt$_2$)$_2$] (1.0 g) was dissolved in TOP (15 ml) and injected into hot TOPO (20 g). A decrease in temperature of 20 - 30 °C was observed. The solution was then allowed to stabilize at 250 °C and heated for 40 min. at this temperature. The pale white solution was cooled to approximately 70 °C and an excess of methanol added, a flocculant precipitate formed. The precipitate was separated by centrifugation and redispersed in toluene

Preparation of Mn doped ZnS quantum dots

In a typical experiment $[Zn(S_2CNEt_2)_2]$ (2.0 g) and $MnCl_2$ (6.9 mg, 1%) was dissolved in TOP (15 ml). The resulting deep red solution was then injected into hot TOPO (20 g) at 250°C. After an initial drop in temperature to 210°C the temperature was stabilised at 240°C and the reaction was allowed to proceed for 40 minutes. The mixture was cooled to 70°C and methanol was then added to flocculate the nanoparticles. After centrifugation, the supernatant solution was discarded and the pale white precipitate of nanoparticles was washed further with methanol to remove the excess TOPO, followed by dissolution in toluene. The above procedure was repeated for 3% and 5% doping, by varying the amount of $MnCl_2$.

RESULTS AND DISCUSSION

The increase in the band gap of semiconductor nanoparticles due to quantum confinement effects is widely reported.[10-13] The size effects are prevalent when the size of particles are comparable to the Bohr exciton diameter which in the case of bulk ZnS is ca. 5 nm. TOPO capped ZnS nanoparticles synthesized from $Zn(S_2CNEt_2)_2$, an air stable single source precursor exhibits a band edge at 315 nm (3.96 eV) with an absorption shoulder at 330 nm. This represents a blue shift of 0.31 eV in comparison to bulk ZnS (340 nm, 3.65 eV). The band edge of Mn doped ZnS is at 324 nm (3.82 eV) for 1%, and is at 320 nm (3.87 eV) for 3 and 5%.

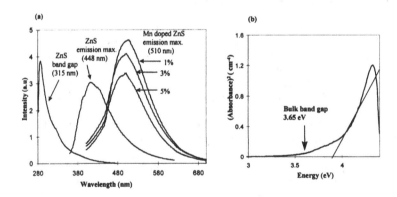

Figure 1. (a) The absorption and emission spectrum of ZnS and Mn doped ZnS (b) Band gap fitted to a direct transition

The relatively small difference in the band gap of the different samples, shows a narrow size distribution which is due to the reaction kinetics of the nucleation and growth phases. The actual level of Mn doping as measured by ICP-AES is 0.72% for the 1% sample, 1.50% for 3% and 4.62% for the 5%.

The diameter of the undoped particles as predicted by the determined by the effective mass approximation (EMA) model is 3.7 nm. The particles are smaller than the excitonic radius in ZnS;

355

the relatively small difference accounts for the absence of any strong excitonic features in the absorption spectrum. The TEM micrograph of the 1% sample shows particles which are monodispersed with an approximate size of 3-4 nm. The lattice fringes visible in the HRTEM micrograph is indicative of the crystalline nature of the particles. The size of the particles is comparable to those particles visible in the TEM and the size calculated from the EMA model.

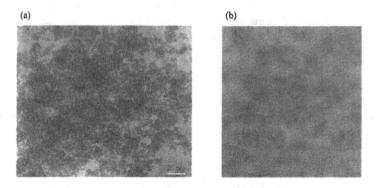

(a)　　　　　　　　　　　　　　　(b)

Figure 2.(a) TEM micrograph showing the Mn doped ZnS (1%) and (b) HRTEM of the same sample showing particles with an approx. diameter of 4 nm.

The XRD pattern of the 1% doped sample shows broad peaks typical for a sample in the nanosize regime. The distinction between the hexagonal and cubic ZnS phases is difficult. However, the peaks in the diffraction pattern can be assigned to the hexagonal phase of ZnS more easily than to cubic with the (002), (110), (103) and (112) planes of the wurtzite phase visible. The selected area electron diffraction pattern exhibits broad diffuse rings typical of particles of these sizes. The (002), (110) and (112) planes are indexed confirming the hexagonal phase. The ZnS nanoparticles synthesized by our method are capped by TOPO which occupies the surface sites resulting in band to band electron–hole recombination. The near band edge emission (448 nm) of the nanocrystalline ZnS is slightly red shifted in relation to its absorption band edge The PL spectrum of all the Mn doped ZnS samples show an emission maximum at 510 nm (λ_{ex} = 380 nm) which is ca. 75 nm blue shifted as compared to those reported previously.[1-7]

There is no change in the wavelength of emission as the levels of the Mn doping changes. The 1% doped sample has the highest intensity emission peak, followed by the 3% and then the 5% sample which is consistent with the results reported by Murphy *et. al.*[5]. There is a clear difference of photoluminescence intensity and emission maximum between ZnS and Mn doped ZnS samples. However the intensity difference between the 5% Mn doped sample and ZnS is negligible. Our PL results are different to those reported by Murphy *et. al.*[5] where in all the samples an emission maximum was observed at 435 nm and 585 nm. In our samples only one emission maximum was observed at 510 nm.

The ESR spectra obtained for the doped samples show (Figure 4) some weak partially resolved signals superimposed on a broad background signal which increases at higher manganese doping (accompanied by a poorer resolution of the weak signals). The weak 6-line spectrum characteristic of the Mn^{2+} can be observed for the 1% sample. The resolution was not greatly improved at low temperatures or for dilute solutions in toluene. It has been suggested that dipole-dipole interactions between the Mn impurities can produce a broadening which gives rise to a poor overall resolution and a broad background signal.[14]

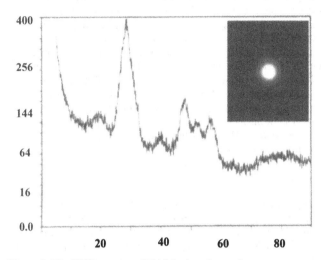

Figure 3. The XRD pattern of 1% Mn doped sample

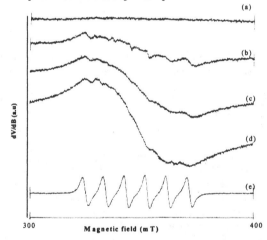

Figure 4. Room temperature EPR spectra of Mn-doped ZnS nanoparticles for ZnS:Mn doping levels of (a) undoped ZnS nanoparticles (b) ZnS:Mn (1%) (c) ZnS:Mn (3%) (d) ZnS:Mn (5%) solution $MnCl_2$ spectrum (in water) to depict some of the hyperfine signals seen for the doped sample.

CONCLUSIONS

Nanoparticles of ZnS and Mn-doped ZnS close to mono-dispersed, capped with TOPO have been prepared by a single source route using *bis*(diethyldithiocarbamato)zinc(II). ZnS nanoparticles show near band-edge luminescence and the optical spectra show quantum size effects in the material. Various levels of Mn-doping were differentiated by the difference in photoluminescence intensity, e.g. ZnS:Mn (1%) showing more intense luminescence than ZnS:Mn (3%) and ZnS:Mn (5%) almost equivalent to ZnS. Photoluminescence show a clear difference between ZnS (448 nm) and ZnS:Mn (510 nm) samples. All samples show a emission max. at 510 nm which is 75 nm lower than those reported previously. ESR spectra and ICP results confirm the presence of Mn in ZnS quantum dots and also correspond to the amount of Mn in each ZnS sample. Transmission Electron Microscopy (TEM), Selected Area Electron Diffraction (SAED), and X-ray diffraction (XRD) pattern show the material to be of the hexagonal phase. The High Resolution Transmission Electron Microscopy (HRTEM) showed the nanosize dots with clear lattice fringes indicating the crystallinity of the material.

ACKNOWLEDGEMENTS

We thank the Royal Society and the FRD (SA) for support to NR and a program of collaboration between UZULU and ICSTM. POB thanks the EPSRC for grant. POB is the Sumitomo/STS Professor of Materials Chemistry and the Royal Society Amersham International Research Fellow (1997/98). We also thank K. Pell (QMW) for Electron microscopy and Dr. A.D. Oduwole (QMW) for EPR spectroscopy.

REFERENCES

1. Y. Wang, N. Herron , K. Moller and T. Bein , Solid State Commun. **77**, p.33 (1991).
2. D. Gallagher, W.E. Heady, J.M. Racz and R.N. Bhargava , J. Crystal Growth **138**, p.970 (1994).
3. Y.L. Soo, Z.H. Ming, S.W. Huang, Y.H. Kao, R.N. Bhargava and D. Gallagher, Phys. Rev. **B50**, p.7602(1994).
4. R.N. Bhargava, D. Gallagher, X. Hong and A. Nurmikko , Phys. Rev. Lett. **72**, p.416 (1994).
5. K. Sooklal, B.S. Cullum, S.M.Angel, C.J.Murphy, J. Phys. Chem. **100**, p.4551 (1994).
6. A.A. Khosravi, M. Kundu, P.D. Vyas, S.K. Kulkarni, G.S. Shekhawat, R.P. Gupta, A.K. Sharma., Appl. Phys. Lett., **67(17)**, p.2506 (1995).
7. I. Yu, T. Isobe and M. Senna, J. Phys. Chem Solids, **57**, p.373 (1996).
8. T. Trindade and P O'Brien, Adv. Mater., **8**, p.161 (1996).
9. T. Trindade and P. O'Brien, Chem. Mater.,**9**, p.523 (1997).
10. D. Duongong, J. Ramsden, M. Gratzel, J. Am. Chem. Soc.,**104**, p.2977 (1982).
11. R. Rossetti, J. L. Ellison, J. M. Gibson, L. E. Brus, J. Chem. Phys., **80**, p.4464 (1984).
12. A. Henglein, Chem. Rev., **89**, p.1861 (1989).
13. M. L. Steigerwald, L. E. Brus, Acc. Chem. Rev., **23**, p.183 (1990).
14. E. E. Schneider and T. S. England, Physica, **17**, p.221 (1951).

IN SITU DIAGNOSTICS OF NANOMATERIAL SYNTHESIS BY LASER ABLATION: TIME-RESOLVED PHOTOLUMINESCENCE SPECTRA AND IMAGING OF GAS-SUSPENDED NANOPARTICLES DEPOSITED FOR THIN FILMS

D. B. GEOHEGAN*, A. A. PURETZKY*, A. MELDRUM*, G. DUSCHER**, and S. J. PENNYCOOK*
*Solid State Division, Oak Ridge National Laboratory, Oak Ridge, TN 37831-6056 odg@ornl.gov
** MPI für Metallforschung, Institut für Werkstoffwissenschaft, Seestr. 92, D-70174 Stuttgart

ABSTRACT

The dynamics of nanoparticle formation by laser ablation into background gases are revealed by gated-ICCD photography of photoluminescence (PL) and Rayleigh-scattering (RS) from gas-suspended nanoparticles. These techniques, along with gated-spectroscopy of PL from isolated, gas-suspended nanoparticles, permit fundamental investigations of nanomaterial growth, doping, and luminescence properties *prior to deposition* for thin films. Using the time-resolved diagnostics, particles unambiguously formed in the gas phase were collected on TEM grids. Silicon nanoparticles, 1–10 nm in diameter, were deposited following laser ablation into 1–10 Torr Ar or He. Three *in situ* PL bands (1.8, 2.6, 3.2 eV) similar to oxidized porous silicon were measured, but with a pronounced vibronic structure. Structureless photoluminescence bands were reproduced in the films (2.1, 2.7, 3.2 eV) only after standared annealing. The ablation of metal zinc into Ar/O_2 is also reported for the preparation of < 10 nm diameter hexagonal zincite nanocrystals, The particles were analyzed by bright field and Z-contrast TEM and high resolution EELS.

INTRODUCTION

Films incorporating silicon nanoparticles are among the most promising optoelectronic materials for applications requiring room-temperature photoluminescence and compatibility with existing silicon processing technology. These 1–10 nm-diameter particles are highly desirable light-emitting materials since quantum confinement (QC) of electron-hole pairs within a nanoparticle can result in bright luminescence from new or previously forbidden optical transitions.[1] Porous silicon (*P*-Si) luminescence is an example of this effect, in which the near-infrared emission from nanometer-sized quantum wires of crystalline Si (*c*-Si) has been traced to QC in quantum dots, i.e. particles of nanocrystalline silicon (*nc*-Si).[2] Thin films containing photoluminescent *nc*-Si and silicon-rich silicon oxide (SRSO) nanoparticles have recently been synthesized by laser ablation[3-7] in variations of the pulsed laser deposition (PLD) technique.[8] Electroluminescent light-emitting diodes (LED's) have now been fabricated from laser ablation-produced *nc*-Si films[9] and surface-emitting LED's based on oxidized *P*-Si have already been successfully integrated into microelectronic circuits.[10]

Controllable gas-phase synthesis of nanocrystals, nanotubes, and nanowires by laser ablation of solid targets into background gases has been hampered by a lack of knowledge of the spatial and temporal scales for clustering and nanoparticle growth. Until recently very little was known about these dynamics, or how the nanoparticles are transported and deposited after their formation. It is often unclear whether nanoparticles found on substrates were grown in the gas phase or from nuclei formed on the substrate surface.[5-7] Only a few *in situ* measurements of gas-formed nanoparticles have recently been employed using different techniques.[3-7,11-13] and considerable controversy still exists regarding the time-scales and dynamics for initial cluster formation, the establishment of critical nuclei, and nanoparticle growth in expanding laser ablation plumes (into vacuum or into background gases).

Here the synthesis and properties of silicon and zinc oxide nanoparticles formed by laser ablation are directly visualized *in situ* with a comparison of measurements involving Rayleigh-scattering (RS) and laser-induced photouminescence (PL) and *ex situ* with a combination of bright-field and Z-contrast transmission electron microscopy (TEM), electron diffraction, and high resolution electron energy loss spectroscopy (HREELS) analysis of individual nanoparticles,[12,13] These techniques form a powerful ensemble to understand nanoparticle formation and doping in gas-phase methods such as laser ablation. Rapid feedback should permit desired PL characteristics to be rapidly surveyed and optimized during synthesis, *prior to deposition*.

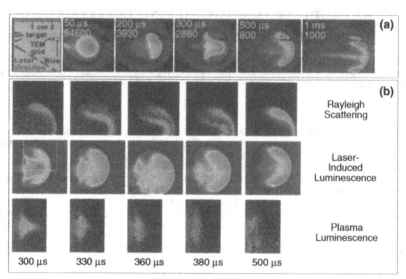

Fig. 1 - (a) Gated-intensified CCD-array (ICCD) photographs of laser-induced luminescence from clusters and nanoparticles as they form and grow following KrF-laser ablation of silicon into 10-Torr He. Photoluminescence from nanoparticles is induced and photographed (3-μs exposures) 50-ns after firing an XeCl sheet beam vertically through the plume at the indicated times after laser ablation. Light from the bright plasma dominates for Δt < 150 μs. For Δt = 200-400 μs, the plasma luminescence extinguishes and two regions of clusters are observed. For Δt >500 μs, surviving nanoparticles are observed on the front of the expanding plume. **(b)** Series of images as above compare spatial distributions of (top panel) Rayleigh scattering (308-nm light, scattered at 90°) from the top half of the plume [see dashed-line box] (middle panel) long-lived laser-induced photoluminescence, and (lower panel) residual plasma luminescence, which is approximately 1/10 as intense as the laser-induced luminescence. The brightest luminescence comes from the core of the plume, from clusters which are too small to scatter significant light by Rayleigh scattering (measured 1-10 nm as collected on the TEM grid at the position in (a) above)

EXPERIMENTAL RESULTS AND DISCUSSION

Polished c-Si wafers or zinc targets were ablated by KrF-laser pulses (λ = 248 nm, focused to energy densities of 5–8 J/cm² in 28-ns FWHM pulses) inside a large, turbopumped vacuum chamber (1 x 10⁻⁶ Torr base pressure) maintained at 1–15 Torr with pure (or mixtures of) 99.9999 % He , 99.9995 % Ar, 99.99% O_2 with variable flow controllers (0–1000 sccm).

After laser ablation, typical laser plasmas emit bright recombination luminescence from excited atoms and molecules which can be photographed at different time delays with the ICCD camera system (5 ns – 3 μs exposures). Figure 1(a) shows a minimal sequence of these digital photographs for the laser ablation of silicon into flowing He at 10 Torr. Within Δt = 20 μs, collisions with the background gas rapidly decelerate the plume of silicon vapor from an initial velocity of 2 cm/μs (kinetic energy/atom = 58 eV) to only 0.01 cm/μs (0.0015 eV).

To investigate Rayleigh scattering (RS) and photoluminescence (PL) from clusters suspended in the gas phase, a sheet of light from another pulsed laser (XeCl, λ = 308 nm, hv = 4.0 eV, 30 ns FWHM pulse, ~ 0.2 J/cm²) illuminated a 1.5-mm-wide cross section of the silicon plume from below.[12] Probe laser light scattered by nanoparticles (with estimated diameters larger than 2 nm)[13] was imaged by gating the camera on *during* the XeCl laser pulse. Alternatively, the camera was gated on *after* the probe laser pulse to capture images of long-lived PL.

In Fig. 1(a) the laser-induced PL is imaged along with the plasma recombination lumines-

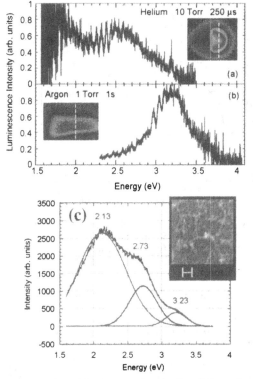

Fig. 2 - **(a,b)** Gated photoluminescence spectra from XeCl-laser-excited, gas-suspended nanoparticles 2-cm away from a laser-ablated Si target in **(a)** 10 Torr Helium, 250 μs after the ablation laser, and **(b)** 1 Torr argon, 1 second delay. Ten-shot averaged spectra, 2.6-nm resolution, 3-μs gate (starting 50-ns after the XeCl excitation pulse) using a gated-intensified diode array (Princeton Inst.) on a 0.3-m monochromator (Acton VM-503, 600 grooves/mm grating). In this case, the sheet of XeCl-light was turned parallel to the target surface.

(c) Strong RT PL spectra (CW, 325 nm, He-Cd laser excitation) from a weblike aggregate nanoparticle film (see inset) prepared in 1-Torr Ar/5%-O_2, collected on L-N_2 cooled Si substrate, and annealed at 800°C in O_2 (no PL) and then 1000°C for 30-min in Ar/5%-H_2 gas. *[spectrum courtesy of T. Makimura group, Univ. of Tsukuba]*

cence (which rapidly drops during the first 400 μs after ablation). By comparing images with and without the excitation pulse in Fig. 1(a), the laser-induced PL could be first discerned at $\Delta t = 150–200$ μs, near the front of the expanding plume. As shown in Figure 1(b), during $\Delta t = 300–500$ μs, the intensity of RS and PL increases at the front of the plume while a central core luminescence suddenly appears as the brightest region of PL in the images at $\Delta t = 300$ μs, and then collapses to the center of the plume, following the extinguishing core of plasma luminescence. This central core of bright luminescence originates from particles smaller than our RS detection threshold (determined by particle collection to be ~ 2 nm diameter), i.e. clusters. The 200 μs onset measured for the appearance of these clusters agrees with laser-induced fluorescence measurements which determined a 200 μs onset for silicon dimerization during ablation into 5 Torr He by Muramoto et al.[11]

For times > 0.5 ms, the nanoparticles segregate into a forward-going (v ~ 10 m/s), rotating ring, resembling a 'smoke ring' with a weak central trail of nanoparticles also visible. These dynamics, and the completely different dynamics resulting in a relatively uniform and stationary cloud of nanoparticles following ablation into heavier argon, are detailed in Ref. 12.

Three broad PL bands (at 1.8, 2.6, and 3.2 eV) were measured for gas-suspended nanoparticles generated by both Si/He and Si/Ar ablation. In general, the three bands occur simultaneously with similar measured decay lifetimes of ~ 1.8 μs.[12] These bands agree with those observed for oxidized nc-Si films and porous silicon.[15] The violet band is clearly linked to the presence of trace impurities of oxygen (from EELS analysis, see below) in the chamber, from the target, or in the ultrahigh purity gases used. This violet luminescence could be maximized by adjusting the flow of gas in the chamber (for Si/Ar, thereby controlling the mixing of the stationary nanoparticle cloud with oxygen-laden gas).[13] In argon, small admixtures of oxygen in the gas flow quenched the violet PL, as did stopping the flow. The maximally luminescent SiO_x stoichiometry was investigated by HREELS (see below). In helium, the turbulent dynamics resulted in

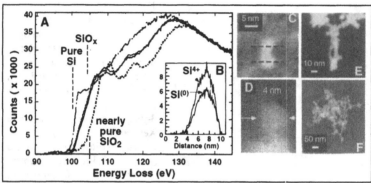

Fig.3 - (a) High-resolution transmission-EELS spectra of individual nanoparticles [collected on TEM grids as in Fig. 1(a)], and imaged by Z-contrast TEM, show a variation of stoichiometries ranging from pure silicon to nearly-pure SiO_2. (b) Translating the 0.27-nm electron beam across individual nanoparticles (as indicated in (c)) revealed that both the pure-silicon ($Si^{(0)}$) and pure-SiO_2 (Si^{4+}) regions are uniformly distributed throughout the 5-nm particles. Z-contrast TEM images showed 1–10 nm particles (c) aggregated in pairs, or (d) single particles for collection early in the expansion as in Fig. 1(a), or aggregated into (e) three-dimensional structures and (f) large networks of ~ 100–200 nm size for collection after many laser shots.

oxidation (and the growth of violet luminescence) for times > 300 μs after ablation, regardless of flow. However, the red and blue-green bands of luminescence appear together early in the expansion, and did not correlate with the extent of oxidation. Figure 2(a) shows these 1.8 and 2.6 eV PL bands measured at 250 μs. Some nearly-pure Si nanoparticles were collected on TEM grids placed at distances from the target which corresponded to the red and blue-green bands (as indicated for the position of the TEM grid in Fig. 1(a)).

This luminescence disappeared after deposition. Weblike films consisting of these aggregated nanoparticles required standard passivation and annealing to exhibit RT PL (Fig. 2(c)). Similar bands (slightly red-shifted red 2.1 eV, blue-green 2.7 eV, and violet 3.2 eV) were retained.

The high-resolution PL-spectra of Fig. 2 reveal a pronounced oscillatory structure decorating the low-energy sides of each of the three PL bands from the gas-suspended nanoparticles. This structure was reproducible at different time delays. In fact, the entire spectrum could be deconvoluted as a sum of Gaussian lineshapes.[13] For the 3.2 eV band in Fig. 2(b), the spacing is quite uniform (62 ± 6 meV) across the spectral region. This roughly agrees with the spacing observed for the 2.6 eV band of Fig. 2(a) (57 ± 5 meV). The 1.8 eV band appeared to contain an additional vibronic series (only an average interpeak spacing of 34 ± 6 meV was derived).[13]

Numerous reports of both irregular and regular spectral features decorating PL from nc-Si and P-Si are reviewed in Ref. 13. However, the most likely explanation for the regular vibronic structure in Fig. 3 is photoexcitation of Si-Si vibrations at localized surface states, or from unassigned transitions in highly vibrationally-excited Si_2 which is weakly bound or photodissociated from the nanoparticle surface. In support of surface state luminescence, Kimura and Iwasaki recently reported clear vibronic structure decorating broad ultraviolet and blue PL bands from isolated (4–10 nm) silicon nanocrystals suspended in organic liquids at room temperature. Unlike the apparent Si-Si vibronic modes in our measurements, Si-O-Si and other modes were found, indicating that the vibronic fine structure originated from a localized surface state.[16] Orii et al recently observed similar vibronic luminescence with increasing fluence during PL in the CdSe nanocrystal system which they attributed to photodissociated Se_2 dimers.[17]

For c-Si ablation into 1 Torr Ar or 10 Torr He, spherical 5 nm diameter nanoparticles were collected (2 cm from the target) with other sizes ranging from 1–10 nm. For deposits collected after 1–10 laser shots, individual and or small groups of nanoparticles (as in Fig. 3(c,d)) were found. For deposits collected after hundreds of laser shots, groups of ~ 5 nm particles were also

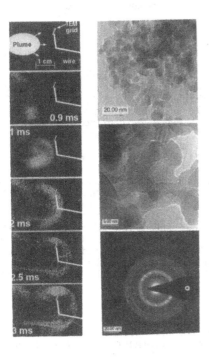

Fig.4 - (left) Gated-ICCD-photographs of XeCl-laser induced luminescence from 5-20 nm ZnO nanoparticles formed when a 2"-target of Zn metal (left of frame) is KrF-laser-ablated into a 7.5 Torr Ar/O_2 mixture (5 Torr Ar / 2.5 Torr O_2). The position of the TEM collection grid is shown in the images. These PL images (and RS images, not shown) indicate that clusters of 5–20 nm size require 1–3 ms to grow.

(right) Bright field TEM images and electron diffraction patterns reveal crystalline ZnO nanoparticles, 5–20 nm in size, of hexagonal zincite.

found in aggregates of up to 100–300 nm dimensions (Figs. 3(e,f)). These aggregates interlinked to form a web-like microstructure very similar to those reported by El-Shall et al.[4] I n d i v i d u a l nanoparticles were examined for composition and nanostructure with HREELS (0.27 nm spatial resolution, energy resolution ~ 0.8 eV). The intensity and ELNES (energy loss near edge structure) were examined at the Si-L absorption edge in Si (99 eV) and SiO_2 (102 eV), and at the O–K absorption edge to determine the nanoparticle composition. As shown in Fig. 3(a), particles found with stoichiometries between nearly-pure Si to nearly-pure SiO_2.

Compositional profiles of individual SiO_x nanoparticles were acquired by translating the STEM beam (as indicated in Fig. 3(c)) while ~100 ELNES spectra were recorded. These HREELS linescans showed (as in Fig. 3(b)) a uniform distribution of $Si^{(0)}$ bonding (as in pure silicon) and $Si^{(4+)}$ bonding (as in pure SiO_2). The particles appear to be composed of a homogeneous mixture of pure and oxidized Si regions, which contrasts with surface-oxidized Si nanoparticles produced and analyzed in other studies.[1] Although the TEM resolution was sufficient to resolve lattice fringes from c-Si by Z-contrast imaging, repeated attempts were unsuccessful, indicating either amorphous silicon or c-Si regions in the sub-1 nm size range (i.e. clusters of c-Si < 25 atoms).

The ex situ stability of as-deposited nearly-pure silicon nanoparticles indicates that oxidation leading to the overall stoichiometry is accomplished in the gas phase. In all cases where violet gas phase PL was observed, SRSO nanoparticles were collected. Introduction of oxygen-containing impurities (oxygen, water vapor) to condensing Si nanoparticles was controlled by increasing the Ar flow. This condition correlated with an overall ex situ $SiO_{1.4}$ stoichiometry.

These same techniques have now been applied for the growth of crystalline Ge, SiGe, ZnO, Y_2O_3, and MgO nanoparticles. Figure 4 shows laser-induced luminescence images which trace the appearance, growth, and propagation of ZnO nanoparticles following KrF laser ablation of Zn metal into a 7.5 Torr Ar/O_2 mixture. Direct imaging of the particle collection process shows that particles were collected on the TEM grid from 2–3 ms after ablation. After 20 laser shots, ex situ bright-field TEM images and electron diffraction measurements shown in Figure 4 reveal that aggregated 5–20 nm diameter hexagonal zincite ZnO nanocrystals were formed.

CONCLUSIONS

In situ imaging of PL and RS from nanoparticles formed by laser ablation into background gases reveals long times for nanoparticle growth (Si: $\Delta t = 3$ ms in 1 Torr Ar, $\Delta t = 200$ μs in 10 Torr He, ZnO: 1-3 ms in 7.5 Torr Ar/O_2). In situ PL spectroscopy of Si nanoparticles reveals a changing spectrum, with the additional 3.2 eV band appearing and growing to dominate the 1.8 eV and 2.6 eV bands according to the extent of oxidation in the plume (as analyzed by ex situ HREELS and ELNES). Maximal violet PL correlated to a $SiO_{1.4}$ stoichiometry in nanoparticles of uniform composition. The brightest gas-phase PL for Si/He in Figs. 1, 2(a) correlated to clusters too small to detect by RS (< 2 nm). The size-independence of the broad PL spectra and apparent vibronic structure, coupled with the measured distribution of particle sizes, suggest a surface luminescence origin. Nanoparticle-aggregate films were deposited which required standard passivation annealing before bright photoluminescence could be obtained. The film PL exhibited bands similar to the gas-suspended nanoparticles, but without vibronic structure. The fact that SiO_x nanoparticles, when isolated in the background gas, did show strong luminescence demonstrates the critical role of isolation.[18] Oxidation of nanocrystals is necessary as a means of isolation in thin films, however partial oxidation accompanied by spatial isolation in the gas phase appears to produce the same result. Crystalline ZnO nanoparticles (< 20 nm) were similarly detected following formation in 7.5 Torr Ar/O_2 and deposited 1-3 ms following laser ablation of Zn metal. The authors gratefully acknowledge valuable research assistance by T. Makimura, C.W. White and B.C. Sales. This research was sponsored by the Oak Ridge National Laboratory, managed by Lockheed Martin Energy Research Corp., for the U.S. Department of Energy, under contract DE-AC05-96OR22464.

REFERENCES

1. W. L. Wilson, P. F. Szajowski, L.E. Brus, *Science* **262**, 1242 (1993).
2. S. Schuppler, S. L. Friedman, M. A. Marcus, D. L. Adler, Y.-H. Xie, F. M. Ross, Y. J. Chabal, T. D. Harris, L.E. Brus, W.L. Brown, E. E. Chaban, P. F. Szajowski, S. B. Christman, and P. H. Citrin, *Phys. Rev. B* **52**, 4910 (1995).
3. (a) L.A. Chiu, A. A. Seraphin, and K.D. Kolenbrander, *J. Electronic Materials* **23**, 347 (1994). (b) E. Werwa, A. A. Seraphin, L.A. Chiu, C. Zhou, and K.D. Kolenbrander, *Appl. Phys. Lett.* **64**, 1821 (1994).
4. (a) M.S. El-Shall, S. Li, and T. Turkki, D. Graiver, U.C. Pernisz, M.I. Baraton, *J.Phys. Chem.* **99**, 17805 (1995). (b) S. Li, S.J. Silvers, and M. S. El-Shall, *J. Phys. Chem. B*, **101**, 1794 (1997).
5. I.A. Movtchan, W. Marine, R.W. Dreyfus, H.C. Le, M. Sentis, and M. Autric, *Appl. Surf. Sci.* **96-98**, 251 (1996).
6. (a) T. Yoshida, S. Takeyama, Y. Yamada, and K. Mutoh, *Appl. Phys. Lett.* **68**, 1772 (1996). (b) Y. Yamada, T. Orii, I. Umezu, S. Takeyama and T. Yoshida, *Jpn. J. Appl. Phys.* **35**, 1361 (1996).
7. T. Makimura, Y. Kunii, and K. Murakami, *Jpn. J. Appl. Phys.*, **35** 4780 (1996).
8. (a) *Pulsed Laser Deposition of Thin Films*, Ed. by D. B. Chrisey and G. K. Hubler, (Wiley-Interscience Publisher), 1994., (b) D.H. Lowndes, D. B. Geohegan, A. A. Puretzky, D. P. Norton, and C.M. Rouleau, *Science* **273**, 898 (1996).
9. T. Yoshida, Y. Yamada, and T. Orii, *Technical Digest of the International Electron Devices Meeting*, San Francisco, CA, Dec. 8-11, 1996, IEEE.
10. K.D. Hirschman, L. Tsybeskov, S.P. Duttagupta, and P.M. Fauchet, *Nature* **384**, 338 (1996).
11. J. Muramoto, Y. Nakata, T. Okada and M. Maeda, *Jpn. J. Appl. Phys.* **36** L563 (1997).
12. D. B. Geohegan, A. A. Puretzky, G. Duscher, and S. J. Pennycook, *Appl. Phys. Lett.* **72**, 2987 (1998) and references cited therein..
13. D. B. Geohegan, A. A. Puretzky, G. Duscher, and S. J. Pennycook, *Appl. Phys. Lett.* **73**, 438 (1998) and references cited therein.
14. H.C. van de Hulst: *Light Scattering by Small Particles* (Dover Publications, New York, 1981).
15. Broad reviews are given by (a) P. M. Fauchet, *J. Lumin.* **70**, 294 (1996). (b) F. Koch, V. Petrova-Koch, *J. Non-Cryst. Solids* **198-200**, 846 (1996).
16. K. Kimura and S. Iwasaki, *Mat. Res. Soc. Proc.*, **452**, 165 (1997).
17. T. Orii, S. Kaito, K. Matsuishi, S. Onari, and T. Arai, J. Phys. D: Condesn. Matter **9**, 4483 (1997).
18. Louis Brus, *J. Phys. Chem.* **98**, 3575 (1994.

THIOL-CAPPED CdSe AND CdTe NANOCLUSTERS: SYNTHESIS BY A WET CHEMICAL ROUTE, STRUCTURAL AND OPTICAL PROPERTIES

A.L. ROGACH[*+], A. EYCHMÜLLER[*], J. ROCKENBERGER[*&], A. KORNOWSKI[*],
H. WELLER[*], L. TRÖGER[%], M.Y. GAO[§], M.T. HARRISON[#], S. V. KERSHAW[#],
M.G. BURT[#]

[*]Institute of Physical Chemistry, University of Hamburg, 20146 Hamburg, Germany
[+]Physico-Chemical Research Institute, Belarussian State University, 220050 Minsk, Belarus
[&]Present address: University of California, Berkeley, Department of Chemistry, Berkeley, CA 94720, USA
[%]Hamburger Synchrotronstrahlungslabor HASYLAB, DESY, 22603 Hamburg, Germany
[§]Max-Planck-Institute of Colloids and Interfaces, 12489 Berlin, Germany
[#]BT Laboratories, Martlesham Heath, Ipswich, Suffolk, IP5 3RE, UK

ABSTRACT

CdSe and CdTe nanoclusters were formed in aqueous solutions at moderate temperatures by a wet chemical route in the presence of thiols as effective stabilizing agents. The nature of the stabilizing agent (thioalcohols or thioacids) had an important influence on the particle size and largely determined the photoluminescence properties. The nanoclusters were characterized by means of UV-vis absorption and photoluminescence spectroscopy, powder X-ray diffraction, high resolution transmission electron microscopy, and extended X-ray absorption fine structure measurements. CdSe and CdTe nanoclusters were crystalline, in the cubic zincblende phase, with mean sizes in the range of 2 to 5 nm depending on the preparative conditions and the post-preparative size-selective fractionation, and showed pronounced electronic transitions in the absorption spectra. Thioglycerol–stabilized CdTe nanoclusters possessed sharp band–egde photoluminescence being tunable with particle size.

INTRODUCTION

The properties of semiconductor nanoclusters whose size is smaller than the dimension of the respective bulk exciton are largely determined by their size and surface chemistry. Semiconductor nanoparticles showing a number of unique structural and optical features [1-4] have become the object of investigation in one of the most rapidly growing branches of chemistry and physics in the last decade. Generally the small particle research includes three main topics: the synthesis of nanoparticles with narrow size distribution and controlled surface properties, their precise structural and optical characterization followed by the utilization of their unique properties in nanotechnology. In this communication we report on the wet chemical synthesis and the structural and optical properties of a series of CdSe and CdTe nanoclusters with extremely small sizes stabilized with thioalcohols or thioacids.

EXPERIMENT

Aqueous colloidal solutions of CdSe and CdTe nanoclusters have been synthesized through the addition of freshly prepared oxygen-free NaHSe or NaHTe solutions to N_2-saturated $Cd(ClO_4)_2 \cdot 6H_2O$ solutions at pH 11.2 in the presence of different thiols (RS) as stabilizing agents following the method of Ref [5]. The molar ratio of $Cd^{2+}:Se^{2-}(Te^{2-}):RS^-$ was chosen as 1:0.5:2.4. The particle size was controlled by the duration of the heat treatment, through post-preparative size-selective fractionation, and by the nature of the stabilizing agent (2-mercaptoethanol, 1-thioglycerol or thioglycolic acid). The size-selective precipitation technique [6] was applied for the post-preparative nanoparticle fractionation into a series of CdSe and CdTe nanocrystals with narrow size distributions. This technique allows the gram scale preparation of redispersible cluster powders which can be handled like ordinary chemical substances.

RESULTS

The crystal structure and particle sizes for CdSe and CdTe samples were obtained from powder X-ray diffractograms and HRTEM images. Figure 1 shows typical XRD patterns of CdSe and CdTe nanoparticles with different sizes. For both CdSe and CdTe nanoparticles a predominant cubic (zincblende phase) crystalline structure could be derived from the diffractograms. The broadness of the diffraction peaks increased gradually with a decrease of particle size. A reflection maximum appeared also in the small-angle region due to a periodicity of the cluster arrangement which is a confirmation of the narrow size distribution of the particles. The mean particle sizes obtained from the diffractograms using the Bragg (small angle region) and Scherrer (wide angle region) equations laid between 2 and 5 nm.

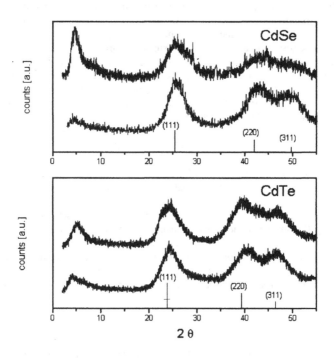

Figure 1. X-ray diffractograms of CdSe and CdTe nanoclusters with different sizes. In each panel: top spectrum – smaller, bottom – larger nanoparticles. The line spectra give the bulk CdSe and CdTe zincblende reflections with their relative intensities.

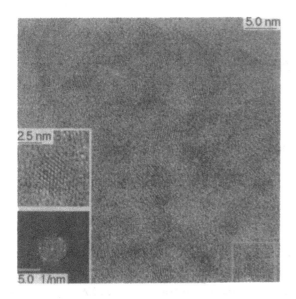

Figure 2. HRTEM image of CdSe nanoclusters. Inserts show a single CdSe particle with a corresponding FFT.

Figure 2 shows a HRTEM image of the nanoparticles (here: CdSe) with a corresponding Fast Fourier Transformation (FFT). The existence of the lattice planes in HRTEM images further confirmed the crystallinity of CdSe and CdTe nanoparticles. The average sizes estimated from HRTEM micrographs is generally larger than those obtained from XRD patterns. The reason could be an overestimation of the relative content of larger particles which are better seen in TEM.

As has been shown recently on a series of CdS colloids and colloidal crystals, EXAFS provides a valuable tool to determine structural and dynamical properties of nanoclusters [7]. Therefore, temperature-dependent EXAFS measurements were performed on 2-mercaptoethanol-stabilized CdTe nanocrystals in order to study properties of the CdTe core and the Cd-SR shell separately. Figure 3 shows the Fourier transforms of the EXAFS spectra of thiolcapped CdTe nanoclusters taken at the Cd K-edge and at the Te K–edge at 8K. The observed splitting of the first coordination shell into two contributions clearly indicates that the coordination of the Cd atom consists of Te atoms (from the cluster core) and S atoms (from the ligand shell). Attempts to fit this splitted peak by replacing sulfur by oxygen or by using an additional oxygen coordination resulted in poor fits with non-reliable fit parameters. In contrast, the respective Fourier transform taken at the Te K-egde shows only one coordination of tellurium which is identified as a Cd shell.

Based on the coordination numbers and the Cd/Te-ratio determined by EXAFS, the formula $[Cd_{54}Te_{32}(SCH_2CH_2OH)_{52}]^{8-}$ was suggested for these extremely small (2.5 nm in diameter) CdTe nanoclusters [8]. The underlying structure is a tetrahedral CdTe core (zincblende) partially coated by a Cd-SR surface layer.

The interatomic distance of the Cd-Te bonds was slightly contracted in CdTe nanoparticles with respect to the bulk material whereas a significant expansion of Cd-S bonds is observed. An isotropic model of elastic strain distribution in a spherical core-shell system of a zincblende CdTe core, heteroepitaxially overgrown by a CdS monolayer described the experimental results reasonably well [8].

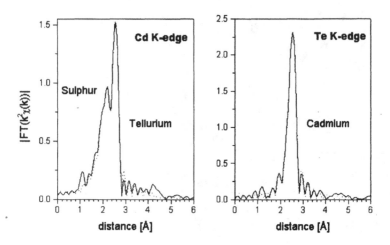

Figure 3. Fourier transforms (measured curves - solid, fits - dashed lines) of the EXAFS spectra of 2-mercaptoethanol-stabilized CdTe nanoclusters at 8 K at the Cd K-edge and at the Te K-edge.

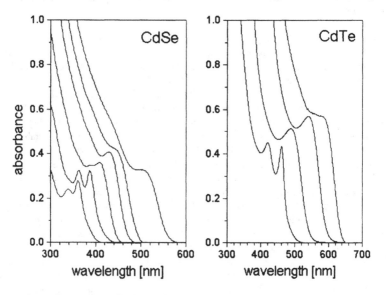

Figure 4. Absorption spectra of CdSe and CdTe nanoclusters with different sizes.

All nanoparticles synthesized were in the size quantization regime and showed a well-developed maximum near the absorption onset which was ascribed to the first excitonic transition (Figure 4). In the case of the smallest CdSe and CdTe nanoclusters transitions at higher energy were also observed. The size dependent shift of the bandgap energies was described reasonably well by a finite depth potential well model in the framework of the effective mass approximation [9].

Both CdSe and CdTe nanoparticles possessed photoluminescence. All CdSe nanoparticles showed only a relatively weak broad emission strongly red-shifted from the absorption onset which is practically independent of the particle size ("trapped" photoluminescence). The smallest (2–mercaptoethanol stabilized) CdTe nanoparticles also showed the trapped photoluminescence only, which coincides with the proposed EXAFS–based model of a CdTe tetrahedral core partially coated by a Cd-SR layer with some uncoordinated Te atoms on the surface providing hole trap states [8].

In contrast, CdTe nanoparticles capped with 1–thioglycerol or thioglycolic acid possessed, besides a weak trapped emisson, a sharp emission near the absorption onset ("excitonic" photoluminescence) which was tunable through the visible spectral range from green to red (Figure 5) with particle size. The quantum yield of this excitonic emission at room temperature was approximately 3–5 %. It was found for CdTe nanoparticles stabilized with thioglycolic acid that the photoluminescence efficiency strongly depends on the pH value of the colloidal solution [10]. The maximum quantum yield at room temperature increased up to 18 % when the pH value of the CdTe solution was brought to 4.5. The optical spectroscopy studies implied that the pH-dependent behavior of the CdTe nanocrystals' photoluminescence was caused by structural changes on the surface. In the acidic range a shell of cadmium-thiol complexes was formed around the CdTe core causing the drastic enhancement of the photoluminescence quantum yield.

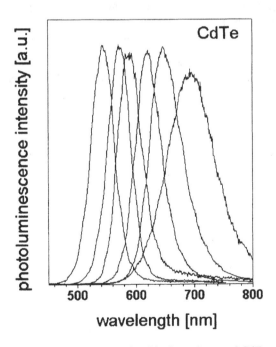

Figure 5. Luminescence spectra of 1–thioglycerol–capped CdTe nanoclusters with different sizes.

CONCLUSIONS

We report on a recently developed method for the synthesis, and on the structural and optical properties of thiol–stabilized CdSe and CdTe nanoclusters in the size–quantization regime. Powder X-ray diffraction, HRTEM and EXAFS measurements provide valuable tools for the structural characterization of nanosized samples. The size–dependent photophysical properties of CdTe nanoclusters make them promising candidates for application in photovoltaic devices.

ACKNOWLEDGMENTS

This work was supported by the Volkswagen Foundation, Hannover, and in part by a Collaborative Research Grant of the NATO Scientific and Environmental Affairs Division.

REFERENCES

1. H. Weller, Angew. Chem. Int. Ed. Engl. **32**, 41 (1993).
2. H. Weller, Adv. Mater. **5**, 88 (1993).
3. A. P. Alivisatos, J. Phys. Chem. **100**, 13226 (1996).
4. J. Z. Zhang, Acc. Chem. Res. **30**, 423 (1997).
5. A. L. Rogach, L. Katsikas, A. Kornowski, D. Su, A. Eychmüller, and H.Weller, Ber. Bunsenges. Phys. Chem. **100**, 1772 (1996).
6. A. Chemseddine and H. Weller, Ber. Bunsenges. Phys. Chem. **97**, 636 (1993).
7. J. Rockenberger, L. Tröger, A. Kornowski, T. Vossmeyer, A. Eychmüller, J. Feldhaus, and H. Weller. J. Phys. Chem. B **101**, 2691 (1997).
8. J. Rockenberger, L. Tröger, A. L. Rogach, M. Tischer, M. Grundmann, A. Eychmüller, and H. Weller. J. Chem. Phys. **108**, 7807 (1998).
9. D. Schooss, A. Mews, A. Eychmüller, and H. Weller, Phys. Rev. B **49**, 17072 (1994).
10. M. Y. Gao, S. Kirstein, H. Möhwald, A. L. Rogach, A. Kornowski, A. Eychmüller, and H. Weller, J. Phys. Chem. B. **102**, 8360 (1998).

A NOVEL ROUTE FOR THE PREPARATION OF CuSe AND CuInSe2 NANOPARTICLES

M. AZAD MALIK,[a] PAUL O'BRIEN,[a*] N. REVAPRASADU[a,b] G. WAKEFIELD[c]
a. Department of Chemistry, Imperial College of Science, Technology and Medicine, Exhibition Road, London, SW7 2AZ, UK. b. Department of Chemistry, University of Zululand, Private Bag X1001, Kwadlangezwa, 3886. SA. c. Department of Engineering Science, University of Oxford, Oxford, OX1 3PJ.UK. Email: p.obrien@ic.ac.uk; m.malik@ic.ac.uk; n.revaprasadu@ic.ac.uk

ABSTRACT

Good quality, close to mono-dispersed, nanoparticles of CuSe and CuInSe2 have been prepared from thermolysis reactions in tri-n-octylphosphine oxide (TOPO). Bis(diethyldiselenocarbamato)-copper(II) ([Cu(Se2CNEt2)2]) was used to prepare CuSe and TOPSe, CuCl and InCl3 were reacted to form CuInSe2. HRTEM images showed that the particles have reasonable monodispersity and are crystalline. Clear lattice fringes were observed in single dots of CuInSe2 (ca. 4 nm) and CuSe (ca.16 nm). The particles are capped with TOPO and the SAED pattern showed cubic CuInSe2 and hexagonal CuSe.

INTRODUCTION

The ternary compound semiconductor, CuInSe2 has potential for use in thin film solar cells because of its high absorption coefficient, band gap and radiation stability.[1] Conversion efficiencies of more than 14% have been reported.[2] The efficiency is influenced by the stoichiometry, defect structure and any impurities present in the material. It has been reported that In-rich CuInSe2 films have more defects than stoichiometric or Cu-rich films.[3] Photoluminescence (PL) has been used to investigate the defects and impurities. Structural features such as the grain size and crystallinity have been studied by X-ray diffraction (XRD), transmission electron microscopy (TEM) and scanning electron microscopy (SEM).

Films of CuInSe2 have been grown using methods such as molecular beam epitaxy (MBE),[3-7] liquid phase epitaxy,[8] and halogen vapour epitaxy (VPE).[9] Polycrystalline CuInSe2 films have also been grown by Metal Organic Chemical Vapour Deposition (MOCVD) using separate Cu and In precursors[10] and more recently by single-source methods using Cu(Se2CNMeHex)2 and In(Se2CNMeHex^n)3.[11]

Copper selenide is also used in solar cells.[12] Several methods have been used to prepare CuSe including the thermolysis of Cu and Se powder mixtures (400 °C - 470 °C under argon),[13] the mechanical alloying of Se and Cu in high-energy ball mill,[14] and the reaction of selenium with elemental Cu in liquid ammonia.[15,16] Wang et al.[17] have reported the synthesis of nanocrystalline Cu2-xSe (particle size ca.18 nm) using a solvothermal method in which CuI and Se were heated at 90 °C in an autoclave using ethylenediamine as a solvent.

In this paper we report the synthesis of TOPO capped CuInSe2 nanoparticles using a two step reaction from InCl3 and CuCl in tri-n-octylphosphine (TOP) followed by injection into tri-n-octylphosphine oxide (TOPO) at 100°C. TOPSe was then added to the reaction mixture at the elevated temperature of 250 °C. Dark green nanoparticles of CuInSe2 were formed after precipitation with methanol. TOPO capped CuSe nanoparticles were synthesized via a single source route as described previously for CdS, CdSe and

ZnSe,[18-20] Cu(Se₂CNEt₂)₂ used as the single source precursor. The nanoparticles were studied by various techniques to investigate their optical and structural properties.

EXPERIMENTAL

Chemicals

Selenium, tri-*n*-octylphosphine oxide; tri-*n*-octylphosphine, indium(III) chloride and copper (I) chloride were purchased from Aldrich chemical Company Ltd and methanol, toluene from BDH. The solvents used for air sensitive chemistry were distilled, deoxygenated under a nitrogen flow and stored over molecular sieves (type 4 Å, BDH) before use.

TOPO was purified by vacuum distillation at *ca.* 250°C (0.1 torr). The solvents used for air sensitive chemistry were distilled, deoxygenated under a nitrogen flow and stored over molecular sieves (type 4 Å, BDH) before use.

Instrumentation: UV/VIS Absorption Spectroscopy: A Philips PU 8710 spectrophotometer was used to carry out the optical measurements of the semiconductor nanoparticles. The samples were placed in silica cuvettes (1 cm path length). Photoluminescence Spectroscopy: A Spex FluoroMax instrument with a xenon lamp (150 W) and a 152 P photomultiplier tube as a detector was used to measure the photoluminescence of the particles. Good spectral data was recorded with the slits set at 2nm and an integration time of 1 second. The samples were quantitatively prepared by dissolving 25 mg in 10 ml toluene. The samples were placed in quartz cuvettes (1 cm path length). The wavelength of excitation was set at a lower value than onset of absorption of a particular sample. X-Ray Diffraction (XRD): X-Ray diffraction patterns were measured using a Philips PW 1700 series automated powder diffractometer using Cu-K (radiation at 40 kV/40 mA with a secondary graphite crystal monochromator. Samples were prepared on glass slides (5 cm). A concentrated toluene solution was slowly evaporated at room temperature on a glass slide to obtain a sample for analysis. Electron microscopy: A Joel 2000 FX MK 1 operated at 200 kV electron microscope with an Oxford Instrument AN 10000 EDS Analyser was used for the conventional TEM (transmission electron microscopy) images. Selected area electron diffraction (SAED) patterns were obtained using a Jeol 2000FX MK 2 electron microscope operated at 200 kV. The samples for TEM and SAED were prepared by placing a drop of a dilute solution of sample in toluene on a copper grid (400 mesh, Agar). The excess solvent was wicked away with a paper tip and completely dried at room temperature. EDAX (Energy dispersion analytical X-ray) was performed on the sample deposited by evaporation on glass substrates by using JEOL JSM35CF scanning electron microscope. For HRTEM a drop of the dilute solution of sample was placed on a holey carbon film and left to evaporate and then examined in a JEOL 4000EX TEM at 400 kV. ICPAES analysis (Cu, In, Se) were recorded on a ARL instrument, Geology Department, Imperial College.

Synthesis

Preparation of CuSe Quantum dots: A modification of the single source route as described by Trindade and O' Brien was used for the preparation of the CuSe nanoparticles.[10,11] Typically Cu(Et₂CNSe₂)₂ (1.0 g) was dissolved in TOP (15 ml). This solution was then injected into hot TOPO (20 g) at 250 °C . A decrease in temperature of 20 - 30 °C was observed. The solution was then allowed to stabilize at 250 °C and heated for 45 min. at this temperature. The solution was cooled to approximately 70 °C and an excess of methanol added, a flocculant precipitate formed. The solid was separated by centrifugation, washed with methanol and redispersed in toluene. The toluene was removed under vacuum to give pale yellow TOPO capped CuSe nanoparticles. The particles were washed three times with methanol and redissolved in toluene. Size selective precipitation was then carried out on the particles to obtain four fractions of varying average particle size.

Preparation of CuInSe$_2$ Quantum dots: TOPO (20 g) was heated to 100 °C, followed by degassing and flushing with nitrogen. An equimolar solution of InCl$_3$ (2.20 g, 0.01 mol) and CuCl (1.0 g, 0.01 mol) in TOP (15 ml) was injected into the hot TOPO at 100 °C. The colourless TOPO turned bright yellow after injection and the temperature dropped to 80 ° C. The reaction was allowed to proceed for an hour after which the temperature was increased to 250 °C, then 1.0 M TOPSe (20 ml) was injected into the TOPO solution. After an initial drop in temperature to 220 °C, the reaction was stabilised at 250 °C and allowed to proceed for 24 hours. The reaction mixture was cooled to 60 °C followed by the addition of excess methanol to flocculate the particles. There was no immediate visible precipitation on the addition of methanol. A fine precipitate did appear after overnight stirring and was separated by centrifugation. The precipitate was washed with methanol to remove the excess TOPO and then dissolved in toluene.

Scheme 1 Stepwise synthesis of CuInSe$_2$ nanoparticles

$$InCl_3 + CuCl \xrightarrow{\text{TOP}} CuInTOP\text{-complex} \xrightarrow[100\ °C,\ 1\ h]{\text{TOPO}} CuInTOPO\text{-complex}$$

$$CuInTOPO\text{-complex} \xrightarrow[250\ °C,\ 24\ h]{\text{TOPSe}} CuInSe_2\ (TOPO\text{-capped})$$

Results and Discussion

The optical properties of semiconductor nanoparticles are different from the corresponding bulk material. The most striking feature of these particles is the increase in the optical band gap due to the quantum confinement effect observed when the size of the particle is comparable to the size of the Bohr exciton radius in the material. This increase in the band gap is observed as a blue shift in the absorption spectra. Properties such as deep trap emission, size distribution, structural characteristics, crystallinity and surface composition have been reported for numerous samples of semiconductor quantum dots.

The absorption spectra of both CuSe and CuInSe$_2$ show dramatic blue shifts in the band edges in relation to the bulk materials. The CuSe nanocrystals show a considerable increase in band gap. Samples were fractionated by stepwise precipitation from toluene solution by adding methanol and their band gaps (using the direct band gap method[21]) are as follows, fraction1, 1.63 eV (760 nm); fraction 2, 1.74 eV (712 nm); fraction 3, 1.77 eV (700 nm) and fraction 4, 1.87 eV (663 nm). The bulk material has a band gap of, 1.05 eV (1180 nm). The broad absorption edge with an excitonic peak at 520 nm observed for fractions 3 and 4 suggests a decrease in the particle size and narrow size distribution. The absorption band edges fitted to a direct transition for fraction 1 and 4 are shown in Figure 1. The average particle size of CuSe ranges from 14 nm to 20 nm. Figure 2 shows the relationship between the absorption spectra and the TEM (images and particle size histograms) for fractions 1, 3 and 4.

A HRTEM micrograph of CuSe (fraction 4) shows monodispersed, spherical nanoparticles with an approximate diameter of (16.1 nm). The average size of the particles decreases from fraction 1 to 4 as seen in the particle size histograms and images. This change corresponds to the shift in the absorption spectra (blue shift of fraction 1, 0.50 eV; fraction 3, 0.72 eV; fraction 4, 0.82 eV). The particles visible in the HRTEM micrograph are approx. (15 nm) in size confirming the size distribution in the TEM image. The lattice fringes visible in the HRTEM image are indicative of the crystalline nature of these

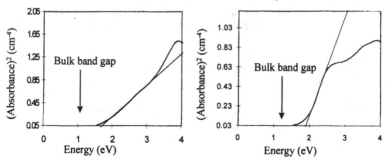

Figure 1. Optical absorption edge of CuSe fitted to a direct transition (a) Fraction 1 showing a band edge at 1.63 eV (b) Fraction 4 showing band edge at 1.87 eV.

particles (Figure 3). The actual spacing of the lattice planes could not be determined from the HRTEM image. However the indexing of the lattice parameters of the SAED pattern correspond to the (110), (200) and (300) lattice planes of hexagonal CuSe. Analysis by EDAX in the electron microscope showed the presence of copper and selenium together with a strong phosphorus peak due to the capping by TOPO. Other low intensity peaks for potassium, calcium and silicon are due to the glass substrate.

The as-prepared TOPO capped $CuInSe_2$ nanoparticles were studied without further fractionation and have a band edge at 420 nm (2.95 eV) with a sharp excitonic peak at 352 nm (Figure 4). The band edge is blue shifted by approximately 1.95 eV in relation to bulk $CuInSe_2$ (1.04 eV). The excitonic feature is due to the first electronic transition occurring in $CuInSe_2$. The shape of the absorption band edge suggests a narrow size distribution of the particles. The photoluminescence spectrum shows a emission maximum at 440 nm (for an excitation at 380 nm). The emission spectrum is red shifted in relation to the band edge by ca. 20 nm (0.02 eV).

The micrograph (TEM) of $CuInSe_2$ shows monodispersed, spherical nanoparticles with an approximate diameter of 4.50 nm. The HRTEM (Figure 5) shows a single quantum dot 4 nm in diameter. Visible lattice fringes correspond to the (220) lattice planes of the cubic phase of $CuInSe_2$ as confirmed by the SAED pattern. The predicted diameter of the $CuInSe_2$ nanoparticles using the effective mass approximation model is 4.04 nm. This value is smaller than the average calculated from TEM (4.50 nm). However such deviations are expected for smaller particles using this model.[22a,b,c] EDAX pattern clearly confirms the presence of copper, indium and selenium. The strong peak for phosphorus is due to the capping of the particles by TOPO. ICPAES analysis (Cu, In, Se) of the dots shows the composition as $Cu_1In_{1.1}Se_{1.8}$. The low selenium content may be due to oxidation during the analysis.

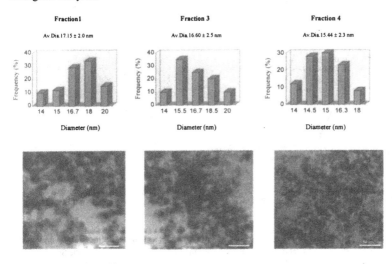

Figure 2. Particle size distribution histograms and TEM images of CuSe (fractions 1, 3 and 4)

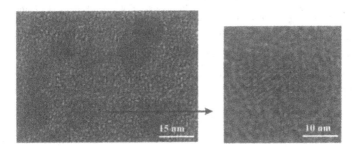

Figure 3. HRTEM images of CuSe quantum dots (*ca.* diameter 16 nm)

Conclusion

Good quality close to mono-dispersed nanoparticles of CuSe have been prepared by thermolysis in TOPO using $Cu(Se_2CNEt_2)_2$ as a single source precursor for the first time. Highly crystalline nanoparticles of $CuInSe_2$ were prepared by reacting TOPSe, CuCl and $InCl_3$ in TOPO. HRTEM images showed that the particles are close to mono-dispersed and clear lattice fringes of a single dot of $CuInSe_2$ confirmed that the nanosize particles are crystalline with a size of *ca.*4 nm. The presence of a strong phosphorus peak in EDAX pattern indicated that the particles are capped with TOPO and the SAED pattern showed a cubic CuInSe₂ and hexagonal CuSe.

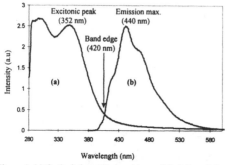

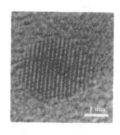

Figure 4. (a) Optical absorption spectrum of CuInSe₂ of showing an edge at 420 nm and an excitonic peak at 352 nm (b) PL spectrum showing an emission max. at 440 nm.

Figure 5. HRTEM image of band a single quantum dot of CuInSe₂ (ca. diameter 4 nm)

ACKNOWLEDGEMENTS: We thank the Royal Society and FRD (SA) for support to NR and a program of collaboration between UZULU and ICSTM. POB thanks the EPSRC for grant. POB is the Sumitomo/STS Professor of Materials Chemistry and the Royal Society Amersham International Research Fellow (1997/98).

REFERENCES

1. V. Nadenau, D. Braunger, D. Hariskos, M. Kaiser, Ch. Koble, M. Ruckh, R. Schaffer, D. Schmid, T. Walter, S. Zwergart, and H. W. Schock, Prog. Photo. Res. Appl., 3, p.363 (1995).
2. L. Stolt, M. Bodegard, J. Kessler, M. Ruckh, K.O. Velthaus and H.W. Schock, Proc. 11ᵗʰ European Photovoltaic Solar Energy Conference, Monteux, Harwood Academic, Chur Switzerland, , p.120 (1993).
3. S. Niki, P.J. Fons, A. Yamada O. Hellman, T. Kurafuji, S. Chichibu and H. Nakanishi, Appl. Phys. Lett., 69, p.647 (1996).
4. F.R White, A.H. Clark, M.C. Graf and L.L. Kazmerski, J. Appl. Phys., 51, p.544 (1979)
5. B. Schumann, T. Tempel, and G. Kuhn, Sol. Cells, 16, p.43 (1986).
6. A.N. Tiwari, S. Blunier, K. Kessler, V. Zelezny, and H. Zogg, Appl. Phys., Lett. 65, p. 2299 (1994).
7. A.N. Tiwari, S. Blunier, M. Filmoser, H.Zogg, D. Schmid and H.W. Schock, Appl. Phys. Lett., 65, p.3347 (1994).
8. H. Takenoshita, Sol. Cells, 16, p.65 (1986).
9. O. Igarashi, J. Cryst. Growth, 130, p. 343 (1993).
10. B. Sagnes, A. Salesse, M.C. Artaud, S. Duchemin, J. Bougnot and G. Bougnot, J. Cryst. Growth, 124, p.620 (1992).
11. J. McAleese, P. O'Brien and D. J. Otway, Mat. Res. Soc. Symp. Proc., 485, p.157 (1998).
12. S.T. Lakshmikvar, Sol. Energy. Mater. Sol. Cells, 32, p.7 (1994).
13. O. Akira, Jpn. Kokai Tokkyo Koho, JP 61, 222, p. 910.
14. T. Ohtani and M. Motoki, Mater. Res. Bull., 30, p. 195 (1995).
15. G. Henshaw, I. P. Parkin and G. Shaw, Chem. Commun., p. 1095 (1996).
16. G. Henshaw, I. P. Parkin and G. Shaw, J. Chem. Soc., Dalton Trans., p. 231 (1997).
17. W. Wang, P. Yan, F. Liu, Y. Xie, Y. Geng and Y. Qian, J. Mater. Chem., 8, p. 2321 (1998).
18. T. Trindade, P.O' Brien, Adv. Mater., 8, p.161 (1996).
19. T. Trindade, P.O' Brien, Chem. Mater., 9, p.523 (1997).
20. N. Revaprasadu, M.A. Malik, P O' Brien, G. Wakefield, J. Mater. Chem., 8, p.1885 (1998).
21. J. I. Pankove, Optical processes in Semiconductors, Dover Publications Inc., New York, (1970).
22. (a) L. E. Brus, J. Chem. Phys., 80, p.4403 (1984) (b) L. E. Brus, J. Phys. Chem., 90, p.2555 (1986) (c) L. E.Brus, J. Chem. Phys., 79, p.5566 (1983).

The Preparation and Characterization of Nanocrystalline Indium Tin Oxide Films

J. AIKENS, H.W. SARKAS, R.W. BROTZMAN, Jr.
Nanophase Technologies Corporation, Burr Ridge, IL 60521

Abstract

Nanocrystalline indium tin oxide (ITO) powder is prepared by gas phase condensation (GPC) and used to produce conductive optical coatings. Materials produced by GPC are commercially available at reasonable cost and bridge the gap between basic science and product development. Nanocrystalline ITO films are transparent (>95%), conductive, and offer distinct device processing advantages. The oxidation state and surface chemistry of ITO may be tailored for particular applications and offers a flexible, affordable alternative to competitive films. The preparation and characterization of nanocrystalline ITO films and device applications will be discussed.

Introduction

Indium tin oxide (ITO) represents a unique ceramic material that when processed can yield semi-conductive conductive transparent films useful in a large array of applications from glass coatings to semi-conductive films for displays and electronics. One of the limitations to the use of ITO in the above mentioned applications is the difficulty and cost associated with producing films of sufficient transparency to prevent haze or other optical aberrations that effect the quality of the overall device. One solution to these problems is the use of sputtering techniques which produce thin films of high optical and conductive quality [1,2]. Sputtering processes are quite costly in equipment and maintenance and can be difficult to scale to large or unusually shaped substrates. An alternative method for ITO film production is to use chemical vapor deposition (CVD) processes to prepare ITO films [3,4]. Like sputtering, CVD processes have equipment expenses and while useful for certain semi-conductive applications, can be difficult to control film uniformity and thus optical quality. To achieve high optical and semi-conductive characteristics at reasonable costs require other methods for ITO preparation and deposition.

Films prepared by particulate deposition offer an attractive option to produce ITO films of high optical quality and reasonable conductivity for a fraction of the cost of other deposition methods. Particles of ITO may be deposited with a variety of methods depending on the desired characteristics of the final film. Particle deposition is sensitive to application methods as well as the nature of the particles used. Major challenges associated with particle deposited films involve control of particle size and stability of ITO dispersion. For optimal transparency and conductivity particles should have an average particle size below 100 nanometers [5]. The difficulties in preparing particle deposited films are dependent upon the methods used to prepare ITO particles in the appropriate size regime used in deposition dispersions.

One method for preparing ultrafine ITO particles is by wet chemical processes such as sol-gel or solution phase precipitation chemistry [6,7]. Resulting particles are of ultrafine size by both methods but particle size stability is difficult to maintain without secondary processes especially when the initial ITO must be formulated into a coating dispersion. Particle dispersions derived from wet chemical processes often contain by-products that must be removed. In addition, resulting films often need substantial thermal processing to "cure" the ITO in order to remove by-products and other components including binders to achieve sufficiently connected semi-conductive films.

A final means of preparing ultrafine ITO particles involves physical methods such as mechanical size reduction or gas phase condensation (GPC) [8]. Mechanical size reduction begins with ITO particles larger than 100nm or large agglomerates of ultrafine particles and by an iterative process of milling or grinding, reduces the particle size to the ultrafine range. A drawback to mechanical size reduction is that it can be labor intensive and costly to produce ultrafine particles with narrow particle size distributions. Small concentrations of larger particles remaining in the sample can settle or otherwise affect film quality to produce hazy, uneven films. The alternative to size reduction is GPC synthesis Figure 1, which produces ultrafine ITO substantially free of larger particles. GPC prepared ITO once in dispersion, can be readily diluted with a variety of solvents to produce particle deposited films.

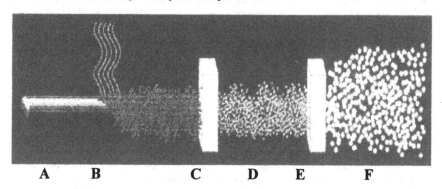

Figure 1. GPC Process for Preparing Ultrafine ITO. Diagram and listing of the basic events in GPC. A. Precursor feed. B. Plasma energy input; precursor vaporized. C. Quench gas input. D. Vapor Condensation. E. Cooling gas input. F. Particle collection.

Unlike sputtering or CVD techniques, particle deposited films require preparation of stable stock dispersions that are often complex mixtures of solvent components. Formulations for film formation must have proper balance of volatile components to allow for even spread of the dispersion across the surface. Once the surface is wet the dispersion in liquid film state must dry to deposit particles with close packing of ITO to achieve optimal optical transparency and low resistivity. Particle deposition techniques require thermal processing to remove residual solvent associated with the film and to organize particulate interfaces to optimize connectivity. In spite of these processing requirements, particulate deposited films offer relatively low cost alternatives to forming transparent conductive ITO films. Particulate deposited films prepared from ITO dispersions of GPC derived ITO are discussed and compared to other commercial ITO particle dispersions.

Experimental

ITO was prepared by feeding ITO precursor into a GPC reactor system as described previously [8]. Collected ITO powder was dispersed in water and thoroughly wetted by mechanical agitation of the slurry either by high shear mixing or media milling. ITO slurry was then allowed to settle in order to remove larger contaminants and undispersed powder. Coating formulations were prepared by diluting sedimented aqueous stock ITO slurries into solvents described previously [5,9] or diluting stocks into a mixture of ethanol and commercially available nonionic surfactants. The basic procedure for preparing particle deposited ITO films is outlined in Figure 2. Not shown is optional application of a barrier topcoat layer composed of hydrolyzed alkoxysilane used to prevent physical damage and atmospheric degradation of the ITO film [5].

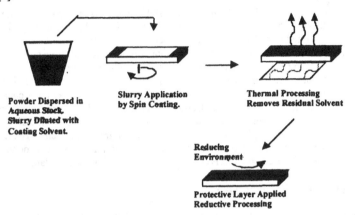

Figure 2. General procedure for Preparation of particle deposited ITO films. ITO coating formulation (2 wt.% ITO) applied by spin coating on a 4"X4" float glass substrate with Ag ink electrodes. Barrier topcoat can then be applied by second spin coating operation followed by thermal treatment at 160°C for 30 minutes. Further processing can be performed in reducing atmosphere at 350°C for 30 minutes.

Resistivity was measured by two point contact across the Ag electrodes on films with or without barrier top coated films after each had cooled to room temperature. In some cases, resistivity was measured prior to thermal treatment, with or without top coating. Transparency was measured by positioning films in a Perkin Elmer Lambda 4 spectrophotometer and scanning through the visible wavelengths versus substrates without ITO films. Surface hardness was determined qualitatively by resistance to film scratching by a metal probe.

Results

ITO Formulations
Previous reports indicate that ultrafine ITO particle dispersions can be applied to substrates to form films with high optical transparency and resistivities approaching $10^4 \Omega/\square$ [5]. ITO particles that comprise these dispersions are found to be spheroidal with surfaces that appear poorly resolved by high resolution TEM analysis. GPC prepared ITO appears rhomboidal with clearly discernable crystal lattice structure even at particle edges. Whether these morphological differences between ITO materials would be manifest in altered performance in deposited thin films became the focal point of the following experiments. To this end, GPC ITO was dispersed according to methods for commercial production of ultrafine ITO dispersions [5,7,9]. In no case

could GPC ITO be dispersed to primary particles in organic solvents directly as determined by settling/flocculation rates and dynamic light scattering. GPC ITO required an aqueous environment to form stock dispersions with a rapid corresponding acidic shift of the measured pH. Attempts to disperse ITO in basic aqueous media resulted in immediate flocculation and settling. Aqueous stock dispersions were found to be sensitive to water quality as settling/flocculation of associated particles occurred in the dispersion as a function of initial ionic character of the water.

Stock dispersions of GPC ITO in water are stable, monodisperse particles with an average size of less than 50 nanometers as determined by dynamic light scattering. Stock dispersions were diluted into alcohol/ketone based coating formulations without significant loss in dispersion stability as determined by both the amount of settling over time and dynamic light scattering. Attempts to prepare aromatic solvent based formulations resulted in rapid settling. Aqueous formulations consisting of small quantities of ethanol and surfactants produced stable dispersions except for cases where ionic surfactants were employed. Under these conditions, immediate flocculation/settling occurred. Commercial samples of ITO dispersions were obtained and used as is in either alcohol/ketone or aromatic solvent based formulations at 2% by weight ITO. Each of these samples were stable for periods of greater than one week if stored below 0°C. No direct comparisons of GPC ITO in aqueous formulation were made to other commercial sources because at present no water based formulations are available.

Particle Deposition and Film Formation

Single layer films of ITO were prepared and evaluated for optical, electrical and surface hardness performance. Both GPC and commercial ITO formulations yielded deposited films with good optical transparency Figure 3. Resistivity was generally found to be $>10^6\Omega/\square$ without some thermal treatment. Surface hardness and stability with respect to resistivity were found to be dependent on the presence of the barrier top coat. GPC formulations typically required modification from commercial formulation to achieve similar film characteristics. Properties of formulations that differed most dramatically were the time required for drying and surface uniformity of the resulting film. Without sufficient drying, resistivity before thermal treatment was poor while after thermal treatment resistivity improved but barrier top coats tended to be brittle to scratching. Water based formulations required the longest drying times and displayed the greatest variance in film quality depending on the combination of surfactants used. Typical optical deterioration appeared as a haze in the film while resistivity ranged from $4 \times 10^4\Omega/\square$ to $>10^6\Omega/\square$.

Further thermal treatment in air at 350°C for 30 minutes resulted in slight reduction in resistivity to $\sim 10^4\Omega/\square$. Further reduction to $\sim 10^3\Omega/\square$ could be affected by thermal treatment of the GPC ITO deposited films at 350°C for 30 minutes under a H_2/Ar atmosphere. Stability of the resulting reduced film was dependent on the presence of barrier top coat as rapid deterioration of the resistivity was observed when deposited films without top coats were exposed to air. The comparative studies with different ITO particles and different coating formulations suggest that in spite of different conditions for dispersing powders, GPC prepared ITO particles produce deposited films with equivalent properties to ITO particles produced by other methods.

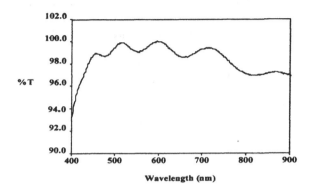

Figure 3. Transparency scan of ITO film. Typical transparency measurement of GPC ITO deposited film on float glass substrate. Film prepared as single layer ITO with barrier top coat described in experimental section.

Conclusions

GPC ITO particles can be dispersed to form aqueous stock dispersions that subsequently formulate into a variety of aqueous or aqueous/organic dispersions. When applied to substrates and processed under similar conditions, particle deposited films of ITO are comparable in optical, electronic and surface hardness independent of the morphological nature of the ITO particle. Unlike other commercial sources of ITO dispersion, GPC prepared ITO stock slurries can be prepared as stable aqueous stocks at weight percentages in excess of 25% by weight ITO. In addition to organic formulations, stock slurries can be diluted to form stable, substantially aqueous formulations. Aqueous formulations display similar film characteristics to organic based formulations but have minimal VOC emissions. GPC prepared ITO particles offer the possibility of producing the same low cost, high quality deposited films as other particle based ITO processes but require less labor to form stock dispersions and can be shipped aqueous, reducing costs associated with shipping final formulations. Ongoing research is focusing on methods to further optimize resistivity of ITO particle based dispersions by taking advantage of the unique morphological and chemical nature of GPC ITO surfaces.

References

1. L. J. Meng, A. Macarico, and R. Martins, Vacuum, **46**, 673-680 (1990).

2. T. J. Vink, W. Walrave, J. L. C. Daams, P. C. Baarslag, and J. E. A. M. Van den Meerakker, Solid Thin Films, **266**, 145-151 (1995).

3. T. Maruyama, and K. Fukui, J. Appl. Phys., **70**, 3848-3851 (1991).

4. T. Maruyama, K. Tabata, Jpn. J. Appl. Phys., **29**, 355-357 (1990).

5. M. Kawata, and M. Yukinobu, U. S. Patent No., 5,662,962 (2 September 1997).

6. T. Furusaki, J. Takahashi, and K. Kodaira, J. Am. Ceram. Soc., **102**, 200-204 (1994).

7. T. Hashimoto, H. Yoshitake, H. Yokoyama, and A. Nakazono, U. S. Patent No., 4,594,182 (10 June 1986).

8. J. C. Parker, M. Ali, and B. B. Lympany, U. S. Patent No., 5,514,349 (7 May 1996).

9. M. Muronchi, T. Hayashi, and A. Nishihara, U. S. Patent No., 5,504,133 (2 April 1996).

ENHANCED PHOTOLUMINESCENCE FOR Zns NANOCRYSTALS DOPED WITH Mn^{2+} CLOSE TO CARBOXYL GROUPS AND/OR S^{2-} VACANCIES

T. ISOBE, T. IGARASHI, M. KONISHI, M. SENNA
Department of Applied Chemistry, Faculty of Science and Technology, Keio University, Hiyoshi, Yokohama 223-8522, JAPAN. isobe@applc.keio.ac.jp

ABSTRACT

ZnS nanocrystals doped with Mn^{2+} ions are prepared by a solution process and subsequent UV irradiation to produce the samples with different S/(Zn+Mn) ratios and/or surface modification by acrylic acids. Coordination states around Mn^{2+} ions were examined at 9 and 35 GHz by electron paramagnetic resonance spectroscopy. The Mn^{2+} sites in the vicinity of S^{2-} vacancies or carboxyl groups are observed at the frequencies more than 9 or 35 GHz, respectively, for nanocrystals, but are not for the bulk sample of 250 nm diameter. Such Mn^{2+} sites enhance the photoluminescence due to d-d transition of Mn^{2+} ions through energy transfer from S^{2-} vacancies or carboxyl groups, excited simultaneously by a light of 350 nm for exciting ZnS.

INTRODUCTION

The photoluminescence (PL) intensity due to the Mn^{2+} d-d transition is enhanced when Mn^{2+}-doped ZnS (ZnS:Mn) nanocrystals are modified by carboxylic acids, such as methacrylic acid and acrylic acid (AA) [1-4]. We focus on the chemical and electronic interaction between Mn^{2+} and carboxyl groups as well as quantum confinement effects, since the carboxyl groups and ZnS are excited simultaneously by a UV light. In this study, Mn^{2+} sites in ZnS:Mn nanocrystals are examined by electron paramagnetic resonance (EPR) spectroscopy at X- and Q-bands, and are compared with that in bulk to discuss roles of carboxyl groups. We also discuss a common role of carboxyl groups and S^{2-} vacancies near Mn^{2+} ions.

EXPERIMENT

The AA-modified nanocrystal was prepared by admixing a solution, containing $(CH_3COO)_2Zn$ and $(CH_3COO)_2Mn$, into an Na_2S solution in the atomic ratio, Zn:Mn:S=1:0.01:1.01, and then by adding AA ($CH_2=CHCOOH$). The details in this solution method is described in ref. [2]. The unmodified nanocrystals with varying S/(Zn+Mn) ratios were prepared in the atomic ratio, Zn:Mn:S=1:0.01:x (0.8 ≤ x ≤ 1.1), by the same procedure without adding AA. The ZnS:Mn bulk sample was synthesized by the following Mn-coating method [5]. ZnS particles of 250 nm diameter were dispersed in 100 cm^3 of 8x10^{-3} M $(CH_3COO)_2Mn$ aqueous solution. A 4 cm^3 of 13 M NH_3 aqueous solution was added into this suspension. The filtered powder was dried at 80°C for 2 h, and then heated in N_2 at 450°C for 2 h.

The samples were observed by transmission electron microscopy (TEM, JEOL, 2000FXII) and scanning electron microscopy (SEM, JEOL, 5200). A PL spectrum for a sample excited by a 350 nm light was measured at room temperature at intervals of 0.5 nm by spectrophotofluorometer (JASCO, FP-777). A PL excitation (PLE) spectrum was measured at room temperature at intervals of 0.2 nm. After dissolving 0.1 g specimen in 10 cm^3 of 12 M HCl aqueous solution, an Mn/(Zn+Mn) atomic ratio was determined from Kα lines of Mn and Zn by x-ray fluorescent analysis (XFA). An S/Zn atomic ratio was measured from Kα lines of S and Zn by energy dispersive spectroscopy (EDS) installed in TEM. The coordination states around Mn^{2+} ions were examined at room temperature by EPR spectroscopy (JEOL, JES-RE3X). A first derivative EPR spectrum was measured at room temperature and X band (9 GHz), where the microwave power was 2 mW; the modulation width, 0.32 mT and the modulation frequency, 100 kHz. An EPR spectrum was measured at Q band (35 GHz), where the microwave power was 0.2 mW; the modulation width, 0.20 mT and the modulation frequency, 100 kHz.

RESULTS

Coordination States around Mn²⁺ Ions in ZnS:Mn Nanocrystals and Bulk

According to the observation by TEM and SEM, the mean particle diameter, d_{av}, is 2~4 nm for nanocrystals, 250 nm for the synthesized bulk sample, as shown in Table 1. The composition, determined by XFA and EDS, is also given in Table 1. All samples exhibit zinc blende (cubic) structure, as confirmed by x-ray diffractometry.

Table 1   Properties of nanocrystals and bulk.

Sample	d_{av} (nm)	PLE (eV)	PL (eV)	Luminance (cd m⁻²)	S/(S+Zn+Mn) (at %)	Mn/(Zn+Mn) (at%)
AA-modified nanocrystal	3.6±0.4	3.77	2.13	52	51	0.19
unmodified nanocrystal	2.6±0.4	3.68	2.14	21	45±3	0.47
synthesized bulk	250	3.54	2.16	5	-----*⁾	-----*⁾

*) not measured yet.

Fig. 1 shows EPR spectra measured at X-band. The sextet signal I with g = 2.0024 and hyperfine coupling constant $|A|$ = 6.9 mT, accompanied with five doublet peaks between each peak of the signal I, is observed in the spectrum (Fig. 1(a)) of the AA-modified nanocrystal. In contrast, the additional sextet signal II with g = 2.0013 and $|A|$ = 9.0 mT is observed in the spectrum (Fig. 1(b)) of the unmodified nanocrystal. The signal I is observed in the spectrum (Fig. 1(c)) of the bulk sample. The signal I is assigned to the Mn²⁺ sites coordinated fully by four S²⁻ ions, whereas the signal II is assigned to the Mn²⁺ sites with S²⁻ coordination number less than 3 [2,6]. The doublet peaks in Fig. 1(a) are ascribed to zero field splitting [7]. When EPR spectra are measured at Q-band, the signals I and II are clearly observed in the spectra of both nanocrystals regardless of surface modification, as shown in Fig. 2(a) and (b). In contrast, only the signal I is observed in the spectrum (Fig. 2(c)) of the bulk sample. We also note that the broad singlet signal due to Mn²⁺ clusters with exchange interaction [8] is overlapped with the hyperfine structure in Fig. 1(b) and (c), and Fig. 2(b) and (c).

The shape of each peak for the signals I and II of all the samples is Lorentzian. The EPR peak height of signal, I_m, measured at X- and Q-bands is given by [9]:

$$I_m = 2 \ - \ A^2(35 \ - \ 4m^2) / [2(g\beta H)^2] \ - \ 5.334D^2 / (g\beta H)^2$$
$$- \ 34.14D^2 (35 \ - \ 4m^2) / (g\beta H)^2 + \ 208D^4(35 \ - \ 4m^2)^2 / (g\beta H)^4 \qquad (1)$$

where m is nuclear spin quantum number; D, the zero filed splitting constant; β, Bohr magneton and H, magnetic field. D was calculated from the intensity ratio, $I_{5/2} / I_{3/2}$ by eq. (1). The values of D are 0.0091 cm⁻¹ and 0.016 cm⁻¹ for the signals I and II, respectively, of the AA-modified nanocrystal, being larger than 0.0044 cm⁻¹ for the signal I of the synthesized bulk sample. D of unmodified nanocrystal could not be determined precisely because of overlapped signals.

Optical Properties of ZnS:Mn Nanocrystals Modified by AA

The PL peak due to d-d transition of Mn²⁺ ions is observed at around 580 nm when excited by a light of 350 nm. The luminance and the peak positions of PL and PLE are summarized in Table 1. The blue shift of PLE peak, ΔE_{PLE}, and the red shift of PL peak are recognized for nanocrystals, with reference to the bulk. In addition, the AA modification increases ΔE_{PLE}. The luminance of the AA-modified nanocrystal is 2.5 times larger than that of the unmodified nanocrystal, although the Mn/(Zn+Mn) ratio of the former is 60% less than that of the latter, i.e., the AA modification increases the luminance, and prevents S²⁻ loss, but leaches out Mn²⁺ ions from ZnS:Mn, as also shown in Table 1. We note that the PL peak due to the conformation of carboxyl groups [10] is observed at around 440 nm for the AA-modified nanocrystal. When the Mn/(Zn+Mn) ratio changes from 0 at% to 0.19 at%, the PL intensity at 580 nm increases, as shown in Fig. 3. At the same time, the PL intensity at 440 nm decreases, in spite of the constant AA amount, 44 wt%, determined by thermogravimetry.

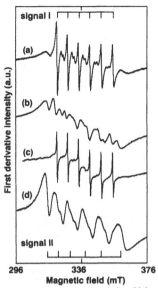

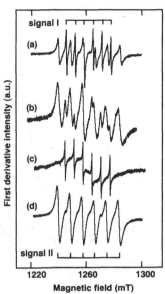

Fig. 1   EPR spectra measured at X-band. (a) the AA-modified nanocrystal, (b) the unmodified nanocrystal, (c) the synthesized bulk sample, (d) the unmodified S^{2-} deficient nanocrystal with S/(S+Zn+Mn) = 42.7 at%.

Fig. 2   EPR spectra measured at Q-band. (a) the AA-modified nanocrystal, (b) the unmodified nanocrystal, (c) the synthesized bulk sample, (d) the unmodified S^{2-} deficient nanocrystal with S/(S+Zn+Mn) = 42.7 at%.

Roles of S^{2-} Vacancies on Optical Properties of Unmodified Nanocrystals

Several unmodified ZnS:Mn nanocrystals were prepared from solutions with different S/(Zn+Mn) ratios.   The signal *II* is predominant in the EPR spectrum of the S^{2-} deficient sample with S/(S+Zn+Mn) = 42.7 at%, as shown in Figs. 1(d) and 2(d).   Fig. 4 shows PL spectra of unmodified nanocrystals, where a 0.05 g sample was compressed at 173 MPa to make a pellet of 6 mm diameter for measurement.   The PL peak due to S^{2-} vacancies [11,12] is observed at around 440 nm in the spectra of the samples with S/(S+Zn+Mn) less than 50 at%, as shown by (a-2) and (b-2) in Fig. 4.   When a comparison is made between the PL spectra of the samples with the same Mn/(Zn+Mn) ratio, the PL intensity at 580 nm is larger for the S^{2-} deficient samples (Fig. 4 (a-2) and (b-2)) than for the stoichiometric and S^{2-} rich samples (Fig. 4 (a-1) and (b-1)).   When the Mn/(Zn+Mn) ratio in the S^{2-} deficient samples increases from 0 at% to 0.47 at%, the PL intensity at 580 nm increases while that at 440 nm decreases, as shown in Fig. 5.

UV Irradiation Effects on ZnS:Mn Nanocrystals

Prolonged excitation of ZnS:Mn samples, shown in Table 1, by the UV light of 350 nm increases the PL intensity of nanocrystals at 580 nm, as shown by curves (a) (b) in Fig. 6.   In contrast, no appreciable change is observed for the commercial bulk sample, as shown in Fig. 6(d).   The increase in the PL intensity at 580 nm by the UV irradiation is smaller for the nanocrystal modified by polyacrylic acid (PAA), shown in Fig. 6(c), than for the AA-modified nanocrystal, shown in Fig. 6(b).   After the PL intensity at 580 nm leveled off, the PL spectra of the nanocrystals without and with AA were measured and are shown in Figs. 7 and 8, respectively.   The saturated PL intensity at 580 nm increases by a factor of 4 for the unmodified nanocrystal, while by a factor of 1.5 for the AA-modified nanocrystal.   We note that the PL intensity at around 430 nm is also enhanced irrespective of modification.   The PL intensity at 430 nm relative to at 580 nm after UV irradiation is larger for the AA-modified nanocrystal than for the unmodified nanocrystal.

385

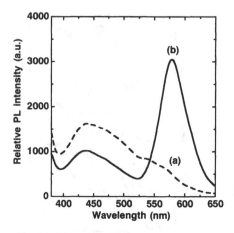

Fig. 3 PL spectra of AA-modified nanocrystals with different Mn/(Zn+Mn) at%: (a) 0, (b) 0.19.

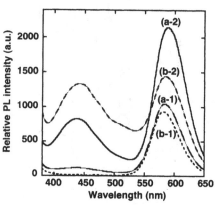

Fig. 4 Change in the PL spectra of unmodified nanocrystals with different S/(S+Zn+Mn) at%: (a-1) 50.5, (a-2) 45.5, (b-1) 52.7, (b-2) 42.7; Mn/(Zn+Mn) at%: (a-1, a-2) 0. 47, (b-1, b-2) 0. 35.

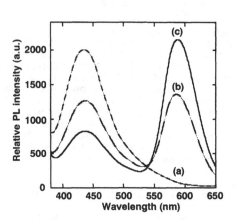

Fig. 5 PL spectra of the S^{2-} deficient unmodified nanocrystals with different Mn/(Zn+Mn) at%: (a) 0, (b) 0.17, (c) 0.47. S/(S+Zn+Mn) at% is 45.5.

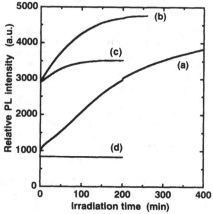

Fig. 6 Change in the PL intensity at 580 nm with UV irradiation time. The wavelength of UV light is 350 nm. (a) the unmodified nanocrystal, (b) the AA-modified nanocrystal, (c) the PAA-modified nanocrystal, (d) the commercial bulk sample of 0.5 μm diameter.

386

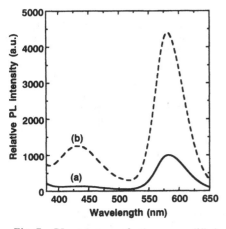

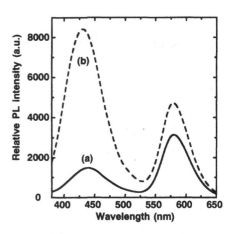

Fig. 7 PL spectra of the unmodified nanocrystal (a) before and (b) after UV irradiation for 24 h.

Fig. 8 PL spectra of the AA-modified nanocrystal (a) before and (b) after UV irradiation for 7 h.

DISCUSSION

The EPR signal II due to Mn^{2+} ions with lower coordination number is observed only for nanocrystals. This is attributed to the larger fraction of surface Mn^{2+} site for nanocrystals than for bulk. D of the signal II for the AA-modified nanocrystal is 3.6 times larger than that of the signal I for the bulk sample, showing that the symmetry of Mn^{2+} crystal field is lower for the former. For the signal I due to Mn^{2+} ions coordinated completely by four S^{2-} ions, D is twice larger for nanocrystals than for bulk. These results could explain a higher probability of Mn^{2+} d-d transition in the PL processes of ZnS:Mn nanocrystals.

The signal II is difficult to observe at X-band for AA-modified nanocrystal, but is detected easily at Q-band, as recognized by comparison of Figs. 1(a) and 2(a). This is attributed to the shorter spin relaxation time by coordination of carboxyl groups to Mn^{2+} ions [13], since the observation time, $t_{ob}=3 \times 10^{-11}$ sec, at Q-band is shorter than $t_{ob}=1 \times 10^{-10}$ sec at X-band. In addition, the broad singlet EPR signal due to exchange interaction between Mn^{2+} ions is weak for the AA-modified nanocrystal, showing the AA modification prevents the formation of Mn^{2+} clusters.

The PL intensity due to Mn^{2+} d-d transition increases by AA modification, although 60% Mn^{2+} ions are leached out by AA, as shown in Table 1. Thus, the PL intensity is not simply determined by the Mn content. The COO^- groups in AA are excited by the same energy as ZnS to exhibit PL at 440 nm. Since the PL intensity due to carboxyl groups is markedly weaker for AA than for PAA, the appearance of PL at 440 nm indicates that the AA monomers are polymerized during drying at 50°C for 1 day. The coating of particles by polymerized AA causes quantum confinement effect, judging from the blue shift of PLE peak. Energy transfer from carboxyl groups to Mn^{2+} ions are also observed for the AA-modified nanocrystal, as shown in Fig. 3. Therefore, the enhanced PL intensity due to Mn^{2+} d-d transition of the AA-modified nanocrystal is attributed to quantum confinement effect and energy transfer from carboxyl groups to Mn^{2+}, in addition to the above-mentioned low symmetry effect.

When the S/(Zn+Mn) ratio decreases, the increase in the PL intensity due to the Mn^{2+} d-d transition is also observed together with the appearance of PL at 440 nm due to S^{2-} vacancies, as shown in Fig. 4. The strong EPR signal II for S^{2-} deficient samples, shown in Figs. 1(d) and 2(d), indicates that S^{2-} vacancies are located in the vicinity of Mn^{2+} ions. The decrease in the PL intensity at 440 nm with increasing Mn/(Zn+Mn) ratio, shown in Fig. 5, is attributed to energy transfer from S^{2-} vacancies to Mn^{2+} ions, being similar to the above-mentioned phenomenon for the AA-modified nanocrystal, shown in Fig. 3. When ZnS:Mn nanocrystals are irradiated by a UV light of 350 nm, the PL intensity at 430 nm and 580 nm increases, regardless of the AA modification. The enhanced PL intensity at 430 nm is attributed to the formation of S^{2-} vacancies and the polymerization of AA monomers. These in turn causes energy transfer by exchange interaction between Mn^{2+} ions and S^{2-} vacancies or carboxyl groups in PAA, leading to the enhanced PL intensity at 580 nm. These results in this work are similar to the enhanced PL intensity by irradiating ZnS:Mn nanocrystals by a 248 nm excimer laser [14] and a 300 nm light [15] or by energy transfer from surface trapped carriers to Mn^{2+} ions in CdS:Mn nanocrystals [16].

CONCLUSIONS

The probability of Mn^{2+} ions with lower coordination number at the near-surface is higher for the nanocrystals than for the bulk sample, as confirmed by EPR. The carboxyl groups, therefore, directly coordinate Mn^{2+} ions for the AA-modified nanocrystal. The strong interaction between Mn^{2+} and carboxyl groups or S^{2-} vacancies near the surface of nanocrystals is one of the important factors to enhance the PL intensity due to d-d transition of Mn^{2+} ions through energy transfer by exchange interaction.

ACKNOWLEDGMENT

This work was supported in part by a grant-in aid for scientific researches (No. 09750764) from the Ministry of Education, Science, Sports and Culture in Japan.

REFERENCES

1. M. Senna, T. Igarashi, M. Konishi and T. Isobe, Proceedings of the Fourth International Display Workshop, (ITE&SID, Nagoya, 1997) pp.613-616.
2. T. Igarashi, T. Isobe and M. Senna, Phys. Rev. **B56**(11), 6444-6445 (1997).
3. T. Isobe, T. Igarashi and M. Senna, in *Microcrystalline and Nanocrystalline Semiconductors* , edited by R.W. Collins, P.M. Fauchet, I. Shimizu, J.C. Vial, T. Shimada and A.P. Alivisatos. (Mat. Res. Soc. Symp. Proc. **452**, Pittsburgh, PA, 1997) pp. 305-310.
4. I. Yu, T. Isobe and M. Senna, J. Phys. Chem. Solids, **57**(4), 373-379 (1996).
5. I. Yu, T. Isobe, M. Senna and S. Takahashi, Mater. Sci. Eng. **B38**, 177-181 (1996).
6. T.A. Kennedy, E.R. Glaser, P.B. Klein and R.N. Bhargava, Phys. Rev. **B52**, R14356-R14359 (1995).
7. S.K. Misra, Physica B, **203**, 193-200 (1994).
8. Y. Ishikawa, J. Phys. Soc. Jpn. **21**(8), 1473-1481 (1966).
9. B. Allen, J. Chem. Phys. **43**(11), 3820-3826 (1965).
10. L.W. Johnson, H.J. Maria and S.P. McGlynn, J. Chem. Phys. **54**(9), 3823-3829 (1971).
11. A.A. Khosravi, M. Kundu, L. Jatwa, S.K. Deshpande, U.A. Bhagwat, M. Sastry and S.K. Kulkarni, Appl. Phys. Lett. **67**(18), 2702-2704 (1995).
12. W.G. Becker and A.J. Bard, J. Phys. Chem. **87**, 4888-4893 (1983).
13. K. Hikichi, T. Hiraoki and N. Ohta, Polymer J. **16**(5), 437-439 (1984).
14. J. Yu, H. Liu, Y. Wang, F.E. Fernandez, W. Jia, L. Sun, C. Jin, D. Li, J. Liu and S. Huang, Opt. Lett. **22**(12), 913-915 (1997).
15. D. Gallagher, W.E. Heady, J.M. Racz and R.N. Bhargava, J. Mater. Res. **10**(4), 870-876 (1995).
16. G. Counio, S. Esnouf, T. Gacoin and J.P. Boilot, J. Phys. Chem. **100**, 20021-20026 (1996).

MICROSTRUCTURE AND SENSING PROPERTIES OF CRYOSOL DERIVED NANOCRYSTALLINE TIN DIOXIDE.

S.M.KUDRYAVTSEVA*, A.A.VERTEGEL*, S.V.KALININ*, L.I.KHEIFETS*, J. VAN LANDUYT**, L.L. MESHKOV*, S.N.NESTERENKO*, E.S. REMBEZA**, A.M.GASKOV*.
[1]Department of Chemistry, Moscow State University, 119899, Moscow, Russia
[2]EMAT, University of Antwerp, 2020 Antwerp, Belgium

ABSTRACT

The powders of nanocrystalline tin dioxide were prepared by two different methods: conventional hydrolysis of $SnCl_4$ in aqueous solution and novel cryosol technique. The microstructure, composition, and electrical properties of the samples were investigated. The sintered pellets obtained by means of the cryosol method are characterized by significantly higher values of electrical resistance as compared to those prepared by conventional technique. A significant effect of the microstructure on the sensing properties of nanocrystalline SnO_2 has been found. The sensitivity to H_2S of the samples synthesized by cryosol method was shown to be higher than that of the samples obtained by traditional precipitation.

INTRODUCTION

Synthesis of nanomaterials is one of the most important challenges in modern materials science. The properties of these species are quite different from that of their large-grained counterparts [1,2]. Decreasing the size of crystallites results in the considerable increase of the surface free energy. This effect contributes to the thermodynamical properties of solids, such as phase transition temperature. On the other hand, nanocrystalline materials may possess some unusual (in comparison with the bulk phase) physical properties due to specific collective interactions in the systems containing from several hundreds to several hundred thousands atoms.

Specific physicochemical properties of nanomaterials are used in some practically important applications. Among those, one should mention gas sensors, catalysts, optical devices, lasers, and so on. However, wide application of functional nanocrystalline materials is hindered by a number of problems. One of the most important difficulties is the instability of nanostructures due to high value of the excessive surface free energy. Thus, synthesis of stable nanostructures is one of the most important issues in the modern materials science.

In Ref. [3] we applied a novel cryosol technique for the synthesis of nanocrystalline tin dioxide. The method is based on the pH-controlled treatment of aqueous solution of sodium stannate by cation exchange resin in H^+-form. The treatment yields colloidal solutions of tin acid stable at the wide range of concentrations and pHs. Freeze-drying of these sols was shown to result in the formation of nanocrystalline SnO_2. In the present paper we investigate the sensitivity of the cryosol-derived samples to gaseous H_2S as compared to that of the samples prepared by traditional precipitation. Influence of the microstructure on the functional properties of materials is discussed.

EXPERIMENT

Powders of SnO_2 were prepared by conventional hydrolysis of $SnCl_4$ (sample SX) and by cryosol technique (sample SK). These methods were described in detail elsewhere [3]. The obtained samples SX and SK were annealed at different temperatures (300-700°C) for 24 h in air.

Phase composition of the samples was studied by X-ray diffraction analysis on the Siemens conventional diffractometer with $Cu_{K\alpha}$ radiation and by electron diffraction method on Philips CM20 electron microscope operated at accelerating voltage of 200 kV. Microstructural studies were carried out in transmission mode on the 400 kV electron microscope JEOL 4000 and in scanning mode on the 200 kV electron microscope JEOL 2000FX-II. Tin dioxide powders for electron microscopy were suspended in ethanol and spread over titanium grid covered with amorphous carbon layer. The average grain size of SnO_2 was also estimated by means of X-ray diffraction broadening analysis of the {110} and {101} lines. The average crystalline size was determined from Scherrer's equation.

The surface composition was investigated using Auger spectroscopy (JAMP-10 CCS JEOL). The primary electron beam diameter was 100 nm and the points were spaced 500-1000 nm apart. Spectra were recorded from 50 to 1000 eV in dN(E)/dE mode. In order to study surface and bulk distribution of the components, the samples were etched by 3 kV Ar^+ beam (etching rate 130 Å/min).

The porosity of the samples was studied by means of capillary absorption method. The data on the nitrogen adsorption isotherms at 77 K have been obtained on the automated unit SORPTMATIC 1900 (Karlo Erba Str.).

The electrical conductance was investigated in the temperature range 77-300 K. The measurements were carried for the powders annealed at 300°C (24 h), compacted in pellets at 700 MPa and sintered at 700°C for 4 h. Metal electrodes were sprayed on the surface of the pellets (the distance between the contacts was choosen to be 2 mm). Sensing properties of the samples (steady state conductance and electrical response) were studied in the homemade automatic measurement unit at controlled temperature and gas phase composition.

RESULTS AND DISCUSSION

X-ray and electron diffraction studies of samples SX and SK annealed at $T \geq 300°C$ revealed the presence of single phase cassiterite (rutile-type structure). The Auger spectra of the samples demonstrated that both SX and SK contained a small amount of chlorine (<1 at.%), which cannot be removed from initial precipitate of α-stannic acid. In addition, sample SK contained sodium on the grain surface, which was probably adsorbed by colloidal particles during the cation exchange. The ratio of the normalized intensities, I_O/I_{Sn}, on the surface and in volume of the samples SX and SK is given in Table I. This ratio in bulk is similar for both samples and agrees with data obtained in Ref. [4]. At the same time, increase of surface I_O/I_{Sn} ratio for sample SK may be ascribed to abnormally high chemisorption of oxygen, which is probably associated with the enhanced surface charge.

Table I. Ratio of the normalized intensities, I_O/I_{Sn}, for samples SX and SK

	SX	SK
surface	0.7±0.1	1.4±0.2
volume	0.5±0.05	0.6±0.04

The change of the grain size D of SnO_2 crystallites upon the annealing temperature was studied by X-ray diffraction broadening analysis (Fig.1a). Crystallite size for sample SX substantially grows with the increase of annealing temperature (from ≈6 nm at 300°C up to ≈40 nm at 700°C). At the same time, grain size of sample SK changes negligibly and does not exceed 6 nm even after annealing at 700°C.

Measurements of the specific surface area as a function of annealing temperature are in a good agreement with X-ray diffraction data. The surface area of samples SX and SK annealed at 300°C were found to be approximately the same (≈ 70 m^2/g). However, the specific surface area of sample SX drastically decreases with the increase of annealing temperature (up to ≈ 5 m^2/g at 700°C) while for sample SK its value remains >50 m^2/g (Fig. 1b).

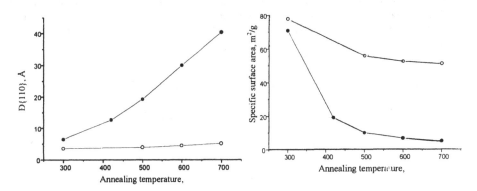

Fig 1. Grain size (a) and specific surface area (b) of samples SX (black circles) and SK (hollow circles) as a function of annealing temperature.

Transmission electron microscopy of samples SX and SK annealed at 700°C showed that crystallites in both cases had irregular spheroid shapes and were connected in aggregates via interparticle necks (Fig. 2 a,b). Average grain sizes calculated by statistical analysis of the dark field images are were 40±10 nm for sample SX and 10±1 nm for sample SK. One can see that these data are in good agreement with the results of X-ray diffraction broadening analysis discussed above

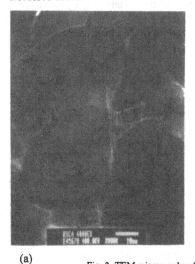 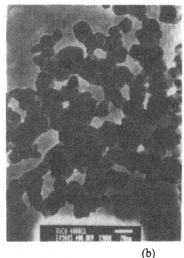

(a)   (b)

Fig. 2. TEM micrographs of samples SX (a) and SK (b) annealed at 700 °C.

391

The large specific surface area and thermal stability of nanocrystalline state for cryosol derived sample SK can be explained from the viewpoint of the theory suggested by Lifshits and Slezov [5-6]. Their idea is that a solid state system with the narrow particles' size distribution develops during the thermal treatment sufficiently more slowly than that characterized by the broad size distribution. In the case of cryosol technique, the uniformity of the particles' sizes is predetermined by the method of the synthesis. Indeed, the ion exchange results in the formation of the restricted number of polymeric hydroxocomplexes with nanometric sizes (primary particles), which are connected in the colloidal aggregates by relatively weak bonds. The freeze-drying enables to fix the structures prepared in a colloidal solution and convert them into solid state without significant change. Therefore, a solid state system consisting of the uniform primary particles should be stable to heat.

Studies of capillary absorption of N_2 at 77 K showed that samples SX are characterized by H2 type hysteresis loop, according to IUPAC nomenclature. (Fig.3a). Such type of absorption

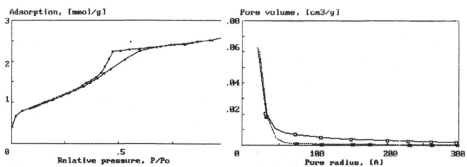

Fig. 3. Absorption (upper)-desorption isotherm (a) and corresponding pore size distribution (b) for sample SX annealed at 700°C. Upper and lower curves in Fig. 3b are integral and differential distribution functions respectively.

behavior indicates that these samples are composed of three-dimensional particles with non-uniform size distribution function. Another important feature of these samples is relatively broad pore size distribution (Fig. 3b).

At the same time, samples SK possess wide H3-type hysteresis loop, characteristic for

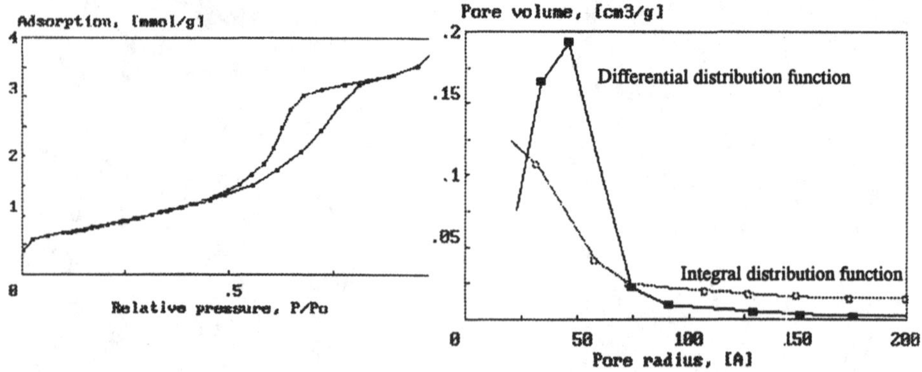

Fig. 4. Absorption-desorption isotherm (a) and corresponding pore size distribution (b) for sample SK annealed at 700°C.

specimens composed of parallel two-dimensional platelets (Fig. 4a). Pore size distribution in samples SK is much more uniform than in case of SX (Fig. 4b) and average pore size is close to 5 nm. These data provides additional argument for the uniformity of the particle size distribution in cryosol derived tin dioxide samples.

The microstructure of the samples was investigated by scanning electron microscopy. SEM micrographs presented in Fig.5 (a,b) demonstrate that the morphology of SX and SK powders is quite different. The sample SX consists of three-dimensional agglomerates with diameter of about 5 μm. On the contrary, the crystallites of the sample SK are agglomerated in

(a)

(b)

Fig. 5. SEM micrographs of samples SX (a) and SK (b) annealed at 700 °C.

thin two-dimensional platelets. From the fact that the platelets are transparent to the electron beam one can conclude that their thickness does not exceed 10 nm. This feature of the microstructure is probably due to peculiarities of freezing of colloidal solutions. Indeed, freezing in liquid nitrogen results in the formation of hexagonal ice that crystallizes in thin platelets during the rapid growth. These crystallites probably act as templates for the following formation of SnO_2. It also should be noted that particles of aluminum and iron (III) oxides obtained by cryosol technique are also characterized by two-dimensional morphology [7].

The steady state conductance of pure samples SX and SK annealed at 700°C was measured during the gradual temperature increase from 100°C to 500°C in oxygen atmosphere. The sample obtained by the cryosol method is characterized by significantly higher values of the electrical resistance. The low value of conductivity for sample SK may have several explanations. First, this sample consists of crystallites with size *ca.* 6 nm. This value is about twice as large as the thickness of the electron depletion layer L determined by the Debye length and the oxygen chemisorption energy [8,9]. That is why the whole volume of sample SK may be considered as the depletion layer and hence concentration of electrons in sample SK is lower than that in sample SX (grain size *ca.* 40 nm). It should be noted that resistance of the pure ceramic samples is mainly due to resistance of the grain boundaries. This suggests that microstructure makes considerable contribution to the conductance behavior [10]. Thus, the second possible reason for the higher resistivity of sample SK is higher amount of barriers (i.e. grain boundaries) per unit volume due to smaller grain size.

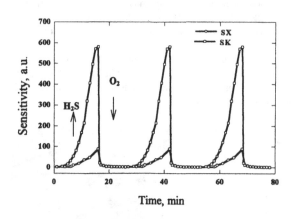

Fig. 6. Sensing response of samples SX and SK during the cyclic change of gas atmosphere O_2/N_2 + 1000 ppm H_2S at $T = 180°C$.

Investigation of conductivity in various atmospheres (O_2 and H_2S+N_2 gas mixture) is of special interest because of its direct relation to the sensing properties. Comparison of the typical response of samples SX and SK to a periodic change of the gas atmosphere at $T = 180°C$ is shown in Fig. 6. It is well seen that the sensitivity of sample SK is essentially higher than that of sample SX. One of the major factors responsible for good sensing properties is high ratio of surface to volume atoms. Ratio of surface to volume atoms should increase with the decrease of the grain size. At the same time, samples SK possess two-dimensional microstructure with slit-like pores, which facilitates gas intrusion into the sample.

CONCLUSION

Sensing properties of tin dioxide-based gas sensors characterized by different types of microstructure are investigated. The sensitivity of the samples prepared by the cryosol method is higher than that of the samples obtained by traditional precipitation. This effect is probably due to both the smaller grain size and two-dimensional microstructure.

REFERENCES

1. J.H. Fendler, I. Dekany, *Nanoparticles in Solids and Solutions*, Kluwer Acad. Publ., New York, 1986, 644p.
2. A.P. Alivisatos, MRS Bull., **20**, p. 23 (1995).
3. S.M.Kudryavtseva, A.A.Vertegel, S.V. Kalinin, N.N.Oleynikov, L.I.Ryabova, L.L.Meshkov, S.N.Nesterenko, M.N.Rumyantseva, A.M.Gaskov, J. Mater. Chem. 7 (11), p. 2269 (1997).
4. G.N. Advani, A.G.Jordan, J.Electronic Mater.9 (1), p. 29 (1980).
5. I.M. Lifshitz and V.V. Slezov, ZETP (Rus.) **35**, p. 479 (1958).
6. I.M. Lifshitz and V.V. Slezov, Solid State Physics, 1, p. 1401 (1959).
7. Kalinin S.V., Metlin Yu.G, Oleynikov N.N., Tretyakov Yu.D, Vertegel A.A., J. Mater. Res., **13** (4), p. 901 (1998).
8. C.Xu, J.Tamaki, N.Miura and N.Yamazoe, J. Mater. Sci. **27**, p. 963 (1992).
9. C.Xu, J.Tamaki, N.Miura and N.Yamazoe, J. Electrochem. Soc. Jpn., **58**, p. 1143 (1990).
10. M.N.Rumyantseva, M.Labeau, J.P.Senateur, G.Delabouglise, M.N.Boulova, A.M.Gaskov, J.Mater. Sci. & Engineering, **B41** , p. 228 (1996).

Chemical Solution Deposited Copper-doped CdSe Quantum Dot Films

N.Chandrasekharan, S.Gorer and G.Hodes
Weizmann Institute of Science, Rehovot, 76100 Israel.

ABSTRACT

Copper doped CdSe quantum dots (Qds) and larger nanocrystals were deposited by chemical solution deposition at two different temperatures (5°C and 80°C) and their morphological , optical and surface properties compared with similarly deposited undoped CdSe. Crystal sizes and structure were investigated by X-ray diffraction and transmission electron microscopy. All the samples exhibited cubic (sphalerite) structure except for the 80°C Cu-doped ones which were substantially hexagonal (wurtzite). For the 5°C samples, the crystal sizes were similar for the doped and undoped samples (ca. 3.4 nm). For the 80°C samples, the crystal size of the doped CdSe was much larger (84 nm) than the undoped one (6.3 nm) and extensive twinning occurred. Optical measurements on the doped samples showed the presence of surface states due to the copper seen as an increase in absorbance in the long wavelength region. Xray photoemission spectroscopy similarly showed tails in the valence band spectra of the doped samples. These tails were dominant in the 5°C samples and less so for the 80°C ones. XPS also showed that the Cu was present predominantly in the univalent state. The copper was found to be located on the QD surface for the 2-3 nm QDs and distributed both on the surface and in the bulk (or at twin boundaries) for the larger nanoparticles.

INTRODUCTION

The energy level structure of semiconductor quantum dots changes gradually from the band structure of the bulk semiconductor to an atomic-like structure as the quantum dot size decreases. This results in an increase in the effective optical band gap as a function of crystal size. Apart from this inherent flexibility in tailoring quantum dot properties, it is of interest to see how doping affects these properties further. For quantum dots in the size range investigated by us, a single dopant atom would drastically change the properties of the dots and therefore doping might be expected to affect the semiconductor properties in a different manner than it does for bulk semiconductors. We have studied Cu-doped CdSe quantum dots, deposited as films by chemical solution deposition and compared their properties to those of undoped CdSe quantum dot films.

EXPERIMENTAL

The CdSe quantum dot films were deposited on either glass or SnO_2-conducting glass from aqueous solutions containing $CdSO_4$, potassium nitrilotriacetate (KNTA) (as complexing agent) and Na_2SeSO_3. The Cu-doped films were prepared by adding 0.8 mM CuCl (1% of the Cd concentration) in KNTA to the deposition solution. The films were ca. 70 nm thick. Detailed information about the synthesis procedure may be obtained from reference [1].

RESULTS

X-ray Diffraction (XRD): XRD patterns of the 4 different films are shown in fig. 1 and the Scherrer formula used to estimate coherence length from the 42° peaks. The low temperature undoped and doped· samples both show an average particle size of 3.2 nm. and a cubic zinc blende structure. The high temperature films show a significant change in particle size on doping: 6 nm for the undoped sample and 7.7 nm (assuming spherical shape – 10.3 nm for hexagonal particle shape) for the doped one.

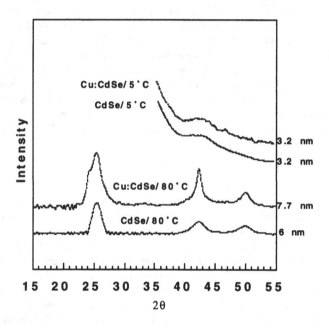

Fig. 1. X-ray diffraction pattern of the copper doped and undoped CdSe deposited at two different temperatures, 5˚C and 80˚C. The 25° peaks are not clearly seen in the 5ºC samples due to the overwhelming glass background in this region and are not shown here.

The undoped CdSe was cubic while the doped sample was predominantly of the wurtzite (hexagonal) structure. While the undoped 80°C sample is also cubic (one peak at 25°), the corresponding doped sample has a large hexagonal (wurtzite) component seen by the unresolved, but clearly visible triplet structure centered around 25°.

Transmission electron microscopy(TEM) : The mean particle sizes calculated from the TEM were in good agreement with those obtained from x-ray diffraction for the 5°C samples as well as the high temperature 80°C undoped films. The copper doped 80°C film however, revealed substantial deviation of the particle size obtained from the two measurements. The table below gives the particle sizes as derived from XRD and those obtained from TEM.

TEMPERATURE	XRD (nm)		TEM (nm)	
	CdSe	Cu:CdSe	CdSe	Cu:CdSe
5°C	3.2	3.2	3.6	3.1
80°C	6	7.7	6.3	84

The bright field images also revealed very large particle sizes for the doped 80°C film , some of the particles being several hundred nanometers. The particle size distribution was in general much wider for the doped films compared to the undoped ones. The electron diffraction revealed cubic symmetry for the low temperature 5°C films and in the case of the high temperature 80°C samples, the undoped CdSe showed cubic symmetry whereas the undoped was indexed within the hexagonal structure. Another interesting point was extensive twinning in the doped Cu:CdSe deposited at 80°C , which was totally absent in the other samples.

Xray photoemission spectroscopy (XPS) : XPS was used both to quantify the copper in the doped samples and to measure shifts in the Fermi level due to copper. The amount of copper (found to be in the univalent state) present in the doped samples was 5% (Cu to Cd ratio) for the 5°C sample and 2.4% for the 80°C one. On sputtering to a depth of ~ 1nm, it was observed that the copper was reduced in the case of the doped sample deposited at 5°C and enhanced for the 80°C one. Another point is that the Se-O invariably present in the undoped sample (5.4%) does not appear in the doped one. Figs. 2 shows the valence band region (VB) of the doped and undoped samples at 5°C and 80°C respectively. Certain common features are present in the two figures. In both, the doped sample shows a pronounced tail above the valence band edge of the undoped film. This effect is more pronounced in the 5°C doped sample. For the 5°C sample, the Fermi level moves from 1.4 eV above the valence band onset for the undoped sample to .0.9 eV for the doped one. The corresponding values for the 80°C samples are 1.2 eV. and 1.0 eV.

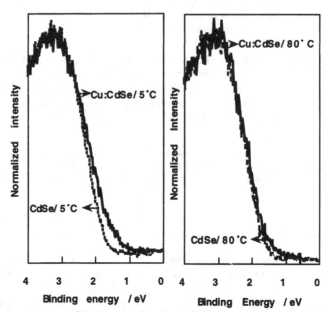

Fig.2. Valence band spectra of the doped and undoped CdSe deposited at 5°C and 80°C. A tail is seen near the valence band edge in the doped sample, which is not present in the undoped sample.

Optical Spectroscopy : The absorbance spectra of the 5°C and the 80°C samples are shown in Fig.3. The band edge of the undoped CdSe (5°C) is at 550 nm. It is difficult to estimate the band edge of the doped sample owing to the presence of the broad tail in the sub-band gap region. On treatment with KCN, the sub-band gap signal disappears completely. The 80°C films are red-shifted compared to the low temperature samples. The onset of the absorption is at 628 nm. A distinct tail in the long wavelength region is seen for the doped film. On treatment with KCN, the tail does not disappear completely, even after prolonged treatment. The tail in the sub-band gap region in the copper doped films presumably arises from surface state absorption induced by the copper doping. In the 5°C film the copper is present on the surface, and dipping in KCN removes all the copper (and thereby all the surface state contributions). In the 80°C film, copper is also present in the 'bulk' (possibly at the twin boundaries) of the crystals, and treatment with KCN removes only the surface copper.

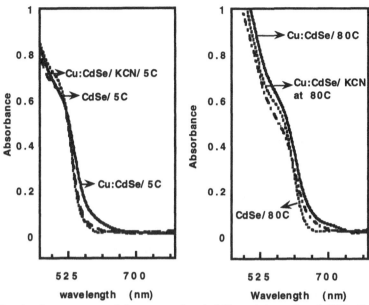

Fig. 3 The absorbance spectra of the copper doped CdSe and the undoped CdSe films at two different temperatures, 5°C and 80°C.

CONCLUSIONS

The structure and morphology of the low-temperature CdSe films are not much affected by copper doping. For the high temperature films, doping greatly increases crystal size and changes the phase from cubic to hexagonal. In the low temperature films, the copper is located at the surface and can be easily removed by cyanide solution as seen in the optical measurements. The high temperature films in contrast, contain at least part of the copper inside the crystals although this may be at twin boundaries.

ACKNOWLEDGEMENT

This research was supported by the Israel Ministry of Science.

REFERENCE

[1] S. Gorer and G. Hodes J. Phys. Chem. **98**, 5338, (1994)

CHARACTERIZATION OF ZnSe NANOPARTICLES PREPARED USING ULTRASONIC RADIATION METHOD

Jianfeng Xu *, Wei Ji *, Sing-Hai Tang *, and Wei Huang **
* Department of Physics, National University of Singapore, Lower Kent Ridge Road, Singapore 119260, Republic of Singapore
**Institute of Materials Research and Engineering, National University of Singapore, Singapore 119260, Republic of Singapore

ABSTRACT

ZnSe nanoparticles with an average size of 15 nm were prepared using the ultrasonic radiation method. The characterization was carried out by means of XRD, TEM, XPS and Raman scattering spectroscopy. The experimental results indicate that the as-prepared powders are composed of ZnSe with zinc-blende structure. The high purity of ZnSe particles was confirmed by XPS analysis. In the Raman spectra, TO and LO phonon modes were observed at 205 and 257 cm^{-1} in the ZnSe nanoparticles.

INTRODUCTION

Nanosized materials, as a type of new quantum solid material, have been subjected to extensive research since the 1980s for unique physical and chemical properties. A wide range of applications could be anticipated in the use of nanometer-sized particles in electronic devices (e.g., single electron tunneling effects) [1]. Semiconductor microdevices produced by epitaxy should offer in particular novel electronic properties. Usually, semiconducting nanoparticles are synthesized by chemical methods. The particle sizes are controlled by using matrices or ligand shells which also prevent agglomeration of the nanometer-sized particles. However, these matrices or ligand shells might influence the electronic properties of the material. It may be therefore more favorable to generate nanometer-sized semiconductor particles which are not in contact with matrices or ligand shells.

ZnSe nanoparticles have the potential for applications in areas such as nonlinear optical devices and fast optical switches. Because most reported ZnSe nanoparticles are prepared and stabilized by capping or passivation with organic materials, the influence of the stabilizer on ZnSe nanoparticles could not be avoided [2, 3]. Here, we present some results about ZnSe nanoparticles generated by a new method [4], which allows the production of particles without any stabilizing shells. First, Zn nanoparticles were prepared by the inert-gas evaporation technique with induction heating. These Zn nanoparticles were mixed with Na_2Se solution under ultrasonic radiation. Normally, Zn can not react chemically with Na_2Se effectively. However, in the nanometer range, this reaction may occur as Zn nanoparticles have high surface reactivity.

EXPERIMENTS

By the inert-gas evaporation technique with induction heating method, Zn nanoparticles were firstly prepared in Ar. From transmission electron microscopy (TEM) analysis, the average particle size can be estimated to be 80 nm.

ZnSe nanoparticles were synthesized in a Na_2Se solution-Zn suspension system under ultrasonic radiation. The schematic diagram of the ultrasonic radiation device has been described elsewhere [4]. Synthesis of ZnSe in the present system may proceed as follow:

$$Zn(s) + 2H_2O(l) \xrightarrow{ultrasound} Zn(OH)_2(s) + H_2(g)$$

$$Zn(OH)_2(s) + Na_2Se(aq) \xrightarrow{ultrasound} ZnSe(s) + 2NaOH(aq)$$

Sodium selenide was dissolved in distilled water, and the concentration of the solution was 0.1 mol/L. The Zn nanoparticles were added into the solution. During reaction, the solution was radiated by an ultrasonic device, and the solution temperature was kept at 50 °C. After a given reaction time of 5h, the sample was washed with distilled water using ultrasonic agitation and centrifuging, and finally washed with ethanol. The product was dried at 50 °C under reduced pressured for 3 h.

The crystal structure of the synthesized powders was analyzed by XRD using a Philips PW 1710 diffractometer equipped with a Cu K_α X-ray generator under ambient conditions. Particle sizes, shape and state of aggregation of the powders were determined from the direct observation carried out on a JEM-200 CX transmission electron microscope. XPS measurements were performed using a VG ESCALAB MK II with a Mg K_α (1253.6 eV) excitation source. For the Raman spectra, a micro-Raman system was used in the experiment. This system consisted of an Olympus microscope, a Spex 1704 spectrometer, and a CCD detector. The laser excitation was provided by an Ar^+ laser operated at a wavelength of 488 nm.

RESULTS

The nanocrystalline ZnSe was first characterized by XRD. Figure 1 shows a typical diffraction pattern from the ZnSe powders. From Figure 1 it becomes obvious that the samples only consist of one phase. This phase has the equilibrium structure of cubic ZnSe.

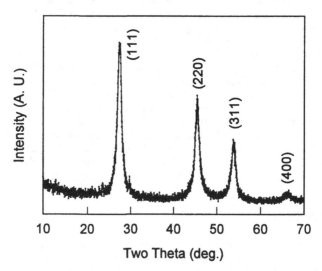

Figure 1. X-ray diffraction spectrum of as-prepared sample

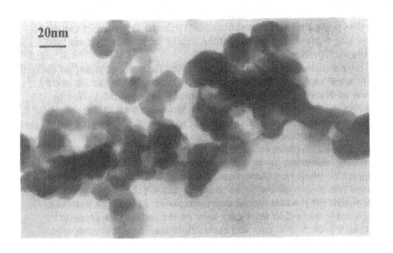

Figure 2. A TEM photo of ZnSe nanoparticles

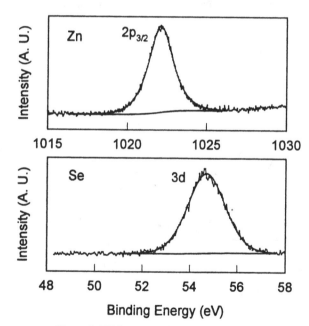

Binding Energy (eV)

Figure 3. XPS spectra of Zn 2p and Se 3d

One of the transmission electron micrographs for the sample is shown in Figure 2. Particle size analysis yields an average of 15 nm for the particles' diameter, somewhat larger than that estimated (12 nm) from XRD Scherrer formula. This larger particle size is probable due to undercounting of the smaller particles, which are hard to distinguish in the micrographs.

Using an X-ray photoelectron spectrometer, the states of atoms on the surface of the ZnSe particles were analyzed. The XPS spectra of the Se 3d and Zn 2p were obtained as shown in Figure 3. The peak position of Zn $2p_{3/2}$ lies at 1022.1 eV, corresponding to ZnSe. A broad Se 3d peak located at 54.7 eV is seen in our sample. This value is the same as that reported for the bulk ZnSe crystal [5]. No peaks for metallic Se and Se in SeO_2 are observed. Eventually, it is concluded that the sample is composed of single phase ZnSe, in agreement with the previous XRD results.

Raman scattering spectroscopy is one of the most important methods for obtaining information on phonons in nano-sized materials. Figure 4 presents the Zn nanoparticles before reaction with Na_2Se solution (a) and the ZnSe nanoparticles prepared after the reaction (b). As mentioned above, Zn particles have high reactivity, when they are exposed to air, a ZnO layer must form on the surface of a Zn particle. The Raman peak at 561 cm^{-1} in Figure 4(a) comes from the surface phonon mode within the ZnO coating [6]. The broad Raman band between 180 and 250 cm^{-1} seems related to the oxygen defects in the ZnO coating [6]. After the reaction, Zn and ZnO are missing, and ZnSe particles with zinc blende phase were prepared, their Raman spectrum is shown in Figure 4(b). The longitudinal optical (LO) and transverse optical (TO) phonon were observed at 257 and 205 cm^{-1}, respectively, in agreement with those reported in the previous work [7].

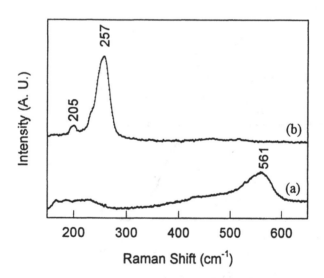

Figure 4. Raman spectra of (a) Zn particles before the reaction with Na_2Se and (b) ZnSe particles after the reaction of 5h

CONCLUSIONS

We have successfully prepared cubic ZnSe nanoparticles using ultrasonic radiation method. The sample is stable under ambient conditions. XPS analysis clearly reveals the formation of high purity ZnSe quantum dots. By using Raman scattering technique, we have observed the TO and LO phonon in the ZnSe nanoparticles at 205 and 257 cm^{-1}.

REFERENCES

1. D. V. Averin, A. N. Korotkov, and K. K. Likharev, Phys. Rev. B **44**, 6199 (1991).

2. G. Li and M. Nogami, J. Appl. Phys. **75**, 4276 (1994).

3. S.W. Haggata, D. J. Cole-Hamilton, and J. R. Fryer, J. Mater. Chem. 7, 1969 (1997).

4. J. F. Xu, W. Ji, J. Y. Lin, S. H. Tang, and Y. W. Du, Appl. Phys. A **66**, 639 (1998).

5. S. Major, S. Kumar, M. Bhatnagar, and K. L. Chopra, Appl. Phys. Lett. **49**, 394 (1986).

6. J. F. Xu, W. Ji, X. B. Wang, H. Shu, Z. X. Shen, and S. H. Tang, J. Raman Spectrosc. **29**, 639 (1998).

7. S. S. Mitra, O. Brafman, W. B. Daniels, and R. K. Crawford, Phys. Rev. **186**, 942 (1969).

NANOPARTICLE-BASED CONTACTS TO CDTE

D.L. Schulz, R. Ribelin, C.J. Curtis, D.E. King, D.S. Ginley
Photovoltaic and Electronic Materials Center and Basic Sciences Center,
National Renewable Energy Laboratory, Golden, CO 80401-3393

ABSTRACT

Our team has been investigating the use of particle-based contacts in CdTe solar cell technologies. Toward this end, particles of Cu-doped HgTe (Hg-Cu-Te) and Sb-Te have been applied as contacts to CdTe/CdS/SnO$_2$ heterostructures. These metal telluride materials were characterized by standard methods. Hg-Cu-Te particles in graphite electrodag contacts produced CdTe solar cells with efficiencies above 12% and series resistance (R_{se}) of 6 Ω or less. Metathesis preparation of Cu(I) and Cu(II) tellurides (i.e., Cu$_2$Te and CuTe, respectively) were attempted as a means of characterizing the valence state of Cu in the Hg-Cu-Te ink. For Sb-Te contacts to CdTe, open circuit voltages (V_{oc}s) in excess of 800 mV were observed, however, efficiencies were limited to 9%; perhaps a consequence of the marked increase in the R_{se} (i.e., >20 Ω) in these non-graphite containing contacts. Acetylene black was mixed into the methanolic Sb-Te colloid as a means of reducing R_{se}, however, no improvement in device properties was observed.

INTRODUCTION

In this paper, we discuss the use of particulate-based contacts for solar cells. The use of particulate systems is attractive for a number of reasons. First, given the existence of proper synthetic routes, particles with variable compositions (e.g., Hg$_{1-x}$Cu$_x$Te) can be synthesized with controlled stoichiometries. This ability allows easy screening and optimizing as compared to, for instance, sputtering where changes in film stoichiometry often require the purchase of a new target. Also, particulate-based inks are amenable to low-cost deposition approaches such as spray deposition, as well as screen and ink-jet printing. The use of small-sized particles could lead to narrower line widths and patterns not easily attainable by conventional approaches. There are some challenges to this particle-based approach. First, small-grained materials have very large surface areas which leads to a large number of grain boundaries (GBs) in sprayed layers. As many device properties are limited by GBs, surface passivation and/or other treatments may be required to produce useful devices. Second, a tradeoff is made between ink stability and film contamination. Coordinating Lewis base species (e.g., diglyme, ethylene glycol, trioctylphosphine oxide) are often added to the solvent/particle mixture as a way of stabilizing the ink. Incorporation of these additives, sometime referred to as surfactants, as well as solvent into the deposited films should be avoided as this often leads to non-optimal device performance.

Our team has been developing particle-based approaches to form back contacts to CdTe solar cells using nanoparticles. The standard NREL back-contact is a graphite-based Cu-doped HgTe material similar to that described by Britt and Ferekides [1]. The use of solution-synthesized Hg-Cu-Te particles in graphite-paste contacts was evaluated to determine feasibility versus the micron-sized particles that are normally used. Sb-Te particles were employed in preliminary studies as a contacting material based on the recent report of a Cu-free contact by the Parma group [2]. We present results on the application of Hg-Cu-Te and Sb-Te particles to CdTe.

EXPERIMENTAL

Attempts were made to produce Cu-doped HgTe, Cu_2Te, CuTe and Sb_2Te_3 materials by metathesis reactions of metal salts with sodium telluride in methanol at reduced temperature under inert atmosphere according to Eqs. 1-4, respectively. These particles were characterized by transmission electron microscopy (TEM), TEM-elemental determination by x-ray spectroscopy (TEM-EDS) and powder X-ray diffraction (PXRD). For the Hg-Cu-Te ink contacts to CdTe, graphite ink was prepared by mixing in air a known amount of Hg-Cu-Te with Electrodag 114 (Acheson Colloids Co., Port Huron, MI). Electrodag is a mixture of a proprietary acrylic resin, carbon black, and graphite in a mixed ketone solvent. Using a paint brush, this mixture was applied to NREL prepared CdTe heterostructures that had been previously treated with $CdCl_2$ and etched in nitric-phosphoric (NP) acid. Contact anneals were then performed under Ar. For the Sb-Te contacts, $CdCl_2$-treated CdTe films were received as a gift from Doug Rose at Solar Cells Inc. (Toledo, OH). Bromine-methanol (Br_2/MeOH) or NP acid surface treatments were performed and the Sb-Te in methanol (Sb-Te/MeOH) ink was immediately spray deposited at 150 °C. In some instances, degassed acetylene black was added to the Sb-Te/MeOH ink. The Sb-Te contacts were then annealed under N_2. In all cases, Ag paint was added via spin coating as the final layer, the entire composite was annealed at 100°C for 1 hr, and the devices were finished according to standard NREL protocol. Electrical properties of these solar cells were characterized by standard light and dark I-V.

$$HgI_2 + Na_2Te \xrightarrow{MeOH,-78°C,-2NaI} HgTe \qquad (1)$$

$$2CuI + Na_2Te \xrightarrow{MeOH,-78°C,-2NaI} Cu_2Te \qquad (2)$$

$$Cu(BF_4)_2 + Na_2Te \xrightarrow{MeOH,-78°C,-2NaBF_4} CuTe \qquad (3)$$

$$2SbI_3 + 3Na_2Te \xrightarrow{MeOH,-78°C,-6NaI} Sb_2Te_3 \quad ?? \qquad (4)$$

RESULTS

Hg-Cu-Te Particles

TEM characterization of as-synthesized Hg-Cu-Te particles showed the samples are micron-sized agglomerates composed of smaller particles ($d \leq 10nm$) with Hg, Cu, and Te observed by TEM elemental determination by x-ray spectroscopy (TEM-EDS). No C or O was observed after the brief exposure to atmosphere during TEM grid sample preparation. Powder X-ray diffraction (PXRD) showed the majority phase was the Coloradoite phase of HgTe (PDF 32-665) with no Cu-Te phases observed.

Cu_2Te and CuTe Particles – Attempted Preparation

In order to better understand the nature of our precursor ink and, further, how the valence state of Cu in the Hg-Cu-Te ink affects the performance of the contacts, attempts were made to form Cu_2Te and CuTe by metathesis reaction. PXRD of product particles showed the formation of Cu_7Te_4 (PDF 18-456) and $Cu_{2.72}Te_2$ (PDF 43-1401) for reaction of Na_2Te with CuI and $Cu(BF_4)_2$, respectively. X-ray photoelectron spectroscopy of sputtered samples of these two product materials showed both samples are approximately 85% Cu^{1+} and 15% Cu^{2+}. This determination was made by curve fitting the data for Cu^{1+} (932.4 eV) and Cu^{2+} (933.6 eV), with a goodness of fit of approximately 40%.

Sb-Te Particles

The metathesis reaction 2 molar equivalents of SbI_3 with 3 molar equivalents of Na_2Te in methanol at $-78°C$ did not produce Sb_2Te_3. Figure 1 shows the PXRD pattern for the product powder consists of a mixture of Sb (PDF 35-732) and Te (PDF 36-1452).

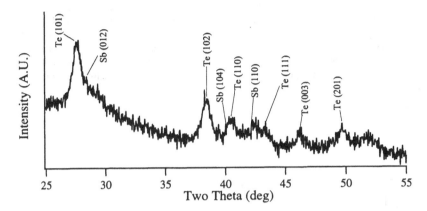

Figure 1. PXRD Pattern of Sb-Te Colloid Sample Dried Under a N_2 Stream.

CdTe Solar Cells Fabricated Using Hg-Cu-Te Contacts

Both the ink composition (i.e., solids loading) and annealing conditions (i.e., time and temperature) were varied during these preliminary studies. The first round compared two different Hg-Cu-Te in Electrodag loadings (i.e., 18 and 55 wt.%) at three different temperature/time schedules (i.e., 220°C/20min, 220°C/40min, and 320°C/20min). IV characterization showed the contacts made with 18 wt.% loaded inks performed better in all categories. Next, 18 wt.% was compared to 37 wt.% using a similar annealing dataset. Table I shows the annealing conditions employed as well as the electrical characterization of the solar cells. While the statistical significance of these data remains to be determined, it is encouraging to note stable contacts that exhibit V_{oc} in excess of 800 mV are observed in several cases with one cell being 12.6% efficient. Furthermore, the 220 °C anneal is approximately 50° lower than normally used for micron-sized Hg-Cu-Te/dag contacts.

Table I. I-V Data for CdTe Solar Cells Contacted with Hg-Cu-Te Dag.

IV Characteristic	220°C / 18 wt.%	20 min 37 wt.%	220°C / 18 wt.%	40 min 37 wt.%	270°C / 18 wt.%	20 min 37 wt.%
V_{oc} (mV)	793	811	813	820	812	799
J_{sc} (mA/cm²)	20.7	22.5	19.9	23.0	21.7	22.5
FF (%)	57.5	63.8	60.8	66.9	67.4	64.2
R_{series} (Ω)	5.8	5.0	6.1	4.4	4.5	5.0
R_{shunt} (Ω)	283	655	354	1063	433	452
Efficiency (%)	9.4	11.7	9.8	12.6	11.9	11.5

Solar Cells Fabricated Using Sb-Te Contacts

A preliminary study of sprayed Sb-Te contacts on CdTe was performed. The contact anneal in nitrogen gas was varied from 200 to 300°C. The IV characteristics of these solar cells were determined with interesting results. Figure 2 shows the V_{oc} and efficiency as a function of contact anneal temperature. The point at 150 °C represents an as-sprayed sample (i.e., a control) that was not subjected to further treatment. V_{oc} is noted to increase from less than 600 mV in the as-sprayed sample up to ~750 mV for samples annealed at 200-250 °C. V_{oc}s in excess of 800 mV are observed when a contact anneal of 300 °C was performed. Efficiencies also increased with contact anneal temperature from 2.4% in the control cell up to 9.1% in one sample annealed at 300 °C for 60 min. This set of solar cells all exhibited $R_{se} > 20$ Ω; a seeming limit to device performance.

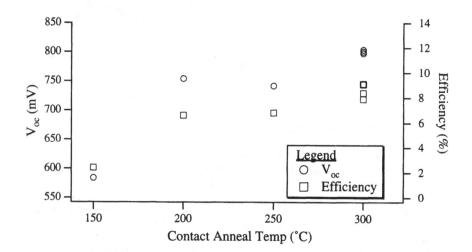

Figure 2. Effect of Contact Anneal Temperature on V_{oc} and Efficiency for Sb_2Te_3-contacted CdTe Devices.

In an attempt to reduce R_{se}, a conducting material, acetylene black, was added to the Sb-Te / methanol mixture. For this round of solar cell preparation, CdTe films from Solar Cells Inc. were subjected to surface treatments in NP and Br_2/MeOH etches prior to spray deposition of the Sb-Te-C mixture. A contact anneal was performed at variable time for 40 min under nitrogen gas. Table II shows the experimental conditions (i.e., surface treatment and annealing conditions) for this dataset.

The addition of C to the contacting layer caused a shunting problem during device isolation owing to flakes of acetylene black shorting directly to the SnO_2 transparent conducting oxide. The problem is not encountered when a graphite-based Electrodag system is employed presumably owing to the structural stability of the acrylic polymer-based dag layer.

The performance of CdTe solar cells contacted with C-containing Sb-Te is not dramatically improved over the C-free Sb-Te contacted films. Except in one case, R_{se} values remain greater than 20 Ω and R_{sh} is less than 400 Ω (Table II). V_{oc} never exceeds 800 mV, J_{sc} always remain less than 20 mA/cm^2, fill factors are less than 53% and efficiencies are limited to 7.2% in these devices.

Table II. Experimental Conditions and IV Characteristics of Sb-Te-C on CdTe Solar Cells

Etch type	Etch Time	Anneal Temp.	Anneal Time	V_{oc} (mV)	J_{sc} (mA/cm^2)	FF (%)	R_{se} (Ω)	R_{sh} (Ω)	Eff. (%)
Br$_2$/MeOH	10sec	300°C	40min	723	18.8	52.8	16.0	316	7.2
Br$_2$/MeOH	60sec	300°C	40min	761	19.1	48.8	25.2	450	7.1
Br$_2$/MeOH	10sec	350°C	40min	729	16.5	46.9	61.8	231	5.7
Br$_2$/MeOH	60sec	350°C	40min	757	17.2	48.7	67.7	356	6.4
NP	10sec	300°C	40min	718	18.2	47.1	30.9	394	6.1
NP*	60sec	300°C	40min	686	18.3	47.8	23.4	378	6.0
NP	10sec	350°C	40min	730	17.3	38.4	343	251	4.9
NP	60sec	350°C	40min	714	16.7	46.2	81.1	269	5.5
None		300°C	40min	660	2.6	41.9	78.6	269	0.7
None		350°C	40min	705	16.7	47.3	59.1	245	5.6

*Note: thickness of Sb-Te-C coating 1/2 that of all other samples.

CONCLUSIONS

Hg-Cu-Te and Sb-Te particle contacts to CdTe solar cells have been investigated. The use of a conducting resin (e.g., Electrodag 114) seems necessary to reduce R_{se} in the contacting layer. CdTe solar cells with good electrical properties have been isolated using Hg-Cu-Te/dag. The contact anneal temperature in these devices is ~50° lower than that used for standard ball-milled Hg-Cu-Te/dag contacts. Addition of Electrodag to our Sb/Te system is planned.

ACKNOWLEDGMENTS

The authors wish to thank Kim M. Jones for TEM characterizations. The gift of CdCl2-treated CdTe heterostructures from Doug Rose at Solar Cells is gratefully acknowledged. This research was funded by the U.S. Department of Energy, Office of Energy Research, Chemical Sciences Division and Materials Science Division and the U.S. Department of Energy National PV Program.

REFERENCES

1. Jeffrey S. Britt and Christos S. Ferekides, U.S. Patent No. 5 557 146 (September 17, 1996).

2. N. Romeo, A. Bosio, R. Tedeschi, and V. Canevari, "High Efficiency and Stable CdTe/CdS Thin Film Solar Cells on Soda Lime Glass," presented at the 2nd World Conference and Exhibition on Photovoltaic Solar Energy Conversion, Vienna, Austria, July 6-10, 1998.

Surface Stoichiometry of CdSe Nanocrystals

Jason Taylor,[I] Tadd Kippeny,[I] Jonathan C. Bennett,[II] Mengbing Huang,[II] Leonard C. Feldman,[II] and *Sandra J. Rosenthal.[I]

[I]Department of Chemistry, [II]Department of Physics, Vanderbilt University, Nashville, TN 37235
*Author to whom correspondence should be addressed: sjr@femto.cas.vanderbilt.edu

ABSTRACT

Rutherford backscattering spectroscopy (RBS) has been applied to determine the constitution of prototypical CdSe nanocrystals[1] synthesized by the high temperature pyrolysis of organometallics in trioctylphosphine oxide (TOPO). The diameter of the nanocrystals was varied from 22 Å to 58 Å. For all nanocrystal sizes the nanocrystals are Cd rich with an average Cd:Se ratio of 1.2 ± 0.1. The Cd:Se stoichiometery is independent of the starting Cd:Se ratio used for nanocrystal preparation, indicating the excess Cd is not associated with the initial abundance but is an intrinsic property of nanocrystals prepared by this method. The size dependence of excess Cd indicates the extra Cd is on the surface of the crystallite. The coverage of the surface passivating TOPO ligands has also been determined and is larger than reported in previous X-ray photoelectron spectroscopy (XPS) studies of Bowen Katari et al.[2] The origin and structural implications of non-stoichiometric Cd are discussed.

INTRODUCTION

Determining the constitution and structure of the surface of chemically synthesized nanocrystals is necessary to understand the influence of the surface on the optical and electronic properties of the nanocrystal, synthesizing complicated nanocrystal architectures, and for successfully implementing these nanocrystals in electronic, opto-electronic, electro-optic and biological imaging applications. CdSe nanocrystals prepared by the high temperature pyrolysis of organometallics in either TOPO or a mixture of trioctylphoshpine and TOPO (TOP/TOPO) have been the workhorse system for II-VI semiconducting nanocrystal research.[1] Nevertheless, an atomic level description of the entire nanocrystal (both core and surface) has not been obtained owing to the difficulty of probing the nanocrystal surface.

Previous studies of CdSe nanocrystals prepared by pyrolysis indicate these crystals are wurtzite[3] with C_{3v} symmetry,[4] prolate with a weakly size dependent aspect ratio,[4,5] and have a permanent dipole moment in the electronic ground state.[6] The dipole moment is thought to originate from the terminating pole planes of Cd and Se which must be present for a stoichiometric, wurtzite crystal (Figure 1). The Cd:Se stoichiometry of these nanocrystals has been assumed to be 1:1 and is not reported in publications describing the preparation of CdSe nanocrystals. Analysis of the stoichiometry of CdSe nanocrystals prepared in TOPO by XPS yielded a value of 1.02 ± 0.14.[2] XPS surface analysis of these nanocrystals indicate that surface Cd atoms are coordinated to TOPO while the majority of surface Se atoms are bare. The TOPO coverage was found to be size dependent with a larger fraction of the surface covered at smaller sizes.[2] Solid state ^{31}P NMR studies of 37 Å diameter CdSe prepared in TOP/TOPO identified both TOP (bonded to Se) and TOPO surface species, with 55% coverage of all surface atoms.[7]

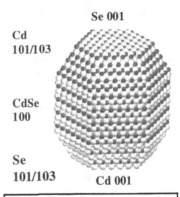

Se 001
Cd 101/103
CdSe 100
Se 101/103
Cd 001

Figure 1. Model of a stoichiometric, wurtzite CdSe nanocrystal[5] obtained from analysis of HR-TEM images.[4]

Thermal gravimetric analysis in conjunction with quantitative ¹H NMR indicated these same 37 Å nanocrystals have only 30% of the surface Cd atoms passivated with either TOPO or TOP bonded to Se.[8]

In the present study we initially set out to use RBS to verify the size dependent TOPO surface coverage of CdSe prepared in TOPO and quantify subsequent ligand exchange. Surprisingly, we found deviation from the assumed 1:1 stoichiometery of CdSe nanocrystals, which is shown to be due to a non-stoichiometeric surface layer.

MATERIALS AND METHODS

CdSe nanocrystals were prepared according to the methods reported by Bowen Katari et al.[2] and the "size focusing" method of Peng et al.[9] Briefly, a solution of dimethyl cadmium [$(CH_3)_2Cd$] and Se metal complexed with tributylphosphine [$P(CH_2CH_2CH_2CH_3)_3$] is pyrolized in TOPO [$OP((CH_2)_7CH_3)_3$, shown right] at 360°C. After addition of the starting materials the temperature of the reaction mixture drops and is maintained at 300 °C. The reaction is stopped after the desired nanocrystal size has been grown as determined by absorption. The nanocrystals are then isolated by precipitation in methanol and washed three times to remove excess starting material. RBS verified that all starting materials were removed after the third wash. Nanocrystals were also prepared by adding additional reagents during the reaction to "focus" the size distribution of larger nanocrystals.[9] Characterization by absorption, luminescence, X-ray diffraction and transmission electron microscopy indicates the expected properties for wurtzite CdSe nanocrystals. For this study the initial Cd:Se mole ratio was varied (1:1, 1.4:1, 1:9). For comparison a nanocrystal sample was prepared by the TOP/TOPO method of Murray et al.[10]

RBS samples were prepared by dropping 0.2 ml of a concentrated nanocrystal/toluene solution onto a 1 cm² graphite substrate and wicking off excess solution. These samples were immediately analyzed or stored under argon to minimize exposure to air. This is significant as Bowen Katari et al. demonstrated SeO₂ can form on the nanocrystal surface if nanocrystals prepared as a monolayer are exposed to air for 24 hrs. This SeO₂ can then desorb, which would lead to erroneous results of the stoichiometry. RBS experiments were performed in a high vacuum chamber (~10⁻⁷ torr) with 1.8 MeV He ion beams at normal incidence. Scattered ions were collected on axis with a solid state detector. No angular dependence (i.e. channeling through the nanocrystal) was observed. The spectrum of a Cd and Se thin film calibration sample yielded the correct 1:1 Cd:Se stoichiometery. The Cd:Se ratios obtained for the nanocrystals were independent of ion beam exposure time, indicating no change in stoichiometery was caused by the ion beam. Good RBS spectra (Cd and Se peaks resolved) are obtained for samples containing 10 monolayers or less of nanocrystals. A typical RBS spectrum is shown in Figure 2. The Cd, Se, P, and O peaks are clearly visible. To obtain the Cd, Se, and P ratios the individual peaks are integrated and normalized by the square of their atomic numbers.

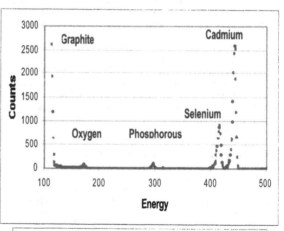

Figure 2. Typical RBS spectrum of a 25Å mean diameter CdSe nanocrystal.

RESULTS AND DISCUSSION

Cd:Se Stoichiometry:

The absolute Cd:Se ratios determined for nanocrystals ranging from 22 to 58 Å in mean diameter are summarized in Figure 3A. The mean diameter (average of the long and short axis diameters) is determined by correlating the wavelength of the first absorption feature with sizes determined by TEM and small angle X-ray scattering.[2] In all cases we find the nanocrystals are Cd rich. The Cd:Se stoichiometry decreases from 1.2 at 22 Å to 1.15 at 56 Å. To convert the Cd:Se ratios in Figure 3A to a number of excess Cd atoms (Figure 3B) it is only necessary to know the

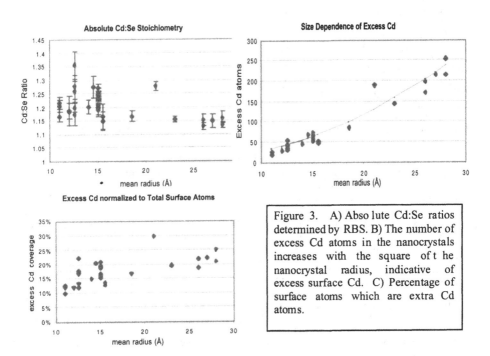

Figure 3. A) Abso lute Cd:Se ratios determined by RBS. B) The number of excess Cd atoms in the nanocrystals increases with the square of t he nanocrystal radius, indicative of excess surface Cd. C) Percentage of surface atoms which are extra Cd atoms.

total number of atoms in a nanocrystal of a given mean diameter. Using the experimentally deter-mined shape[4] (figure 1) and size dependent aspect ratio (long/short axis)[5] we have determined the number of atoms in a hexagonal prism capped with two frustums. The fit to the data in Figure 3B assumes a simple spherical model for the nanocrystal for which the addition of a shell of surface atoms contains a fraction of excess Cd atoms. The results of this fit indicate the extra Cd is on the surface of the nanocrystal.

This is consistent with our characterization of these nanocrystals. X-ray diffraction indi-cates the core is stoichiometric wurtzite and the luminescence is dominated by the band edge fluo-rescence, indicating the absence of vacancy defects in the nanocrystals. In Figure 3C the excess Cd is displayed as a fraction of the surface atoms. Figure 3C is obtained by converting the number of excess Cd atoms using the surface density of atoms for a hexagonal prism capped with two frustrums. Values are in the 15-20% range, consistent with a substantial surface enrichment. These values indicate that approximately 1 in every 5-6 surface atoms is an excess Cd atom.

Origin of Excess Cd:

What is the origin of the Cd rich nanocrystal surface indicated by the data presented in figure 3? The presence of excess Cd is not due to contamination from unreacted starting material as RBS analysis of the supernatants used in the washing procedure verified that unreacted Cd and Se are removed with the third methanol wash. Two hypothes remain. The excess Cd is either a result of the initial excess Cd used in the synthetic reaction mixture or is a result of the surface passivating ability of the TOPO ligand. To test the first hypothesis we varied the Cd:Se ratio used in the reaction mixture and synthesized 30 Å nanocrystals. Figure 4 displays the average volume stoichiometry obtained for 12 different 30 Å samples prepared with 1:1, 1.4:1, and 1.9:1 Cd:Se mole ratios of reactants. We find the Cd:Se stoichiometry is independent of initial Cd and Se abundance, therefore the excess Cd must be due to the stabilization of Cd dangling bonds by the passivating TOPO.

Dangling bonds from unpassivated Cd and Se surface atoms lead to charged, higher energy surfaces. As indicated by XPS,[5] surface Se atoms are bare while surface Cd atoms are coordinated to TOPO. It is therefore consistent that excess Cd would passivate surface Se atoms. However, not

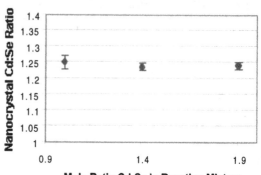

Mole Ratio Cd:Se in Reaction Mixture

Figure 4. Final nanocrystal stoichiometry in 30 Å nanocrytals vs. the initial Cd:Se mole ratios used in the synthetic preparation.

all surface Se atoms are equivalent. For example, Se atoms on the 101/103 facets (see figure 1) are ideal sites for excess Cd, as one Cd atom can tie up two Se dangling bonds without the addition of strain energy or requiring surface reconstruction. The Se atoms on the top surface Se 001 and equatorial CdSe 100 planes however have only one dangling bond. If excess Cd atoms passivated these Se surface atoms, one dangling Se bond would be passivated but two Cd dangling bonds would be created, so this is a less favorable site for excess Cd. Indeed, the result that one in every ten surface atoms is an excess Cd atom is consistent with preferential coverage of the Se atoms at the 101/103 facets. For example, a perfectly stoichiometric 32 Å nanocrystal has 54 Se atoms on the 101/103 facets. We find experimentally that 32 Å diameter nanocrystals have an excess of 50 Cd atoms. Preferential passivation of the Se facets is also consistent with the observation that CdSe nanocrystals prepared in TOPO have a permanent dipole moment,[6] as the Se 001 terminating plane retains its partial negative charge.

To test the hypothesis that the excess Cd is driven by TOPO passivation, we prepared a nanocrystal sample in a mixture of TOP/TOPO,[10] as TOP passivates surface Se atoms.[7] For a 30 Å diameter TOP/TOPO nanocrystal we also find the nanocrystals are Cd rich. The NMR results of Beccera et al.[7] indicate that of the 55% coverage of surface TOPO and TOP species, 70% is TOPO and 30% is TOP. These results are also consistent with TOPO driven excess Cd, and indicate TOPO coordination to Cd is stronger than the TOP passivation of surface Se. The stoichiometry of TOP/TOPO coated CdSe nanocrystals requires more detailed investigation.

TOPO Surface Coverage:

The absolute P:Cd stoichiometry determined by RBS is displayed in Table 1. The absolute P:Cd ratios determined by RBS are extremely similar to those determined by XPS.[5] The percentage of phosphorous atoms covering the surface atoms determined by RBS is presented in Figure 5. As XPS analysis indicated that there is no tributylphosphine coordinated to the nanocrystals,[5] each phosphorous atom represents one TOPO molecule. From this data we obtain an average TOPO surface coverage of 70%. This is consistent with TOPO driven excess Cd, as each surface Cd atom is passivated by a TOPO molecule. This TOPO surface coverage is however larger than that reported by XPS.[5] The difference originates from considering the atomic surface density of a hexagonal prism capped with two frustums.

Table 1. P:Cd Ratios

Radius(Å)	P:Cd ratio
11	0.34
12	0.33
12.5	0.33
14	0.39
15	0.36
15.5	0.24
18.5	0.34
21	0.3
23	0.27
26	0.20
27	0.21
28	0.21

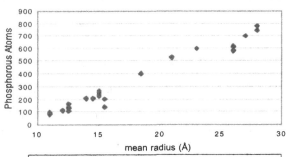

Figure 5. Size dependence of phosphorous atoms in TOPO molecules on the surface of CdSe nanocrystals.

CONCLUSIONS

Our results indicate CdSe nanocrystals prepared by the common pyrolysis method have an average Cd:Se stoichiometry of 1.2:1 ± 0.1. The size dependence of the excess Cd indicates the Cd is on the surface of the nanocrystals. On average, 70% of the nanocrystal surface is Cd, while 30% is Se. The stoichiometry is independent of the initial Cd:Se ratio in the starting material. The origin of the excess Cd is therefore attributed to the superior passivating ability of TOPO for surface Cd atoms, leading to an intrinsic excess of surface Cd atoms on CdSe nanocrystals. Analysis of the TOPO passivating ligand indicates that there is one TOPO molecule for every surface Cd atom, consistent with TOPO driven excess Cd. The absolute P: Cd ratios are in good agreement with those determined previously by XPS.[5] We find however a larger TOPO coverage (%70) than previously reported due to considering the atomic surface density of atoms for a hexagonal prism capped with two frustums. We have also demonstrated that RBS is a quantitative, sensitive tool for determining nanocrystal composition, surface stoichiometry, and surface ligand coverage at the atomic level.

AKNOWLEDGEMENTS

Jason Taylor acknowledges the Vanderbilt University Undergraduate Summer Research Program for sponsorship. We would also like to thank Andreas Kadavanich for helpful discussions and providing the nanocrystal model in Figure 1. Finally, we acknowledge the assistance of J. Keay and W. M. Augustyniak.

REFERENCES

1. The optical and structural properties of CdSe nanocrystals prepared by pyrolysis in TOPO and TOP/TOPO have been extensively studied. For review see A. P. Alivisatos, J. Phys. Chem. **100**, 13226 (1996).
2. J. E. Bowen Katari, V. L. Colvin, and A. P. Alivisatos, J. Phys. Chem. **98**, 4109, (1994).
3. S. H. Tolbert and A. P. Alivisatos, Science, **265**, 373 (1994).
4. J. J. Shiang, A. V. Kadavanich, R. K. Grubbs, and A. P. Alivisatos, **99**, 17418 (1995).
5. A. V. Kadavanich, The Structure and Morphology of Semiconductor Nanocrystals" Ph.D Thesis, University of California, Berkeley, 1997.
6. S. A. Blanton, R. L. Leheny, M. A. Hines and P. Guyot-Sionnest, Phys. Rev. Lett. **79**, 865 (1997).
7. L. R. Becerra, C. B. Murray, R. G. Griffin, and M. G. Bawendi, J. Chem. Phys. **100**, 3297 (1994).
8. M. Kuno, J. K. Lee, B. O. Dabbousi, F. V. Mikulec, and M. G. Bawendi, J. Chem. Phys. **106**, 9869 (1997).
9. X. Peng, J. Wickham, and A. P. Alivisatos, J. Am. Chem. Soc. **120**, 5343 (1998).
10. C. B. Murray, D. J. Norris, and M. G. Bawendi, J. Am. Chem. Soc. **115**, 8706 (1993).

FEMTOSECOND INTERFACIAL ELECTRON TRANSFER DYNAMICS OF CdSe SEMICONDUCTOR NANOPARTICLES

C. BURDA, T. C. GREEN, S. LINK and M. A. EL-SAYED*
Laser Dynamics Laboratory, Georgia Institute of Technology, Department of Chemistry, Atlanta, GA 30332-0400

ABSTRACT

The effect of the adsorption of an electron donor (thiophenol, TP) on the surface of CdSe nanoparticles (Nps) on the emission and electron-hole dynamics is studied. It is found that while the emission is completely quenched, the effect on the transient bleach recovery of the band gap absorption is only slight. This is explained by a mechanism in which the hole in the valence band of the NP is rapidly neutralized by electron transfer from the TP. However, the excited electron in the conduction band is not transferred to the TP cation, i. e. the electron does not shuttle via the organic moiety as it does when naphthoquinone is adsorbed [1]. The excited electron is rather trapped by surface states. Thus the rate of bleach recovery in the CdSe NP system is determined by the rate of electron trapping and not by hole trapping. Comparable conclusions resulted previously [2] for the CdS NP when the CdS-MV^{2+} system is studied. A comparative discussion of the electron-hole dynamics in these systems (CdSe-NQ, CdS-MV^{2+} and CdSe-TP) is given.

INTRODUCTION

In recent years the charge carrier dynamics of semiconductor nanostructures has drawn increasing interest [3-6]. The electronic properties of nanoparticles (NP) depend largely on their size and shape as the NP diameter reaches the dimensions of the exciton Bohr radius [7]. This usually occurs on the nanometer length scale. For such NPs charge carrier trapping by surface states becomes very important in determining the electron-hole dynamics and recombination, and thus the emission properties of these particles. For this reason, the controlled preparation and surface modification of semiconductor NP systems yields materials with new optical and electronic properties.

The electron transfer from the laser-excited CdSe NP across the NP interface to an adsorbed quinone (Q) was recently investigated [1]. Subsequent back electron transfer into the NP valence band occurred in the NP-Q system five orders of magnitude faster than the inherent electron hole recombination in CdSe NP (without added acceptors). In a study on the femtosecond dynamics of CdS NPs by Logunov et al. [2], the hole dynamics was determined by the electron removal from the excited CdS NP to the surface electron acceptor methylviologen (MV^{2+}). From the results, it was concluded that the trapping of the excited electrons is the rate limiting step (τ_{trap} = 30 ps) for the bleach recovery in CdS NPs.

In this paper we report the new preparation and investigation of CdSe NP composites with an electron donor on the surface. The addition of electron donors should lead to neutralization of the holes in the valence band of the excited NP. If the back electron transfer from the reduced NP to the donor cation is slow enough, it should result in the spectroscopic determination of the electron relaxation processes in the NP-electron donor system. In fact, the addition of thiophenol (TP) to the NP suspension led to efficient quenching of the CdSe NP emission, while the bleach recovery kinetics were not accelerated. The results suggest that in CdSe NPs, as in CdS NPs, electron trapping is the rate determining process (τ_{trap} = 40 ps) of the bleach recovery.

EXPERIMENT

The femtosecond transient absorption experiments were carried out as follows: An amplified Ti-Sapphire laser system (Clark MXR CPA 1000) was pumped by a diode-pumped, freqeuncy-doubled Nd:Vanadate laser (Coherent Verdi). This produced laser pulses of 100 fs duration (HWFM) and an energy of 1 mJ at 790 nm. The repetition rate was 1 kHz. A small part (4%) of the fundamental was used to generate a white light continuum in a 1 mm sapphire plate. The remaining laser light was split into two equal parts in order to pump two identical OPAs (Quantronix TOPAS). Each produced signal and idler waves with a total energy of 150 μJ. Tunable excitation wavelengths in the visible range were then produced by SHG and SFG of the signal wave. The excitation beam was modulated by an optical chopper (HMS 221) with a frequency of 500 Hz. The second OPA was used to generate tunable probe wavelengths outside the continuum range. The probe light was split into a reference and a signal beam. After passing the monochromator (Acton Research) both beams were detected by two photodiodes (Thorlab). The kinetic traces were obtained using a sample-and-hold unit and a lock-in-amplifier (Stanford Research Systems). The typical measured optical density (OD) changes were in the range of 50 mOD. For spectral measurements a CCD camera (Princeton Instruments) attached to a spectrograph (Acton Research) was used. The group velocity dispersion of the white light continuum was compensated. All measurements were carried out at lowest possible laser excitation powers. Emission spectra were measured on a PTI Quantum Master spectrofluorometer.

CdSe nanoparticles were synthesized according to the procedure developed by Murray, et al. [8]. Preparation of the CdSe NP composites was carried out by adding an excess of freshly sublimed 1,2-naphthoquinone or thiophenol (Aldrich, used as purchased) to the CdSe NP toluene solution, followed by shaking for several minutes. The direct adsorption of the electron acceptor and donor was confirmed by FT-IR studies on KBr pellets which will be presented elsewhere. We took UV/vis absorption, emission spectra and TEM pictures of our samples to check for possible degradation. We found no photodegradation during our experiments but the samples with addional electron acceptors or donors degraded after several days. As a consequence only fresh prepared samples were used in our studies.

RESULTS AND CONCLUSIONS

By addition of the thiophenol (10 μl per ml NP solution), the steady-state emission of CdSe NP was completely quenched. With femtosecond transient spectroscopy we monitored the bleach recovery of the CdSe NPs in the presence of TP. In Figure 1, the transient pump-probe spectra of the CdSe NP-TP system are shown.

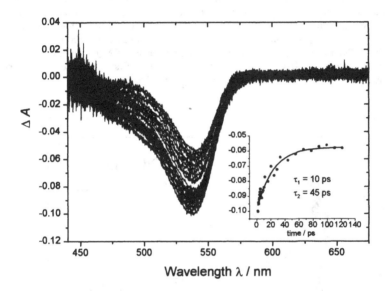

Figure 1. The time dependence of the bleach spectra of the CdSe NP in colloidal solution with thiophenol adsorbed on its surface. The inset shows the decay of the observed bleach at its maximum 550 nm.

The measured bleach recovery time, observed at 550 nm, became slightly longer than in the unperturbed NP (τ_1 = 10 ps, τ_2 > 45 ps versus τ_1 = 2.5 ps, τ_2 > 40 ps). The quenching of the CdSe NP emission suggests that the thiophenol efficiently reduces the holes in the valence band of the photoexcited NP and thus inhibited electron-hole recombination.

The bleach recovery in semiconductor NPs is determined by the dynamics of the excited charge carriers. From the fact that the CdSe NP emission appeared mainly as near band gap emission, we conclude that trapping of the charge carriers by deep traps can be neglected as an early relaxation process. The multi-exponential decay traces of the pure

NP indicate rather competitive kinetics between internal state relaxation and shallow surface trapping. Our results suggest, that in our time window of < 150 ps, mainly trapping of the electrons leads to the recovery of the bleach.

The addition of 1,2-naphthoquinone (NQ) to the CdSe NP suspension also led to efficient quenching of the steady-state near band gap emission. With femtosecond transient spectroscopy we monitored the bleach recovery of the CdSe NPs in the presence of NQ [1]. The resulting transient femtosecond spectra of CdSe NP in the presence of NQ showed the formation of an absorption between 600 and 680 nm. This was assigned to the previously reported absorption of the radical anion of NQ [5]. We determined that the rate of formation for this radical anion absorption (200 fs, observed at 650 nm) had the same rate constant as the formation of the bleach (200 fs, observed at 550 nm. In addition, the decay times for both the absorption of the NQ anion and the bleach recovery were the same (2.8 ps). The bleach recovery time was reduced from > 100 ns in the bare CdSe NP (without quinones) to less than 3 ps in the presence of NQ. This was attributed to the electron shuttling effect of the surface quinones which first accept the electron and subsequently shuttle it back to the hole in the NP valence band.

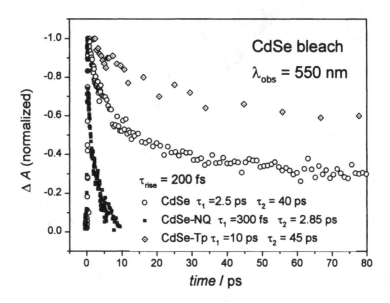

Figure 2. The effect of the adsorption of the electron acceptor naphthoquinone (dark squares) and the electron donor thiophenol (diamonds) on the transient bleach of the CdSe NP band gap absorption (circles) at 550 nm.

If the bleach decay of Cdse-TP is compared with the one of the pure CdSe NP, it is interesting to note that mainly the fastest observed lifetime increased (from 2.5 ps to 10 ps). This might suggest, that the fastest time component is due to electron-trapping processes which are eliminated by the addition of TP.

In summary, comparison of the effects of the addition of the electron donor (TP) with the effect of the addition of the electron acceptor (NQ) to the NP on the electron-hole dynamics leads to the following conclusions: 1.) electron transfer via the CdSe NP interface is demonstrated in both directions 2.) the added organic components on NP surfaces can act either as an efficient electron shuttle (NQ), as electron robber (MV^{2+}), or as electron donor (TP). 3.) The forward electron transfer rates in all cases are orders of magnitudes faster than the back electron transfer.

Figure 3 summarizes the proposed electron transfer dynamics for the different composite systems.

a) Naphthoquinone as Electron Shuttle   b) MV^{2+} as Electron Acceptor   c) Thiophenol (TP) as Electron Donor

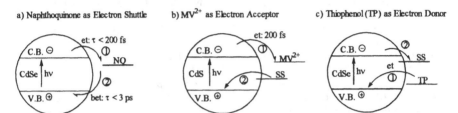

Figure 3. The electron-hole dynamics in CdS (b) and CdSe NP in presence of electron acceptor (a) or electron donor (c). For (b) and (c) it is proposed that the electron-hole dynamics is determined by trapping into surface states (SS).

In (a), the relaxation of the excited electron (step 1) and its combination with the hole (step 2) in the valence band (which leads to bleach recovery) occurs via the NQ and by-passes surface trapping [9,10] and/or changes in the state multiplicity [11,12]. As a result an acceleration of the bleach recovery is observed.

The electron-hole dynamics in the CdS-MV^{2+} system [2] is summarized in (b). In this case, the electron is rapidly transferred from the conduction band of the NP to the electron acceptor (MV^{2+}) (step 1). MV^+ does not shuttle the electron back to the hole in the valence band. The latter is thus trapped in 7 ps by the surface states (step 2). This led to the conclusion [2] that the observed bleach recovery time of 30 ps in the bare NP must be rate determined by the electron trapping and not by the hole trapping.

In (c), the hole in the valence band of the excited CdSe particles is first neutralized by the electron transfer from the electron donor (step 1). The removal of the excited electron (and thus the bleach recovery) takes place by surface trapping (step 2). The fact that the bleach recovery is not greatly affected by the addition of TP suggests that surface trapping in the CdSe NP is faster than the back electron transfer from the NP conduction band to the TP. It is then concluded, that the bleach recovery in the CdSe NP, like that in the CdS NP, is rate determined by the electron and not by the hole trapping.

ACKNOWLEDGEMENTS

C.B. wishes to thank the German Academic Exchange Service (DAAD) for a postdoctoral fellowship, T.G. thanks the Molecular Design Institute (ONR) at Georgia Tech and S.L. the German Fonds der Chemischen Industrie and BMBF for a Ph.D. fellowship. The continued support of this work by the Office of Naval Research (ONR grant No. N00014-95-1-0306) is greatly appreciated.

REFERENCES

1. Burda, C.; Green, T.C.; Link, S.; and El-Sayed, M. A. submitted to *J. Phys. Chem. A.*
2. Logunov, S.; Green, T.; Marguet, S.; and El-Sayed, M.A. *J. Phys. Chem. A* **1998**, *102*, 5652.
3. Alivisatos, A.P.; Harris, A.L.; Levinos, N.J.; Steigerwald, M.L.; Brus, L.E.; *J. Chem. Phys.* **1988**, *89*, 4001.
4. Bawendi, M.G.; Wilson, W.L.; Rothberg, L.; Caroll, P.J.; Jedju, T.M.; Steigerwald, M.L.; Brus, L.E.; *Phys. Rev. Lett.* **1990**, *65*, 1623.
5. Hunsche, S.; Dekorsy, T.; Klimov, V.; and Kurz. H. *Appl. Phys. B* **1996**, *62*, 3.
6. Efros, Al. L.; Rosen, M.; Kuno, M.; Nirmal, M.; Norris, D.J.; and Bawendi, M.G. *Phys. Rev. B* **1996**, *54*, 4843.
7. Brus, L.E. *J. Chem. Phys.* **1984**, *80*, 4403.
8. Murray, C.B.; Norris, D.J.; and Bawendi, M.G. *J. Am. Chem. Soc.* **1993**, *115*, 8706.
9. Bawendi, M.G.; Wilson, W.L.; Rothberg, L.; Caroll, P.J.; Jedju, T.M.; Steigerwald, M.L.; Brus, L.E.; *Phys. Rev. Lett.* **1990**, *65*, 1623.
10. Nirmal, M.; Murray, C.B.; and Bawendi, M.G. *Phys. Rev. B* **1994**, *50*, 2293.
11. Nirmal, M; Norris, D.J.; Kuno, M.; Bawendi, M.G.; Efros, Al.L.; and Rosen, M. *Phys. Rev. Lett.* **1995**, *75*, 3728.
12. Efros, Al. L.; Rosen, M.; Kuno, M.; Nirmal, M.; Norris, D.J.; and Bawendi, M.G. *Phys. Rev. B* **1996**, *54*, 4843.

Part VII

Microcrystalline and Polycrystalline Semiconductors

STRUCTURAL AND ELECTRONIC PROPERTIES OF LASER CRYSTALLIZED SILICON FILMS

T. Sameshima
Tokyo A&T University, Tokyo, 184-8588, JAPAN, tsamesim@cc.tuat.ac.jp

ABSTRACT

Fundamental properties of silicon films crystallized by a 30-ns-pulsed XeCl excimer laser were discussed. Although crystallization of 50-nm thick silicon films formed on quartz substrates occurred through laser heating at the crystalline threshold energy density of 160 mJ/cm^2, a higher laser energy density at 360 mJ/cm^2 was necessary to crystallize silicon films completely. Analyses of free carrier optical absorption revealed that phosphorus-doped silicon films with a carrier density about 2×10^{20} cm^{-3} had a high carrier mobility of 20 cm^2/Vs for irradiation at the crystallization threshold energy density, while Hall effect measurements gave a carrier mobility of electrical current traversing grain boundaries of 3 cm^2/Vs. This suggested that the crystalline grains had good electrical properties. As the laser energy density increased to 360 mJ/cm^2 and laser pulse number increased to 5, the carrier mobility obtained by the Hall effect measurements markedly increased to 28 cm^2/Vs because of improvement of grain boundary properties, while the carrier mobility obtained by analysis of free carrier absorption increased to 40 cm^2/Vs. A post annealing method at 190°C with high-pressure H$_2$O vapor was developed to reduce the density of defect states. Increase of carrier mobility to 500 cm^2/Vs was demonstrated in the polycrystalline silicon thin film transistors fabricated in laser crystallized silicon films.

INTRODUCTION

The pulsed laser heating method has been developed to make crystalline silicon films at a low processing temperature. Crystallization of silicon films and activation of dopant species are achieved by laser irradiation with a small energy density because of the rapid and local heating of silicon. The pulsed-laser crystallization has been widely applied to fabrication of various devices, for example, polycrystalline silicon thin film transistors (poly-Si TFTs) and solar cells in order to reduce fabrication costs or to use inexpensive substrates with low heat resistivity such as glass [1-5]. A carrier mobility higher than 500 cm^2/Vs has been achieved in TFTs fabricated with pulsed laser crystallized polycrystalline silicon films [6,7]. This indicates that electrical circuits with a high operation speed can be fabricated with poly-Si TFTs. Equipments for pulsed laser crystallization of silicon with a large area (>50 cm x 50 cm) have been developed for commercial production of poly-Si TFTs. However, crystalline properties have not been well understood. An optimum crystallization method have not been established.

This paper discusses fundamental properties of pulsed laser crystallized silicon films. Evolution of crystalline structures with increasing laser energy density is presented with experimental results of Raman scattering and optical reflectivity spectra. We also discuss the carrier mobility and grain boundary properties of doped polycrystalline silicon films basing on free carrier absorption and Hall effects measurements. We propose and demonstrate a post annealing method with H$_2$O vapor to reduce the density of defect states in polycrystalline silicon films. An increase in carrier mobility of poly-Si TFT is presented.

CRYSTALLIZATION OF SILICON FILMS

Crystallization of silicon films formed on quartz glass substrates occurred when silicon films were heated by 30-ns-pulsed XeCl excimer laser with a wavelength of 308 nm and a laser energy density above 160 mJ/cm². The laser light was effectively absorbed at the silicon surface within 10 nm in depth because of a large optical absorption coefficient at 10^6 cm⁻¹ at 308 nm, which results in heating of the silicon surface to high temperatures. Because the heating energy diffuses into the underlying substrate during laser irradiation, annealing properties of silicon films such as the threshold energy density for crystallization are governed by the thermal diffusivity, the density and the specific heat of underlying substrates if silicon films are thin enough compared with a heat diffusion length in Si of about 1 μm during laser irradiation [8,9]. The low heat diffusivity at ~1.4 W/mK of quartz glass [10] results in localization of heating energy at surface region about 300 nm during laser irradiation and the low threshold energy density for crystallization at 160 mJ/cm². Pulsed laser irradiation above the threshold energy density causes melting of the silicon surface. Surface melting is observed as increase in the optical reflectivity of the silicon surface in the visual and infrared ranges due to phase change from solid silicon to liquid silicon [11, 12]. Crystallization occurs according to solidification of liquid silicon after laser irradiation. The solidification (crystallization) velocity was governed by heat diffusion into substrate [13]. It was experimentally estimated to be approximate 1 m/s by change in electrical conductivity of silicon films during and after laser irradiation using transient conductance measurements [14,15], which takes an advantage of the high electrical conductivity ~10^4 S/cm of liquid silicon [16].

Figure 1 shows stokes Raman scattering spectra of 50-nm-thick amorphous silicon (a-Si) films on quartz glass substrates fabricated by low pressure chemical vapor deposition (LPCVD) at 425°C followed by 30-ns pulsed XeCl excimer laser crystallization. Laser irradiation was conducted in vacuum at 2×10^{-4} Pa at room temperature. Multiple-step energy irradiation was carried out. The laser energy density was increased step-by-step from 160 mJ/cm² to 360 mJ/cm² with a 40 mJ/cm² increment. Five pulses were irradiated at each step. The spectrum of initial a-Si films showed a broad peak around 450 cm⁻¹. A small and sharp peak around 516.8 cm⁻¹ associated with the transverse optical (TO) phonon of crystalline silicon appeared in the spectrum for the sample heated by laser with a laser energy density of 160 mJ/cm², the crystallization threshold. There was a residual broad spectral line in the lower wave number region than 516.8 cm⁻¹ owing to the optical phonon of amorphous silicon. The Raman spectrum shows that the silicon film annealed at the crystalline threshold energy density is a mixture of crystalline and amorphous phases. As the laser energy density increased from 160 to 360 mJ/cm², the peak intensity associated with the TO phonon of crystalline silicon increased and its full width at half maximum (FWHM) decreased from 20 cm⁻¹ to 4.9 cm⁻¹ under an instrumental resolution of 4 cm⁻¹, as shown in Fig. 1. This means that both the crystalline volume ratio and the crystalline grain size increased[17]. Transmission electron microscopy (TEM) measurement revealed that very fine crystalline grains with an average grain size about 10 nm were formed for laser irradiation at the crystallization threshold energy density. The grain size increased to about 100 nm as the laser energy density increased to 360 mJ/cm². No substantial defects were observed in crystalline grains for each laser irradiation case. Low crystalline volume ratios for irradiation at low laser energy densities probably resulted from the disordered amorphous state at grain boundaries and Si/SiO₂ interfaces.

The crystalline volume ratio was estimated from Raman integrated intensities of the crystalline and amorphous TO phonon peaks, which were separated assuming Gaussian-type spectral line shapes, as shown in Fig. 2. The crystalline volume fraction was about 0.4 for laser irradiation at the crystallization threshold energy density. Although the intensity of the broad spectral band associated with optical phonon of amorphous silicon was reduced and the crystalline volume ration increased as the laser energy density increased, a high energy density of 360 mJ/cm², much higher than the crystallization threshold, was required to reduce the optical phonon peak of amorphous silicon to the instrumental noise level. The large laser energy is necessary to melt 50-nm thick silicon films completely because of the large latent heat for melting silicon, 1810 J/g, in spite of the low specific heat, ~ 0.97J/gK [8]. If the temperature of the silicon surface reaches to the melting point of silicon for laser irradiation at 160 mJ/cm², a simple estimation shows that a high laser energy density at 370 mJ/cm² is necessary for melting the entire silicon films.

Samples crystallized with different laser energy densities had the same peak wave number of 516.8 cm^{-1}, which is lower than that of single crystalline silicon, 519.5 cm^{-1}. The red shift probably resulted from a large tensile stress caused by film formation on quartz substrates using LPCVD. The stress of the crystallized films was estimated as 5×10^8 Pa from this peak shift[18].

Measurements of optical reflectivity spectra in ultraviolet region are also useful for estimation of crystalline state of silicon films because crystalline silicon has a peak around 276 nm (E_2 peak) in optical reflectivity spectra which is caused by large joint density of states at the X point in the Brillouin zone [19], while amorphous silicon has no peak around 276 nm. Because the optical absorption coefficient is large, $\sim 10^6$ cm^{-1}, in the ultraviolet region, crystalline states at the surface and the bottom Si/SiO$_2$ interface regions can be investigated with 10-nm resolution in the depth direction. The E_2 peak heights divided by the peak reflectivity ($\Delta R/R$) were estimated for the top surface and the bottom interface of the samples crystallized by laser irradiation with different energy densities. The E_2 peak heights ($\Delta R/R$) were normalized by that of single crystalline, 0.172, as shown in Fig. 2. The reflectivity spectra at the bottom interface are different from those at the top surface because the refractive index of SiO$_2$ is larger than that of air (=1). In order to correct the substrate effect, the reflectivity spectra of the bottom interface were multiplied by a correction factor 0.73, which was obtained by calculations of reflectivities for SiO$_2$ substrate/Si interfaces with different refractive indexes and extinction coefficients between amorphous and single crystalline silicon.

Small E_2 peaks were detected in spectra of the top surface as well as the bottom interface for sample irradiated at 160 mJ/cm^2, the crystallization threshold energy density. This result means that the crystallization occurred throughout film thickness for energies just at the threshold energy density. The E_2 peak heights ($\Delta R/R$) at the both silicon surfaces was lower than that of single crystalline silicon. This shows that there were a substantial amount of disordered amorphous regions in the films even at the top surface. The E_2 peak heights ($\Delta R/R$) at the both surfaces increased as the laser energy density increased. The E_2 peak height ($\Delta R/R$) at the top surface became almost the same as that of single crystalline silicon for laser energy densities higher than 320 mJ/cm^2. Crystalline state was dominant at the top surface region for irradiation at these high energy densities. On the other hand, the E_2 peak height ($\Delta R/R$) at the bottom interface was lower than that of the top surface for all samples. This means that disordered amorphous states remained at the Si/SiO$_2$ interface. The crystalline nucleation probably initiates at the bottom interface and the crystalline growth proceeds from the bottom interfaces to the top surface because heat diffuses into the underlying substrate. If the density of nucleation sites is high and the average grain size is small at the bottom interface, disordered states may remain among small crystalline grains. The small broad band observed in Raman spectra is mainly due to the amorphous state located at the bottom interfaces especially for irradiation with large laser energy densities at 280 ~ 360 mJ/cm^2.

CARRIER MOBILITY AND GRAIN BOUNDARY PROPERTIES

An analysis of optical reflectivity or transimissivity spectra of doped silicon films gives the average carrier mobility and the carrier density in crystalline grains because free carrier optical absorption occurs via excitation induced by the electrical field of incident photons followed by energy relaxation in the crystalline grains [20,21]. On the other hand, the Hall effect measurement provides the effective carrier mobility of the electrical current which traverses many grain boundaries in polycrystalline silicon films, so it strongly depends on grain boundary properties. The free carrier absorption causes change in the refractive index as well as in the extinction coefficient, as in the following equations [21,22],

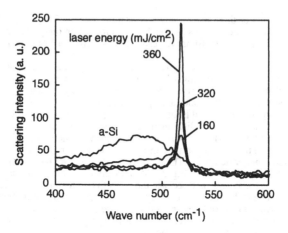

Fig.1 Stokes Raman scattering spectra for undoped 50-nm-thick silicon films on quartz substrate. An a-Si film was deposited by LPCVD, and was crystallized by laser irradiation whose energy density increased to the final energy densities, as shown in the figure, with a 40 mJ/cm² step. 5 pulses of laser-irradiation were supplied at each energy density step.

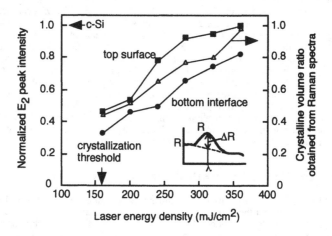

Fig. 2 E_2 peak heights divided by the peak reflectivity ($\Delta R/R$) at the top surface and the bottom interface as functions of the final laser energy density. The intensitiy was normalized with respect to $\Delta R/R$ for single crystalline Si. The samples were undoped 50-nm-thick silicon films crystallized by laser irradiation whose energy density increased to the final energy density with a 30 mJ/cm² step and with 5 pulses at each energy density step. The figure also shows the crystalline volume ratio estimated from integrated intensity of crystalline and amorphous TO phonon peaks in Raman scattering spectra.

$$n_f = \frac{1}{\sqrt{2}} \left[n_{Si}^2 - A + \left\{ \left(n_{Si}^2 - A \right)^2 + \frac{A^2 e^2}{4\pi^2 m^2 c^2 \mu^2 K^2} \right\}^{0.5} \right]^{0.5}$$

$$k_f = \frac{1}{\sqrt{2}} \left[A - n_{Si}^2 + \left\{ \left(n_{Si}^2 - A \right)^2 + \frac{A^2 e^2}{4\pi^2 m^2 c^2 \mu^2 K^2} \right\}^{0.5} \right]^{0.5}$$

$$A = Nm\mu^2 \varepsilon_0^{-1} (1 + 4\pi^2 m^2 \mu^2 c^2 e^{-2} K^2)^{-1} \tag{1}$$

, where n_{si} is the refractive index of undoped crystalline silicon, c is the velocity of light in vacuum, e is the electrical charge, m is the effective mass of the carrier, whose dependence on the carrier density was determined by Miyao et al. [23], K is the wave number, μ is the carrier mobility and N is the carrier density. The optical reflectivity was measured in the wave number range between 400 cm⁻¹ and 4000 cm⁻¹ using a conventional Fourier transform infrared spectrometry (FTIR) to analyze optical absorption caused by free carrier in the doped silicon films. In the infrared region, the reflectivity spectra for the air/Si/substrate system are modulated by the optical interference effect as given in the following equation[22],

$$R = \left| \left(r_0 + r_1 \exp \left(i4\pi \tilde{n}_f d K \right) \right) \left(1 + r_0 r_1 \exp \left(i4\pi \tilde{n}_f d K \right) \right)^{-1} \right|^2$$

$$r_0 = \left(1 - \tilde{n}_f \right) \left(1 + \tilde{n}_f \right)^{-1}, r_1 = \left(\tilde{n}_f - \tilde{n}_{SiO2} \right) \left(\tilde{n}_f + \tilde{n}_{SiO2} \right)^{-1},$$

$$\tilde{n}_f = n_f + ik_f, \; \tilde{n}_{SiO2} = n_{SiO2} + ik_{SiO2} \tag{2}$$

, where d is the film thickness, K is the wave number, $\tilde{n}_f$ and $\tilde{n}_{SiO2}$ are the complex refractive indexes of Si films and SiO₂ substrate, respectively, which consist of the real refractive indexes (n_f and n_{SiO2}) and the extinction coefficients (k_f and k_{SiO2}). Undoped crystalline silicon is transparent (k_f=0) in the infrared range. On the other hand, SiO₂ has a substantial absorption coefficient for wave numbers lower than 2000 cm⁻¹ so that the reflectivity includes optical absorption effect of the quartz substrate. The experimental spectra were compared to the spectra obtained by the interference calculation given by the eq.(2) with the refractive index and the extinction coefficient given by the eq. (1) with changing parameters of the carrier mobility and the carrier density until best agreement of the two spectra was obtained.

Figure 3 shows the carrier density (a) and the carrier mobility (b) obtained by the analyses of free carrier absorption and by Hall effect measurements. The samples were 50-nm-thick silicon films formed on quartz substrates implanted with phosphorus atoms. They were crystallized by laser irradiation with increasing the laser energy density step by step with single and 5 pulses at each energy density step. The both analyses of free carrier absorption and Hall effect measurements gave approximately the same carrier density, as shown in Fig.3 (a). The carrier density was slightly increased to 2.5×10^{20} cm⁻³ as the laser energy density increased from 160 mJ/cm² to 280 mJ/cm² and it leveled off for laser energy density higher than 280 mJ/cm². Dopant atoms were effectively activated and the high density of carriers were generated. The increase in the carrier density with increasing laser energy density means that the activation did not complete over the entire film in the low energy density region.

The analysis of free carrier absorption gave a large carrier mobility about 20 cm²/Vs for samples annealed at 160 mJ/cm², although the average grain size was small, about 10 nm as obtained by TEM measurements. The carrier mobility increased to 40 cm²/Vs as the laser energy density increased. Irradiation with a single pulse and five pulses for each energy density step resulted in approximately the same carrier mobilities. These results indicate that laser

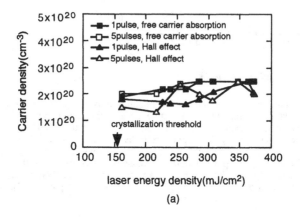

(a)

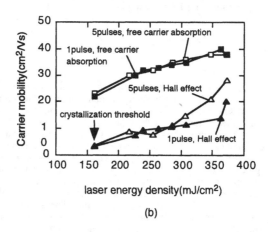

(b)

Fig. 3 Carrier density (a) and carrier mobility (b), which were obtained by analyses of free carrier absorption and by Hall effect measurements, as functions of the final laser energy density. The samples were phosphorus-doped 50-nm thick silicon films crystallized by laser irradiation whose energy density increased to the final energy density with 20~30 mJ/cm² steps and with single and 5 pulses at each energy density step.

irradiation formed crystalline grains with good electrical characteristics even for lower energy densities near the crystallization threshold energy. Irradiation with high laser energy densities improved the carrier mobility, which is ascribed to an increase in the grain size and reduction in the density of defect states. On the other hand, no change in the carrier mobility with increasing pulse number means that the laser pulse number is not an important process parameter for improvement of electrical properties of the crystalline grains. The Hall effect measurements resulted in lower carrier mobilities than those obtained by the free carrier optical absorption analysis, especially for the samples irradiated with low laser energy densities. The carrier mobility obtained by the Hall effect measurements markedly increased from 3 cm^2/Vs to 28 cm^2/Vs as the laser energy density was increased. This increase was much larger than the change of the mobility obtained by the free carrier absorption analysis. This result is interpreted as improvement of the grain boundary properties by laser irradiation with high energy densities because the carrier mobility obtained by Hall effect measurements is the drift mobility which is affected by carrier trap states and a high potential energy barrier at the grain boundaries. Disordered states with a high density of the dangling bonds at the grain boundary were reduced by laser irradiation with high energy densities because of the long melt duration and the low quenching rate [24]. Moreover, the carrier mobility obtained by Hall effect measurements for samples crystallized with 5 pulses was higher than that for samples crystallized with a single pulse for high laser energies of 310 mJ/cm^2 ~ 375 mJ/cm^2, which is contrasting to the above-mentioned observation that the carrier mobility analyzed by free carrier absorption did not depend on the laser pulse number. Multiple pulse irradiation is important for reduction of the average barrier height at grain boundaries. The maximum carrier mobility of 40 cm^2/Vs obtained by analysis of free carrier optical absorption was close to that of single crystalline doped silicon reported by Irvin [25]. This means that the crystalline grains formed by the laser crystallization method have approximately the same electrical properties as doped single crystalline silicon. The average energy barrier height (ΔE) at grain boundaries can be roughly estimated by simply assuming the mobility obtained by the Hall effect measurements (μ_H) as,

$\mu_H = \mu_A \exp(-\Delta E/kT)$, where μ_A is the average carrier mobility in crystalline grains obtained by the free carrier optical absorption, k is the Boltzmann constant and T is the absolute temperature ~300K. The estimated barrier height is 9 meV for the maximum carrier mobility (μ_H) obtained by Hall effect measurement was 28 cm^2/Vs.

Figure 4 shows the carrier mobilities as a function of temperature for 50-nm-thick phosphorus doped silicon films crystallized at 160 mJ/cm^2 and 280 mJ/cm^2. The carrier mobility obtained by the analysis of free carrier optical absorption increased monotonously as the temperature decreased from 473 K to 77 K for both samples. There was no change in the carrier density, 2×10^{21}cm^{-3}, with temperature. The mobility increase is interpreted as reduction of carrier scattering caused by the lattice vibration. On the other hand, the Hall effect measurements revealed that the mobility slightly decreased as the temperature decreased from 473 K to 77 K for the sample crystallized at 160 mJ/cm^2, while the poly-Si films formed at 280 mJ/cm^2 showed little temperature dependence of the mobility. When the grain boundary properties are poor and the average energy barrier at the boundaries is high, the carrier mobility obtained from the Hall effect measurements decreases as the temperature decreases, because the thermal excitation energy for carriers to cross the boundaries is reduced. No change in the carrier mobility with temperature for 280 mJ/cm^2-crystallized silicon films means that the energy barrier height was lower than the case for the 160 mJ/cm^2-sample.

POST ANNEALING WITH H$_2$O VAPOR

Reduction of defects by post annealing is important for a variety of device applications. Several techniques have been investigated. For example, hydrogen plasma and hydrogen radical treatment have been widely investigated [26-33]. We have recently proposed simple heat treatment with high-pressure H$_2$O vapor at low temperatures to reduce densities of defect states in silicon as well as SiO$_2$ films [34-37]. Figure 5 shows the effect of the annealing on the dark conductivity and the photoconductivity of the undoped laser-crystallized polycrystalline films.

433

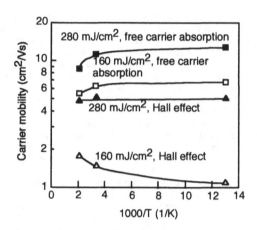

Fig. 4 Carrier mobilities, which were obtained by analyses of free carrier absorption and by Hall effect measurements, as functions of temperature. The samples were phosphorus-doped 50-nm thick silicon films crystallized by laser irradiation whose energy density increased to 160 mJ/cm^2 and 280 mJ/cm^2 with a 20 mJ/cm^2 step and with single and 5 pulses at each energy density step. The carrier density was about 2×10^{21} cm^{-3} for both samples.

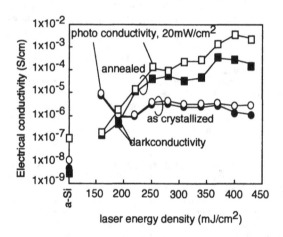

Fig. 5 Effects of heat treatment under H$_2$O vapor on dark and photo-electrical conductivities. The samples were 50-nm-thick LPCVD a-Si films and crystallized by laser irradiation whose energy density increased to the final energy densities with 20~30 mJ/cm^2 steps and with 5 pulses at each energy density step. Heat treatment was carried out under 3.6×10^5-Pa-H$_2$O vapor for 1 h at 190°C. Photoconductivity was measured under illumination of an 1600-K-black-body radiation with an intensity of 30 mW/cm^2.

The heat treatment continued for 1 h at 190 °C with 3.6×10^5-Pa-H_2O vapor pressure. The dark conductivity for as crystallized samples ranged from 10^{-7} to 10^{-6} S/cm when the crystallization laser energy density was varied from 160 mJ/cm^2 to 430 mJ/cm^2. Very low photoconductivities under illumination of 1600-K-black body radiation at 30 mW/cm^2 were observed for these samples. This means that the as crystallized poly-Si films have high densities of defect states which cause recombination of photo induced carriers for. After heat treatment, the dark conductivity increased to ~10^{-4} S/cm as the laser energy density increased to 430 mJ/cm^2. The photoconductivity also increased to 10^{-3} S/cm for sample crystallized at the laser energy density above 360 mJ/cm^2. The increase in the dark conductivity and photoconductivity means the density of defect states was reduced and the carrier mobility thereby increased after heat treatment. The silicon dangling bonds were probably terminated by hydrogen atoms or oxygen atoms from the H_2O incorporated into the films. The H_2O annealing was effective especially for the silicon films crystallized at high laser energy densities above 360 mJ/cm^2, because for these samples the disordered amorphous regions are little and the regions with high densities of dangling bonds are only localized near the grain boundaries, as the results of Figs 1~4 revealed. No change in the optical absorption coefficient spectra in a photon energy range from 1.2 to 2.5 eV was observed within the detection limit, 100 cm^{-1}, after heat treatment [36]. FTIR measurement revealed that the increase of Si-H bond was less than ~2×10^{20} cm^{-3} (~0.4 at. %) [36]. These results indicate that only a small amount of hydrogen or oxygen atoms were incorporated in the films, but they effectively terminated the dangling bonds.

Heat treatment with high-pressure H_2O vapor at 190°C for 1 h was applied to annealing of n-channel-top-gate-type-poly-Si TFTs fabricated using laser crystallized poly-Si films [37]. A 20-nm-thick hydrogenated amorphous silicon films doped with 2 atom % phosphorus was first formed on a glass substrate at 250°C using conventional plasma-enhanced chemical vapor deposition (PECVD) and then islands as source and drain regions were defined with plasma etching. Undoped a-Si:H films 20 nm thick were subsequently deposited using PECVD over the entire substrate. Island patterning was conducted by etching the amorphous layers. The silicon layers were crystallized by laser irradiation in vacuum to form undoped and doped poly-Si regions. In order to realize a smooth surface by slowly desorbing hydrogen from silicon films, irradiation with multiple energy steps was performed. The final laser energy was controlled to 210 mJ/cm^2 and 235 mJ/cm^2, which was below the amorphization threshold (240 mJ/cm^2) for 20-nm-thick films. Laser heating induced diffusion of dopant atoms from the underlying doped region vertically through the entire thickness of the Si layers so that doped poly-Si regions as well as undoped poly-Si regions were formed. A triode-type remote plasma CVD was used to form 160-nm-thick SiO$_2$ films[38]. Through the decomposition of SiH$_4$ gas by oxygen and helium radicals, SiO$_2$ films were formed on a substrate heated at 250 °C. Then, Al gate, source and drain electrodes were formed. The TFT samples were then heated at 190°C with H_2O vapor pressures from 1.8×10^5 Pa to 7.2×10^5 Pa for 1 h.

Figure 6 shows output characteristics before and after heat treatment at 190°C with 1.8×10^5-Pa H_2O vapor for 1 h for a poly-Si TFT which showed an initial carrier mobility of 320 cm^2/Vs. The heat treatment increased the drain current at any gate voltage. This means that heat treatment increased the carrier mobility in the channel region. Figure 7 shows changes in the carrier mobility for TFTs whose initial carrier mobility were 320 cm^2/Vs and 50 cm^2/Vs. The difference in the initial mobility was mainly caused by differences in laser irradiation energy density. The carrier mobility was increased to 500 cm^2/Vs after the heat treatment for both samples. The subthreshold slope was decreased after heat treatments. These results indicate that heat treatment with high-pressure H_2O vapor improved properties of SiO$_2$/Si interfaces as well as polycrystalline silicon films. This result demonstrated that the heat treatment with high-pressure H_2O vapor is effective even when it is carried out after all the device-fabrication steps are completed. H_2O probably diffused into silicon and SiO$_2$ effectively during heating. This is an advantage for establish a simple fabrication process.

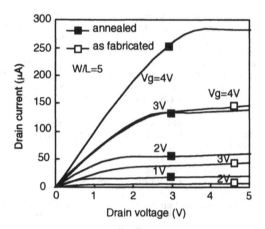

Fig.6 Output characteristics of the TFTs as-fabricated and annealed at 190°C with 3.6x10⁵ Pa H_2O vapor. The ratio of channel width to channel length (W/L) was 5. The thickness of the gate SiO_2 insulator was 160 nm. Initial carrier mobility was 300 cm²/Vs.

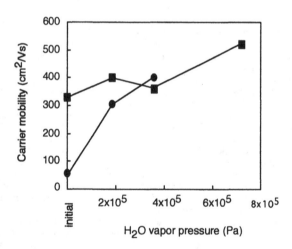

Fig.7 Carrier mobility of TFTs as a function of H_2O vapor pressure for heat treatment at 190°C for 1h. The initial carrier mobilities of the TFTs were 50 cm²/Vs and 320 cm²/Vs.

CONCLUSIONS

Irradiation with a 30-ns-pulsed-XeCl excimer laser with a wavelength of 308 nm caused crystallization of 50-nm thick silicon films formed on quartz substrates by LPCVD with the crystalline threshold energy density of 160 mJ/cm². Measurements of TO phonon peaks in Raman scattering spectra and E_2 peak in reflectivity spectra revealed that silicon films had mixed states of crystalline and disordered amorphous silicon. The crystalline volume ratio increased to approximately 1 as the laser energy density increased from 160 mJ/cm² to 360 mJ/cm² in a step-up laser energy irradiation method with 5 pulse each energy density step. From measurements of the E_2 peak, there are disordered state at the bottom SiO_2/Si interface even for high laser energy density cases at 280 ~ 360 mJ/cm². Analyses of free carrier optical absorption revealed that phosphorus doped silicon films with a carrier density about 2×10^{20} cm⁻³ had a high carrier mobility at 20 cm²/Vs for irradiation at the crystallization threshold energy, while Hall effect measurements gave a carrier mobility of electrical current traversing grain boundaries of 3 cm²/Vs. This results suggested that the crystalline grains had good electrical properties although the crystalline grain size was as small as at ~10 nm and that there was substantial disordered states between crystalline grain, which reduced the carrier mobility of electrical currents traversing grain boundaries. The carrier mobility obtained by analysis of free carrier absorption increased to 40 cm²/Vs as the laser energy density increased to 360 mJ/cm². This increase in carrier mobility probably resulted from increase in the grain size. The carrier mobility obtained by analysis of free carrier absorption did not depend on the laser pulse number. The carrier mobility obtained by the Hall effect measurements markedly increased to 28 cm²/Vs as the laser energy density increased to 360 mJ/cm² because of improvement of grain boundary properties. Irradiation with multiple laser pulse number was effective to improve grain boundary properties which resulted in increase in the carrier mobility especially for laser energy density higher than 300 mJ/cm². Heat treatment with 3.6x10⁵-Pa-H_2O vapor at 190°C increased the dark electrical conductivity from 10⁻⁶ S/cm to 10⁻⁴ S/cm and increased photoconductivity to 10⁻³ S/cm under illumination of 1600-K black body radiation with 30 mW/cm² for the samples crystallized at laser energy density higher than 340 mJ/cm². These results indicates that the heat treatment reduced the density of defect states and increased the carrier mobility. Increase in the carrier mobility by heat treatment high-pressure H_2O vapor was demonstrated for the polycrystalline silicon thin film transistors fabricated in laser crystallized silicon films. The carrier mobility was increased to 500 cm²/Vs after heat treatment because of defect reduction at the SiO_2/Si interfaces as well as in the silicon films.

ACKNOWLEDGMENTS

The author thanks M.Kondo, A.Matsuda, S.Yamasaki, S.Higashi, S.Inoue, H.Ohshima for their support.

REFERENCES

1. T. Sameshima, S.Usui and M.Sekiya, IEEE Electron Dev. Lett. **EDL-7,** 276 (1986).
2. K. Sera, F. Okumura, H. Uchida, S, Itoh, S. Kaneko and K. Hotta, IEEE Trans. Electron Devices **36,** 2868 (1989).
3. T. Serikawa, S. Shirai, A. Okamoto and S.Suyama, Jpn. J. Appl. Phys. **28,** L1871 (1989).
4. T.Noguchi and Y.Kanaishi, IEEE Trans. Electron Device Lett. **10,** 543 (1989).
5. H. Kuriyama, T.Nohda, Y.Aya, T.Kuwahara, K.Wakisaka, S.Kiyama and T.Tsuda, Jpn. J. Appl. Phys. **33,** 5657 (1994).
6. S.Shirai and T.Serikawa, IEEE Trans.Electron Devices **39,** 450 (1992).
7. A.Kohno, T.Sameshima, N.Sano, M.Sekiya and M.Hara, IEEE Trans. Electron Devices **42,** 251 (1995).
8. *Lanbolt-Boernstein,* eds. O. Madelung, M. Schultz and H.Weiss (Springer-Verlag, Berlin

1984) Vol.17-c, Chap 6.
9. H. S. Carslaw and J. C. Jaeger, *Conduction oh Heat in Solid* (Oxford University Press, Oxford 1959) Chaps. 2 and 10.
10. A. Goldsmith, T.E. Waterman and H.J.Hirschhorn, *Handbook of Thermophysical Properties of Solid Materials* (Pergamon, New York, 1961) Vols. 1 and 3.
11. D. H. Auston, J. A. Golovchenko, P. R. Smith, C. M. Surko and T. N. C. Venkatesan, Appl. Phys. Lett. **48**, 33 (1982).
12. T. Sameshima, S. Usui and H. Tomita, Jpn. J. Appl. Phys. **26**, L1678 (1987).
13. R.F.Wood and G.E.Giles, Phys. Rev. **B23**, 2923 (1981).
14. G.J.Galvin, M.O.Thompson, J.W.Mayer, R. B. Hammond, N. Paulter and P.S. Peercy, Phys. Pev. Lett. **48**, 33 (1982).
15 T. Sameshima, M. Hara and S. Usui, Jpn. J. Appl/ Phys. **28**, 1789 (1989).
16 V. M. Glazov, S. N. Chizhenvskaya and N. N. Glagoleva, *Liquid Semiconductors*, (plenum Press, New York, 1969) p60.
17. Z. Iqbal and S.Veprek, J. Phys.C: Solid State Phys. **15**, 377 (1982).
18. B. A. Weinstein and G. J. Piermarini, Rhys. Rev. **B12**, 1172 (1975).
19. J. R. Chelikowsky and M. L. Cohen, Phys. Rev. **B10**, 5095 (1974).
20. H. Engstrom, J. Appl. Phys. **51**, 5245 (1980).
21. T.Sameshima, K.Saitoh, M.Satoh, A.Tajima and N.Takashima, Jpn. J. Appl. Phys. **36**, L1360 (1997).
22. M. Born and E. Wolf, *Principles of Optics* (Pergamon, New York, 1974) Chaps 1 & 13.
23. M.Miyao, T. Motooda, N.Natuaki and T. Tokuyama, *Proc. of Laser and Electron-Beam Solid Interactions and Material* (Elsvier, North Holland, 1981) p.163.
24. T. Sameshima and S.Usui, J. Appl. Phys., **74**, 6592 (1993).
25. J. C. Irvin, Bell System Tech. J. **41**, 387 (1962).
26. M.Rodder and S.Aur, IEEE Electron Device Lett. **12**, 233 (1991).
27. R.A.Ditizio, G.Liu, S.J.Fonash, B.-C.Hseih and D.W.Greve, Appl. Phys. Lett. **56**, 1140 (1990).
28. I-W. Wu, A.G.Lewis, T-Y, Huang and A.Chiang, IEEE Electron Device Lett. **10**, 123 (1989).
29. K.Baert, H.Murai, K.Kobayashi, H.Namizaki and M.Nunoshita, Jpn. J. Appl. Phys. **32**, 2601 (1993).
30. T. C. Lee and G. W. Neudeck, J. Appl. Phys. **54**, 199 (1983).
31. U.Mitra, B.Rossi and B.Khan, J. Electrochem. Soc. **138**, 3420 (1991).
32. D. Jousse, S. L. Delage and S. S. Iyer, Philos. Mag. **B63**, 443 (1991).
33. I. Yamamoto, H. Kuwano and Y.Saito: J.Appl. Phys. **71**, 3350 (1992).
34. T.Sameshima and M.Satoh, Jpn. J. Appl. Phys. **36**, L687 (1997).
35 T.Sameshima, K.Sakamoto and M. Satoh: to be published in Thin Solid Films (1998).
36. T. Sameshima, M. Satoh, K. Sakamoto, K. Ozaki and K.Saitoh, Jpn. J.Appl. Phys. **37**, L1030 (1998).
37. T. Sameshima, M. Satoh, K. Sakamoto, K. Ozaki and K.Saitoh, Jpn. J. Appl. Phys. **37**, 4254 (1998).
38. N.Sano, K. Kohno, M. Hara, M. Sekiya and T. Sameshima, Applied Phys. Lett., **65**, 162 (1994).

NUCLEATION PROCESSES IN Si CVD ON ULTRATHIN SiO_2 LAYERS

T. Yasuda*, D. S. Hwang**, K. Ikuta*, S. Yamasaki*, and K. Tanaka*

* Joint Research Center for Atom Technology (JRCAT)-National Institute for Advanced Interdisciplinary Research (NAIR), Tsukuba, 305-8562, JAPAN, tyasuda@jrcat.or.jp
** Joint Research Center for Atom Technology (JRCAT)-Angstrom Technology Partnership (ATP), Tsukuba, 305-0046, JAPAN

ABSTRACT

We investigate nucleation densities in UHV-CVD of Si on ultrathin SiO_2 layers (0.2-2 nm) which were prepared by three different oxidation methods: thermal, UV-ozone, and plasma oxidation. The experiments changing the Si_2H_6 pressure in UHV-CVD indicate that these oxide surfaces have preferred sites for nucleation. Among the three oxidation methods, the nucleation density, N_s, on the thermal oxide is the lowest, while the plasma oxide shows the highest N_s. These results suggest that strained bonds and ion-induced damages in the oxide layers assist nucleation. For UV-ozone and plasma oxides N_s is independent of orientation, reconstruction, and morphology of the initial Si surface.

INTRODUCTION

Device application of nanocrystals often requires control of their lateral positions on the substrate. When one forms nanocrystals by taking advantage of three-dimensional island growth, controlling the position of each dot is inherently difficult. We have so far demonstrated that deposition of Si microstructures at the pre-assigned positions on Si wafers is possible through an in-vacuo selective-area processing scheme that consists of (i) formation of an ultrathin SiO_2 or SiO_2/SiN_x mask layer on Si, (ii) definition of growth area by direct electron-beam irradiation on the mask, and (iii) selective-area deposition of Si by ultrahigh-vacuum chemical vapor deposition (UHV-CVD) [1,2]. Si nucleation in Step (iii), however, takes place not only at the beam-defined positions but also on the other parts of the SiO_2 mask surface. To suppress such uncontrolled nucleation, understanding the nucleation mechanism is indispensable. Nucleation at low densities (10^9 - 10^{10} cm^{-2}) [2,3] and with induction time [3,4] is commonly observed in Si CVD on SiO_2, however, there has been only a limited amount of information on the mechanism of the nucleation process.

In this paper we investigate the Si nucleation processes on ultrathin SiO_2 layers. Our approach is to prepare the SiO_2 layers by different oxidation methods and compare the nucleation processes on them. In particular we focus on the variation of the Si nucleation densities, N_s. Oxidation methods we employed are thermal, UV-ozone and plasma oxidation. In thermal oxidation, molecular O_2 is the only source of oxygen. UV-ozone oxidation involves O_3 and atomic O, while atomic O and O_2^+ ions are the oxidizing species for plasma oxidation. These three kinds of oxides are presumably different from one another in defect densities and network structures. Since nucleation of the SiO_2 surfaces takes place on preferred sites (defects, impurities, etc.) as shown later, the oxide dependence of the nucleation processes provides insights into the identity of the nucleation centers.

EXPERIMENT

Formation of ultrathin SiO_2 layers

Since the nucleation phenomena are anticipated to be sensitive to surface contamination and imperfection, much attention was paid to preparation of the sample surface. For the same reason the oxidation and CVD processes were carried out successively in a multichamber UHV environment [5]. Cleaning of Si(001) samples started with the standard RCA cleaning and formation of a sacrificial oxide layer (~30 nm) by furnace oxidation. Next a H-terminated surface was prepared by etching it off in a 1.6 % HF solution followed by a water rinse. Si (111) wafers were first degreased in acetone and methanol, then were cleaned by the RCA method. H-termination of the (111) surface was accomplished by etching in a 40% NH_4F solution for 5 minutes to prepare an atomically flat (111) surface [6].

The H-terminated Si sample thus prepared was immediately introduced into the multichamber system for oxide formation and Si CVD. The loadlock chamber of the system is equipped with an oil-free pumping system and a conductance-controlling valve for slow pump-down. UV-ozone and plasma oxidation processes were carried out in the loadlock chamber (base pressure ~ 10^{-7} Torr) with the sample at room temperature, while a separate UHV chamber in the same vacuum system was used for thermal oxidation.

(a) Thermal oxidation: Prior to oxidation a Si(001) sample terminated by chemical oxide was heated to 900°C to obtain (2x1) reconstruction as confirmed by reflection high energy electron diffraction (RHEED). The sample was then cooled down below 400°C, and O_2 (1 - 8 mTorr) was introduced into the UHV chamber pumped by a turbomolecular pump. Oxidation was initiated by raising the sample temperature to >700°C. At no time during oxidation an ion gauge nor an ion pump was turned on, so that oxidation by the excited oxygen was not allowed to take place.
(b) UV-ozone oxidation: The chamber was filled with a pure O_2 gas (1 - 500 Torr), and a UV light from a low-pressure Hg lamp (40 W) was introduced through a sapphire window.
(c) Plasma oxidation: Parallel-plate electrode configuration was employed for plasma generation. An RF power of 0.5 W at 13.56 MHz was applied to an electrode of 55 mm in diameter. The plasma gas was 0.02 % O_2 (He balance), and the pressure was kept at 300 mTorr.

UHV-CVD of Si

The as-oxidized sample was analyzed by Auger electron spectroscopy (AES) for thickness determination, then was transferred to a Si CVD chamber. The source gas was Si_2H_6, and its pressure was typically 0.4 mTorr. At this pressure the gas-phase reaction of Si_2H_6 is negligible, and molecular Si_2H_6 is the only gas-phase species. The growth rate at a typical sample temperature of 580°C was 0.04 nm/s. At this temperature the growth rate is limited by desorption of H_2 from the surface [7]. The absolute value of the growth rate and its temperature- and pressure-dependences were consistent with those reported in the literature [7]. Below 700°C desorption of the thin oxide layer was negligible. After CVD processing for 200 s, the Si deposits, which are nanocrystals as observed by RHEED, were observed by ex-situ atomic force microscopy (AFM; tapping mode).

RESULTS AND DISCUSSION

Nucleation on preferred sites

Figure 1 schematically illustrates two mechanisms for nucleation on a substrate surface [8]. In the case of nucleation on the planar sites (i.e., intrinsic nucleation), a stable nucleus is formed through collision of a migrating monomer (Si_2H_6 in this study) with another monomer or a cluster (Si_nH_m, n > 2) of the sub-critical size. In this case, the nucleation rate is proportional to N_1^k, where N_1 is the surface monomer concentration, and k is related to the critical size of the cluster but always equal to or larger than 2. In the desorption-limited condition, which is the case of this study, N_1 is related to the monomer flux, R, as $N_1 = R \cdot \tau$ where τ is the surface residence time of a monomer. After all, the nucleation rate at the planar sites should be proportional to R^k.

The other case shown in Fig. 1 considers a finite number of preferred nucleation sites on the substrate surface (i.e., extrinsic nucleation). Once a monomer is adsorbed on such a site, it is stabilized and does not desorb or migrate any more. The nucleus grows by capture of migrating monomers at such sites. If we define N_{pr} as concentration of the preferred sites, the initial nucleation rate is proportional to $N_{pr} \cdot R$, and the nucleation density N_s saturates at N_{pr}.

The relative importance of these two mechanisms can be assessed from the dependence of the nucleation rate on R. Figure 2 shows the change of N_s as a function of the pressure of Si_2H_6. In this experiment, the CVD time was inversely proportional to the Si_2H_6 pressure so that the integrated dose of Si_2H_6 was the same for all the samples. On both plasma and UV-ozone oxide surfaces, the observed dependence of N_s on the Si_2H_6 pressure is weaker than expected from the planar-site mechanism. This indicates that preferred sites do exist on these surfaces. The difference between the UV-ozone and plasma oxides is due to different densities of nucleation centers, as examined in more detail in the next section. Experiments changing growth temperature (not included in this paper) showed that N_s is independent of temperature, which is also consistent with dominance of nucleation at preferred sites. We point out that a quantitative assessment of the relative importance of the two mechanisms from the Fig. 2 is not straightforward since the surface energy of the Si clusters varies by a large amount depending on the H coverage.

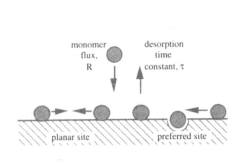

Figure 1 Schematic illustration of nucleation on a planar and a preferred site. The surface monomer density, N_1, is given by $N_1 = R \cdot \tau$ in the desorption-control limit.

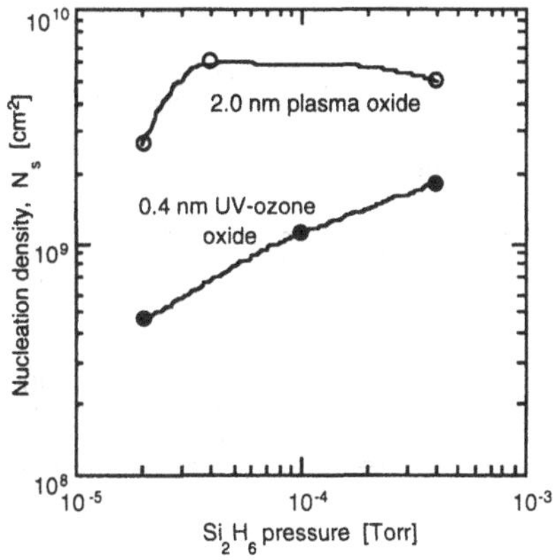

Figure 2 Dependence of N_s on the Si_2H_6 pressure. Two sets of experiments were carried out on UV-ozone and plasma oxides. The total Si_2H_6 dose was 2.1×10^{19} [molecules cm^{-2}] for all the samples.

Nucleation densities on various oxides

In what follows, the effects of the oxidation method on N_s are examined in detail. For all the experiments discussd below, the Si_2H_6 pressure and deposition time were fixed at 0.4 mTorr and 200 s, respectively. Nuclei height was typically 5-10 nm.

(a) Thermal oxides

Table 1 summarizes the values of N_s on thermal oxides. Several samples were prepared by in-situ oxidation under different conditions, which resulted in variation in the oxide thickness from < 0.1 nm to 0.6 nm. There is a clear trend that N_s decreases with increasing oxide thickness. To obtain N_s below 1×10^9 cm^{-2} an oxide thickness larger than 0.2 nm is required [9]. For UV-ozone and plasma oxides, the critical thickness for nucleation suppression was also about 0.2 nm [2]. The in-situ oxide of 0.6 nm in thickness showed N_s of 3×10^8 cm^{-2} which is the lowest value we observed in this study.

Table 1   Nucleation densities on ultrathin SiO_2 mask layers formed by thermal oxidation. The Si_2H_6 pressure and CVD time were fixed at 0.4 mTorr and 200 s for all the experiments.

oxidation method	oxidation conditions			oxide thickness [nm]	nucleation density [$\times 10^9$ nuclei/cm^2]
	temperature [°C]	time [s]	O_2 pressure [Torr]		
in situ	700	120	2×10^{-3}	< 0.1	> 100
	720	120	1×10^{-3}	~ 0.15	5.4
	775	600	5×10^{-3}	0.2	1.0
	775	1800	8×10^{-3}	0.6	0.3
furnace	900	1800	760	210	1.8*
	900	1800	760	210	4.5**

*The oxidized sample was transferred to the CVD reactor within 30 minutes after unloading from the furnace.
** The oxidized sample was kept in air for 2 days before the CVD processing.

Table 2   As Table 1 for UV/ozone oxidation

initial Si surface	oxidation conditions		oxide thickness [nm]	nucleation density [$\times 10^9$ nuclei/cm^2]
	time [s]	O_2 pressure [Torr]		
H-terminated (001)	14400	1	0.5	2.1
	3600	500	0.45	1.8
	3600	20	0.4	1.8
	1800	10	0.4	1.3
H-terminated (111), NH$_4$F etch	3600	3600	0.45	1.3

Table 3   As Table 1 for plasma oxidation

initial Si surface	oxidation time [s]	post-oxidation annealing in UHV	oxide thickness [nm]	nucleation density [$\times 10^9$ nuclei/cm^2]
H-terminated (001)	300	-	2.0	4.5 ± 1.5
	300	580°C, 1800 s	2.0	3.9
	300	700°C, 30 s	2.0	4.4
	30	-	1.0	2.5
	5	-	0.2	4.2
(2x1)-(001)	300	-	2.0	3.5
	10	-	0.45	5.0
H-terminated (111), NH$_4$F etch	300	-	2.0	3.0
HF etch	300	-	2.0	3.7

In order to compare the present results with previous studies in the literature, N_s on furnace oxides were also studied. RCA-cleaned Si samples were oxidized in a furnace in a separate clean room, then were transferred to our CVD system. The sample exposed to air for the minimum time for transfer (< 30 minutes) showed N_s of 1.8×10^9 cm^{-2} which is higher than those of the in-situ oxides thicker than 0.2 nm. A longer exposure to air (2 days) resulted in further increase of N_s. It is evident from this experiment that airborne surface contaminants act as a nucleation center. Hu et al reported a systematic study of Si nucleation on thermally grown oxide of 20-50 nm [3]. They reported that N_s was in the range from 5×10^9 to 1×10^{10} cm^{-2} for growth to nuclei height of 2-20 nm when Si_2H_6 was used for the source gas. This density is comparable to our results for the furnace oxide.

(b) UV-ozone oxides

Observed N_s values for UV-ozone oxides are summarized in Table 2. The oxide thickness was in the range of 0.4 - 0.5 nm, and N_s was reproducible in the range of $1 \times 10^9 \sim 2 \times 10^9$ cm^{-2}. The oxide layer formed on a NH_4F-treated (i.e., atomically-flat) Si(111) surface showed N_s in the same range, which suggests that neither surface orientation nor morphology of the initial Si surface is an important factor for the nucleation processes. Such insensitivity to the structure of the initial Si surface was also observed for plasma oxide (see Table 3).

Compared to the thermal oxides thicker than 0.2 nm, the ozone oxides show higher N_s. We tentatively attribute this to the strained bonds in the UV-ozone oxides. Since ozone oxides were formed by strongly oxidizing species at room temperature, they presumably contain more strained bonds than the thermal oxides. Contribution of the surface-OH groups is less likely judging from the results for the plasma oxidation on H-free (2x1) surface as discussed below.

(c) Plasma oxides

Table 3 summarizes the results for plasma oxides formed on various Si surfaces. The oxidation rate for the plasma method is fast even at a very low partial pressure of O_2 employed (6×10^{-5} Torr). We repeated experiments on 2 nm-thick plasma-oxide layers, and have found that N_s scatters in the range of $(4.5 \pm 1.5) \times 10^9$ cm^{-2}, significantly higher than those on UV-ozone and thermal oxides. N_s on thinner layers (0.2 and 1.0 nm) is also in this range. Effects of surface orientation and morphology were investigated by preparing H-terminated Si(111) surface of different smoothness (i.e., NH_4F and HF treatments for smooth and rough morphology, respectively). As shown in the last two lines in Table 3, neither the surface orientation nor morphology appears to affect N_s significantly.

Since the plasma oxidation involves ions, ion-induced damages and contamination by the sputtering mechanism are among the possible causes for the high values of N_s. In this regard, Hu et al reported that H^+ ion irradiation of the thermal oxide enhances Si nucleation [3]. Fast oxidation by atomic O and oxygen ions may generate strained bonds as in the case for UV-ozone oxides. In order to see whether annealing is effective to reduce N_s, we prepared two samples for which annealing was inserted between the oxide formation and CVD steps. One was kept at the CVD temperature for 30 minutes, and the other was annealed in UHV at 700℃ for 30 s. The N_s value, however, was hardly changed by such annealing.

When one oxidizes H-terminated Si surfaces by a plasma or UV-ozone method at temperature below 300℃, the surface-H atoms are most likely converted to OH groups and a part of them may remain even at the CVD temperature (580℃). To check whether the surface OH groups are effective nucleation centers, we carried out experiments on the oxide layers that were formed by plasma oxidation of H-free (2x1)-reconstructed Si(001) surfaces. Because the processing time is shorter and the gas pressure is lower for the plasma oxidation than the UV-ozone method, the former is better suited for oxidation of the (2x1) surface which is susceptible to contamination by the impurities in the process gas. As show in Table 3, the H-free oxide layers formed on the (2x1) surface show N_s of $\sim 4 \times 10^9$ cm^{-2}. Thus no positive evidence was obtained for the OH groups as a nucleation center.

Microscopic studies

The results of Tables 2 and 3 indicate that surface orientation and morphology are not important factors, at least for low-temperature oxides. Related to this, we show an AFM image in Fig. 3 which was obtained for the Si nuclei formed on the plasma-oxide layer on an NH_4F-treated Si(111) surface. The step-and-terrace structure of the H-terminated surface prior to oxidation is well preserved even after the oxide formation, which ensures that the SiO_2 surface is atomically flat [10]. In many film-growth experiments we often encounter preferred nucleation at step or kink sites. A close examination of Fig. 4, however, reveals that most of the nuclei are formed randomly on the terrace. Similar microscopic investigation of nucleation on UV-ozone and thermal oxides is in progress.

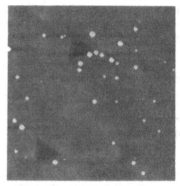

Figure 3 A 1 μm × 1 μm AFM image of Si nuclei formed on a 0.5 nm-thick plasma oxide layer on NH_4F-treated Si(111).

SUMMARY

Nucleation densities in UHV-CVD of Si on ultrathin SiO_2 layers are investigated for three different oxidation methods: plasma, UV-ozone, and thermal oxidations. The minimum thickness for nucleation suppression is found to be approximately 0.2 nm. The observed nucleation densities are explained by considering preferred nucleation sites. The oxide dependence of nucleation suggests that strained bonds and ion-induced damages are effective nucleation centers, while surface orientation and morphology were found to be unimportant factors.

ACKNOWLEDGMENT

This study, partly supported by NEDO, was carried out at JRCAT under the joint research agreement between NAIR and ATP.

REFERENCES

1. D. S. Hwang, T. Yasuda, K.Ikuta, S. Yamasaki, and K. Tanaka, Jpn. J. Appl. Phys. 3 7, L1087 (1998).

2. T. Yasuda, D.S. Hwang, J.W. Park, K. Ikuta, S. Yamasaki, K. Tanaka, Appl. Phys. Lett. 7 4 (in press)

3. Y.Z. Hu, D. J. Diel, C.Y. Zhao, C.L. Wang, Q. Liu, E.A. Irene, K.N. Christensen, D. Venable, D.M. Maher, J. Vac. Sci. Technol. B 1 4, 744 (1996); Y.Z. Hu, C.Y. Zhao, C. Basa, W.X. Gao, E.A. Irene, Appl. Phys. Lett. 6 9, 485 (1996).

4. K.F. Kelton, A. L. Greer, and C. V. Thompson, J. Chem. Phys. 7 9, 6261 (1983).

5. J.W. Park, T. Yasuda, K. Ikuta, L.H. Kuo, S. Yamasaki, and K. Tanaka, Mater. Res. Soc. Symp. Proc. 4 4 8, 271 (1997).

6. G. S. Higashi, R. S. Becker, Y. J. Chabal, and A. J. Becker, Appl. Phys. Lett. 5 8, 1656 (1991).

7. T.R. Bramblett, Q. Lu, T. Karasawa, M.-A. Hasan, S.K. Jo, and J.E. Greene, J. Appl. Phys. 7 6, 1884 (1994).

8. B. Lewis and J. C. Andersono, *Nucleation and Growth of Thin Films* (Academic Press, London, 1978), p.287.

9. K. Fujita, H. Watanabe, and M. Ichikawa, Appl. Phys. Lett. 7 0, 2807 (1997).

10. K. Ikuta, J.W. Park, L.H. Kuo, T. Yasuda, S. Yamasaki, and K. Tanaka, Mater. Res. Soc. Symp. Proc. 4 5 2, 743 (1997).

Gaas MICRO CRYSTAL GROWTH ON A As-TERMINATED Si (001) SURFACE BY LOW ENERGY FOCUSED ION BEAM

Toyohiro Chikyow and Nobuyuki Koguchi
National Research Institute for Metals,
1-2-1 Sengen, Tsukuba Ibaraki 305, JAPAN
e-mail: tchikyo@momokusa.nrim.go.jp

ABSTRACT

Ordered Gaas micro crystal growth on a As-terminated Si (001) surface was demonstrated using a low energy focused ion beam. Si (001) surface was terminated by Arsenic. The surface showed a (2X1) structure with As dimers. The As layer was sputtered periodically with low energy focused Ga ion beam. Supplied Ga atoms migrated on the surface and trapped at the As removed region, forming Ga droplets. Gaas micro crystals were grown from Ga droplets by As molecule supply. The proposed method was shown to be effective as a fabrication method.

INTRODUCTION

Recently fabrication methods of ordered semiconductor structure have been of interest. Ordered nano scale micro crystals are applicable for low dimension carrier confinement, such as quantum dots[1,2]. Semiconductor crystals ordered on a 1.0 micron scale are needed for the photonic crystal structure [3,4]. In both cases, a conventional fabrication method is required. As a sophisticated fabrication method, " Droplet Epitaxy ", where GaAs micro crystals grown from Ga droplets, has been proposed and successful results have been reported[5,6]. A combination of " Droplet Epitaxy" and low energy focused ion beam (LE-FIB) method have shown a possibility of ordered sub-micron scale micro crystal growth[7]. In this combination, stable surface with low surface free energy and nucleation by low energy ion bombardment were essential factors.

An As-terminated Si (001) surface is known as a stable surface. The surface has a dimer structure. The dangling bonds are filled, which gives a lower surface free energy [8]. As shown in Fig.1, As dimer has 0.3 eV/bond, while Si dimmer has 1.48 eV/bond. This huge surface free energy difference gives differences in surface diffusivity and reactivity at the surfaces.

In this paper, a fabrication of ordered GaAs micro crystal on a As-terminated Si (001) surface is demonstrated and an application for photonic crystal is discussed.

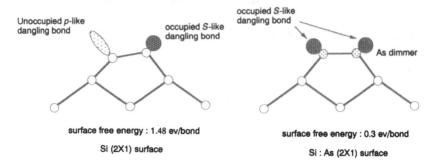

surface free energy : 1.48 ev/bond

Si (2X1) surface

surface free energy : 0.3 ev/bond

Si : As (2X1) surface

Fig.1 Atomic structires of Si-Si- dimer and As-As dimer on Si (001) surface

EXPERIMENTS

Low energy focused ion beam system (LE-FIB)

The LE-FIB system with a liquid Ga ion source had six static lenses. The first lens was used to control the ion current for 100 pA and the second one was used for focusing. Extracting voltage was 8.5-8.7 KV and the accelerating voltage was 10 KV at the second lens end. The other four lenses are retarding ones to reduce the ion kinetic energy down to 100 eV by applying 9.90 KV voltage. In this condition, the final beam size was 1.0 micron on the sample. Reflectors in the middle of the column scanned the beam according to step bias for X and Y movement. For a combination of the LE-FIB and " Droplet Epitaxy ", a conventional molecular beam epitaxy system (MBE), the LE-FIB system and a scanning tunneling microscope (STM) were integrated. The sample holder was transferred between the three systems without breaking vacuum. Sample heating was carried out by direct current.

As-termination and Ga droplets formation

After the conventional cleaning process, n-type Si (001) was dipped into HF solution (HF:H_2O=1:9) and loaded into the MBE system. The sample was heated up to 1200 °C and flashed for 3 s by direct current in an As atmosphere (10^{-4} Pa). Next, the sample temperature was reduced to 400 °C to achieve an As termination of Si surface. Ga molecules were supplied to the surface to form Ga droplets.

446

Nucleation sites formation and Growth of GaAs

Another As-terminated Si sample was transferred to the LE-FIB system. Ga ions 100 eV in kinetic energy was irradiated to the surface to remove the As layer periodically. We have ever reported that Ga ion 100 eV in kinetic energy was sufficient to sputter surface atoms[9]. The spacing between the sites was 2.5 micron. At this energy, a few monolayers of the surface is sputtered. Namely, As layers were removed periodically.Then the sample was transfer to the MBE system again. Ga molecules were supplied to the surface at 400 °C and subsequently As molecules were supplied to grow GaAs at the same temperature. The reason why we choice the Ga ion in 100 eV was to sputter the surface atoms

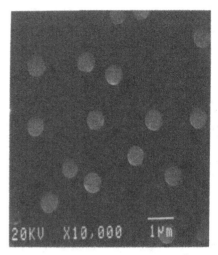

Fig.2 Ga droplets on As-terminated Si (001) surface.

The sputtered surface was observed by STM and characterized by scanning tunneling spectroscopy (STS).

RESULTS AND DISCUSSION

On the As-terminated Si (001) surface, Ga molecules being supplied, Ga droplets 1.0 micron in size were observed at 400 °C as shown in Fig.2. This means that Ga did not react with the As-terminated surface. Actually the As-Si bonding is reported stable up to 500 °C [8]. The average distance between Ga droplets were 3.8 micron in this case. However, the droplet size and spacing were controlled by substrate temperature.

After the nucleation sites formation by LE-FIB, Ga molecules being supplied to the surface, 5 or 6 Ga droplet groups were observed as shown in Fig.3. The average distance between Ga droplets was 2.5 micron. Between the droplet, Ga droplets were not observed. This means that the Ga atoms could migrate on the As-terminated area and be stopped at the ion irradiated area due to the surface diffusion length of 3.8 micron at 400 °C.

On the clean Si surface, Ga molecules form one atomic layer [10]. From this report, the supplied Ga migrates on the surface on the As-terminated Si and is trapped at the ion irradiated area due to the active Si dangling bonds. The size of droplets were 200 - 300 nm. The droplets seemed to

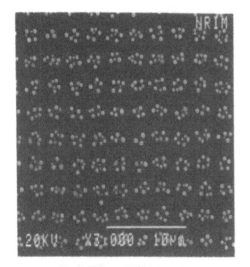 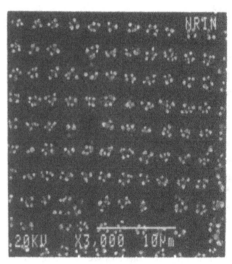

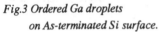

Fig.3 Ordered Ga droplets
on As-terminated Si surface.

Fig.4. GaAs micro crystals
on As-terminated Si surface.

be formed near the irradiated area. On the As-terminated surface, the diffusion length is thought to be longer than that on the Ga -terminated Si surface. At the edge of the ion irradiated area ,Ga atoms migrating from the As-terminated area go into the irradiated area and have more chance to collide, being nuclei of Ga droplets.

After the As molecule supply, the Ga droplets were converted to GaAs micro crystals as shown in Fig.4.

The As-terminated Si surface was investigated by STM. The surface shows a (2X1) reconstructed structure. Also a rod or triangular shape structure was observed as shown in Fig.5. These structures were speculated to be an additional As layer. Measured spacing in the As-As dimer was estimated to be 0.25 nm. This distance is longer than that of a Si-Si dimer, which is 0.235 nm. The spacing between As-As dimers agrees with a calculated value of 0.253 nm [11].

This surface was found to have little surface states because significant band vending was not observed by STS. Due to the strong Si-As bonding, a possibility of defect formation seems reduced, leading to few surface states.

After the Ga ion irradiation, white spots on an atomic scale were observed as indicated in Fig.6. These images were similar to white spots observed on (2X1) Si surface, which are identified as A or B defects. These white spots were thought to be formed by Ga ion irradiation. A characteristic

Fig.5 STM image of
As-terminated Si (001) surface.

Fig.6 STM image
after Ga ion bombardment.

point of the spots were each spot bridges between two dimers. It means that two dimers in another row were removed simultaneously with a single ion collision. A strong interaction between adjacent dimers may exist. STS analysis showed the Fermi level pinning occurred due to increased surface states. This means that the generated surface defects have electrically active dangling bonds. These dangling bonds seem to capture Ga atoms migrating from the As-terminated area.

GaAs array on Si has an advantage for a photonic crystal. GaAs has a direct optical transition, with 1.54 eV band gap. Si has an indirect transition and the value of band gap is 1.1 eV. Also the reflactive index of GaAs is 3.69,while that of Si is 3.44. These means that there is a particular light which can go through GaAs but not for Si. Actually, a light of 900 nm wave length can only go though the GaAs and can not penetrate into Si.

In addition,light can be reflected at the GaAs/Si interface due to the reflactive index when the light comes from a particular incident angle.

GaAs micro crystal array on Si has a potential to realize an ideal two dimensional photonic crystal. It also have a possibility of integrating optical devices and Si LSIs.

In summary, an array of GaAs micro crystals were fabricated on a As-terminated Si (001) using a low energy focused ion beam. Si (001) surface was terminated by As and it showed (2X1) dimer structure. The dimers were sputtered periodically by low energy Ga ion beam. After the Ga molecule supply, periodic Ga droplet groups were observed. On the surface Ga molecules migrated on the As-terminated surface and trapped at the As removed region, forming Ga droplets. GaAs micro crystals were grown from Ga droplets by As molecule supply.

The proposed method was proved to be convenient for GaAs micro crystal array formation and it gives a effective method of photonic crystal design.

References

1. H.Sakaki, Surface Science **267**,623 (1992).

2. Y.Arakawa and H.Sakaki, Appl. Phys. Lett **40** ,939 (1982) .

3. M.Ohtaka, J. Appl. Phys. **34**, 4502 (1995).

4. Yablonovitch, Phys. Rev. Lett. **58**, 2059 (1987).

5. T.Chikyow and N.Koguchi, Appl. Phys. Lett.,**61**, 2431 (1992).

6. N.Koguchi and K.Ishige, Jpn. J. Appl. Phys.,**32**, 2052 (1993) .

7. T.Chikyow and N.Koguchi, J.Vac.Sci.Technol.**B16**(**4**), 2538 (1998).

8. D.KBiegelsen,R.D.Bringans,J.E.Northrop, M.C.Schabel and L.E.Swartz
   Phys.Rev.**B47**,9589 (1993).

9.T.Chikyow and N.Koguchi, Surface Science **386**, 254 (1997).

10. Y.Nakada and H.Okumura, J.Vac. Sci. Technol. **B16**, 645 (1998).

11. P.Kreager and J.Pollman, Phys.Rev.**B47**,1898 (1993).

AMORPHOUS/MICROCRYSTALLINE PHASE CONTROL IN SILICON FILM DEPOSITION FOR IMPROVED SOLAR CELL PERFORMANCE

Joohyun KOH *, H. FUJIWARA **, Yeeheng LEE *, C. R. WRONSKI *, R. W. COLLINS *
* Center for Thin Film Devices, The Pennsylvania State University, University Park, PA 16802.
** Electrotechnical Laboratory, 1-1-4 Umezono, Tsukuba-shi, Ibaraki 305-8568, JAPAN.

ABSTRACT

Real time optical studies have provided insights into the growth of hydrogenated amorphous silicon (a-Si:H) and microcrystalline silicon (μc-Si:H) thin films by plasma-enhanced chemical vapor deposition as a function of the H_2-dilution gas flow ratio R=$[H_2]/[SiH_4]$, the accumulated film thickness d_b, and the substrate material. Results pertinent to the optimization of a-Si:H-based solar cells have been obtained in studies of Si film growth at moderate to high R on dense amorphous semiconductor film surfaces. For depositions with 15$\leq$R$\leq$80 on freshly-deposited a-Si:H, initial film growth occurs in the amorphous phase. Upon continued growth, however, a transition is observed as crystallites begin to nucleate from the amorphous film. The thickness at which this amorphous-to-microcrystalline (a$\rightarrow$$\mu$c) transition occurs is found to decrease with increasing R. Based on these results, a deposition phase diagram has been proposed that describes the a$\rightarrow$$\mu$c transition as a continuous function of R and d_b. We find that the optimum stabilized a-Si:H p-i-n solar cell performance is obtained in an i-layer growth process that is maintained as close as possible to the phase boundary (but on the amorphous side) versus film thickness.

INTRODUCTION

Recently, important gains in the stabilized performance of a-Si:H solar cells have been realized by incorporating i-layers prepared using moderate hydrogen dilution of silane, typically with R= $[H_2]/[SiH_4]$~10-15 [1,2]. Under these conditions, R lies just below values generally recognized to yield μc-Si:H film growth [3]. This approach has been proposed as the key to obtaining a-Si:H with optimum photoelectronic properties and stability for solar cells [4,5]. The position of the amorphous-to-microcrystalline (a$\rightarrow$$\mu$c) phase boundary in preparation parameter space is expected to be very sensitive to the structure of the underlying material, i.e., the substrate or accumulated film [3]. Comparing two extreme cases, film deposition on a H-terminated crystalline Si (c-Si) surface is expected to lead to an a$\rightarrow$$\mu$c transition at lower R than deposition on dense a-Si:H. In addition, the position of the boundary is expected to depend on the film thickness -- as material accumulates, the memory of the substrate may be gradually erased. Thus, films prepared on H-terminated c-Si at lower R may evolve from substrate-induced crystallinity to an amorphous structure characteristic of the growth conditions. In contrast, films prepared on a-Si:H at higher R may evolve from a substrate-induced amorphous structure to microcrystallinity.

Such expectations suggest two important conclusions.

(1) *It is invalid to correlate materials properties and device performance unless the films used in the materials studies are deposited on similar substrates to similar thicknesses as in the device structure.* For example, if one is interested in applying infrared (ir) spectroscopy to characterize the bonding in an i-layer material used in the steel/Ag/ZnO/n-i-p solar cell configuration, one can simulate the solar cell configuration by depositing a thin n-layer on a c-Si substrate followed by an i-layer, the latter under identical conditions with similar thickness as in the solar cell. Then, one can perform ir transmission on the c-Si/n/i structure. Even this approach can be dangerous for a μc-Si:H n-layer, since n-layer growth on the c-Si can lead to enhanced crystallinity compared with growth on ZnO, and this would then affect the structural development of the overlying i-layer.

(2) *Because of the thickness dependence of the component layer properties, optimization of these layers should be considered in terms of multistep processes.* For example, in the glass/SnO$_2$:F/p-i-n structure, the p-layer is usually amorphous. Because of this, the initial i-layer material can be prepared at higher R than would ordinarily be possible for c-Si, glass, or silica substrates, while still retaining an amorphous film structure. To avoid a gradual transition to microcrystallinity with accumulating thickness, R is then decreased later in the i-layer deposition. Although this approach defines a two step process, a multistep or even a continuous variation in R may lead to important further improvements in cell performance and stability.

Mat. Res. Soc. Symp. Proc. Vol. 536 © 1999 Materials Research Society

The motivation for this study is to demonstrate the validity of these conclusions based on real time spectroscopic ellipsometry (RTSE) of Si film growth in correlation with device performance.

EXPERIMENTAL

For RTSE studies of the effect of substrate on the position of the a→μc phase boundary, Si films were deposited in a single-chamber system on c-Si at 200°C. H_2-dilution ratios R from 0 to 40 were obtained by fixing the SiH_4 partial pressure at 0.07 Torr with a 5 sccm flow and increasing the H_2 flow from 0 to 200 sccm. For the applied rf power flux of 70 mW/cm^2, the deposition rate varied from 1.3 (R=0) to 0.2 Å/s (R=40). For comparison, a similar series from R=10 to 80 was deposited on a-Si:H substrate films. These experiments simulate i-layer growth on the p-layer of a p-i-n solar cell. The a-Si:H substrate films were prepared with R=0 under otherwise identical conditions. The structural evolution and optical properties of the resulting a-Si:H and μc-Si:H films were measured by RTSE using a multichannel ellipsometer [6]. The acquisition time for full spectra (1.4–4.5 eV) ranged from 0.16 s (R=0) to 0.8 s (R=80).

In order to assess the device performance of the Si films obtained using H_2-dilution, p-i-n solar cells were fabricated in a multichamber system on specular SnO_2:F-coated glass at 200°C. The ~250 Å p-type a-$Si_{0.91}C_{0.09}$:H:B window layer was deposited first using flow ratios of [SiH_4]:[CH_4]:[B(CH_3)$_3$]:[He]=10:10:0.1:5 (in sccm). The ~4000 Å i-layer was deposited next using H_2-diluted SiH_4 in one or two step processes with variable R. Finally, the ~350 Å μc-Si:H:P n-layer was deposited using ratios of [H_2]:[SiH_4]:[PH_3]=100:2:0.023.

RESULTS AND DISCUSSION

Materials Studies

RTSE is effective in identifying the a→μc transition as a function of the H_2-dilution ratio R and bulk layer thickness d_b owing to its abilities (i) to measure the nucleating and surface roughness layers with sub-monolayer sensitivity [7] and (ii) to provide the dielectric function (ϵ_1, ϵ_2) in films even a few monolayers thick [8]. As a first example, Fig. 1 shows the roughness layer thickness d_s versus bulk layer thickness d_b for the Si films deposited on native oxide-coated c-Si substrates at R values from 0 to 40. Figure 2 shows the bulk layer ϵ_2 spectra for these films when d_b is within the range of 150-200 Å. Finally, Fig. 3 (open points) provides the crystalline Si volume fraction estimated from effective medium theory (EMT) analyses of (ϵ_1, ϵ_2) for d_b=100 and 2000 Å.

Figure 1 shows that for depositions with R ranging from 0 to 20, the nuclei make contact after a thickness of d_n~17 Å. The nucleating layer of thickness d_n is transformed into a surface roughness layer of thickness $d_s=d_n$ after the nuclei make contact and the first bulk monolayer forms (at d_b~2.5 Å). For the depositions with R=0-20, the films grow initially as a-Si:H, as indicated by their (ϵ_1, ϵ_2) spectra (see, e.g., Fig. 2). The relatively stable surface roughness layers for the films prepared with R=0 and 10 are indicative of continued growth of a-Si:H, whereas the rapid increase in roughness starting at 200 Å for the film with R=20 is consistent with the development of μc-Si:H. For R=40, the nuclei make contact at a thickness of 58 Å. The resulting lower nucleation density and the bulk film (ϵ_1, ϵ_2) spectra for R=40 are both consistent with immediate nucleation of μc-Si:H on the c-Si. Thus, for Si film growth on the oxide-covered c-Si substrate, Figs. 1 and 3 suggest that the a→μc phase boundary occurs just above R=20 for a film thickness of 100 Å. For a much thicker film with d_b=2000 Å, the transition occurs close to R=15.

Figure 4 shows results for ϵ_2 corresponding to those in Fig. 2, but for Si film growth on a-Si:H substrate films. In this case, the a→μc phase boundary occurs near R=50 for a film thickness of 100 Å, as shown in Fig. 3 (solid points). As was suggested in the introductory remarks, we conclude that the a→μc phase boundary depends not only on the film thickness, but also on the substrate. Figure 3 suggests that the a-Si:H substrate significantly suppresses μc-Si:H development by imposing its structure on the growing film in the initial stages. In contrast, the native oxide on the c-Si substrate only slightly suppresses the nucleation of μc-Si:H. This can be seen by the small shift to lower R in the a→μc transition with increasing thickness in Fig. 3.

Figure 5 shows a schematic phase diagram versus R and d_b for Si film growth on a-Si:H. Above and below the diagonal solid line, the films are accumulating as μc-Si:H and a-Si:H, respectively. Several features are noted from the phase diagram. (i) For R=10, the structure

452

remains amorphous throughout the growth of a 1 μm film. (ii) For R=15 to 20, the films develop measurable microcrystallinity starting from d_b=3000 to 2000 Å, respectively. This is the range of typical i-layers in solar cells. (iii) For R=40 to 50, the films develop microcrystallinity starting from d_b=200 to 100 Å, as has been demonstrated in Figs. 3 and 4. This is the range of typical "buffer layers" used at cell interfaces. (iv) Even for R>50, we find that a relatively low-density amorphous transition region must form on the a-Si:H substrate before μc-Si:H can nucleate.

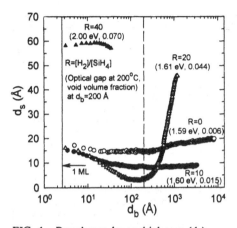

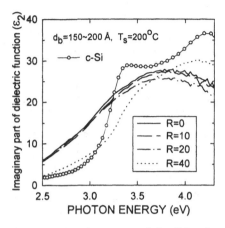

FIG. 1   Roughness layer thickness (d_s) versus bulk layer thickness (d_b) for Si films deposited on c-Si at different H2 dilution ratios. Also given are the $\varepsilon_2^{1/2}$ gaps at 200°C and the relative void volume fractions, both deduced at d_b=200 Å.

FIG. 2   Imaginary parts of the dielectric functions (at 200°C) after bulk layer thicknesses of 150-200 Å for Si films deposited on c-Si substrates at different H$_2$ dilution ratios. The data for single-crystal Si at 200°C are also shown.

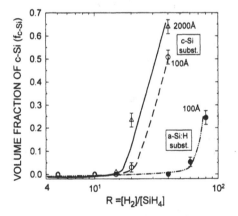

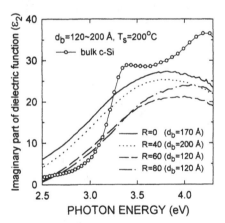

FIG. 3   Dependence of crystal Si volume fraction on H$_2$-dilution ratio for Si depositions on c-Si wafers at thicknesses of d_b=100 (open circles) and 2000 Å (open triangles) and for depositions on a-Si:H at a thickness of d_b=100 Å (closed circles).

FIG. 4   Imaginary parts of the dielectric functions (at 200°C) after bulk layer thicknesses of 120-200 Å for Si films deposited on a-Si:H substrate films at different H$_2$ dilution ratios. The data for single-crystal Si at 200°C are also shown.

453

Device Studies

The higher-R/lower-d_b (R≥15 / d_b<3000 Å) films that nucleate first as a-Si:H, but evolve to a microcrystalline structure with accumulating thickness are potentially useful for incorporation in solar cells. We have called this the "protocrystalline regime" since the amorphous phase serves as an underlying precursor from which the crystalline phase nucleates [5]. The key question, then, is whether the thin a-Si:H prepared versus R in the protocrystalline regime exhibits improvements in material properties that extend to the a→µc phase boundary. Here, the microstructural evolution of the a-Si:H films on oxide-covered c-Si in Fig. 1 provides insights. Previous RTSE results for a-Si:H depositions on smooth c-Si have shown a correlation between the microstructural evolution and the photoelectronic properties [7]. It was found that conditions leading to the maximum surface smoothening (Δd_s) during coalescence (d_b<100 Å) are precisely those leading to optimum photoelectronic properties. Thus, the results in Fig. 1 suggest that the film with R=20, which has been prepared at the a→µc boundary and has exhibited the largest smoothening effect during coalescence yet observed for a-Si:H (Δd_s=12 Å), should also exhibit the best photoelectronic properties among the 100 Å amorphous films of Fig. 1 prepared on c-Si. Further evidence for improved materials in the protocrystalline regime comes from recent analyses of $(\varepsilon_1, \varepsilon_2)$. Among a-Si:H films, those prepared closest to the a→µc boundary exhibit the smallest broadening parameter Γ in fits to $(\varepsilon_1, \varepsilon_2)$ and hence the longest lifetimes $\tau=\hbar/\Gamma$ for band excitations [9].

In order to explore these suggestions further, we have fabricated a-Si:H p-i-n solar cells using one and two step i-layer growth processes with different H_2 dilution ratios R. An example of the two step device structure is shown in Fig. 6. The layer nearest the p/i interface is deposited first to a thickness of 100 or 200 Å, and the bulk layer is deposited next, leading to a total i-layer thickness of ~4000 Å. Figure 7 shows the initial (annealed state) performance parameters of open circuit voltage (V_{oc}) and fill factor (FF) for the solar cells prepared in the one step i-layer process plotted versus the R value used to prepare the i-layer (open circles). Results are also shown for cells prepared in the two step process plotted versus the R value used in the first step of the i-layer adjacent to the p/i interface (filled points). In the latter case, R is fixed at 10 for the second step. Figure 8 shows V_{oc} and FF for selected solar cells as a function of AM1.5 light-soaking time. In the following paragraph, we interpret the results in Figs. 7 and 8 in terms of the schematic phase diagram of Fig. 5. We expect this diagram to be relevant to the i-layer fabrication process as well since, in both the materials and solar cell studies, the underlying substrate material is a dense amorphous film, in one study pure a-Si:H and in the other p-type a-Si$_{1-x}$C$_x$:H (x~0.09).

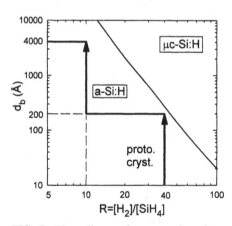

FIG. 5   Phase diagram for preparation of Si films on a-Si:H vs. R and d_b. The diagonal line separates a-Si:H and µc-Si:H growth regimes. The bold lines depict an optimized two step i-layer process.

Metal contact (Cr)	~1000 Å
µ c-Si:H:P n-layer	350 Å
a-Si:H i-layer (R=10)	3800~ 3900 Å
a-Si:H p/i interface layer (R=40)	100~ 200 Å
a-Si$_{0.91}$C$_{0.09}$:H:B p-layer	250 Å
SnO$_2$:F	~7500 Å
Glass	

FIG. 6   Schematic structure of a p-i-n solar cell with a two step i-layer. The H_2-dilution ratios shown are those that optimize performance in the initial and stabilized states.

454

The results in Fig. 7 for the one step i-layer suggest that optimum initial performance of $V_{oc}xFF$ is obtained at R=10. For lower R, V_{oc} is reduced whereas for higher R, FF declines. After ~200 hours of light-soaking (Fig. 8), the performance optimum at R=10 is accentuated since the highest stability is also obtained at R=10. In fact, for the solar cell prepared with an R=0 i-layer, continuous degradation of V_{oc} and FF occur, to values of 0.78 V and 0.53, whereas for R=10, V_{oc} and FF have stabilized at 0.86 V and 0.61, respectively. We attribute the increase in stabilized performance with increasing R to the beneficial effects of gas phase H_2 dilution on the bonding and microstructure of the i-layer. Such effects may include (i) greater structural relaxation of the network owing to the lower deposition rate, (ii) elimination of strained Si-Si bonds due to the penetration of a possibly higher concentration of H from the plasma, and (iii) elimination of high sticking coefficient precursors such as SiH and SiH_2 or polysilanes that limit structural coalescence and generate localized defects [10]. We attribute the rapid reduction in FF above R=10 in Fig. 7 to the formation of microcrystalline material within the top ~1000 Å of the 4000 Å i-layer in accordance with the predictions of the phase diagram of Fig. 5. Thus, a prescription to obtain the optimum performance of the solar cell with a one step i-layer is to prepare the i-layer at the highest possible R while avoiding the transition to microcrystallinity.

The initial performance of the solar cells prepared using two step i-layers with a fixed second step R value of 10 shows a similar optimum with a first step R value of 40. As for the one step i-layer, V_{oc} is reduced at lower R whereas FF declines for higher R. The improvement with increasing R to R=40 may have similar underlying reasons as the improvement from R=0 to 10 for the one step i-layer. In fact, the films in the protocrystalline regime exhibit measureably lower Si-Si bond packing densities than the highest density films prepared with R=0 and 5. This effect is evident in Figs. 2 and 4. Here the films prepared with R=20 on c-Si and with R=40 on a-Si:H exhibit lower ε_2 magnitudes, best fit assuming void fractions of ~0.05-0.07 relative to the densest a-Si:H with R=5. After 200 hr of light soaking, the performance optimum at R=40 is accentuated since the highest stability is also obtained near R=40. For the cell prepared with a one step R=10 i-layer, V_{oc} and FF stabilize at 0.86 V and 0.61, whereas for the two step (R=40, 10) i-layer, V_{oc} and FF stabilize at 0.92 V and 0.63, respectively. The improved fill factor stability is reproducible

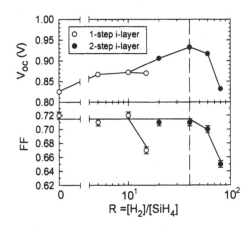

FIG. 7   Initial open circuit voltages and fill factors of a-Si:H p-i-n solar cells vs. the bulk i-layer H_2-dilution ratio R for one step i-layers and vs. the near-interface R for two step i-layers. For the two step i-layers, R=10 for all bulk i-layer regions.

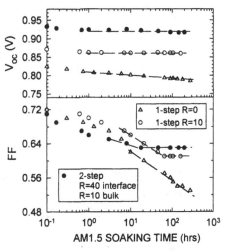

FIG. 8   Time dependence of the open circuit voltages and fill factors during light soaking under AM1.5 illumination for p-i-n solar cells fabricated with one and two step i-layers. The lines are guides for the eyes.

and is attributed to a slight shift in the deposition process for the bulk R=10 i-layer toward the a→μc phase boundary when growth occurs on an R=40 a-Si:H film. This is further evidence of a substrate dependent phase boundary. For example, if the higher void fraction in the underlying R=40 film is columnar in nature, ellipsometric studies show that these voids can propagate across the R=40/R=10 boundary. A higher void fraction may lead to an enhanced ability of H to enter the network and promote relaxation to a more ordered or possibly microcrystalline state.

The decline in performance of the solar cells with two step i-layers observed when the first step dilution ratio is set above R=40 is attributed to microcrystallite formation in the first layer. It should be noted that for the first step of the two step i-layers in Fig. 7, the thickness was 200 Å for R=20 and 40, but was 100 Å for R=60 and 80. For a 200 Å first layer with R=60 and 80, the performance decline was even more pronounced. These overall results are consistent with the phase diagram of Fig. 5. For R=60, crystallites are expected to be present only in the top 25% of the layer for a 100 Å film whereas they would occur in the top 60% for a 200 Å film. For R=80, only the first ~30 Å is amorphous so the properties of the first layer are dominated by the crystalline phase irrespective of thickness. On the basis of these results, we conclude that a prescription to obtain the optimum performance of solar cells with a two step i-layer is to prepare both layers as close as possible to the a→μc phase boundary while ensuring that the boundary is not crossed. The optimum two step process is shown on the phase diagram of Fig. 5 (bold lines).

SUMMARY

Recently, substantial evidence has been put forward to suggest that for optimum a-Si:H solar cells, the deposition process for the i-layer should be maintained as close as possible to the regime in which microcrystallinity is obtained, without actually passing into this regime. In this study, we clarified this issue by establishing a deposition phase diagram that identifies the regimes of film thickness and H_2-dilution ratio over which a-Si:H and μc-Si:H thin films are obtained for a given substrate material. Two key observations with respect to the phase diagram have been made that are pertinent to solar cell optimization. (i) The H_2 dilution ratio at which the amorphous-to-microcrystalline (a→μc) transition occurs is a sensitive function of the substrate material. (ii) For a given substrate material, the H_2-dilution ratio $R=[H_2]/[SiH_4]$ at which the a→μc transition occurs depends on the accumulated thickness d_b. Both observations demonstrate that correlations between film characteristics and solar cell performance for materials prepared near the a→μc boundary must be established using films deposited to similar thicknesses on similar substrates as in the devices. The second observation suggests that i-layer optimization for solar cells should not be viewed in terms of a single step process with fixed parameters but rather a multistep or continuous process that is maintained near the a→μc boundary versus thickness.

ACKNOWLEDGMENTS

This research was supported by the National Renewable Energy Laboratory (Subcontract No. XAN-4-13318-03) and the National Science Foundation (Grant No. DMR-9622774).

REFERENCES

[1] X. Xu, J. Yang, and S. Guha, J. Non-Cryst. Solids 198-200, 60 (1996).
[2] Y. Lee, L. Jiao, H. Liu, Z. Lu, R.W. Collins, and C.R. Wronski, 25th IEEE Photovoltaic Specialists Conference, May 1996, Washington DC, USA (IEEE, 1996), p. 1165.
[3] C.C. Tsai, in Amorphous Silicon and Related Materials, Vol. 1, edited by H. Fritzsche (World Scientific, Singapore, 1988), p. 123.
[4] D. Tsu, B. Chao, S. Ovshinsky, S. Guha, and J. Yang, Appl. Phys. Lett. 71, 1317 (1997).
[5] J. Koh, Y. Lee, H. Fujiwara, C.R. Wronski, and R.W. Collins, Appl. Phys. Lett. 73, 1526 (1998).
[6] R.W. Collins, Rev. Sci. Instrum. 61, 2029 (1990).
[7] Y. Li, I. An, H. Nguyen, C. Wronski, and R.W. Collins, Phys. Rev. Lett. 68, 2814 (1992).
[8] H. Nguyen, Y. Lu, M. Wakagi, and R.W. Collins, Phys. Rev. Lett. 74, 3880 (1995).
[9] J. Koh and R.W. Collins (unpublished results, 1998).
[10] R.W. Collins and H. Fujiwara, Current Opinion: Solid State & Mater. Sci. 2, 417 (1997).

PREPARATION AND CHARACTERIZATION OF MICROCRYSTALLINE AND EPITACTIALLY GROWN EMITTER LAYERS FOR SILICON SOLAR CELLS

K. LIPS, J. PLATEN, S. BREHME, S. GALL, I. SIEBER, L. ELSTNER, AND W. FUHS

Hahn-Meitner-Institut, Abt. Photovoltaik, Rudower Chaussee 5, 12489 Berlin, Germany

ABSTRACT

We have deposited thin B- and P-doped Si layers by electron cyclotron resonance CVD on c-Si (4 Ωcm, CZ) and on quartz glass substrates at T=325 °C. Films grown on quartz glass are of microcrystalline nature with crystalline volume fractions of about 70 % and a resistivity ranging from 0.01 - 10 $(\Omega\text{cm})^{-1}$ depending on doping concentration. The doping efficiency is close to unity with the carrier mobility being independent of doping concentration for both B- and P-doping. Films grown on c-Si, on the other hand, exhibit perfect homoepitaxial morphology when the gas phase doping concentration exceeds 1000 ppm and 5000 ppm for P- and B-doping, respectively. The quality of the films is tested by preparing thin film emitter solar cells. We find efficiencies above 11 % for cells without ARC. The result are compared to cells with diffused emitters, otherwise prepared with the same technological steps.

INTRODUCTION

One of the truly challenging aspects of silicon technology is the development of a polycrystalline thin-film solar cell using low-cost substrates such as glass. This limits the process temperature to the softening temperature of the glass (T < 600 °C). The basic problem is the development of deposition processes which allow to deposit thin crystalline Si films (2-3 µm) and have the potential for large scale production. Encouraging results with solar cells of microcrystalline Si (µc-Si) with efficiencies in the range 7 – 10 % have been obtained in various laboratories using plasma-enhanced chemical vapor deposition (PECVD) at process temperatures of about 200 °C [1, 2]. A major concern, however, is the deposition rate which is still by far too low for a production process. Deposition processes which hold promise for higher deposition rates are ion assisted techniques and in particular ECR-CVD allowing high density plasma excitation at pressures as low as 10^{-3} Torr [3]. Using this technique, even low-temperature homoepitaxy of Si has been achieved at T = 450 °C with a rate of 25 nm/min [4]. In this paper we present a study on the deposition of doped films at a deposition temperature of 325 °C using quartz glass and Si as substrates. We found that in spite of the low temperature the addition of dopants strongly enhances epitaxial growth on the Si-wafers in excellent quality.

SAMPLE PREPARATION AND EXPERIMENT

The thin-film silicon deposition was carried out by ECR-CVD (Electron Cyclotron Resonance – Chemical Vapor Deposition) as described in Ref. 3. We simultaneously deposited on quartz glass and on crystalline silicon (c-Si) substrates. The p- and n-type c-Si wafers (4 Ωcm, (100), CZ) were polished on the front side and had a back-surface field driven in by diffusion from solid sources. The c-Si wafers were RCA-cleaned and then etched in dilute HF solution and rinsed with DI water just before they were introduced into the load-lock. The substrates were then transferred to the ECR-CVD deposition chamber were they were heated to 325 °C. No in-situ surface cleaning was performed prior to the deposition. For all depositions H_2 was used as excitation gas and a mixture of SiH_4, B_2H_6, PH_3 and H_2 was used as source gas. The hydrogen dilution ratio, $X = H_2/(H_2+SiH_4)$ remained unchanged through out all depositions at 97 % for P- and 98 % for B-doping, respectively. The process pressure was kept constant at 10 mTorr and the microwave power was set to 1000 W. A divergent magnetic field extracted the plasma to the substrate holder in the reaction chamber.

The samples were characterized by Raman backscattering, scanning electron microscopy (SEM), optical spectroscopy, secondary ion mass spectroscopy (SIMS), electrical conductivity and Hall effect measurements. Additionally, the layers deposited on c-Si were characterized by high resolution cross-sectional transmission electron microscopy (HRTEM), Rutherford backscattering spectroscopy (RBS) and high resolution X-ray diffractometry (XRD). The film thickness was measured on quartz glass or SiO_2 covered c-Si substrates by a step-height profiler.

For the preparation of solar cells we simply deposited a highly doped thin film silicon onto c-Si wafers. After the deposition, the solar cells were patterned by mesa etching to define the area (0.16 cm^2 or 4 cm^2). Afterwards the front and backside were metalized by a 500nm thin Al layer. The front grid was structured by photolithography. In a few cases, an additional silicon nitride antireflection coating (ARC) was deposited on top of the metal grid. It has to be emphasized that for the solar cells none of the above mentioned deposition and preparation procedures were optimized. Yet to be able to classify our solar cell performance, we also prepared solar cells with in-diffused emitters using otherwise the same technological steps as above.

RESULTS AND DISCUSSION

When depositing doped silicon on quartz substrates with ECR-CVD at T = 325°C under the present conditions, the films grow in a polycrystalline structure with grain sizes in the order of 10 nm. This material will in the following be referred to as microcrystalline silicon (μc-Si). Only in a few cases, at high doping levels and poor microwave coupling to the reactor, the films grew amorphous. The deposition rate for P- and B-doped μc-Si films decreases with increasing doping concentration with the deposition rate for B-doping being generally higher than for P-doping (see Fig. 1). This behavior is due to the catalytical properties of diboran for dissociation of silan. On quartz glass the films grow with columnar structure preferentially oriented in the (110) direction independent of the gas phase doping level. From Raman measurements we obtain a crystallinity of about 70% with the peak of the Raman spectrum slightly shifting with increasing doping concentration from 520 cm^{-1} to about 517 cm^{-1}. The crystallinity is within experimental certainty independent of doping. For n-type doping the carrier concentration determined from Hall measurements increases linearly with the gas phase doping concentration (Fig. 2 for P doping). SIMS data which are also included in Fig. 2 show that the doping efficiency in the thin films is close to unity. We found similar behavior also for B-

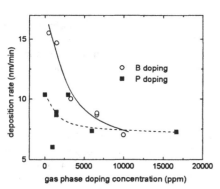

Fig. 1: Deposition rate as a function of the gas phase doping concentration (PH3/SiH4 and B2H6/SiH4). The lines are guides to the eye.

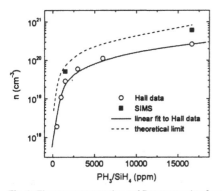

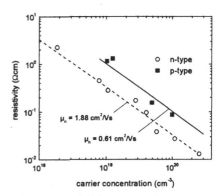

Fig. 2: Electron concentration and P concentration from SIMS versus PH$_3$/SiH$_4$. The dotted curve is the theoretical limit of P-doping.

Fig. 3: Dark resisitivity versus carrier concentration. The lines represent fits to the data assuming a constant mobility.

doping (not shown here). We conclude from these results that almost all dopant atoms are electrically active. Comparing these results with the theoretical limit of phosphorus incorporation (Fig. 2, dashed line) we conclude that the P- and B-atoms occupy substitutional sites inside the crystallites and do not segregate at grain boundaries even at the highest doping levels.

This conclusion is further supported by the observation that the dark resistivity measured in a coplanar van der Pauw configuration decreases linearly with increasing carrier concentration and hence with doping concentration (Fig. 3). We find minimal resistivities (film thickness 90-200 nm) between 10^{-2} Ωcm and 10^{-1} Ωcm for n-type and p-type doping, respectively. Although the concentration of dopant atoms is varied over 2 orders of magnitude, we observe that mobilities of electrons ($\mu_n = 1.9 \pm 0.2$ cm^2/Vs) and holes ($\mu_h = 0.6 \pm 0.1$ cm^2/Vs) are independent of doping concentration. This behavior is not found in Si single crystals and leads us to conclude that presumably disorder-induced effects or barriers at the grain boundaries determine the carrier mobility. The above results are in good agreement for what is reported for highly doped thin films deposited by PECVD [5], and HW-CVD [6]. Summarizing the results for films deposited on quartz glass, we conclude that ECR-CVD is capable of producing state-of-the-art highly doped thin films at relatively high deposition rates.

For highly P and B-doped layers grown on c-Si wafers the film morphology is distinctively different from that observed above for films deposited on quartz. When the doping concentration exceeds 1000 ppm in case of P-doping and 5000 ppm in case of B-doping, we find a dramatic change from columnar growth which is typical for μc-Si with the interface always being clearly resolved (Fig. 4a) to an extremely compact growth with smooth surfaces where the interface is not observable (Fig. 4b). The morphology of the highly doped films was investigated by low and high resolution TEM (Fig. 5). The films are perfectly crystalline as is shown by the selective area diffraction pattern (SAD) in the inset of Fig. 5a. Although the interface with the c-Si appears as a thin dark line in Fig. 5a, in high resolution TEM (Fig. 5b) no crystal imperfections like stacking faults or dislocations are observable. To further prove the excellent growth quality, we additionally used the Rutherford backscattering (RBS) channeling technique performed with 1.4 MeV He$^+$ ions. The minimum yield was determined to be clearly lower than 4 % for the best samples. This is very close to the theoretically predicted value of 3.4 % for crystalline silicon and gives prove of the excellent quality of the homoepitaxially grown films. Homoepitaxial growth of undoped Si at low temperature with ECR-CVD has been reported before [7-9]. It was found that

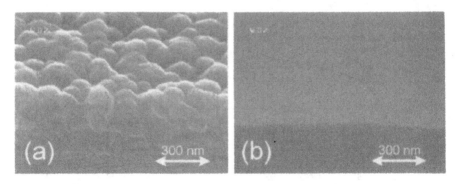

Fig. 4: SEM micrographs of a p-type microcrystalline (a) and epitaxially (b) thin film grown on c-Si. The doping concentration was 500 ppm (a) and 6700 ppm (b), respectively.

for T < 400°C the films grow with a high number of crystal defects [10, 11]. Our experiments clearly show that by adding dopant gases to the reactor epitaxial growth with high crystalline quality is possible under conditions were usually good high quality μc-Si films are grown on glass. XRD measurements suggest that with increasing doping concentration the stress in the epi-layer is strongly reduced which enables perfect film growth. Details will be published elsewhere.

SOLAR CELLS

In order to test the electronic quality of the n- and p-type thin films deposited on the Si wafers we prepared solar cell by using such layers as emitters. From each deposition (total of 26) about 30 cells with A = 0.16 cm^2 or 4 cells with A = 4 cm^2 were prepared on a single 3" c-Si wafer. We find good lateral homogeneity and only little scatter in the solar cells parameters for both p- and n-type emitters. In Fig.6 the main solar cell parameters are summarized for n-type emitters as a function of the gas phase doping concentration. The general behavior reflects the change in morphology and the increasing quality of the epitaxial emitter with increasing doping level. Similar result were found for p-type emitters, however, with the cell performance changing at a

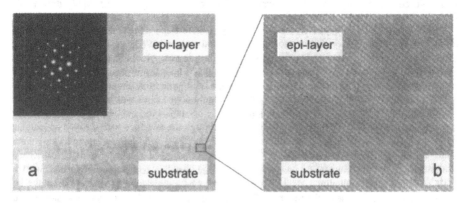

Fig. 5: (a) TEM micrograph of a highly B-doped film grown on c-Si. The inset shows a SAD pattern. (b) HRTEM image of the indicated area in (a) showing the high quality interface.

somewhat higher doping level of 5000 ppm. The best as deposited cells have efficiencies of about 11 %. When additionally an ARC is added, the efficiency is increased to about 14%. The cells with A = 4 cm^2 usually have a slightly lower efficiency which is presumably due a higher series resistance of the cell. These results are similar to what has been found for emitter layers deposited by ECR-CVD at T = 450 °C [4].

The dramatic change in solar cell performance with change in growth morphology is also reflected in the external quantum efficiency as plotted in Fig. 7 for n-type emitters; p-type emitters show qualitatively the same behavior. The blue response increases dramatically with increasing P-doping which gives evidence of the improved structural and electronic quality of the emitter. Although part of this loss in blue response is due to the fact that the emitter layer thickness increases with decreasing P-doping, the main effect is understood in terms of a poor quality emitter with high interface recombination. Since the red response of the cells, on the other hand, is mainly determined by the properties of the absorber, only little variation is observed with doping.

Generally, the open circuit voltage and for the large area cells also the fill factor are still too low. The fill factor may be limited by a non-optimized grid metalization. When we compared the results to cells with diffused emitters (see Fig. 6) prepared otherwise under the same condition, (same substrate, simple technology) we find similar V_{OC} and FF. The clearly lower I_{SC} is due to the fact that diffused emitters are by a factor of 5 wider that the thin film emitters. This comparison demonstrates the potential of low temperature deposited thin film emitters

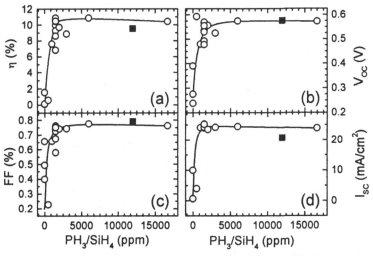

Fig. 6: Efficiency η (a), open-circuit voltage V_{OC} (b), fill factor FF (c), and short circuit current I_{SC} (d) of n-type emitters (0.16 cm^2) versus gas phase concentration. The solid lines are guides to the eye. The solid squares represent data from a diffused emitter. For details see text.

CONCLUSION

We have deposited highly doped Si films on glass and on c-Si by ECR-CVD. Films on glass are μc-Si whereas films on c-Si grow perfectly homoepitaxial. We have prepared thin film emitter solar cells with efficiencies above 11% (14% with ARC). The cell performance is better compared to cells with diffused emitters prepared under the same conditions.

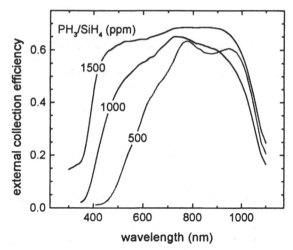

Fig. 7: External collection efficiency of n-type emitters with various gas phase doping concentrations.

ACKNOWLEDGEMENT

We greatly appreciate the support of G. Keiler, J. Krause, B. Rabe, and D. Patzek for technical assistance, M. Kunst and O. Abdallah for depositing the ARC and W. Henrion for optical measurements. We are also indebted to B. Selle for RBS measurements, U. Zeimer (Ferdinand-Braun-Institut) for XRD and N. Wanderka for TEM measurement. The work was partially supported by the BMBF under contract # 0329773.

REFERENCES

1   H. Keppner, U. Kroll, P. Torres, J. Meier, D. Fischer, M. Goetz, R. Tscharner and A. Shah, in *25th IEEE PVSC*, Washington D.C., 1996, p. 669.

2   K. Yamamoto, M. Yoshimi, T. Suzuki, Y. Tawada, T. Okamoto and A. Nakajima, in *2nd World Conf. on PVSEC*, Vienna, Austria, 1998, to be publ.

3   P. Müller, I. Beckers, E. Conrad, L. Elstner and W. Fuhs, in *25th IEEE PVSC*, Washington D.C., 1996, p. 673.

4   E. Conrad, L. Elstner, W. Fuhs, W. Henrion, P. Müller, B. Selle and U. Zeimer, in *26th PVSC*, Anaheim, Ca, 1997, p. 755.

5   R. Carius, F. Finger, U. Backhausen, M. Luysberg, P. Hapke, L. Houben, M. Otte and H. Overhof, in *Mat. Res. Soc. Symp.*, edited by S. Wagner, M. Hack, E. A. Schiff, R. Schropp and I. Shimizu (MRS, San Francisco, 1997), Vol. 467, p. 283.

6   A. R. Middya, J. Guillet, R. Brenot, J. Perrin, J. E. Bouree, C. Longeaud and J. P. Kleider, in *Mat. Res. Soc. Symp.* (MRS, San Francisco, 1997), Vol. 467, p. 271.

7   J. L. Rogers, P. S. Andry, W. J. Varhue, P. McGaughnea, E. Adams and R. Kontra, Appl. Phys. Lett. **67**, 971 (1995).

8   D. S. L. Mui, S. F. Fang and H. Morkoç , Appl. Phys. Lett. **59**, 1887 (1991).

9   W. J. Varhue, J. L. Rogers, P. S. Andry and E. Adams, Appl. Phys. Lett. **68**, 349 (1996).

10 H. Yamada and Y. Torri, Appl. Phys. Lett. **50**, 386 (1987).

11 H.-S. Tae, S.-H. Hwang, S.-J. Park, E. Yoon and K.-W. Whang, J. Appl. Phys. **78**, 4112 (1995).

PROBING THE ELEMENTARY SURFACE REACTIONS OF HYDROGENATED SILICON PECVD BY *In-situ* ESR

Satoshi Yamasaki, Claus Malten, Takehide Umeda, Jun-ichi Isoya, and Kazunobu Tanaka
Joint Research Center for Atom Technology (JRCAT),
1-1-4, Higashi, Tsukuba, 305-8562, Japan.

ABSTRACT

The dynamic change of the dangling bond (db) intensity in hydrogenated amorphous silicon (a-Si:H) during H_2 and Ar plasma treatments was observed using an *in-situ* electron-spin-resonance (ESR) technique. The experimental results show that the time to reach the steady state between gas-phase H atoms and the a-Si:H surface is less than 1 sec, and Ar plasma treatments create a top-surface region with an extremely high db density.

INTRODUCTION

Microcrystalline silicon (μc-Si:H) can be grown by several techniques, such as PECVD (plasma enhanced chemical vapor deposition) using a high hydrogen-dilution silane-plasma, a flow modulation technique of silane and H_2, a hot-wire technique, etc. To address the μc-Si:H and a-Si:H formation mechanism, several groups have reported on the surface morphology and bonding configurations during film growth using real-time spectroscopic ellipsometry [1], ultrahigh-vacuum scanning-tunneling-microscopy[2], and infrared absorption reflection spectroscopy[3]. These measurements, however, give us information on the macroscopic change of surface structures. On the other hand, in any deposition technique used for μc-Si:H formation atomic hydrogen plays a crucial role in modifying the surface reactions and in growing μc-Si:H. Also, for the growth of μc-Si:H as well as hydrogenated amorphous silicon (a-Si:H), dangling bonds (dbs) in the surface region play a crucial role as reaction sites for precursors to bond to the surface. From the viewpoint of surface dbs, hydrogen assumes two competing roles in the reactions on a growing surface; annihilation of dbs and creation of dbs. Creation of dbs takes place due to breaking of Si-Si bonds by H insertion reactions (H + Si-Si $\rightarrow$ Si-H + Si—) and by hydrogen abstraction reactions (H + Si-H $\rightarrow$ H_2 + Si—). For db annihilation, H atoms terminate Si dbs (Si— + H $\rightarrow$ Si-H). In addition, due to structural reconfiguration, two dbs can move during and after reaction, and intimate pairs of dbs annihilate through rebondings (Si— + Si— $\rightarrow$ Si-Si), although this process is not directly related to H atoms.

In this study, to clarify the role of hydrogen atoms in the growth of Si:H, *in-situ* electron spin resonance (ESR) measurements [4-6] have been made during hydrogen plasma treatment of a-Si:H films, as well as Ar plasma treatment as a reference system, using a remote plasma technique. A dynamic change of the Si db ESR intensity was observed during and after the plasma treatment. Although, in both cases dbs were created during the plasma treatments, the db intensity of rise after plasma-on and its decay after plasma-off are different from one another. On the basis of these experimental results surface microchemical reactions of hydrogen atoms and excited Ar atoms on Si:H films are discussed.

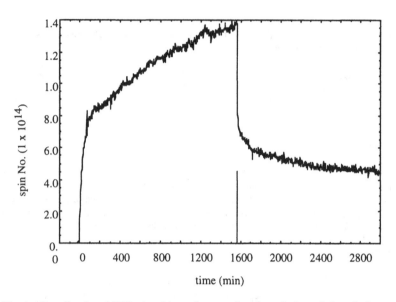

Fig. 1. Dangling bond ESR signal intensity as a function of elapsed time during a-Si:H deposition.

EXPERIMENT

For *in-situ* ESR measurements, a remote plasma system was used [4]. A plasma discharge with a power of 50W was sustained within a microwave cavity (2.45 GHz) for H_2 and Ar plasmas. It was located at the position above the X-band frequency (9 GHz) of the ESR cavity. H_2 (100 SCCM) flowing through a high-purity vitreous-silica (HPVS) tube, with an outer diameter of 6 mm, was dissociated into H atoms upon passing through the discharge cavity. Ar atoms (70 SCCM) were excited to Ar* (11.6 eV) in the discharge cavity with the same plasma conditions as the H_2 plasma. For film deposition, silane molecules (10 SCCM) were reacted with the atomic H in the ESR cavity and the primary reaction taking place is $SiH_4 + H \rightarrow SiH_3 + H_2$ [7]. As a result, an a-Si:H film was deposited on the inner wall (inner diameter ; 8 mm) of the outer HPVS tube. After deposition H_2 and Ar plasma treatments were performed. All deposition and plasma treatments were performed near 100°C. The ESR measurements were made using an ESP BRUKER 300E system. To get the time-resolved change of the db ESR intensity, the magnetic field was set at the peak position of the db ESR spectrum in the derivative form.

RESULTS

The db ESR intensity was measured as a function of time during a-Si:H deposition, as shown in Fig. 1. A quick increase of the signal intensity was observed when the deposition was started, followed by a relatively slow increase of signal intensity over the entire deposition time due to an increase of the film thickness. A rapid decrease occurred when the deposition was terminated by stopping the plasma. From a detailed analysis, it was clarified that during the film growth there exists a surface region where the db density is much higher than that in the bulk region [5].

After deposition an Ar plasma treatment was performed followed by a H_2 plasma treatment. Figure 2 shows the change of the db intensity as a function of time. Each point was taken as an average of 1024 points collected during an 84-sec time scan. For both cases there are excess dbs during plasma treatment. The db ESR intensity during H_2 plasma treatment was almost constant. On the other hand, for the case of the Ar plasma the db density did not saturate within 30 min, but continued to increase with time. For both cases long decays of db ESR intensities were observed after plasma-off.

To see the transient behavior, the data obtained for the 84-sec scans of plasma-on and plasma-off for H_2 and Ar plasma treatments are shown in Fig. 3. Although the Ar plasma created an excess db intensity like the H_2 plasma, the time constants of the rise after plasma-on and the decay after plasma-off were quite different from the case of the H_2 plasma. After the H_2 plasma was started (Fig. 3 (a)), the db ESR intensity quickly increased with a time constant shorter than the time resolution in this experiment (< 1 sec). The rapid increase is followed by a slow decrease to a constant intensity approximately 10 sec after plasma-on. The time evolution of the db density after Ar plasma-on (Fig. 3 (c)) shows an increasing trend with two components: First, the db intensity increases after plasma-on and appears to saturate after around 5 sec in the short time regime of Fig. 3(c). Second, a slow increase is observed in the long time regime of Fig. 2. On the other hand, after plasma-off, dbs created by the H_2 plasma decay continuously from the intensity measured during plasma treatment in the short time regime (Fig. 3 (b)), and connect continuously to the long time regime in Fig. 2. For the case of the Ar plasma (Fig. 3 (d)), a part of the db ESR intensity decays quickly (< 1 sec), although in the long term measurement another slow-decay component is also observed, as shown in Fig. 2.

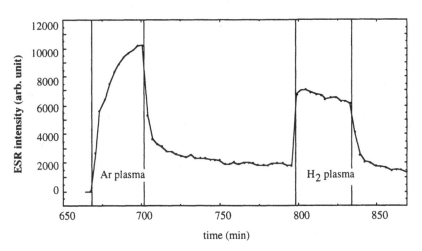

Fig. 2 Time evolution of dangling bond intensity during Ar and H_2 plasmas. Each point was taken as an average of 1024 points during an 84-sec scan with the magnetic field fixed at the peak of the db ESR spectrum.

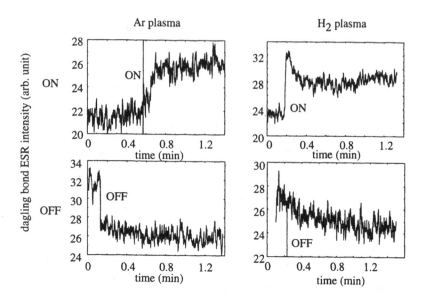

Fig. 3. Transient dangling bond intensity for plasma-on (a) and plasma-off (b) for H_2 plasma treatments, and for plasma-on (c) and plasma-off (d) for Ar plasma treatments.

DISCUSSION

Here, we take into account the primary reaction, although other reactions related to secondary reactions or to impurities can be considered. With respect to the role of hydrogen two competing reactions on the growing surface can occur; annihilation of dbs and creation of dbs. Creation of dbs takes place due to breakings of Si-Si bonds by H insertion reactions (H + Si-Si → Si-H + Si—) and hydrogen abstraction reactions (H + Si-H → H_2 + Si—). For annihilation, H atoms terminate Si dbs (Si— + H → Si-H). In addition, in a process not directly related to H atoms structural reconfiguration allows, two dbs to diffuse during and after reaction. Then, intimate pairs of dbs annihilate through rebonding (Si— + Si— → Si-Si).

For the H_2 plasma treatment, evolution of the db density is expressed as,

$$dN/dt = k_{cH}FS - k_{tH}FN - 2k_{aH}N^2,$$ (1)

where N is the surface db density, F is the flux of H atoms, S is the number of Si-Si and Si-H bonds in the reacting region, k_{cH}, k_{tH}, and k_{aH} are the rate constants of db creation, db annihilation due to a termination by H atoms, and db annihilation due to thermal reconstruction, respectively. The first term describes the creation of dbs due to the breaking of Si-Si bonds and Si-H bonds, the second annihilation term describes hydrogen termination, and the last term describes the annihilation due to rebonding between two dbs.

For the Ar plasma, different reactions can be considered. Excited Ar atoms, Ar*, create dbs through bond breaking (Ar* + Si-Si → Ar + Si— + Si— and Ar* + Si-H → Ar + Si— + H). The time evolution of dbs for Ar plasma is expressed as,

$$dN/dt = 2k_{cAr}FS - 2k_{aAr}N^2, \qquad (2)$$

where k_{cAr} and k_{aAr} are the rate constant of db creation and db annihilation due to thermal reconstruction, respectively.

When the H_2 plasma is started, the db ESR intensity rises within the time resolution in this experiment (< 1 s). Following the quick rise of the signal, we observe the decrease of the db intensity and it becomes a constant approximately 10 sec after plasma-on. This decrease is explained by the change of plasma conditions caused by the plasma ignition. This explanation is checked by direct H density measurements using in-situ ESR observation of the H hyperfine structure [4]. The intensity of the H hyperfine structure also showed the same behavior; a quick rise of the signal after plasma-on and a slow decay with a time constant of 10 sec. Therefore, it is concluded from results of the time evolution for the H_2 plasma that the time to reach a steady state between gas-phase H atoms and the a-Si:H surface is shorter than 1 sec.

When the plasma is stopped, the decay of db intensities is expressed as,

$$N = N_o/(2N_o k_{aH} (t-t_e) +1), \qquad (3)$$

where t_e is the time when the plasma is stopped. The time constant of the signal decay is determined by k_{aH}. When the plasma is started, the db density increases with a time constant that is a function of k_{cH}, k_{iH}, F, S. Since the time constant of the signal decay for the H_2 plasma is much longer than that of the signal rise, as seen from Fig. 2, the term $2k_{aH}N^2$ in eq. (1) is thought to be smaller than the term $k_{iH}FN$, except in the case of extremely small F. If we ignore the third term, the rise of the db density is expressed as,

$$k_{iH}F >> 2k_{aH}, \qquad (4)$$

$$dN/dt = k_{cH}FS - k_{iH}FN, \qquad (5)$$

$$N = N_o [1 - \exp(-k_{iH}Ft)]. \qquad (6)$$

The signal rises with a time constant of $1/k_{iH}F$.

On the other hand, for the Ar plasma it takes a relatively longer time for the db ESR intensity to reach a steady state after plasma-on in comparison with the case of H_2 plasma. Although the ESR intensity appears to be saturated after approximately 5 sec in the short time regime in Fig. 3, even approximately 30 min after plasma-on the db intensity continues to increase in the longer time regime in Fig. 2. The different behavior of the rise times between the H_2 and Ar plasma may originate from the difference in reaction mechanism as mentioned above. For the case of the H_2 plasma, db termination by H atoms occurs in addition to the recombination process of db pairs due to thermal reconstruction. Whereas, for the Ar plasma, only the later process annihilates a db. Therefore, the annihilation probability under Ar plasma exposure should be smaller than that under H_2 plasma exposure. The slow increase of db density after plasma-on may be related to the small annihilation probability during plasma exposure. Also, as is mentioned above, Ar* creates two dbs simultaneously or a complex of a db and a H atom. Therefore, it should be easy to restore these defects to the previous bonding structures (Si— + Si— → Si-Si and Si— + H → Si-H). This suggests that the effective probability for an Ar* atom to create a db may be small. A small annihilation probability and a small creation probability may be the origin of the slow increase after plasma-on for the Ar plasma.

For the Ar plasma, we have observed a db component with a quicker rise after plasma-on than the subsequent slow increase of db intensity. The former component also exhibits a quick decay with a time constant shorter than 1 sec after plasma-off in Fig. 2. This suggests the existence of a high db density region for the Ar plasma, because such a region has a shorter annihilation time for db pairs. It is reasonable that the top-surface (including a couple of atomic layers) has such a high db density. Using an in-situ ellipsometry technique, Collins and Cavese[8] and Fukutani et al. [9] reported that the Ar plasma treatments on an a-Si:H film change the structure only at the top surface. The present data are consistent with their results, because it is reasonable for the modified layer to have a higher db density than the bulk region.

CONCLUSIONS

In-situ ESR measurements have been made during hydrogen and Ar plasma treatments of a-Si:H films, and dynamic changes of the Si dangling-bond intensity were observed during and after the plasma treatment. From the experimental results, it is concluded that the time to achieve the steady state between gas-phase H atoms and an a-Si:H surface is less than 1 sec, and it is suggested that for the Ar plasma there exists a top-surface with a relatively higher db density compared with the case of the H_2 plasma

ACKNOWLEDGMENTS

This work, partly supported by NEDO, was performed in the Joint Research Center for Atom Technology (JRCAT) under the joint research agreement between the National Institute for Advanced Interdisciplinary Research (NAIR) and the Angstrom Technology Partnership (ATP).

REFERENCES

1. R. W. Collins, I. An, H. V. Nguyen, Y. Li, and Y. Lu, in *Physics of Thin Films, Optical Characterization of Real Surfaces and Films*, edited by M. H. Francombe, and J. L. Vossen, (Academic Press, San Diego, 1994) pp 49-125.

2. K. Ikuta, K. Tanaka, S. Yamasaki, K. Miki, and A. Matsuda, Appl. Phys. Lett. **65**, 1760 (1994).

3. Y. Toyoshima, K. Arai, A. Matsuda, and K. Tanaka, J. Non-Cryst. Solids **137&138**, 765 (1991).

4. S. Yamasaki, T. Umeda, J. Isoya, and K. Tanaka, Appl. Phys. Lett. **70**, 1137 (1997).

5. S. Yamasaki, T. Umeda, J. Isoya, and K. Tanaka, J Non-Cryst. Solids **227-230**, 83 (1998).

6. S. Yamasaki, T. Umeda, J. Isoya, and K. Tanaka, in Amorphous and Microcrystalline Silicon Technology - 1997, edited by S. Wagner, M. Hack, E. A. Schiff, R. Schropp, and I. Shimizu (Mat. Res. Soc. Proc. 467, Pittsburgh, PA, 1997), p. 507.

7. K. Y. Choo, P. P. Gaspar, and A. P. Wolf, J. Chem. Phys. **79**, 1752 (1975).

8. R. W. Collins and J. M. Cavese, J. Non-Cryst. Solids **97&98**, 1439 (1987).

9. K. Fukutani, M. Kannbe, W. Futako, B. Kaplan, T. Kamiya, C. M. Fortmann and I. Shimizu, J. Non-Cryst. Solids **227-230**, 63 (1998).

EVAPORATED POLYCRYSTALLINE GERMANIUM FOR NEAR INFRARED PHOTODETECTION

L. COLACE, G. MASINI, F. GALLUZZI, and G. ASSANTO
Department of Electronic Engineering, University 'Roma TRE' & Unità INFM
Via della Vasca Navale, 84 - I-00146- Rome - Italy

ABSTRACT

We present low cost near infrared photodetectors based on polycrystalline Ge film thermally evaporated on a silicon substrate. We demonstrate that, by proper choice of deposition conditions and device configuration, a responsivity of 16mA/W and a response time of a few nanoseconds can be achieved at the wavelength of 1.3micron. The device can operate up to 1.55micron. We also describe the fabrication and the operation of a 16 pixel linear detector array, with pitch of about 100 micron.

INTRODUCTION

SiGe based photodetectors have been demonstrated over a wide range of wavelengths, from the near to the mid infrared[1]. The most important features are the band gap engineering of the strained SiGe alloys and quantum confined heterostructures and the technological compatibility with the well established silicon VLSI. Due to the widespread use of fiber optic communications in the 1.3-1.55 micron range, Ge-rich alloys and pure-Ge are the best candidates. Unfortunately, the large lattice mismatch between germanium and silicon (about 4%) introduces the fundamental limitation known as critical thickness (few monolayers for pure Ge and 100Å for $Ge_{0.5}Si_{0.5}$ alloy)[2]. If the epitaxial growth proceeds above the critical thickness, the strain is relaxed through the formation of defects and dislocations. A Ge epilayer suitable for efficient detection of near infrared light is much thicker than the critical value, and approaches such as the introduction of graded buffers (usually combined with surfactants [3] or chemical-mechanical-polishing[4]) or low temperature grown layers [5] have been proposed. Regardless the type of buffer, it has been demonstrated that two factors are determinant: the cleaning procedure of the substrate, usually performed at high temperatures (800- 1100 °C), and the vacuum in the growth chamber (10^{-10} - 10^{-11} Torr). Both have severe implications in terms of cost and compatibility with silicon technology. Moreover, in order to relax the entire amount of strain, the buffers should be graded (from Si to 100% Ge) at a rate of about 10% Ge μm^{-1}, which implies long and expensive growth sessions. To overcome these problems, we have adopted a novel approach based on evaporation of polycrystalline Ge film on silicon. Deposition uniformity, low cost fabrication, and reduced thermal budget are fully compatible with and well suited for large area wafer production. In this work we demonstrate that polycrystalline Ge films allow the fabrication of efficient near infrared photodiodes with good speed and sensitivity. In the following we discuss the fabrication process, the material characterization and the device performances.

SAMPLE GROWTH

Ge films were deposited by thermal evaporation using a 99.999% purity commercial source and a tungsten crucible in a vacuum with a background pressure of 10^{-6} Torr. Different samples were grown on n- and p- type (resistivity 2-3 Ωcm) <100> silicon as well as on Borofloat™ glass substrates at temperatures in the 25-500°C interval, with evaporation rates of 1.5Å/s. Films of different thickness from 0.12 to 1.8 μm were deposited, as determined with a piezoelectric crystal balance. Silicon substrates were chemically cleaned just before the introduction in the vacuum chamber by dipping into a buffered HF bath for two minutes and successive rinsing in deionized water. The whole cleaning process was performed at room temperature. All the samples exhibited smooth specular surfaces, homogeneous over several square centimeters. For growth temperatures above 250°C the surface color changed from light gray to greenish.

RESULTS AND DISCUSSION

Absorption spectra were determined by measuring transmitted and reflected light intensity and correcting for the substrate absorption and reflection. Thin (0.12μm) and thick (1.8μm) samples were used for measurements in visible and NIR range, respectively. Typical absorption spectra are shown in Fig.1. The transition from amorphous to poly-crystalline structure is evidenced by the neat appearance of absorption edges at 1.54 and 0.58μm, related to the direct transitions at the Γ and Λ points of the crystalline Ge band structure, respectively.[6] The Raman spectra (see the inset of fig.1) confirm the occurence of the amorphous-polycrystalline transition[7] characterized by the displacement of the broad band peaked around 270 cm^{-1} into the sharp transverse optical phonon line at 300 cm^{-1}. For the latter measurements we used microprobe Raman spectroscopy (spot size 2μm) with a laser at 633nm with intensity below 10 mW to prevent local recrystallization.

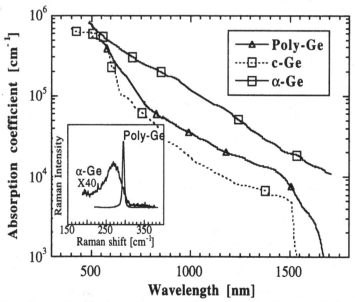

Fig.1 Absorption spectra of poly- and amorphous Ge compared to crystalline Ge. In the inset the Raman spectra of the materials are shown.

The resistivity of the polycrystalline and amorphous Ge films deposited on both Si and glass substrates was 0.1 and 1000Ωcm, respectively, to be compared with 50-100 Ωcm values for undoped Ge single crystals [8]. The higher resistivity of amorphous samples is reasonably due to reduced carrier mobility, whereas the large resistivity decrease of polycrystalline films should be attributed to unintentional metal impurity incorporation during Ge evaporation. Indeed metal atoms such as Cu and Ag introduce shallow acceptor-like states and exhibit high diffusivity values even at 300°C [6].

Dark current-voltage measurements were employed to characterize the Si/Ge heterojunction. To minimize parasitic effects, we fabricated vertical mesa diodes of different areas ranging from 0.1 to 4x10^{-4} cm^2. Contacts to the (top) Ge layer were realized with a 0.4μm silver layer evaporated and lithographically defined on top of the mesas. Silver contacts proved to be ohmic through current-voltage measurements performed between a couple of coplanar electrodes. Diodes fabricated on either n-type or p-type silicon exhibited a strong rectifying behavior such as shown in Figs.2 and 3, respectively. The heterojunctions were forward biased when the Ge was at positive voltages with respect to the silicon for n-type substrates, and the opposite for p-type Si. The current increase in forward bias is limited by the series resistence of the substrate when the voltage exceeds a few hundred millivolt.

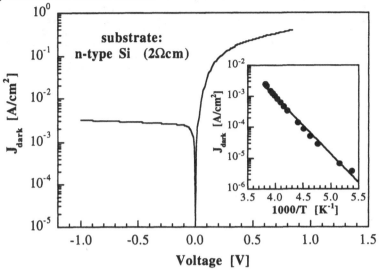

Fig.2 Current-voltage characteristic of a poly-Ge/Si heterojunction. In the inset the dependence of the dark current is plotted

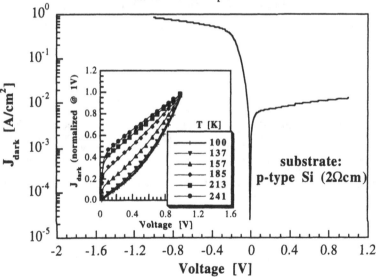

Fig.3 Current-voltage characteristic of poly-Ge/Si p-type heterojunction. In the inset the reverse current characteristics are plotted for different temperatures: experimental data (symbols) are fit by the image charge barrier lowering model.

The analysis of reverse current dependence on bias and temperature allowed to discriminate two different current transport mechanisms in the diodes. In particular, while both currents evidence a temperature activated process, heterojunctions grown on n-type substrates exhibited a flat reverse current independent of bias (inset, fig.2) with an activation energy of 0.37 eV. On the contrary, the reverse current of diodes grown on p-type substrates slowly increases with the applied bias, following an exponential dependence on the fourth root of the voltage [9] (inset, fig.3). This dependence and the extrapolated barrier height (0.4 eV) are consistent with a model in which the current is limited by the barrier at the valence band discontinuity between Si and Ge [10]. The presented results suggest a band alignment between Ge and Si as sketched in Fig.4.

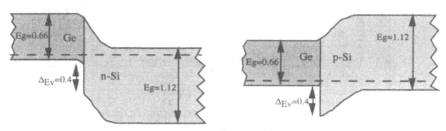

Fig.4 Band alignment for the poly-Ge/Si heterojunction: left, n-type Si; right, p-type Si substrate.

As for diodes photosensitivity, devices based on amorphous Ge on either p- or n-type silicon do not exhibit any NIR photoresponse above 1.2μm. Similarly isotype heterojunctions between poly-Ge and p-type silicon present a NIR photoresponse completely similar to that of pure silicon devices. Indeed, according to the suggested band alignment (fig.4, right), the collection of carriers photogenerated within the Ge layer is completely hindered by the Si-Ge interface barrier in these structures. On the contrary hetero-type junctions between poly-Ge and n-type silicon show a strong photoresponse up to 1.55μm.

A typical photocurrent spectrum of such devices, recorded by shining the monochromatic light through the sample, is plotted in fig.5.

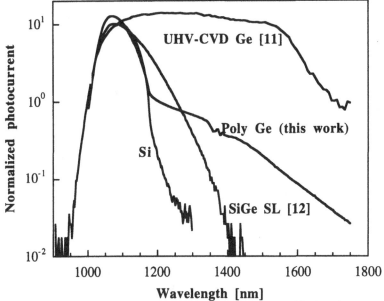

Wavelength [nm]

Fig.5 Photocurrent spectrum of a poly-Ge/Si device compared with photodetectors made in silicon, SiGe superlattice and UHV-CVD Ge-on-Si. The cut around 1000 nm is due to a silicon filter used to suppress the lower order of the monochromator.

In the same figure, the spectra of a monocrystalline Ge-on-Si[11], a SiGe superlattice (SL) [12] and a Si p-i-n photodiode are shown for comparison. It is evident that the poly-Ge based device extends its photoresponse well over that of Si and SiGe SL, up to the third window of fiber optical communications (1.55 μm). At the same time the response of the crystalline Ge detector is higher at the cost of a more complex and less silicon-technology compatible fabrication process.

The minority carriers lifetime was measured by a Shockley-type experiment[8] : light pulses of 100 ps duration at a wavelength of 1.32 μm (uniformly absorbed in the Ge layer) were shined on the diodes and the photocurrent transient was recorded. The duration of the photocurrent exponential decay was independent on the applied bias, as expected for a lifetime-limited collection. A lifetime of 5ns was measured in polycrystalline samples.

The responsivity of the fabricated devices is independent on light intensity in a wide interval and reaches 16mA/W at 1.32μm for 0.2V reverse bias. Due to the high conductivity of the poly-Ge film (corresponding to an extremely narrow depletion depth of a few Ångstrom), the mechanism suggested for photocarrier collection is the diffusion in the quasi-neutral zone rather than the drift in the space charge region. The diffusion length L_D, estimated by the approximated relationship $Y \propto \left(1 - e^{-\alpha L_D}\right)$ (with Y photocurrent quantum yield, α absorption coefficient) is in the range 20 - 30 nm.

To exploit the performances of the poly-Ge/Si system we designed and fabricated a linear array of 16-elements, as sketched in Fig.6 (right inset). Each Ge-on-Si pixel is a planar Metal-Semiconductor-Metal photoelement. However, due to the reduced thickness of the Ge layer compared with the interelectrode spacing (10μm), it results in a double backward Ge/Si heterojunction. The active area of the pixels (100μm x 100μm) is defined by the multiple bends of the central ground electrode, fabricated on a mesa-etched Ge stripe. Metal contacts are insulated from the substrate by 2μm thick photoresist, properly windowed over the sensitive Ge region.

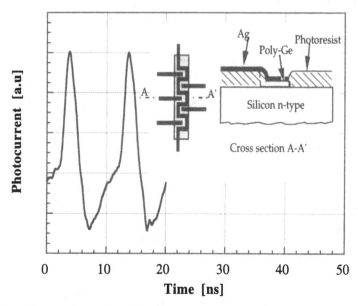

Fig.6 Time response of one pixel excited by a train of 100ps pulses at 1.32 μm. On the right the top view and section of the array are sketched.

To assess the quality of the process through uniformity in photoresponse and pixel definition, we performed Light Beam Induced Current (LBIC) evaluation of each pixel. The measurements were carried out by scanning the array with a 1.32μm laser beam focused to a spot size of about 20 μm, and recording the photocurrent at a bias of 0.6V. A quantitative analysis of the experimental data indicates a ten-fold reduction in the signal from the addressed pixel with respect to the adjacent ones. This guarantees low cross-talk between neighbouring channels, of primary importance in most applications. When scanning the 16 pixels of the array, via LBIC we evaluated a response fluctuation of less than 2% from element to element.

We also measured the speed of photoresponse when each pixel was excited by an optical pulse (100ps) train (100MHz) from a Nd:YAG laser operating at λ=1.32 μm. The device, biased at 3V through a 50Ω resistor, responded with a FWHM duration of about 2 ns.

CONCLUSIONS

Evaporated poly-Ge photodiodes exhibit interesting characteristics as NIR detectors integrated on silicon. The low-cost deposition process and its compatibilty with silicon technology, together with the good responsivity and speed, make this approach a valid alternative to MBE and CVD techniques.

REFERENCES

1. T.P.Pearsall, "Electronic and optical properties of Ge-Si superlattices", Progr. Quantum Electron. 18, 97,1994.

2. R.People, "Physics and applications of GeSi/Si strained layer heterostructures", IEEE J. Quantum Electron. 22, 1696, 1986.
3. H. Presting, H. Kibbel, "Buffer concept of ultrathin SiGe superlattices", Thin Solid Films 222, 215 (1992).
4. M.T. Currie, S.B. Samavedam, T.A. Langdo, C.W. Leitz, E.A. Fitzgerald, "Controlling threading dislocation densities in Ge on Si using graded SiGe layers and chemical-mechanical polishing", Appl. Phys. Lett. 72, 1718 (1998).
5. L. Colace, G. Masini, F. Galluzzi, G. Assanto, G. Capellini, L. Di Gaspare, and F.Evangelisti, "Near infrared light detectors based on UHV-CVD epitaxial Ge on Si (100)", 1997 MRS Fall Meeting, Boston, Dec.1-5, 1997.
6. M. Neuberger, 'Handbook of electronic materials' (IFI/Plenum Press New York 1971)
7. F. Evangelisti, M. Garozzo, and G. Conte, "Structure of vapor-deposited Ge films as a function of substrate temperature", J. Appl. Phys. 53, 7390 (1982)
8. S. M. Sze, 'Physics of Semiconductor Devices', Wiley & Sons, New York (1981)
9. R.S. Muller and T.I. Kamins, 'Device Electronics for Integrated Circuits', Wiley & Sons, New York p. 139, (1986)
10. L. DiGasapare, G. Capellini, C. Chudoba, M. Sebastiani, and F. Evangelisti, Appl. Surf. Sci., 104-105, 595 (1996)
11. L. Colace, G. Masini, F. Galluzzi, G. Assanto, G. Capellini, L. Di Gaspare, E. Palange, and F. Evangelisti, "Metal-Semiconductor-Metal Near Infrared Light Detector Based On Epitaxial Ge/Si", Appl. Phys. Lett. 72, 3175 (1998)
12. G. Masini, L. Colace, G. Assanto, T. P. Pearsall, "Voltage-tunable near-infrared photodetector: Versatile component for optical communication systems", J. Vac. Science & Tech. B 16, 2619 (1998)

"Mirrorless" UV lasers in ZnO polycrystalline films and powder

H. Cao*, Y. G. Zhao*, H. C. Ong**, J. Y. Dai**, X. Liu**, E. W. Seelig**, and R. P. H. Chang**
* Department of Physics and Astronomy, Materials Research Center, Northwestern University, Evanston, IL 60208-3112, h-cao@nwu.edu
** Department of Materials Science and Engineering, Materials Research Center, Northwestern University, Evanston, IL 60208-3116

Abstract

We report the first observation of ultraviolet lasing in ZnO powder and ZnO polycrystalline films grown on amorphous fused silica substrates. In the absence of any fabricated mirrors, laser action occurs in the closed loops formed by multiple optical scattering of light.

Introduction

Throughout the development of semiconductor lasers, disordered semiconductor microstructures have not attracted much attention because of strong optical scattering. Light scattering has traditionally been considered detrimental to laser action since it removes photons from the lasing mode of a conventional cavity. In this letter, we report the first observation of ultraviolet (UV) laser action in zinc oxide powder and polycrystalline films. We will show that when the optical scattering is sufficient, it actually facilitates lasing by providing feedback.

Experiment

ZnO powder with an average particle size of 100 nm was deposited onto ITO coated glass substrates using the electrophoretic process. Thickness of the films of ZnO powder varied from 6 to 15 μm. The samples were optically pumped by a frequency-tripled mode-locked Nd:YAG laser (355 nm, 10 Hz repetition rate, 15 picosecond pulse width). The pump beam was focused to a spot or a stripe on the film surface with normal incidence. Figure 1 shows the evolution of the emission spectra as the pump power was increased. At low excitation intensity, the spectrum consisted of a single broad spontaneous emission peak., the emission peak became narrower due to the preferential amplification at frequencies close to the maximum of the gain spectrum. When the excitation intensity exceeded a threshold, very narrow peaks emerged in the emission spectra. The linewidth of these peaks was less than 0.3 nm, which was more than 30 times smaller than the linewidth of the amplified spontaneous emission peak below the threshold. When the pump power increased further, more sharp peaks appeared. Figure 2 shows the spectrally integrated emission intensity as a function of the pump power. A threshold behavior was observed: above the pump power at which multiple sharp peaks emerged in the emission spectrum, the integrated emission intensity increased much more rapidly with the pump power. The emission above the threshold was strongly polarized. These data suggest that laser action has occurred in the ZnO powder.

The characteristics of lasing in semiconductor powder exhibits remarkable differences from that of a conventional laser. First of all, the laser emission from the ZnO powder could be observed in all directions. The laser emission spectrum varied with the observation angle. Secondly, the pump intensity required to reach lasing threshold depends on the excitation area. As the excitation area decreases, the lasing threshold density increases. When the excitation area was increased, more lasing peaks emerged in the emission spectra. On the other hand, when the excitation area was reduced to below a critical size, laser oscillation stopped.

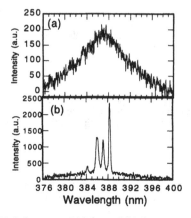

Fig. 1 Emission spectra (a) below and (b) above the threshold

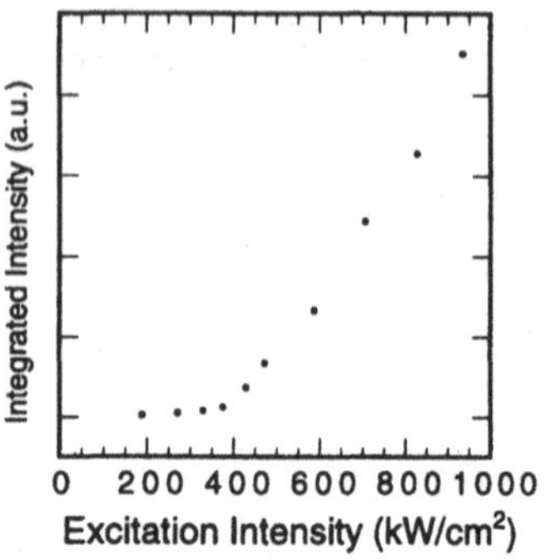

Fig. 2: Integrated emission intensity versus excitation intensity

We have observed similar lasing behavior in ZnO polycrystalline films deposited on amorphous fused silica substrates by laser ablation. TEM images illustrate the polycrystalline grain structure of the ZnO films. The grains appear in irregular shapes, and their sizes vary from 50 nm to 150 nm.

One essential component of a laser is the resonator, which provides the coherent feedback and determines the lasing frequencies. How the laser resonators are formed in the semiconductor powder and polycrystalline films in the absence of mirrors?

The highly disordered structure of semiconductor powder results in strong optical scattering. From the coherent backscattering measurement, we characterized the scattering mean free path in ZnO powder and polycrystalline films is close to the emission wavelength of ZnO. Therefore the emitted light is strongly scattered. Due to the short scattering mean free path, the emitted light may return to the a scatterer from which it was scattered before, thereby forming a closed loop path (see the inset of figure 1) [1,2]. There are many such closed loop paths for light in the powder and polycrystalline films. These loops could serve as ring cavities for light. Along different closed loop paths, the probability of a photon scattered back to its starting point is different. In other words, the ring cavities formed by scattering have different loss. When the pump power increases, the gain exceeds the loss first in the low-loss cavities. Then laser oscillation occurs in these cavities, and the lasing frequencies are determined by the cavity resonances. Laser emission from these resonators results in a small number of discrete narrow peaks in the emission spectrum, as observed in figure 1. As the pump power increases further, the gain increases and it exceeds the loss in the lossier cavities. Laser oscillation in those resonators add more discrete peaks to the emission spectrum.

Since different laser cavities could have different output directions, lasing modes observed at different angles may be different. Hence, some lasing modes could be observed only in certain directions. The others could be observed in several directions because some cavities could have strong output into several directions.

The increase of the number of lasing modes with the excitation area is due to the increase of the number of closed loop paths for light with area. Hence, in a larger excitation area, laser action could occur in more cavities formed by multiple scattering. On the other hand, when the excitation area is reduced to below a critical size, laser oscillation stops because the closed loop paths are too short, and the amplification along the loops is not high enough to achieve lasing. That is why the pump intensity required to reach lasing threshold increases with a decrease of the excitation area.

To actually observe the laser resonators in the ZnO films, we have taken amplified images of the excitation area using a microscope objective (50x) and a UV sensitive CCD camera. Below the lasing threshold we could not observe any patterns. However, above the lasing threshold, we observed the closed loops along which laser action occurred, as shown in figure 3 (a). Since the lateral dimension of the excitation area is much larger than the film thickness, laser cavities are in the plane of the film. The shape and size of the two-dimensional laser cavities changed as we moved the excitation spot across the film. When the excitation spot was close to the edge of the film, a tail was observed in figure 3 (b) due to the reflection of the emission from the edge of the film. The size and shape of the laser resonators are determined by the optical gain coefficient, the grain size and distribution, the scattering cross section, the boundary conditions, etc. Hence, the lasing frequencies vary across the film. This observation provides additional support to the formation of laser resonators through multiple optical scattering.

(a) (b)

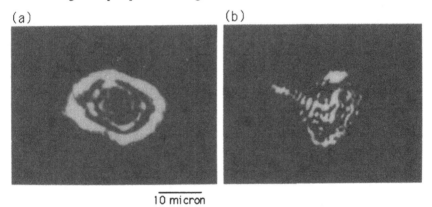

10 micron

Fig. 3: Amplified images of the laser cavities in ZnO films.

In summary, we have observed UV laser action in ZnO powder and polycrystalline films. Laser resonators are formed through multiple optical scattering. This observation may open up the possibility of utilizing disordered semiconductor microstructures as alternative sources of coherent light emission.

References

1. Z.-Q. Zhang, Phys. Rev. B **52**, 7960 (1995).
2. D. S. Wiersma, and A. Lagendijk, Phys. Rev. E **54**, 4256 (1996).

479

PHASE SEGREGATION IN SIPOS: FORMATION OF Si NANOCRYSTALS

A. VILÀ[1], J.R. MORANTE[1], B. CAUSSAT[2], P. BARATHIEU[3], E. SCHEID[3]
[1] Electronic Materials and Engineering, EME. Department of Electronics. University of Barcelona. C/ Martí i Franquès,1. Barcelona 08028. Spain. anna@el.ub.es
[2] LGC, UMR CNRS 5503, 18 chemin de la Loge. 31078 Toulouse Cedex, France.
[3] LAAS, UPR CNRS 8001, 7 av. du Colonel Roche, 31077 Toulouse Cedex, France.

ABSTRACT

A characterization of Semi-insulating Polycrystalline Silicon (SIPOS) layers deposited from SiH_4 on SiO_2 is presented, as a function of growth and annealing conditions (time and temperature), in order to better understand the processes involved in nucleation of silicon nanocrystals. Correlation between optical and XPS measurements allows determination of the starting composition of the amorphous material. After annealing, Fourier transform infrared (FTIR), ultraviolet-visible and Raman spectroscopies have been used to determine the structural and optical characteristics of the resulting material. Thermal treatment promotes a phase separation, modifying the layer properties and degrading the electrical insulation characteristics. Concentration of the silicon dioxide phase increases, whereas elemental silicon precipitates into nanocrystals which nucleate near the interface with the underneath SiO_2. Their density depends of the initial silicon content in the SIPOS layer, and some directions such as <111> and <220> grow preferentially whereas other directions such as <311> show a slower growth. As the percentage of oxygen increases, the formation of precipitates is less marked.

INTRODUCTION

Semi-insulating polycrystalline silicon (SIPOS) is a generic name for nonstoichiometric silicon oxide (SiO_x) usually obtained by chemical vapor deposition (CVD) from silane and oxygen-bearing species such as nitrous oxide[1]. The most important property of this material is that the SIPOS conductivity can be monitored by changing its oxygen content, which depends on the gas-flow rate ratio. This gives it utility in a variety of electronic and optoelectronic devices as a protective, filtering or antireflection coating, or as a dielectric material [2,3]. The composition of the films can vary from x=0 (polycrystalline or amorphous silicon) to x=2 (silicon dioxide). The structure, optical and electrical properties of this material are dependent on the oxygen concentration[4]. Moreover, the influence of deposition conditions and successive annealing procedure on the SIPOS films properties can not be neglected[5].

Despite the large number of studies on SIPOS, some basic questions referring to its structural properties still remain unclear. How the oxygen is distributed in SIPOS is still an open question, since the electronic and optical properties of the films are often linked inextricably to their structure. Currently, there are two different models for the structure of SiO_x: 1) the microscopic random-bond model[6] in which the Si-Si and Si-O bonds are randomly distributed throughout the SiO_x, and 2) the random-mixture model[7] in which the tetrahedrally bonded units of Si and SiO_4 are randomly dispersed and have domain sizes of a few tetrahedral units. A modification of the random-mixture model is the "shell" model, in which each Si grain is surrounded by a thin SiO_2 layer so that Si grains are always isolated from each other[8]. This "shell" model has successfully been utilized to explain the strong dependence of electrical conductivity of SIPOS films on oxygen content[9].

Moreover, usually the average composition of the silicon oxide is kept below x=1 to get a high dielectric constant, which is rather a silicon-rich suboxide. This situation is quite different from that reported as silicon rich SiO_2, for which x is near 2^{10}. In this case, the excess silicon atoms undergoing thermal treatments first nucleate into clusters and then these crystalline nuclei grow. As the diffusion coefficient of interstitial silicon in SiO_2 is quite low, one can expect a slow growth for such nuclei. This mechanism explains the formation and evolution of silicon nanocrystals whose dimensions are in the range of a few nanometers which are responsible of light production above the silicon band gap due to quantum effects. Only when the concentration of silicon is high, the system weakens by forming big silicon precipitates and relaxing the whole system. On the contrary, in SIPOS there is a predominant amorphous silicon lattice distorted by the presence of an important oxygen concentration which causes a substantial modification of the material properties. In this case, it is possible to think that the oxygen trend is to form SiO_2 precipitates and, as a consequence, Si precipitates too, probably giving amorphous clusters that could crystallize after adequate thermal treatment.

In this framework, the present study is devoted to the structural properties of SIPOS films deposited on SiO_2 and undergoing thermal treatments for inducing phase separation, in order to achieve a better understanding of the processes involved in nanocrystal formation.

EXPERIMENTAL

Substrates were one-side polished silicon wafers with thickness of about 380 μm, which had been thermally oxidized to an oxide thickness of about 100 nm. The SIPOS film deposition was performed in a tubular horizontal hot-wall low-pressure chemical vapor deposition (LPCVD) system by the pyrolysis of silane (SiH_4) and nitrous oxide (N_2O). The growth temperature was 635°C, the total pressure during deposition was adjusted to 0.24 Torr. The silane flow rate was 200 sccm (standard cubic cm/min) and the $\gamma=Q(N_2O)/Q(SiH_4)$ flow rate ratio was kept at 30%. The thickness of the SIPOS layers was adjusted to about 300nm. After deposition, the SIPOS layers were annealed at temperatures ranging from 700 to 900°C for times up to 1 hour under a continuous flow of pure nitrogen.

The oxygen content of the starting layers was determined by X-ray photoelectron spectroscopy (XPS) achieved with a Perkin Elmer ESCA spectroscope using a non-monochromatic MgKα X-ray source at a power of 300 W. The spectra interpretation involves some difficulties due to the lack of homogeneity of the material, especially when the phase separation is induced. As the surface of these layers presents a higher oxidation degree than the bulk, XPS measurements do not report the right values. As an alternative, it is possible to subject the samples to an ion gun treatment. However, the different sensitivity of the silicon-oxygen bonds in the suboxides to the sputtering process prevents XPS results yielding absolute values but they can be used in a comparative way.

The thickness, texture and possibility of precipitates in the films were checked by high-resolution transmission electron microscopy (HREM) using a Philips CM30-ST microscope operated at 300 kV, allowing a resolution of 1.9 Å. X-ray diffraction (XRD) was used to determine any preferential orientation, by means of a high resolution Philips instrument in configuration of low incidence angle, to allow more sensitivity to crystallite orientation. Finally, Raman and infrared spectrometries were also applied in order to follow the evolution, respectively, of the silicon and SiO_2 signals with the thermal treatments. The Raman scattering measurements were performed in a Jobin-Yvon T64000 spectrometer coupled with an Olympus microscope, in backscattering geometry using an exciting wavelength of 488 nm from an Ar^+ laser. The excitation power was 1.2 mW and the spot diameter in the sample around 0.6 μm, giving a power density of ~0.4 MW/cm^2 on the sample. The infrared measurements were performed in a Fourier transform BOMEM DA3 spectrometer in diffuse reflection configuration. From the optical point of view, the lack of homogeneity is also a difficulty because it is not possible to simulate the optical medium as formed by two different materials in order to obtain models for the refractive index. Nevertheless, as FTIR and Raman techniques analyze a volume sufficient to average the contribution of the different phases, one can assume the signal as corresponding to an homogeneous material.

RESULTS

After deposition, XPS measurements showed that under the superficial SiO_2 layer, the SIPOS signal is rather constant. The spectra obtained have been deconvoluted considering the five possible oxidation states for the silicon: Si^0, Si^+, Si^{2+}, Si^{3+} and Si^{4+}. Each oxidation state has been fitted using peaks constituted by a symmetric gaussian, their energy positions fixed at those previously reported in the literature[1]. Finally, the oxygen content x=O/Si has been determined by considering the contribution of the area corresponding to each oxidation state to the total area of the XPS spectrum in the Si2P region. Results show the dominance of Si, Si_2O and even SiO states and the presence of very few Si_2O_3 and SiO_2 ones, what gives a composition of x=0.2 rather constant over the whole depth. This value is far away from the stoichiometric (x=2) situation, giving a very silicon-rich starting material. On the other hand, HREM observations indicated a completely amorphous textured structure for the deposited layer.

After annealing, a variation in the distribution of oxidation states can be deduced from IR measurements. Figure 1 presents to Si-O stretching region of the absorbance spectra from

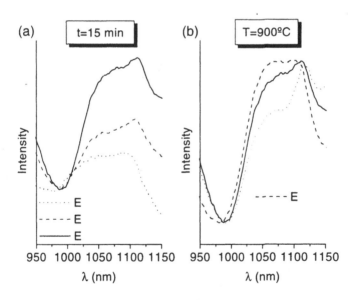

Fig.1. *Si-O stretching region of the IR absorbance spectra from annealed samples normalized against the as-deposited ones, as a function of annealing temperature (a) and time (b).*

samples annealed at different temperatures or different times, all of them normalized against the unannealed ones. Different relative heights can be seen for the areas usually related to the different oxidation states. Peaks are related to states with smaller density in the sample than in the reference (less absorption) whereas valleys are associated to larger-density states (more absorption). It is clear that as annealing temperature increases, more contribution is observed for the SiO_2 band ($\lambda \cong 1100$ nm) and less for the region related to suboxides ($\lambda > 1100$ nm). The same behavior can be observed for increasing annealing times at the right side. This observation supports the hypothesis of the random-mixture model, for which Si and SiO_2 tend to segregate into separated regions inside the amorphous matrix.

The previous changes in stoichiometry are related to changes in crystallinity, as the analysis of Raman scattering measurements indicates. The Raman spectra from amorphous silicon is characterized by the presence of four broad bands centered at 480 cm^{-1} (TO), 380 cm^{-1} (LO), 310 cm^{-1} (LA) and 150 cm^{-1} (TA), while crystalline silicon shows a first order line with a lorentzian shape centered at 521 cm^{-1}. The crystalline mode in the Raman spectra measured on the films annealed at 700°C is not significant compared to the amorphous ones (Fig. 2), whereas for 800 and 900°C an increasing contribution arises in the spectra, which reflects the partial crystallization achieved during thermal treatments. In all cases the center mode is displaced towards wavelengths smaller than the prediction for crystalline silicon (521 cm^{-1}). This effect can mainly be due to reduced grain dimensions (less than around 20 nm), even thought a minor strain contribution can also exist. The tail at the left of the crystalline peak is compatible with a distribution of grain dimensions inside the amorphous matrix and the corresponding surface vibration modes associated to nanograins. Variations diminish with annealing time, as shown in the figure (b), as material becomes more and more crystalline.

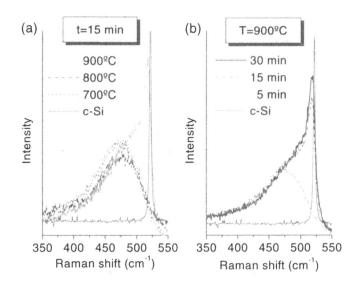

Fig. 2. *Raman spectra of the annealed films in comparison with the one corresponding to crystalline silicon for different annealing temperatures (a) and times (b).*

Once verified the formation of nanocrystalline precipitates in the samples annealed at 800 or more degrees, XRD experiments have been used to obtain some information about preferential orientation of the crystallites. The three densest planes in crystalline silicon are {111}, {220} and {311}. Therefore, the diffraction peaks corresponding to these reflections are the most visible in XRD, and should appear centered at 28.44°, 47.30° and 56.12°, respectively. From the number of counts measured for these reflections as compared to the background level associated to the amorphous matrix, the quantity of material oriented along these directions can be inferred. The XRD spectra obtained from some annealed samples are plotted in figure 3 as a function of the incidence angle. For long annealing times, there can be a randomly distributed orientation of the nanocrystals with a preference for <111> orientation, as deduced from the large area of the peak associated to its reflection, while the <220> and <311> are present too. Figure 4 presents the HREM image of some of those nanoparticles of silicon near the SIPOS/SiO₂ interface, obtained after 1 hour of annealing at 700°C.

Accordingly to Raman results, an increasing annealing temperature results in a better crystallinity, as peaks associated to crystalline silicon become more important. Although spectra suggest that the {111}, {220} and {311} reflections increase in a similar proportion with temperature, the most frequent reflection for long thermal treatments is always the {111} one. However, for short thermal treatments, the reflection {311} is more frequent than after further annealing, indicating a nearly random orientation at the first stages of nucleation which progresses towards {111}. This can be related to the fast growth of silicon in the <111> direction, and could be understood as a consequence of migration of silicon atoms inside the amorphous material.

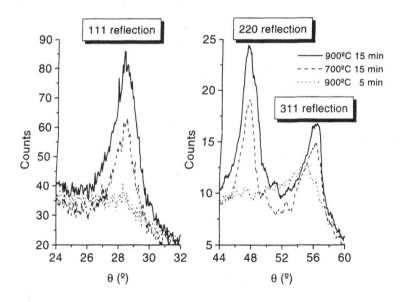

Fig. 3. *XRD spectra from annealed SIPOS layers around the {111}, {220} and {311} reflections. Crystalline contributions are superimposed on the amorphous background.*

485

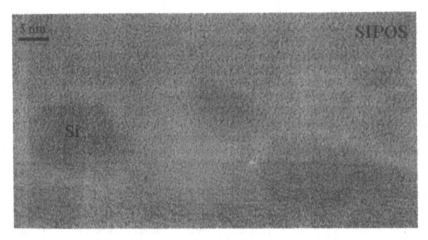

Fig. 4. HREM image of the SIPOS film near the SiO₂ interface annealed at 700°C for 1 hour.

CONCLUSIONS

Results of the present study indicate that for annealing temperatures under 700°C the SIPOS layers deposited under the conditions used for this work need long thermal treatments to form precipitates and to change appreciably their structural properties.

For annealing temperatures higher than 700°C segregation effects are evident, being less effective as the oxygen concentration increases. In this process, silicon precipitates, forming crystals with nanometric dimensions, as shown by HREM and Raman experiments. At the first stages of nucleation, nanocrystals are rather randomly oriented with no clear preferred orientation. However, they tend to {111} orientation for long annealing times. On the other hand, FTIR spectra indicate a major concentration of SiO₂ phase as either annealing temperature or time increase.

These two features corroborate the existence of a phase separation process in SIPOS layers undergoing thermal treatments. Such thermodynamic instabilities could limit their applications.

REFERENCES

[1] M.L. Hitchman and J. Kane, J. Cryst. Growth **55**, 485 (1981); M.L. Hitchman and A.E. Widmer, ibid. **55**, 501 (1981).
[2] A.J. Bennet and L.M. Roth, Phys. Rev. B **4**, 2686 (1971).
[3] G. Fortunato and D. Della Salla, J. Non-Cryst. Solids **97-98**, 423 (1987).
[4] A. Kucírkova, K. Navrátil, L. Pajasova and V. Vorlícek, Appl. Phys. A **63**, 495 (1996).
[5] G. Compagnini, S. Lombardo, R. Reitano and S.O. Campisano, J. Mat. Res. **10**, 885 (1995).
[6] W.Y. Ching, Phys. Rev. B **26**, 6610 (1982).
[7] G.A. Niklasson and C.G. Granqvist, J. Appl. Phys. **55**, 3382 (1984).
[8] J. Ni and E. Arnold, Appl. Phys. Lett. **39**, 554 (1981).
[9] K.T. Chang, C. Lam and K. Rose, Mat. Res. Soc. Symp. Proc. **105**, 193 (1988).
[10] Y. Kanemitsu, Phys. Rev. B **53** (20), 13515 (1996).
[11] O. Benkherourou and J.P. Deville, J. Vac. Sci. Technol. A **6** (6), 3125 (1988).

LOW TEMPERATURE DEPOSITION OF POLYCRYSTALLINE SILICON THIN FILMS PREPARED BY HOT WIRE CELL METHOD

M. ICHIKAWA, J. TAKESHITA, A. YAMADA AND M. KONAGAI
Department of Electrical and Electronic Engineering, Tokyo Institute of Technology, 2-12-1 O-okayama, Meguro-ku, Tokyo, Japan

ABSTRACT

Hot wire (HW) cell method has been newly developed and successfully applied to grow polycrystalline silicon films at a low temperature with a relatively high growth rate. In the HW-cell method, silane is decomposed by reaction with a heated tungsten wire placed near the substrate. It is found that polycrystalline silicon films can be obtained at substrate temperatures of 175-400°C without hydrogen dilution. The film crystallinity is changed from polycrystalline to amorphous with decreasing the total pressure. The X-ray analysis clearly showed that the films grown at the filament temperature of 1700°C have a very strong (220) preferential orientation. The films consist of large grains as well as small grains, and it was found from cross-sectional SEM that the films have columnar structure. These results suggested that the HW-cell method would be a promising candidate to grow device-grade polycrystalline silicon films for photovoltaic application.

INTRODUCTION

In our previous experimental and theoretical works, it was demonstrated that the radical flux ratio ($[SiH_3]/[H]$) on the growing surface was a key factor to grow high quality amorphous and crystal silicon at a low temperature[1]. Therefore, we propose a HW-cell method as a new deposition process. In this process, a heated tungsten filament induces catalytic or pyrolytic dissociation and produces, in addition to silyl radicals, atomic hydrogen, which reacts with silane molecules to produce SiH_3 radicals. This process was almost same as Catalytic Chemical Vapour Deposition (Cat-CVD)[2] and HW-CVD[3-5]. These authors reported the preparation of high-quality poly-Si films deposited at a highly hydrogen-diluted condition resulting in deposition rates of 1.0nm/s, while we successfully deposited polycrystalline silicon films without a hydrogen dilution. The significant difference between our HW-cell method and the previous HW-CVD works is the layout of the HW filament. In other methods, the HW filament was spread over and kept parallel to the substrate holder. On the contrary, the filament was perpendicular to the substrate holder in this work. Because of the layout of the filament, the reactant gas was decomposed effectively while traveling in the filament and the effective decomposition rate of the reactant gas increase. In HW-cell method, atomic hydrogen was also supplied from decomposition of the silane. Therefore polycrystalline silicon film can be obtained at the condition of no hydrogen dilution in our process. In this paper, we present some results on the structural characterization of the silicon films deposited without a hydrogen dilution.

EXPERIMENT

A schematic of the deposition chamber with the HW-cell is shown in Fig. 1. The HW cell consists of gas inlet and a tungsten filament. The axis of the gas inlet into the chamber is perpendicular to the plane of the substrate holder. The tungsten wire with a diameter of 0.3mm is used as the filament. The filament, coiled with a diameter of 4mm and length of 1.5cm, is arranged lengthwise with gas inlet. The distance between the filament and the substrate is about

6cm. The temperature of the filament is measured using an optical pyrometer through a window of the chamber. The filament is heated by supplying electric power directly to it and is kept constant throughout the experiment by controlling the power. The pure silane (SiH₄) is used as a reactant gas. The temperature of the substrate is measured by a calibrated thermocouple attached to the substrate holder. The calibration for substrate temperature is previously obtained using a second thermocouple attached to the substrate. The gas pressure is controlled by adjusting a main valve between the chamber and the mechanical booster pump. Corning 7059 glass is used as a substrate. The main deposition parameters are listed in Table I.

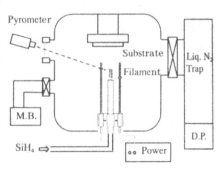

Fig. 1. Schematic of the deposition chamber

The films were morphologically and structurally characterized by Scanning Electron Microscopy (SEM), X-ray diffraction (XRD) and Raman spectroscopy. The Raman spectrum was deconvoluted in their integrated crystalline, I_c (~520cm^{-1}), amorphous, I_a (~480cm^{-1}) and intermediate, I_m (~510cm^{-1}) peaks. The crystalline fraction, X_c, was calculated from the following equation[6]:

Table I. Growth Conditions

Substrate Temperature	175 - 400°C
Filament Temperature	1700 - 2000°C
Total Pressure	0.015 - 0.2Torr
Flow Rate (SiH₄)	5 – 15sccm

$$X_c = (I_c + I_m) / (I_c + I_m + I_a). \qquad (1)$$

The thickness of the films was measured using profilometer.

RESULTS

Deposition Rate

Deposition rate is significantly influenced by the SiH₄ flow rate and the total pressure. Figure 2 shows the dependence of the deposition rate on the SiH₄ flow rate at a total pressure of 0.1Torr. The thickness of poly-Si film was measured at the edge of the sample. However, we checked the uniformity and found that the thickness was almost same within the substrate size (2 x 2cm). It shows that the deposition rate increases linearly with SiH₄ flow rate, and reaches 1.0nm/s at a SiH₄ flow rate of 15sccm. At this point, the deposition rate does not saturate implying that higher growth rate is achievable at higher SiH₄ flow rates. This result

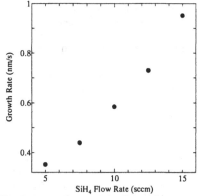

Fig. 2. Dependence of the deposition rate on the SiH₄ flow rate.

suggests that the deposition rate may be limited by the gas phase concentration of the radicals. Figure 3 shows the deposition rate as a function of the total pressure. It shows that the growth rate increases monotonically from 0.35nm/s at the pressure of 0.015Torr to approximately 0.7nm/s at the pressure of 0.2Torr. The deposition rate is also dependent on the substrate temperature, as shown in Fig. 4. The figure shows that the substrate temperature hardly influences deposition rate. These results are consistent with the conclusion that the limiting step in the growth of polycrystalline silicon is the dissociation of SiH_4 molecules at the filament and the gas phase reaction. The decrease of deposition rate on the substrate temperature may be due to the change of the sticking coefficient of the radicals at the growing surface.

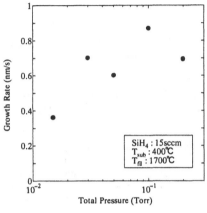

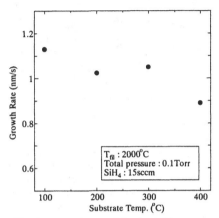

Fig. 3. Dependence of the deposition rate on the total pressure.

Fig. 4. Dependence of the deposition rate on the substrate temperature.

Structural Properties

Crystallinity of the film is significantly affected by the total pressure of the deposition chamber. Figure 5 shows the dependence of the Raman spectra of the films on the total pressure. For the film deposited at the pressure of 0.015Torr, the spectrum shows only a broad low-intensity peak centered around $480cm^{-1}$ which is a characteristic of amorphous phase. When the total pressure increases to the 0.03Torr, the peak centered at $520cm^{-1}$ which is a characteristic of crystalline phase is observed, and the crystalline fraction becomes 40%. As the total pressure increases, the crystallinity of the film increases. Therefore, when the total pressure reaches 0.1Torr, only the narrow peak centered at $520cm^{-1}$ is observed and crystalline fraction of the film reaches about 90%. However, at the total pressure of 0.2Torr, the crystalline fraction of the film decreases to 70%. Thus the structural transition of silicon is quite sensitive to the total pressure. As a result, we obtain polycrystalline silicon films by choosing the total pressure without hydrogen dilution. When the total pressure is kept constant as 0.1Torr, the film crystallinity hardly changes although the substrate temperature decreases. When the substrate temperature decreases from 400°C to 175°C, the crystalline fraction of the film decreases to 80% and the peak of amorphous phase is undetectable. It is noticed that the polycrystalline silicon was obtained at a low temperature of 175°C without hydrogen dilution, which is mainly due to the filament configuration of our system and the high decomposition rate of silane.

Figure 6 shows the XRD spectrum of the films deposited at the total pressure of 0.1Torr. The

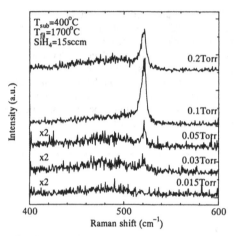

Fig. 5. Dependence of the Raman spectra on the total pressure.

Fig. 6. XRD spectrum of the film deposited at $P_{total} = 0.1$Torr, $T_{sub} = 400°C$.

polycrystalline silicon films are mainly consisted of grains oriented in (111), (220) and (311) directions perpendicular to the substrate surface. The X-ray diffraction spectrum indicating that the film has a very strong (220) preferential orientation. The preferential orientation was found to depend on the filament temperature. As the filament temperature increases from 1700 to 2000°C, the intensity of the (220) diffraction peak decreases and the spectrum shows randomly orientation in the film.

Figure 7 shows the SEM surface (a) and cross-sectional (b) image of the film deposited at the substrate temperature of 325°C. In Figure 7a, it is observed that grains having different size in major axis (0.3-0.5μm) are spread all over the surface of the films. The surface morphology of the samples deposited at substrate temperatures over 250°C has almost same surface morphology.

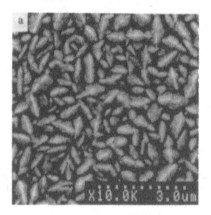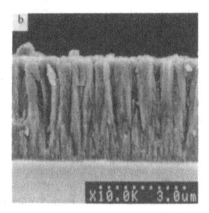

Fig. 7. SEM surface (a) and cross-sectional (b) image of the film deposited at T_{sub}=325°C

490

The cross-sectional view of the film has a columnar structure, which is general feature of all the polycrystalline samples. At the first stage of the growth, the sample has small-sized grains, and the existence of amorphous layer whose thickness is about $0.3\mu m$ is confirmed by a cross-sectional TEM observation in this region. In the near surface region, the grain growth is observed.

Deposition Mechanism

In the deposition process of polycrystalline silicon by HW-cell method, silane is passed through a heated tungsten filament. The dependence of the film structure on the total pressure suggests that the transition from amorphous to polycrystalline depend on the concentration ratios of the radicals near the substrate strongly. The gas phase reactions and the reaction between the reactant gas and the heated filament can form the radicals.

In order to understand the HW-cell deposition process well, simulations with a kinetic model have been performed. The main assumptions of our model are summarized as follows.

(1) Only a certain percentage of the inlet gases react with the filament. The exact reactions of mono silane molecules occurring on the surface of the filament are not clear. In our model, we assume the dissociation reactions as follows:

$$SiH_4 \longrightarrow SiH_2 + H_2 \qquad (2)$$

$$H_2 \longrightarrow 2H \qquad (3)$$

(2) The radicals produced by the dissociation reaction react with the remaining molecules and the reaction continues as the mixed gas flow throughout the chamber towards the substrate. In this step we consider the following reactions[7,8]:

$$H + SiH_4 \longrightarrow H_2 + SiH_3 \qquad (4)$$

$$SiH_3 + SiH_3 \longrightarrow SiH_2 + SiH_4 \qquad (5)$$

$$SiH_2 + SiH_4 \longrightarrow Si_2H_6 \qquad (6)$$

$$H + SiH_3 \longrightarrow H_2 + SiH_2 \qquad (7)$$

$$SiH_2 + H_2 \longrightarrow SiH_3 + H \qquad (8)$$

Calculations are performed by solving the balance equations for the radicals involved, as a function of the total pressure.

Figure 8 shows the calculated concentrations of SiH_3 and SiH_2 and H at the growing surface as a function of the total pressure. At high-pressures, the SiH_2 radical whose reaction rate with SiH_4 is higher than SiH_3 radical deactivates during the travel from the filament to the substrate by gas phase reaction. As a result, SiH_3 radical can preferentially impinge upon the surface at a high-pressure region, and SiH_2 radical is preferential radical at a low-pressure region. This result may well explain our experimental result that the crystallinity of the film is significantly influenced by the total pressure.

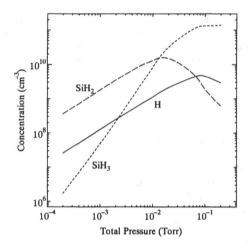

Fig. 8. Calculated concentratiòns of the radicals at the growing surface as a function of the total pressure.

CONCLUSIONS

The structural studies performed with Raman spectroscopy and SEM show that polycrystalline films can be deposited at a substrate temperature of 175-400°C by a HW-cell method when the pure silane is used as a reactant gas. A columnar crystalline structure is observed with preferential orientation in direction (220) on glass substrates when the gas pressure is kept at 0.1Torr. As the gas pressure becomes lower, deposited films show an amorphous phase. The growth rate of the deposited films ranged from 1.1nm/s to 0.4nm/s.

The results of this study shows that the HW-cell method appears very promising for depositing of polycrystalline silicon for photovoltaic applications.

ACKNOWLEDGMENTS

This study was supported in part by NEDO as a part of the New Sunshine Program under the Ministry of International Trade and Industry.

REFERENCES

1. T. Oshima, A. Yamada and M. Konagai, Jpn. J. Appl. Phys. **36**, 6481 (1997).
2. H. Matsumura, Jpn. J. Appl. Phys. **30**, L1522 (1991).
3. J. Cifre, J. Bertomeu, J. Puigdollers, M. C. Polo, J. Andreu and A. Lloret, Appl. Phys. A **59**, 645 (1994).
4. M. Heintze, R. Zedlitz, H. N. Wanka and M. B. Schubert, J. Appl. Phys. **79**, 2699 (1996)
5. P. Brogueira, J. P. Conde, S. Arekat and V. Chu, J. Appl. Phys. **79**, 8748 (1996)
6. T. Kaneko, M. Wakagi, K. Onisawa and T. Minemura, Appl. Phys. Lett. **64** , 1865 (1994)
7. T. Fuyuki, B. Allain and J. Perrin, J. Appl. Phys. **68**, 3322 (1990)
8. J. Perrin and T. Broekhuizen, Appl. Phys. Lett. **50**, 433 (1987)

ENHANCEMENT OF THE AMORPHOUS TO MICROCRYSTALLINE PHASE TRANSITION IN SILICON FILMS DEPOSITED BY SiF_4-H_2-He PLASMAS

G. CICALA[1], M. LOSURDO[1], P. CAPEZZUTO[1], G. BRUNO[1], T. LIGONZO[2], L. SCHIAVULLI[2], C. MINARINI[3] and M.C. ROSSI[4]
[1]Plasma Chemistry Research Center, MITER-CNR, cscpgc07@area.ba.cnr.it, Dipartimento di Chimica-Università di Bari, Via Orabona, 4 70126-Bari, Italy;
[2]Dipartimento di Fisica-Università di Bari;
[3]ENEA C.R. Portici Napoli;
[4]Dipartimento di Ingegneria Elettronica Università di Roma3, Roma Italy.

ABSTRACT

Hydrogenated microcrystalline silicon (μc-Si:H) thin films have been obtained by plasma decomposition of SiF_4-H_2-He mixtures at low temperature (120 °C). The size of crystalline grain and their volume fraction with respect to the amorphous phase have been found dependent on the r.f. power as evaluated by grazing incidence X-ray diffraction, microRaman and ellipsometry measurements. Chemical and electrical properties change according to the microcrystallinity. Pure and/or highly microcrystalline silicon has been obtained at temperature and r.f. power as low as 120 °C and 15 Watt.

INTRODUCTION

Since the demonstration that microcrystalline silicon films exhibit good properties for optoelectronic and photovoltaics applications [1 - 3], there has been a renewed interest in the growth of such silicon films by plasma enhanced chemical vapor deposition (PECVD) [4, 5].

Recently, several methods have been reported for the preparation of thin films of microcrystalline silicon embedded in a-Si:H and, among these, the hydrogen dilution of silane has been considered the "conventional" method to be used as reference [4-6]. However, the main subject of these investigations has been on the enhancement of the amorphous-to-microcrystalline phase transition through the selective etching mechanism of the amorphous phase by H-atoms [4, 6-7].

In the present paper, the feasibility of the SiF_4-H_2-He plasma system in promoting the growth of microcrystalline silicon films is demonstrated. This enhancement has been related to the high etching selectivity of amorphous against crystalline phase and results from the competition between the etching of both F- and H-atoms [8] and deposition assisted by H-atoms [9]. This contribution is concerned with an investigation of structural, compositional and electrical properties of hydrogenated Si films deposited at very low temperature (around 120 °C) and, here, the emphasis is on the deposition of highly crystallized material at low r.f. power. Data from grazing incidence X-ray diffraction (GIXRD), microRaman, spectroscopic ellipsometry (SE), Fourier transform infrared spectroscopy (FTIR), and temperature dependence of conductivity are reported to confirm the effectiveness of the SiF_4-H_2-He plasma system in the production of pure microcrystalline silicon films with properties of some interest.

EXPERIMENT

The microcrystalline silicon films have been grown in a parallel plate ultra high vacuum plasma reactor by an r.f. glow discharges of SiF_4-H_2-He mixtures. The r.f. powered and the grounded electrodes are 10 cm in diameter and 10x10 cm in dimension, respectively, with an interelectrode gap of 3 cm. The grounded electrode can be heated and the film growth on it is followed by laser reflectance interferometry technique for the *in situ* measurement of deposition rate. Silicon films on different substrates (Corning 7059 glass, c-Si (100)) have been deposited at a temperature of 120 °C, pressure 300 mTorr, flow rates of SiF_4, H_2 and He of 20, 10 and 45 sccm, respectively, and r.f. power variable in the range 15-45 Watt.

XRD measurements have been carried out on a diffractometer (PHILIPS X'PERT Multiple Purpose Diffractometer System) using a Cu K_α radiation source. The crystallite size has been calculated from XRD data obtained in standard θ-2θ configuration. To acquire the 2θ spectra a typical θ-2θ goniometer mounted on the line shape radiation source and equipped by a monochromator has been used. The acquisition parameters have been: a degree step of $0.005°$, a measure time of 60 sec for step. The crystalline fraction has been determined from XRD data in grazing angle configuration (with fixed incident angle of $0.7°$), with a low divergence Soller slit and a graphite monochromator, to enhance the thin film signal with respect to the Corning glass substrate.

MicroRaman spectra have been acquired by a Raman ISA spectrometer (LabRAM) in a confocal back-scattering geometry with a spectral resolution of 1 cm^{-1}. The Raman measurements (in micro mode) have been done with a cw He-Ne laser beam (λ = 6328 Å, nominal power 16 mWatt) focused by Olympus optics onto the film surface down to a spot size of 1 μm. A neutral density filter D = 1 has been used in order to avoid sample heating.

Ellipsometric spectra (SE) have been acquired ex-situ in the energy range 1.5-5.5 eV using a phase modulated spectroscopic ellipsometer (UVISEL by ISA-Jobin Yvon). The Bruggeman effective medium approximation (BEMA) has been used to calculate the effective dielectric functions of the deposited samples. And, for the fitting of the SE spectra, optically different types of silicon (a large-grain and a fine-grain polycrystalline silicon p-Si [11] and an amorphous silicon a-Si [12]) have been mixed with voids in the appropriate ratio.

Infrared spectra have been obtained on silicon films deposited on double polished c-Si (100) substrates and acquired by a FTIR Spectrometer (BOMEM, Michelson 102).

Electrical measurements have been performed on silicon films grown on Corning 7059 glass after evaporation of Ag strips 1 cm long, 3 mm wide with 1 mm gaps for coplanar dark- and photo-conductivity, whereas mobility measurements have been carried out by Van der Paw configuration at room temperature.

RESULTS

Three typical samples, grown at different r.f. power values are listed in Table I, where chemical (H-content and Si-H stretching values by IR spectroscopy), electrical (dark conductivity, σ_D, and photosensitivity ratio, σ_{PH}/σ_D) and structural (grain size δ, and crystalline fraction f_c, by GIXRD, microRaman and SE techniques) properties are reported.

TABLE I: Chemical (H-content, c_H, and Si-H stretching mode values, ω^s), electrical (dark conductivity, σ_D, and photosensitivity ratio, σ_{PH}/σ_D) and structural (grain size, δ, and crystalline fraction, f_c, estimated by GIXRD, Raman and SE), properties of some typical samples of silicon films grown at different r.f. power values.

Sample	r.f. power (Watt)	c_H (at. %)	ω^s (cm^{-1})	σ_D (Ωcm)$^{-1}$	σ_{PH}/σ_D	δ (Å) XRD	Raman	SE^a	f_c (%) GIXRD	Raman	SE
Sihe19	15	6	2046	3x10^{-7}	8x10^1	230	180	L	100	83	93
Sihe15	30	11	2028	3x10^{-8}	5x10^2	150	150	F	75	79	83
Sihe18	45	13	2015	3x10^{-9}	5x10^3	140	130	F	56	67	59

^{a} L and F are for Large and Fine grain.

Figure 1 shows the GIXRD patterns of μc-Si:H films having the same thickness (t=410 nm) grown under r.f. power conditions listed in Table I. The crystalline grains of μc-Si:H exhibit three different orientations (111) (2θ= 28.2°), (220) (2θ=47.2°) and (311) (2θ=55.8°). The width data processing of the (220) and (311) Bragg reflections acquired in θ-2θ configuration has been done to evaluate the grain size, δ, by means of the Debye-Sherrer's formula [12]. The comparison of the peak width (220) of the three samples evidences that SiHe15 and SiHe18 exhibit about the same grain size (150-140 Å) and the SiHe19 has the larger grain size of 230 Å. The crystalline fraction has been calculated by the ratio between the diffraction spectra area in the 2θ range of 50-54 degrees of the samples under test and that of a pure amorphous silicon.

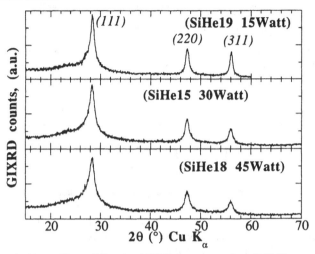

Fig.1 Grazing incidence X-ray diffraction (GIXRD) patterns of μc-Si:H films on Corning 7059 glass at three different r.f.powers (see Table I).

Material structure has also been investigated by Raman spectroscopy allowing the evaluation of the order present in the solid through phonon lineshape analysis. The measured

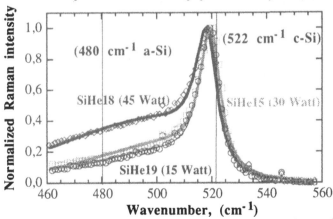

Fig.2 Measured (data points) and calculated (line) Raman intensity vs Raman frequency for a two-phases silicon (microcrystallites embedded in an amorphous matrix) films grown on Corning 7059 glass at different r.f. powers (see Table I).

spectra (data points), shown in Fig.2, are characterized by a rather sharp peak around 520 cm^{-1} with a low-frequency tail, typical for microcrystalline silicon. Such lineshapes are related to the phonon confinement in small domain sizes which induces a partial relaxation of single crystal phonon selection rules, allowing the involvement of a large range of phonons with different momentum to the scattering process. It has been reported by several authors [13, 14] that such a model gives a reasonable estimation of the confinement size and hence of the average crystallite size. However, assuming a single size of confinement, the predicted asymmetric broadening and frequency downshift have not been consistent with the experimental findings. This result indicates that the material structure is not homogeneous but crystallites of different sizes are formed during the growth. In particular, it is reasonable to assume that large-sized silicon nanocrystallites are surrounded by small-sized ones, or embedded in an amorphous tissue. In such a case a bimodal crystallite size distribution function

$$G(R) = (1-y) \, d(R-R_1) + y \, d(R-R_2) \tag{1}$$

can be assumed indicating that the fractions y and (1-y) of the sample are constituted by R_2-sized and R_1-sized nanocrystallites, respectively. The intensity profile of the Raman band can be then written:

$$I(\omega) = C \int dR \, G(R) \, S(\omega, R) \tag{2}$$

where C is a proper constant and $S(\omega, R)$ is the crystallite lineshape function. Assuming a Gaussian confinement and a spherical Brillouin zone [15], the latter function can be written

$$S(\omega, R) = \int_0^1 \frac{dq \, 4\pi q^2 e^{-(qR)^2/4}}{[\omega - \omega(q)]^2 + (\gamma/2)^2} \tag{3}$$

where q is the normalized phonon momentum, $\omega(q)$ is a one-dimensional average of the optical phonon branch dispersion curves and γ is the phonon line-width. Raman lineshapes can be reproduced fairly well by eqs.(2), (3) (continuous line in Fig. 2) assuming an effective dispersion curve $\omega(q) = 480 + 40 \cos (pq)$ [cm^{-1}]. Such a curve is able to predict the Raman lineshapes the broad Raman feature at 480 cm^{-1}, characteristic of amorphous silicon, for a confinement size of 22 Å, in agreement with previous reports [16]. All the investigated samples have been analyzed in the frame of this model. Only for the sample SiHe15 it was necessary to take into account also stress effects in order to obtain a satisfying fit of data. In particular a compressive hydrostatic stress was assumed, leading to a rigid shifting of the phonon dispersion curve by 1 cm^{-1}.

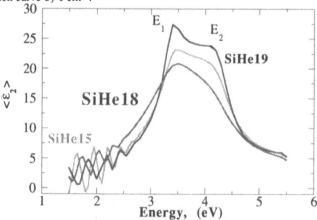

Fig. 3 The imaginary part, $\langle \varepsilon_2 \rangle$ of the pseudo-dielectric function *vs* the energy for μc-Si films obtained at three different r.f. powers.

Figure 3 shows the imaginary part, $\langle\varepsilon_2\rangle$ of the pseudo-dielectric function, uncorrected for surface roughness, recorded for the SiHe19, SiHe15 and SiHe18 samples. It is evident a decrease of the $\langle\varepsilon_2\rangle$ spectra and a broadening of the E_1 and E_2 interband critical points which suggest a reduction of both the crystalline fraction and the grain size, going from the SiHe19 sample to the SiHe18 sample. This behaviour is confirmed by the best fit BEMA models, which consist of a bulk μc-Si layer, including a mixture of μc-Si + a-Si + voids, and a rough surface layer simulated by (μc-Si + voids). The resulting data of the crystalline volume fraction of the bulk-layer for the three samples are listed in Table I. In the Table, it has also been specified that according to XRD and Raman data large-grain polycrystalline silicon component applies for the SiHe19 sample, whereas fine-grain polycrystalline silicon applies for SiHe15 and SiHe18 samples.

The specific features found in the formation of microcrystalline silicon from the He addition to the SiF_4-H_2 mixtures are: the enhancement of crystallization and the production of unstressed crystalline material under mild plasma conditions such as low r.f. power and low deposition temperature (120 °C). At this low value temperature grain size, δ, and crystalline volume fraction, f_c, increase by decreasing the power and, in particular, a pure or highly crystalline material ($f_c \approx 100$ %) is deposited at r.f. power as low as 15 Watt.

Figure 4 shows the crystalline fraction, f_c, as a function of net deposition rate r_D at which the silicon films are grown. In this plot, we can distinguish three regions, the first at deposition rate values less than 0.3 Å/sec, in which pure microcrystalline phase exists in the material, the second at deposition rate values in the range 0.3-0.9 Å/sec where the amorphous and microcrystalline phases coexist at variable percentages; in particular the microcrystalline

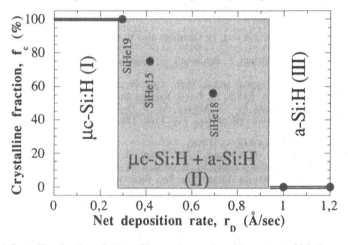

Fig. 4 Crystalline fraction of silicon films *vs* the net deposition rate at which the material is grown.

phase decreases by increasing the deposition rate. The third at deposition rate greater than 1 Å/sec, where pure amorphous phase is obtained [9]. In pure microcrystalline region, the grain size could furtherly increase at lower deposition rate, since the deposition rate and the crystallite size are inversely proportional, as found by G. Bruno et al. [17] in a similar halogenated system. The net deposition rate value, r_D, which is the difference between the growth rate, r_G, and the etching rate, r_E, is determinant in establishing the material structure. At low values of r_D, the selective etching of amorphous phase with respect to the crystalline by F- and H-atoms becomes operative as reported in ref. 8, by enhancing the amorphous/crystalline transition.

Microcrystallinity also has a characteristic signature in the FTIR: when the silicon films become more and more microcrystalline, the total hydrogen content decreases and the residual

H atoms are principally bonded in dihydride structure indicating accumulation to the grain boundaries (see Table I). The hydrogen and fluorine atoms play a key role by promoting a structural order in the Si network and are incorporated in the material at values between 13 and 6 at.% and less than 1 at.%, respectively.

The achievement of highly crystallized films is furtherly demonstrated by the Arrhenius dependence of the dark-conductivity on the temperature for the three samples with different f_c and the same thickness; as the crystalline volume fraction, f_c, increases, the dark-conductivity and the mobility increase. Samples having high crystallinity, e.g. sample SiHe19, exhibit mobility values around 60 cm^2V^{-1}.

CONCLUSIONS

An enhancement of the amorphous to microcrystalline phase transition in silicon occurs when helium is added to SiF_4-H_2 mixture. Helium promotes a high concentration of F-atoms which are etchant species selective for amorphous phase. We found experimental conditions in which high crystalline fraction is obtained at very low deposition temperature (120 °C) and at very low r.f. power (15 Watt), as evaluated by XRD, Raman and ellipsometry techniques.

ACKNOWLEDGMENTS

The financial supports from Progetto Finalizzato CNR "Materiali Speciali per Tecnologie Avanzate II" and European Joule Contract (JOR3-CT97-0145-NEST) are acknowledged.

REFERENCES

1. T. Toyama, T. Matsui, Y. Kurokawa, H. Okamoto and Y. Hamakawa, Appl. Phys. Lett. **69**, 1261 (1996).
2 G. Cicala, P. Capezzuto, G. Bruno, L. Schiavulli, G. Perna and V. Capozzi, J. Appl. Phys. **80**, 6564 (1996).
3. J. Meier, R. Fluckiger, H. Keppner and A. Shah, Appl. Phys. Lett. **65**, 860 (1994).
4. S. Veprek, M. Heintze, F.-A. Sarott, M. Jurcik-Rajman and P. Willmott, in *Amorphous Silicon Technology*, edited by A. Madan, M. J. Thompson, P. C. Taylor, P. G. LeComber and Y. Hamakawa (Mater. Res. Soc. Proc. **118**, Pittsburgh, PA, 1988) pp. 3-17.
5. A. Matsuda, Thin Solid Films **332**, xx (1998)
6. C. C. Tsai, R. Thompson , C. Doland, F. A. Ponce, G. B. Anderson, B. Wacker in *Amorphous Silicon Technology*, edited by A. Madan, M. J. Thompson, P. C. Taylor, P. G. LeComber and Y. Hamakawa (Mater. Res. Soc. Proc. **118**, Pittsburgh, PA, 1988) pp. 49-54.
7. M. Otobe, M. Kimura and S.Oda, Jpn. J. Appl. Phys. **33**, 4442 (1994).
8. G. Cicala, P. Capezzuto and G. Bruno, Thin Solid Films **332**, 1 (1998).
9. G. Bruno, P. Capezzuto, G. Cicala, J. Appl. Phys. **69**, 7256 (1991).
10. H. P. Klug, L. E. in *Alexander, X-ray Diffraction Procedures*, (New York John Wiley &Sons, 1974).
11 G.E. Jellison, M.F. Chisholm, S. M. Gorbatkin, Appl. Phys. Lett. **62**, 1493 (1993).
12. B. G. Baglay, D. E. Aspnes, A. C. Adams, C. J. Mogab, Appl. Phys. Lett. **38**, 56 (1981).
13. I. H. Campbell, P. M. Fauchet, Solid State Commun. **58**, 739 (1986).
14. J. Gonzalez-Hernandez, G. H. Azarbayejani, R. Tsu, F. H. Pollack, Appl. Phys. Lett. **47**, 1350 (1985).
15 H. Richter, Z. P. Wang, L. Ley, Solid State Commun. **39**, 625 (1981).
16 C. Messana, B. A. De Angelis, G. Conte, C. Gramaccioni, J. Phys. D **14**, L91-4 (1981) and refs. therein.
17. G. Bruno, P. Capezzuto and F. Cramarossa, Thin Solid Films **106**, 145 (1983).

DECHANNEALING STUDY OF NANOCRYSTALLINE Si:H LAYERS PRODUCED BY HIGH DOSE HYDROGEN IRRADIATION OF SILICON CRYSTALS

V.P. Popov*, A.K. Gutakovsky*, I.V. Antonova*, K.S. Zhuravlev*, E.V. Spesivtsev*, I.I. Morosov**, G.P. Pokhil***
*Institute of Semiconductor Physics, 630090 Novosibirsk, popov@isp.nsc.ru
**Institute of Nuclear Physics, 630090 Novosibirsk, Russia,
***Research Institute of Nuclear Physics, MSU, Moscow, Russia.

ABSTRACT

A study of Si:H layers formed by high dose hydrogen implantation (up to $3 \times 10^{17} cm^{-2}$) using pulsed beams with mean currents up 40 mA/cm^2 was carried out in the present work. The Rutherford backscattering spectrometry (RBS), channeling of He ions, and transmission electron microscopy (TEM) were used to study the implanted silicon, and to identify the structural defects (a-Si islands and nanocrystallites). Implantation regimes used in this work lead to creation of the layers, which contain hydrogen concentrations higher than 15 at.% as well as the high defect concentrations. As a result, the nano- and microcavities that are created in the silicon fill with hydrogen. Annealing of this silicon removes the radiation defects and leads to a nanocrystalline structure of implanted layer. A strong energy dependence of dechanneling, connected with formation of quasi nanocrystallites, which have mutual small angle disorientation (<1.5°), was found after moderate annealing in the range 200-500°C. The nanocrystalline regions are in the range of 2-4 nm were estimated on the basis of the suggested dechanneling model and transmission electron microscopy (TEM) measurements. Correlation between spectroscopic ellipsometry, visible photoluminescence, and sizes of nanocrystallites in hydrogenated nc-Si:H is observed.

INTRODUCTION

In the last several years, considerable interest of the researchers has been directed to systems containing high concentration of nanometer-sized particles (nanostructured materials). The increased interest in such systems is stimulated basically by two reasons. From the point of view of fundamental research, it is important to clarify the mechanism of visible photoluminescence recently found from nanometer-sized crystals of non-direct semiconductors, such as Ge and Si [1,2]. Hence the research in this area is conducted rather intensively, yet there are no exact representations about the source and mechanism of this phenomenon so far. As a possible reason for occurrence of visible photoluminescence, some authors point out the recombination of quantum-confined excitons in nanocrystals [3,4]. Recombination of electron-hole pairs on defect centers inside the nanocrystals or on interfaces between the crystals and surrounding matrices may be indicated as an alternative reason [5]. Some authors connect the occurrence of visible photoluminescence with the presence of hydrogen [6]. H atoms may be trapped at defects or interfaces thus forming Si-H complexes with dissolution energies of ~2.6 eV or siloxene groups $\equiv$Si-OH. On the other hand, in the literature there are some data suggesting that the presence of hydrogen should not play a decisive role in the formation of luminescening centers, and can only promote an increase in the probability of radiative transitions. This increase can be realized, for example, by the saturation of broken bonds thus lowering the density of non-radiative recombination centers.

The objective of the present research is the preparation and examination of nanocrystalline silicon, synthesized by implantation of a high dose of hydrogen ions in silicon. In this research the nanocrystalline structure is established in silicon using hydrogen ion implantation and subsequent heat treatments, and the preparation process is correlated with optical properties. Hydrogen solubility in crystalline silicon is about 10 ppm but ion implantation allows one to introduce of $10^{22} cm^{-3}$ of the hydrogen atoms or even more.

EXPERIMENT

The <100> p-type Cz-Si wafers were used for experiments. Hole concentration was varied from $10^{14} cm^{-3}$ to $10^{15} cm^{-3}$. Hydrogen pulse implantation was carried out at room temperature by protons at energies of 18 keV. The pulse duration was 4 µs, the interval between pulses was 30 s, and the ion beam density in the pulse was 40 mA*cm^{-2}. The ion projected range was 0.22 µm. Isochronal and isothermal annealings were carried out in argon atmosphere up to 1050°C.

Structural changes (appearance of amorphous regions, nanocrystals, and atomic displacements in crystalline lattice) in implanted layers were observed by optical microscopy with 400 times magnification, TEM, and by RBS and channeling (RBS/C) at different He ion energies. Spectroscopic ellipsometry (SE) and visible photoluminescence were also employed for the analysis of optical properties. The SE energy range was 2 – 4.8 eV at an angle of incidence of 70.5°. PL was excited by the 488 nm line of an argon-ion laser.

RESULTS

The RBS/C spectra for hydrogen as-implanted samples and annealed ones are

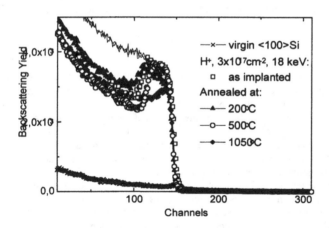

Figure 1. RBS/channeling spectra for hydrogen as-implanted, and annealed samples.

presented in Fig.1. Annealing in the temperature range of 200-500°C leads to removal of the point defects beneath the irradiated layer. Moreover the RBS yield for 200°C annealed samples is only 90-95% of spectrum for initial 'amorphous like' state.

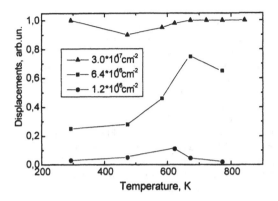

Figure 2. RBS normalized peak yield as a function of annealing temperature for
samples with different implantation dose.

This result suggests intermediate range ordering inside highly disordered layer (Fig.2). The
thickness of the layer is estimated from RBS spectra as 0.28 μm. Annealing at high temperature
(1050°C) does not cause the recovery of the structure (Fig.3).

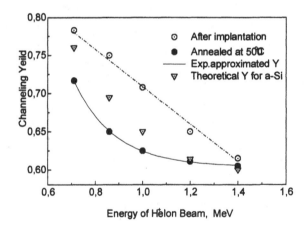

Figure 3. Dechanneling measurements as a function of He beam energy
used for RBS

A detailed study of dechanneling by irradiated layers was made by varying the He beam
energy (Fig.3). In the case of as irradiated layer, this dechanneling is linear and higher than the
theoretical one for a-Si. The dechanneling is lower for 500°C annealed sample than for
experimental and theoretical values of a-Si for energy >1.0 MeV and highly increases with
energy decrease to 0.7 MeV.

A model of disk like platelets of nanopores with small deviation of initially perpendicular to the surface crystalline axis in local regions between such pores was used to describe this dependence. Good results were obtained for the crystalline regions with average sizes a of 30 nm, and disorientation $\theta \sim 1.5°$ to the normal. To study the structure, TEM have been made for samples as irradiated and annealed. TEM images with small defects like vacancy loops, defects, and crystallites are presented in Fig. 4. An image of as irradiated sample has dark spots with average size a of 3 nm, and 30 nm after annealing at 500°C 1h.

Figure 5 shows the changes in visible PL intensity upon annealing. The maximum in intensity corresponds to quantum confinement recombination of electron-hole pairs inside small crystalline regions with 2-4 nm dimensions [7], and visible PL is absent in the case of 10 nm sizes, that is very close to the TEM data (Fig.4 a). The effective medium approximation (EMA) was used to model the SE spectra of the measured samples. Figure 6 presents an experimental and best-fit of real parts ε_1 and imaginary parts ε_2 of pseudo-dielectric function spectra for the sample with hydrogen dose 3×10^{17} cm^{-2} annealed at 500°C 1h.

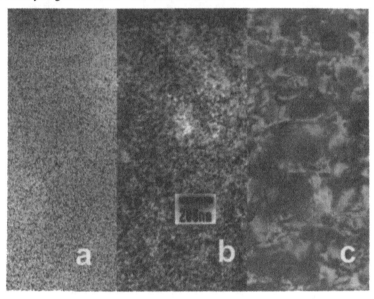

Figure 4. Weak beam TEM images for samples irradiated with H dose of 3×10^{17} cm^{-2} a), and annealed during 1 h at : b) 500°C; c) 1050°C

For analysis a two-layer optical model was employed: substrate is crystalline silicon (c-Si); layer 1 is a mixture of 30% c-Si, 30% a-Si, and 40% oxide or nc-Si:H. Layer 2 is more porous and has 24% a-Si, 8% SiO$_2$ and 68% voids. The thicknesses are 0.28 and 0.02 μm for layers 1 and 2, respectively. According to RBS measurements, however, oxygen atoms are not contented inside layer 1 and 40% of nc-Si:H is more real explanation of SE results.

It is known that most hydrogen-defect complexes are annealed near 250°C [8] but some complexes are stable up to 550°C [9]. One can suppose that VH$_n$ and IH$_n$ hydrided defects are stable up to 500°C, can diffuse rapidly to platelets [10], and annihilate at their walls according to following reaction:

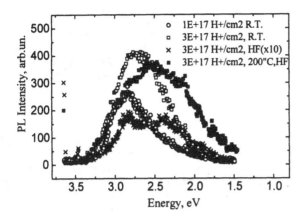

Figure 5. PL spectra for as-implanted samples, covered with SiO₂, and after HF etching and annealing at 200°C during 1h

$$\{V,H,...,H\} + \{I,H,...,H\} = n\,H_2 \qquad (1)$$

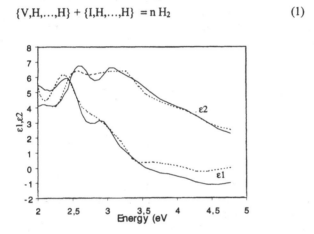

Figure 6. Experimental (solid line) and EMA calculated (broken line) real parts ε_1 and imaginary parts ε_2 of pseudo-dielectric function spectra for the sample with hydrogen dose $3 \times 10^{17}\,cm^{-2}$ annealed at 500°C 1h.

This mechanism can lead to an increase in H_2 gas content inside the nanopore platelets and the growth of their thickness d to d = _*a*sinθ ≃ 0.4 nm upon annealing at constant gas overpressure. Such a mechanism has never been observed on an atomic level. We speculate that such growth is manifested as "negative" annealing (Fig. 2) leading to the increase in the disorientation of the crystalline regions between the nearest platelets for higher dose. If we assume a linear relation between RBS/C yield and strain field caused by local hydrogen accumulation, as was suggested in [9], we obtain the same activation energy $E_a = 0.5 \pm 0.1$ eV for the lower H fluence $1.2x10^{16}cm^{-2}$. The higher value 0.8 ± 0.1 eV was obtained from Fig.2 for samples irradiated with medium dose ($7x10^{16}cm^{-2}$). The first value is closed to the migration barrier for the diffusion of atomic hydrogen in the Si crystal, but the second one shows the more complicated nature of this process including the reaction between defects according to (1) and some extended defects such as platelets or crystalline grain boundaries for higher dose ($3x10^{17}cm^{-2}$).

CONCLUSIONS

The conditions of irradiation and annealing for the production of quasi nanocrystalline and amorphous layers have been found. Pulsed implantation of the hydrogen ions allowed us to produce the layers which contain hydrogen as well as defects in concentrations high enough to generate an amorphous state (hydrogen content higher than 15 at. %). We find a relationship between visible PL and the structure of the hydrogenated nanocrystalline Si:H and amorphous Si:H.

1. Unusual strong dechanneling beneath the high dose hydrogen irradiated layers is observed for nc-Si:H samples annealed at temperatures >350°C for analysis beam energies lower than 1.0 MeV.
2. A model of quasi nanocrystalline Si structure based on channeling measurements is proposed. The minimal sizes of crystallites between the nearest platelets are 2-4 nm that grow to 30 nm after 500oC 1h, with small 1.5° disorientation to the surface normal.
3. This model coincides with TEM images showing the presence of smallest stressed crystalline regions <5 nm and explain the features of spectroscopic measurements by PL and SE.

ACKNOLEDGMENTS

This work was done at the support of ISTC grant no.563.

REFERENCES

1. T. Shimizu-Iwayama, K. Fujita, S. Nakao, K. Saitoh, T. Fujita, N. Itoh. J.Appl.Phys. **75**, 7779 (1994).
2. H.A. Atwater, K.V. Shcheglov, S.S. Wong, K.J. Vahala, R.C. Flagan, M.I. Brongersma, A. Polman. Mater.Res.Soc. Symp.Proc. **316**, 409 (1994).
3. P. Mutti, G. Ghislotti, S. Bertoni, Z. Bonoldi, G.F. Cerofolini, Z.Meda, E. Grilli, M.Guzzi. Appl.Phys.Lett. **66**, 851 (1995).
4. T. Shimizu-Iwayama, Y. Terao, A. Kamiya, M. Takeda, S. Nakao, K. Saitoh. Nucl.Instr.Meth., **B112**, 214 (1996).
5. G.A. Kachurin, I.E. Tyschenko, K.S. Zhuravlev, N.A. Pazdnikov, V.A. Volodin, A.K. Gutakovsky, A.F. Leier, W. Skorupa, R.A. Yankov. Nucl.Instr.Meth. **B122**, 571 (1997).
6. L.S. Liao, X.M. Bao, X.Q. Zheng, N.S. Li, N.B. Min. Appl.Phys.Lett. **68**, 850 (1996)
7. P. F. Trwoga, A. J. Kenyon, and C. W. Pitt. J.Appl.Phys., **83**, 3789 (1998).
8. S.J. Perton, J.W. Corbett, T.S. Shi. Appl.Phys., **A 43**, 153 (1987).
9. M. Bruni, D. Bisero, R. Tonini, G. Ottaviani, G. Queirolo, R. Bottini. Phys. Rev., **B 49**, 5291 (1994).
10. P. Leary, R. Jones, S. Oberg. Phys.Rew.B, **57**, 3887 (1998).

OPTICAL ANALYSIS OF PLASMA ENHANCED CRYSTALLIZATION OF AMORPHOUS SILICON FILMS

L. MONTÈS*, L. TSYBESKOV**, P.M. FAUCHET**, K. PANGAL***, J.C. STURM***, S. WAGNER***,
* Laboratoire de Spectrométrie Physique, CNRS (UMR 5588), 38402 Saint-Martin d'Hères Cedex, France, Laurent.MONTES@ujf-grenoble.fr
** ECE Department, University of Rochester, Rochester, NY 14627.
*** ECE Department, Princeton University, Princeton, NJ 08544.

ABSTRACT

Low-temperature crystallization of a-Si is important for display and Silicon-On-Insulator (SOI) technologies. We present optical characterization (Raman scattering and photoluminescence) of H_2 and O_2 plasma enhanced crystallization of a-Si:H films. H_2 plasma treatment is shown to be the most efficient, leading to larger grain sizes, and both H_2 and O_2 plasma lead to visible photoluminescence (PL). Recently, the PL of re-crystallized a-Si films has been explained in terms of quantum confinement [1]. The mean size of the crystallites in our re-crystallized films is determined by Raman scattering for different treatments parameters. No correlation between size and the photon energy of the visible emission is found. However, we can clearly distinguish between the PL from purely amorphous and re-crystallized a-Si:H films : Their PL temperature dependence and spectra are very different. The origin of the visible PL in re-crystallized thin Si films is discussed.

INTRODUCTION

Silicon nanocrystals (nc-Si) with sizes in the order of the nanometer are now elaborated using different fabrication techniques [1-7]. For silicon grains with size below 10 nm, the quantum size effects emerge as their size is comparable to the diameter of the bulk exciton 4-5 nm, which results in a widening of the band gap and a collateral increase of the probability for optical transitions that can be useful for optoelectronic applications. Porous silicon has been widely studied for its luminescent properties at room temperature in the visible [2,3], yet the mechanism contributing to the emission is still uncertain. In comparison with porous silicon, re-crystallized a-Si:H films are more stable and can be deposited over large area glass substrates, therefore they are of potential use in thin solar cells, as active layers in thin film transistors arrays for flat panel displays, with a better conduction than polycrystalline device. However the origin of the PL in these films is not well known but is of significance to contribute to a better control of the structures in order to increase the efficiency and the stability of the emitting material. Different mechanisms for excitation and radiative recombination are suggested, based on quantum confinement (excitation between the quantized levels inside the nanocrystallites) [1,4], on spatial confinement [3], and luminescent compounds or defect states [5]. Liu et al [1] reported the room-temperature photoluminescence of nc-Si crystals embedded in a large fraction of amorphous tissue, and proposed quantum confinement to be responsible for the PL peak position.

To address the origin of the light emission, a comparative study of samples with different sizes are analyzed through Raman scattering and PL temperature dependence to uncover the relation, if any exists, between the size of the crystals and the photoluminescence.

EXPERIMENT

Hydrogenated amorphous silicon films (a-Si:H) were deposited by PECVD using pure silane, on 7059 glass substrates at substrate temperature T_s of 150°C (set A) and 250°C (set B), and RF power ~5 W. A subsequent RF plasma exposure was realized in a parallel plate Reactive Ion Etcher (RIE) at room temperature with hydrogen or oxygen. Annealing in a furnace at 600°C in N_2 for 3 hours was completed to obtain the re-crystallized samples, using UV reflectance to monitor the crystallization process. The thickness of the film varied from 110 to 180 nm.

The Raman spectra of the transverse optical (TO) mode of the material were acquired with a Jobin Yvon U1000 instrument by exciting the sample with the 514.5 nm line of an Ar^+ laser. The power on the sample was of about 1 mW and the light was collected trough a microscope in the backscattering configuration. For the photoluminescence acquisition, visible emission was detected though a conventional optical multichannel analyzer set-up, while the infra-red region was detected with a North Cost high purity germanium detector.

RESULTS AND DISCUSSION

Raman analysis

The Raman spectrum of a crystalline silicon reference sample is dominated by a sharp lorentzian feature centered at 520 cm⁻¹, with a measured width at half maximum of 4 cm⁻¹. The feature corresponding to a-Si:H has a gaussian shape centered at 480 cm⁻¹, broad because the momentum conservation rule is relaxed by structural disorder. The Raman spectra of all the studied samples presented amorphous bands, while only the re-crystallized ones presented an additional TO-like silicon crystalline band. An example of spectrum is shown in Fig. 1 where both contributions clearly appears.

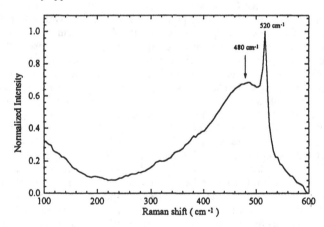

Fig. 1 : Raman spectra after re-crystallization for sample B2 (growth temperature Ts=150°C and hydrogen plasma treatment) : both a-Si:H and nc-Si TO-lines are present.

The crystalline band in our samples is highly asymmetric in the low frequency side, which is a clear indication of the stress of the structure. As the mean crystallite size decreases, the Raman peak broadens and shifts to lower frequency. The Raman spectra in the nc-Si peak region is represented in Fig. 2 for several treated samples.

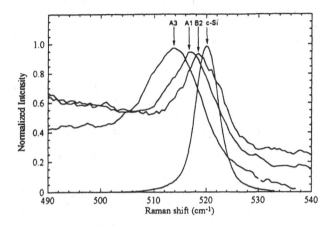

Fig. 2 : Raman shift TO line of re-crystallized samples A1, A3, B2 (see Table 1 for details) with comparison to bulk crystalline silicon as reference.

The correlation length of the nc-Si crystallites was deduced from a fitting with lorentzian curves and comparison with the values calculated by Fauchet [8], assuming a spherical shape of the grains. The resulting values of the peak shift $\Delta\omega$ and width are represented in Fig. 3 for the re-crystallized samples, leading to estimated sizes varying from 5 to 10nm, so that quantum size effects might be expected for these films.

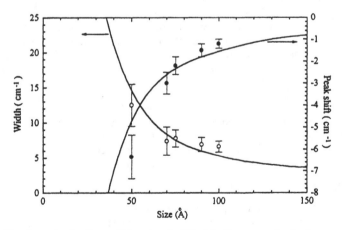

Fig. 3 : Size determination from width and peak shift of the Raman spectra for the re-crystallized samples (from left to right : A3, B1, A2, B3, B2) with comparison to [8].

The mean grain size d can also be deduced from Fig. 1 using the formula :

$$d = 2\pi\sqrt{\frac{B}{\Delta\omega}} \quad \text{with B=3 nm}^2.\text{cm}^{-1} \qquad (1)$$

The volume fraction of nanocrystallites is defined by $X_c = I_c/(I_c + yI_a)$ [9], where y is the ratio of the Raman cross section of the TO mode of crystalline and amorphous, I_c is the crystalline component of the Raman spectra and I_a is the amorphous component. The scattering factor y can be obtain with the formula [9]:

$$y(d) = 0.1 + e^{-(d/250)} \qquad (2)$$

We find sizes d of the crystallites range from 5 to 10 nm in good agreement with our previous plot and a volume ratio of crystalline in the order of 55% for almost all the re-crystallized samples is obtained. The results for the nc-Si samples are reported in Table 1.

Table 1 : nc-Si samples : growth temperature of amorphous film (Ts), plasma treatment before re-crystallization, mean size d and crystalline fraction Xc from Raman spectra, PL peak wavelength at room temperature, characteristic temperature T_0 and energy shift with temperature.

Sample	T_s (°C)	Plasma Treatment	Mean size (A)	Xc (%)	PL (nm)	PL (eV)	T_0 (K)	$\Delta E/\Delta T$ (meV.K^{-1})
A3	150°C	O_2	50	60	803	1.54	64	-0.04
A1	150°C	-	70	55	780	1.59	65	-0.05
B1	250°C	-	70	54	804	1.54	33	-0.13
A2	150°C	H_2	75	54	802	1.54	69	-0.04
B3	250°C	O_2	90	54	737	1.68	42	-0.06
B2	250°C	H_2	100	54	660	1.88	34	-0.12

The effect of the plasma treatment before annealing differs depending on the element (hydrogen or oxygen) and the temperature of the substrate T_s during the deposition. With increasing T_s, the effect of the plasma treatment is more perceptible, as it increases the size of the grains, while the crystalline fraction in the films remains almost constant. Hydrogen plasma is obviously the most efficient, as size is increased from 70Å to 100Å, for $T_s = 250$°C. On the contrary the oxygen plasma treatment for low T_s has a tendency to decrease the mean diameter of the grain while increasing the volume fraction. The fitting of sample A3 in Fig. 3 is unsatisfactory as the width of the Raman line is especially reduced for this size : this is an indication of the stress of the material and is probably correlated with the high volume ratio. The nucleation sites are more distributed in the film, leading to numerous nucleation of smaller crystallites. Higher defects density in the structure may also explain this result, because the temperature T_s corresponding is low and consequently the quality of a-Si:H is affected. Moreover, the presence of defects within the grains can confine the phonons, resulting in smaller correlation length (higher shifts) than the actual grain size. Compared to oxygen (or argon) plasma, hydrogen plasma has the largest effect on grain size, with the crystallization time reduced by a factor of five relative to non-plasma treated samples [7].

Photoluminescence and temperature dependence

Visible PL is obtained for plasma treated samples, persisting even at room temperature for the re-crystallized ones. In the later case, PL was stable over 6 months and no fatigue was observed.

An increase of hydrogen or oxygen content due to the plasma treatment leads to a widening of the band gap and increases the PL quantum efficiency when no crystallization is done. The PL peak energy is blue shifted (1.5eV) relative to the low-temperature PL in the near infrared region of standard a-Si:H films, while the temperature dependence parameters are very close to those of a-Si:H (see Fig. 4).

The values of the PL peak position at room temperature are reported in Table 1 for the re-crystallized films. No size correlation can be deduced from our data : this disconnection eradicates any quantum size effect related phenomenon. The PL temperature dependence was studied to investigate the mechanism involved in the recombination process. In standard a-Si:H films, for temperatures higher than a critical temperature T_c (50-100K for our samples), the recombination process is dominated by the dissociation of electron-hole pairs by thermal activation and non-radiative recombination through defects, which results in the quenching of the PL. Although the parameters are different, we observed PL dependency similar to that of a-Si:H : the temperature quenching of PL follows an exponential decrease with temperature,

$$\frac{I(T)}{I_0} = e^{-\frac{T}{T_0}} \qquad (3)$$

where I_0 is the maximum photoluminescence intensity (corresponding to T_c), and T_0 is a characteristic temperature (~23K for a-Si:H [10]). This ratio is related to the probability of radiative and non-radiative recombination, respectively W_r and W_{nr},

$$\frac{I(T)}{I_0} = \frac{W_r}{W_{nr} + W_{nr}} \qquad (4)$$

Thus Fig. 4 is a plot showing the dependence $\ln(W_{nr}/W_r)=\exp(T/T_0)$ versus T, from which T_0 can be extracted. From Table 1 it is clear that T_0 is strongly dependent on the growth temperature parameter T_s : lower deposition temperature corresponds to higher T_0 values. According to the luminescence model of band-tail states proposed by Street [11], T_0 in a-Si:H is proportional to the width of exponential band tail states. High T_0 values corresponds to deeper extending of the band tail states towards the middle band gap with a less pronounced slope of the exponential decay in the density of states, which reflects an increase of the disorder in the material and a weaker temperature dependence of the PL quenching. Changes in the structure due to the lattice mismatch between the amorphous and crystalline phases, together with

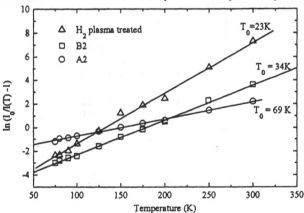

Fig. 4 : PL temperature dependence of for samples A2, B2 and a H_2 plasma treated film without re-crystallization.

abundant hydrogen or oxygen concentration induced by plasma treatment is responsible for those high T_0 values. With increasing temperature, carriers trapped in the band tail states move towards the middle of the band gap, leading to a redshift of the PL peak, as suggested by Street et al. [11]. This shift $\Delta E/\Delta T$ is found to be insignificant for our samples (in the order of $10^{-5} eV.K^{-1}$ instead of $10^{-3} eV.K^{-1}$ for a-Si:H). The weak temperature dependence of luminescence can be interpreted in terms of passivation due to the reduction of the density of mid-gap defects. As a consequence a PL signal is still efficient at room temperature.

CONCLUSIONS

Hydrogen plasma treatment reveals the important role of hydrogen in re-crystallization of a-Si:H films, with an enhancement of the nucleation rate leading to larger grain sizes. Comparison of the different crystallite sizes, with an almost constant crystalline fraction in the films, exclude any correlation between the size of the crystals and the PL wavelength that was observed : the luminescence has no relation with quantum size effect. Plasma treatment lead to PL with the temperature dependence parameters characteristic of standard a-Si:H films, while re-crystallization processes weaken this dependence. The temperature quenching of the PL is diminished and visible light emission is maintained even at room temperature.

REFERENCES

1. S. Tong, X.-n. Liu, and X.-m. Bao, Appl. Phys. Lett. 66 (4), 469 (1995); S. Tong, X.-n. Liu, T. Gao, X-m Bao, Y. Chang, W-z Shen and W.-g. Tang, Solid State Commun. 104 (10), 603 (1997).
2. L.T. Canham, Appl. Phys. Lett. 57 (10), 1046 (1990)
3. I. Solomon, R.B. Wehrspohn, J.-N. Chazalviel, F. Ozanam, J. Non-crystalline Solids 227-230, 248. (1998); M.J. Estes and G. Moddel, Appl. Phys. Lett. 68 (13), 1814 (1996).
4. E. Bustarret, E. Sauvain, and M. Rosenbauer, Thin Solid Films 276(1/2), 134 (1996).
5. K. Luterová, P. Knápek, J. Stuchlík, J. Kocka, A. Poruba, J. Kudrna, P. Malý, J. Valenta, J. Dian, B. Hönerlage, I. Pelant, J. Non-crystalline Solids, 227-230, 254 (1998).
6. D.J. Lockwood,, Z.-H. Lu, and J.M. Baribeau, Phys. Rev. Letters 76 (3), 539 (1996); L. Tsybeskov , G.F. Grom, K.D. Hirshman, L. Montès, and P.M. Fauchet, T.N. Blanton, J.P. McCaffrey, J.M. Baribeau, G.I. Sproulc, H.J. Labbé and D.J. Lockwood, MRS Fall Meeting 1998 (Symp. F)
7. K. Pangal, J.C. Sturm and S. Wagner, Proc. Symp. Mat. Res. Soc. 507 (1998) (to be published)
8. P.M. Fauchet, *Light Scattering in Semiconductors Structures and Superlattices*, edited by D.J. Lockwood and J.F. Young, (Plenum Press, New York, 1991), p.229
9. R. Tsu, J. Gonzalez-Hernandez, S.S. Chao, S.C. Lee, and K. Tanaka, Appl. Phys. Lett. 40, 534 (1982); E. Bustarret and M.A. Hachicha, M. Brunel, Appl. Phys. Lett. 52 (20), 1675 (1988)
10. R.W. Collins, M.A. Paesler and W. Paul, Solid State Commun. 34: (10), 833 (1980)
11. R.A. Street, *Semiconductors and Semimetals*, Vol. 21B, edited by J.I. Pankove (Academic Press, Orlando, 1984), p.197

LEDS BASED ON OXIDIZED POROUS POLYSILICON ON A TRANSPARENT SUBSTRATE

C.C. Striemer, S. Chan, H.A. Lopez, K.D. Hirschman, H. Koyama, Q. Zhu, L. Tsybeskov, and P.M. Fauchet
Department of Electrical and Computer Engineering, University of Rochester, Rochester, NY 14627

N.M. Kalkhoran and L. Depaulis
Spire Corporation, Bedford, MA 01730

ABSTRACT

Light emitting devices (LEDs) based on porous polysilicon (PPS) have been fabricated on a transparent quartz substrate. Several structures have been developed, each consisting of a backside contact (ITO or p$^+$ polysilicon), a light emitting PPS layer, a capping layer, and a metal top contact. Photoluminescence (PL) from PPS is similar to that of etched crystalline Si, peaking near 750 nm and showing degradation during 515 nm laser excitation with intensity <100 mW/cm^2. This degradation disappears if PPS is oxidized after formation. Visible electroluminescence (EL) has been achieved in both oxidized and non-oxidized PPS devices with voltages under 10 V and current densities <200 mA/cm^2.

INTRODUCTION

Extensive research in the area of light emitting porous crystalline silicon (PSi) has been conducted for nearly a decade, yielding several promising optoelectronic applications for this material [1-3]. However, in the area of active matrix display technology PSi has limited potential because it is formed on a silicon wafer. For large displays, the cost and availability of the necessary substrates would be prohibitive. Noting this disadvantage, several groups have started to look toward polysilicon or microcrystalline silicon as possible solutions to the size limitations imposed on traditional PSi [4-6]. Polysilicon can be deposited using low-pressure chemical vapor deposition (LPCVD) or plasma-enhanced chemical vapor deposition (PECVD) on a variety of substrates, such as glass which can be made arbitrarily large to accommodate large area displays. Light emission from PPS is not confined within the porous layer, but emits in all directions, making it ideal for thin film displays. The use of a transparent substrate allows the back side of the PPS layer to be used as the emitter, while driver circuitry can be stacked on top of the film. Using this geometry, driver circuitry no longer inhibits light emission and the active area is optimized. In this investigation, we experimented with several PPS LED structures on transparent quartz substrates, and constructed the first "all-silicon" PPS LED on this transparent substrate.

EXPERIMENT

In the first portion of this work, Corning 7059 glass substrates coated with a thin layer of indium-tin oxide (ITO) were used. The sheet resistivity of the ITO layer was approximately 4×10^{-4} Ω-cm, and its optical transmission was > 85% in the visible spectrum. This clear conductive layer must be present to provide a path through which current can be passed for PPS

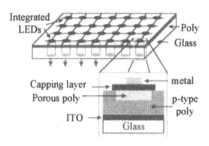

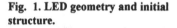

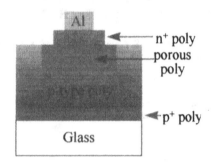

Fig. 1. LED geometry and initial structure.

Fig. 2. Structure of the "all-silicon LED"

formation and for subsequent EL emission. A 1.0 μm layer of polysilicon was deposited onto the ITO by LPCVD with a deposition time of 2 hours at 610°C. The thickness was determined using a Nanospec interferometry system. Ion implantation of boron was then used to dope the polysilicon p-type. This film was anodized at room temperature in the dark for 1-5 minutes using current densities <10 mA/cm^2, in a 49% HF:ethanol=1:5 solution. Some samples were also allowed to stand (open-circuited) for 0-2 minutes under illumination by a 100W tungsten halogen lamp for additional light-assisted etching. The anodization conditions were optimized based on PL output characteristics. After anodization, the as-etched PPS layer was then capped with a layer of spin-on conductive polymer (polyanaline), or in-situ doped n$^+$ a-Si deposited by PECVD. Finally, a metal contact was deposited, completing the device structure shown in Figure 1.

Encouraged by the success of the previous device structure, we then started to work with an improved "all-silicon" structure pictured in Figure 2. In this design, the ITO layer was replaced by a 0.25 μm, heavily doped p$^+$ polysilicon layer. This layer was also prepared using LPCVD followed by ion-implantation of boron (~2x10^{20}/cm^3). An additional 1.0μm polysilicon layer was then deposited, followed by a p-type up-diffusion step in which the device was annealed at 950°C for 30 minutes. A layer of PPS was then formed using a stain etching method in which the substrate was successively dipped in an HF:HNO$_3$:H$_2$O=1:3:5 solution until a strong photoluminescence output could be detected under illumination with a UV lamp. The sample was then oxidized via a furnace anneal at 950°C with N$_2$ flowing. The structure was further stabilized with a 0.25 μm capping layer of n$^+$ polysilicon formed by LPCVD followed by As$^-$ implantation. Three layer Ti:Pd:Au contacts were then sputtered onto the capping layer, completing this structure.

All devices were characterized by PL spectra, current-voltage (I-V) analysis, and finally by EL spectral output. PL was taken with excitation from the 515 nm line of a Coherent Innova 300 Ar$^+$ laser and spectral data was collected using a Thermo Jarrell Ash spectrometer attached to a Princeton Instruments thermo-electrically cooled optical multichannel analyzer (OMA). EL spectra were collected using the same spectrometer/OMA setup. I-V analysis was conducted with a computerized data acquisition system which uses a Keithley programmable current source and autoranging voltmeter in its measurements.

RESULTS

The preparation of PPS is considerably different from that of PSi and much time was spent optimizing these procedures using photoluminescence efficiency as our benchmark. As has been noted by several groups, the presence of grain boundaries [7,8], defects within the Si grains [9], and the random crystallographic orientations of these grains in polysilicon [10] add considerable complexity to the etching process. Preferential etching at grain boundaries and defect locations could potentially lead to the formation of isolated PPS islands on the insulating substrate which could destroy the EL efficiency of these devices. To limit this effect, a fairly dilute solution of HF:ethanol (1:5) and low current levels (<10 mA/cm^2) were used during the anodization process. In addition, the etching time was closely monitored to avoid etching through the polysilicon layer, while trying to leave as thin a layer of unetched polysilicon as possible. Samples with areas that began to etch through to the substrate could be easily identified by voltage changes across the electrochemical cell and holes in the polysilicon layer could be seen if the substrate was held up to the room light.

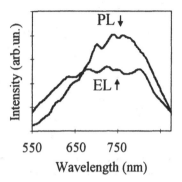

Fig. 3. Characteristic EL and PL spectra from a polymer capped PPS LED.

Our first working LED consisted of an ITO back contact, a PPS active layer, a conductive polyanaline capping layer, and a metal top contact. The PL and EL spectra for this LED are plotted in Figure 3. The EL is broader and slightly blue shifted compared to the PL. The EL from this device was clearly observable at a voltage of 15.9 V and current density of 880 mA/cm^2, with an EL threshold of 8.5 V and 100 mA/cm^2. The emission is also uniform over an area of approximately 0.1 cm^2.

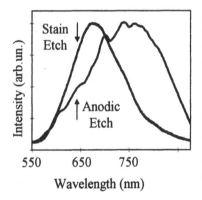

Fig. 4. Comparison of the resulting PL from samples prepared using anodic and stain etching of PPS.

In an effort to use standard silicon processing methods, we then replaced the polymer capping layer with an in-situ doped n$^+$ a-Si layer with an Al top contact. These samples were identical to the previous devices after the anodization process and thus the PL spectra were similar. After the a-Si deposition, the PL remained unchanged. However, the resulting EL from these devices exhibited a considerable red-shift, pushing the spectral peak above 800nm and leaving only a small portion of the spectra in the visible range. Thus, as a visible LED this device was less efficient, although emission was still clearly observable at 8.1 V and 1240 mA/cm^2 (~1 Watt per 0.1 cm^2 device) with an observable EL threshold at 6.5 V, 320 mA/cm^2. The decreased operating voltage of this device was a promising result, despite the higher current levels that were needed to make the visible portion of its EL spectrum comparably bright.

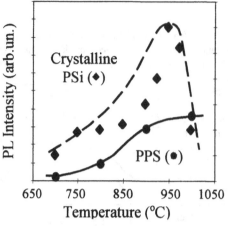

Crystalline
PSi (♦)

PPS (●)

PL Intensity (arb. un.)

650   750   850   950   1050
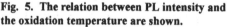
Temperature (°C)

Fig. 5. The relation between PL intensity and the oxidation temperature are shown.

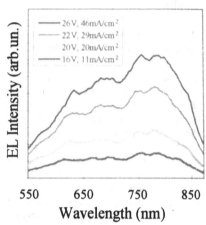

EL Intensity (arb. un.)

—— 26 V, 46mA/cm^2
---- 22 V, 29mA/cm^2
—— 20 V, 20mA/cm^2
—— 16 V, 11mA/cm^2

550   650   750   850
Wavelength (nm)

Fig. 6. EL Spectra of an LED based on oxidized PPS.

The final portion of this study culminated in the development of an "all-silicon" LED based on *oxidized* PPS on the same transparent glass substrate. This device incorporated many processing simplifications in its design. First, the ITO backside contact was replaced with a heavily doped p$^+$ polysilicon layer which could be easily fabricated in-house. Anodic etching of PPS was also replaced with stain etching. Using stain etching, the PPS was formed purely chemically by immersion of the sample in a solution of $HF:HNO_3:H_2O=1:3:5$. The resulting PL of these devices was substantially narrower and significantly blue shifted, as can be seen in Figure 4. Another feature of this device was the incorporation of an oxidation step which is believed to provide additional passivation and stability to the PPS nanostructure. This improvement was obvious by comparing the PL emission stability of oxidized and non-oxidized samples. Samples that have not been oxidized show a steady decrease in PL intensity over time at excitation intensities less than 100 mW/cm^2, while oxidized samples exhibit no such temporal degradation even at much higher excitation intensities. In the oxidation process described in the previous section, the annealing temperature is a critical parameter that needed to be optimized. In Figure 5, the PL intensity is plotted versus annealing temperature for both PPS and PSi. Although the PL appears to steadily increase over the temperature range of 700°C-1000°C, we chose to anneal our samples at 950°C, corresponding to the peak efficiency of oxidized PSi. This decision was made based

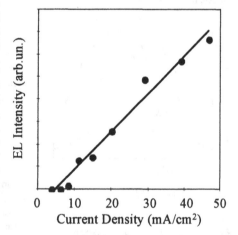

EL Intensity (arb. un.)

0   10   20   30   40   50
Current Density (mA/cm^2)

Fig. 7. Integrated EL intensity of an oxidized PPS LED as a function of current density.

on the similarity in luminescence mechanism that we believe exists between PPS and PSi. After the oxidation process, an n^+ polysilicon capping layer was deposited and implanted with As. Finally, a high quality three layer Ti:Pd:Au contact was sputtered onto the capping layer. Resulting EL spectra measured on this device are plotted in Figure 6. This device has an EL threshold at 9.2 V, with a driving current density of 4 mA/cm^2. The emission increases linearly with current as depicted in Figure 7, and is clearly observable in a dimly lit room at 26 V and 46 mA/cm^2. The higher voltage operation of this device can be attributed to a loss in the continuity of the backside contact due to cracking of the thin polysilicon layer. This cracking is probably due to stress induced during high temperature processing steps (implantation and oxidation anneals) by different coefficients of thermal expansion in the film and substrate. Future work will focus on improving the quality of this device by redesigning our fabrication procedure to avoid some of these stress related problems.

CONCLUSIONS

In this study, several light emitting devices were fabricated based on a PPS active layer. Preferable device geometries were employed using a transparent glass substrate which allows light to propagate from the back side of the PPS layer. This work culminated in the fabrication of the first "all-silicon" PPS LED on a transparent substrate. We believe that the incorporation of standard silicon processing technology with a highly stable oxidized PPS active layer has potential applications in the area of active matrix display technology.

ACKNOWLEDGMENTS

This work was supported in part by SBIR grant #DMI-9522054 and STTR grant #DMI-9712262 from the National Science Foundation. C.C.S. is supported under a fellowship from the Office of Naval Research.

REFERENCES

1. K.D. Hirschman, L. Tsybeskov, S.P. Duttagupta, and P.M. Fauchet, Nature, **384**, p.338 (1996).
2. R.T. Collins, P.M. Fauchet, and M.A. Tischler, Physics Today, **50**, 24 (1997).
3. P.M. Fauchet, in *Light Emission in Silicon from Physics to Devices*, edited by D.J. Lockwood (Semiconductors and Semimetals 49, New York, 1998), p. 206.
4. N. Koshida, E Takizawa, H. Mizuno, S. Arai, H. Koyama, and T. Sameshima, in *Materials and Devices for Silicon-Based Optoelectronics*, edited by A. Polman, S. Coffa, and R. Soref (Mater. Res. Soc. Symp. Proc. **486**, Boston, MA, 1998) p. 151.
5. N. M. Kalkhoran, F. Namavar, and H.P. Maruska, Appl. Phys. Lett. **63**, p.2661 (1993).
6. F. Chane-Che-Lai, C. Beau, D. Briand, and P. Joubert, Applied Surface Science, **102**, p.399 (1996).
7. P. Guyader, P. Joubert, M. Guendouz, C. Beau, and M. Sarret, Appl. Phys. Lett. **65**, p.1787 (1994).
8. P.G. Han, M.C. Poon, P.K. Ko, and J.K.O. Sin, J. Vac. Sci. Technol. B, **14**, p.824 (1996).
9. L. Haji, Y. Le Thomas, F. Chane Che Lai, and P. Joubert, in *Advances in Microcrystalline and Nanocrystalline Semiconductors - 1996*, edited by R.W. Collins, P.M. Fauchet, I Shimizu, J. Vial, T. Shimada, and A.P. Alivisatos, (Mater. Res. Soc. Symp. Proc. **452**, Boston, MA, 1997) p. 421.
10. W.N. Huang, K.Y. Tong, and P.W. Chan, Semicond. Sci. Technol. **12**, p. 228 (1997).

EFFECT OF HYDROGEN PLASMA TREATMENTS AT VERY HIGH FREQUENCY ON p-TYPE AMORPHOUS AND MICROCRYSTALLINE SILICON FILMS

E. CENTURIONI*, A. DESALVO*, R. PINGHINI*, R. RIZZOLI**, C. SUMMONTE**, and F. ZIGNANI*
*Dip. di Chimica Applicata e Scienza dei Materiali, Bologna University, viale Risorgimento 1, I-40136 Bologna, Italy
**CNR-LAMEL, via Gobetti 101, I-40129 Bologna, Italy

ABSTRACT

Very high frequency (100 MHz) hydrogen plasma treatments on a-Si:H deposited by plasma enhanced chemical vapour deposition were studied. Ex–situ optical measurements have shown that etching and chemical transport occur as competing phenomena. The one that prevails is determined by the plasma conditions. In particular, very efficient chemical etching is observed for high H_2 flow rates, while, for low H_2 flow rates, the equilibrium is shifted toward deposition, and a structural modification of the sample is observed. The experimental conditions were stressed in the direction of utilizing chemical transport from the cathode to deposit very thin (7 nm) microcrystalline films with a pure H_2 plasma.

INTRODUCTION

To produce p-type microcrystalline silicon layers at 13.56 MHz the layer-by-layer technique is normally employed [1,2]: deposition steps are alternated with treatments in a pure H_2 plasma. Several models have been proposed to describe the transition from the amorphous to the microcrystalline phase, and a general agreement has not yet been reached, as none of them is fully acceptable to describe the observed results [2]. The role of hydrogen in the plasma gas mixture is still controversial. In ref. [3], some experimental evidence is reported that, upon hydrogen plasma treatment, there is no phase change in an already deposited amorphous layer, and that the microcrystalline deposition entirely comes from chemical transport from the cathode, in a diode type RF apparatus. On the other hand, in ref. [1] it was found that hydrogen treatments induce extreme 'porosity' in the deposited layer and produce eventually a phase modification in the bulk of the material. In Electron Cyclotron Resonance systems it was shown that hydrogen treatment can modify the structure of films [4]. Moreover, the role of silicon contamination of the walls of the deposition chamber [3] has been ignored in many experiments in the literature, and probably affects the experimental results. The role of hydrogen, though accepted as a source of possibly selective etching, was not considered in combination with its ability to cause chemical transport from the cathode.

In this paper, a set of experiments designed to distinguish carefully between deposition, etching and contamination from the chamber, is reported. The experiments were intended to study the dependence of etching rates on the plasma parameters, and to verify the role of chemical transport in the microcrystalline silicon deposition.

EXPERIMENTAL DETAILS

The depositions were performed using a four chamber diode type Plasma Enhanced Chemical Vapour Deposition (PECVD) apparatus. The pre-vacuum is in the 10^{-8} hPa range. The plasma is ignited by a spark. This allows accurate fixing of the starting plasma conditions, and a precise control of the process time. p-type samples were deposited on 100 cm^2 Corning glass

517

under the following conditions: 100 MHz, 0.3 hPa, 170°C, 23 mm electrode distance, with the SiH_4 (3 sccm) + 2% B_2H_6 diluted in H_2 (0.4 sccm) + H_2 (196.7 sccm) gas mixture. The samples were characterized ex-situ. To get reproducible results, the deposition and the set of etching steps were performed within the same day.

Reflectance, R, and transmittance, T, spectra were measured with a UV-visible spectrophotometer. Analysis of the optical data was performed with a computer program which makes use of the conventional matrix Fresnel formulation.

As we observed that, under some experimental conditions, a deposition is obtained with a pure H_2 plasma in a Si-contaminated chamber, accurate cleaning of the chamber and cathode sandblasting were carried out before etching experiments.

RESULTS

Optical measurements

All deposited samples were optically characterized by means of transmission and reflection spectroscopy. The method has the double advantage of being sensitive to extremely thin layers (of the order of nm) and, at the same time, to structural arrangement. In particular, the crystallinity is well detected by reflectivity in the UV (200 to 400 nm range) where c-Si shows two peaks that do not appear in a-Si. Neither Raman spectroscopy nor X-ray diffractometry have the same sensitivity, and, over transmission electron microscopy and spectroscopic ellipsometry, transmission and reflection spectroscopy has the advantage of being far easier to perform.

The experimental spectra were simulated by one or more layers, each constituted either by pure components, or by a mixture of components computed by the Bruggeman Effective Medium Approximation (EMA).

The pure components used here are a-Si:H (ref. [5]), fine grained micro-crystalline silicon ('c-Si FG' in the following) [6], and voids. The a-Si:H was not simulated by a mixture of a-Si and voids as is normally done [1], because no such mixture could reproduce our experimental data. In particular, the best simulation is obtained with a mixture of 92% a-Si + 8% voids, that nonetheless is too absorbing at wavelengths longer than about 500 nm. In fact, our films are very

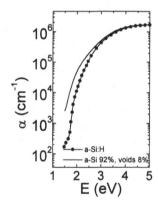

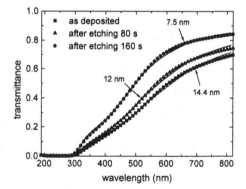

Fig.1 – Absorption coefficient spectra of a-Si:H (ref. [5]) and of a (92% a-Si + 8% voids) mixture computed by EMA.

Fig. 2 – Experimental T spectra of the same sample after deposition and after subsequent etching steps (symbols). Best fitting curves (lines) are simulated with an a-Si:H single layer of the indicated thickness.

thin (maximum thickness 30 nm), and were deposited under conditions for which larger thicknesses produce micro-crystalline films. Their structure is therefore expected to be different from that of standard bulk a-Si:H. A comparison between the absorption of a-Si:H [5] and that of the mixture 92% a-Si + 8% voids is reported in Fig. 1.

Spectral transmittance is used in this paper to determine the film thickness. As an example, in Fig.2 a set of T spectra measured on the same sample at different stages of etching is reported, along with the T curves simulated with pure a-Si:H, in which only the film thickness was allowed to vary. The resulting computed thickness is also indicated. Note that a 2 nm difference is easily resolved.

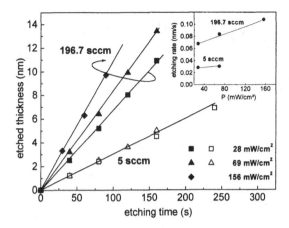

Fig. 3 – Sample etched thickness versus etching time with varying H_2 flow rate and RF power density. The deduced etching rates are reported in the inset. Lines are reported as a guide for the eye.

Etching rates

A set of p-type samples was deposited and rapidly etched in a pure H_2 plasma at 170°C, 100 MHz, 0.2 hPa, 23 mm electrode distance. The H_2 flow rate was either 5 or 196.7 sccm, and the RF power density, P, was 28, 69, or 156 mW/cm^2. The thickness after each etching was evaluated by fitting the transmittance measurements by a single layer / substrate model, as described above. All layers are well simulated by a single layer of a-Si:H, and this continues to be true as the etching time increases.

Fig. 3 shows the etched thickness versus etching time and the deduced etching rates are reported in the inset. For high H_2 flow rate, the etching rate increases as P increases, while little dependence on power is detected for low H_2 flow rate.

An interesting feature is observed in the case of 156 mW/cm^2, 5 sccm H_2 flow rate, which is not reported in the figure due to its anomalous behaviour. In this case, contrary to the other etching conditions, the R spectra show an evolution (upper part of Fig. 4) while the T spectra (not shown) do not change. As can be seen in the upper part of Fig. 4, the reflectance decreases for wavelengths in the range from 200 to 500 nm and the spectra can be no longer simulated by a single layer of a-Si:H. The appearance of a peak at about 4.6 eV, with increasing exposure time to the H_2 plasma, indicates the presence of a certain amount of c-Si FG. The lower part of Fig. 4 displays the results of the fitting procedure of experimental T and R spectra after each "etching" step. The sample structural evolution is well simulated using three layers (surface, bulk and a film/substrate interface layer) and varying their thicknesses and the percentages of a-Si:H, c-Si FG and voids in each layer. For the as-deposited sample bulk and surface layers coincide and show a mixture of c-Si FG and a-Si:H within 4 nm from the surface. The very thin surface layer is introduced in the simulation only after the first etching step and contains only c-Si FG and voids. The presence of a high concentration of voids in the surface layer can also indicate the

519

existence of a noticeable surface roughness, which increases with etching time. This possibly indicates some H induced selective etching of the a-Si:H. With increasing etching time, the H induces strong changes in the bulk layer: the thickness and its void content increase, while the a-Si:H percentage remains constant after the first etching step, and the c-Si FG component disappears after the following step. The interface layer thickness and the corresponding a-Si:H content decrease, while the fraction of c-Si FG increases and the void fraction remains approximately constant.

The simulated structure increases in volume but decreases in density and some induced crystallization seems to occur in the interface layer.

Chemical transport

Basing on the experiments reported in ref. [3], we checked the role of chemical transport deposition, CTD, in a PECVD process. We deposited standard a-Si:H at 13.56 MHz on a 10x10 cm² stainless steel plate, at two different substrate temperatures (50° and 170° C). This plate was then placed on the cathode of a different clean deposition chamber, in order to provide a controlled solid silicon source. A 10 min pure H₂ plasma treatment (100 MHz, 5 sccm, 0.2 hPa, 170°C, 156 mW/cm²) was then performed with this configuration on a clean Corning glass substrate. We obtained the following results:

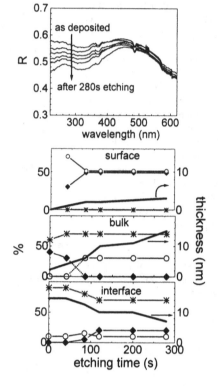

Fig. 4 – Experimental R spectra of the same sample after deposition and subsequent H₂ etching steps at 5 sccm, 156 mW/cm² (upper part). Best fit result versus etching time: thickness (—), and fraction of a-Si:H (✳), c-Si FG (♦), and voids (o) for surface, bulk and substrate interface layers (lower part).

- No deposition on the sample was observed using the plate deposited at 170°C.
- Deposition was observed using the plate deposited at 50°C.

This result is reproducible. Reflectance and transmittance spectra of one film obtained in this way are reported in Fig. 5. It is very interesting to note that the R spectrum shows the two UV peaks that are characteristic of the μc-Si phase. The same figure also shows the best fit obtained from the simulation of T and R spectra, using a very high (about 80%) c-Si FG fraction, and a thickness of about 7 nm. The fitting curves obtained with pure amorphous and pure c-Si FG films are also reported for comparison.

The different behaviour observed using plates deposited at different temperatures is explained by considering that the a-Si:H obtained at low temperature has a much less dense structure, and is apparently more easily etched by the H₂ plasma, thus allowing chemical transport from the cathode to the anode. We point out that the film reported here has a much higher degree of microcrystallinity than films of the same thickness produced by the conventional layer by layer technique.

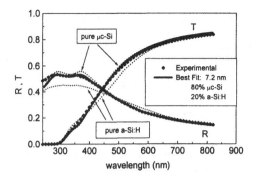

Fig. 5. Experimental and best fit R and T spectra of the sample obtained by CTD. Computed spectra for a-Si:H/glass and μc-Si/glass structures are reported for comparison.

DISCUSSION

Etching of a-Si:H in pure H_2 was performed. This process is actually a reactive etching type process. We observed that for high H_2 flow rate the etching rate increases with power density, which is to be expected as an increasing power density increases the dissociation of the gas [7]. As a consequence, the H radical concentration increases, thus enhancing the rate of transfer of silicon to the gas phase as SiH_4, SiH_3, etc., via surface reactions such as [8]:

$$Si(s) + (4/m)H_m \rightarrow SiH_4 \qquad (1)$$

On the contrary, for an H_2 flow rate as low as 5 sccm, no dependence of the etching rate on power density is observed for P up to 69 mW/cm^2. Moreover, for 156 mW/cm^2, some evidence of structural modification is observed: as the etching time increases, spectral reflectance shows a sign of nucleation of small c-Si grains, associated with an increase of film thickness and a reduction of density. We note that, if the H_2 flow rate is decreased from 196.7 to 5 sccm, the molecule residence time τ_{dwell} is increased by a factor 40, and is estimated to be about 15 s under our experimental conditions. On the other hand, the reaction half time $\tau_{1/2}$ [8] decreases with increasing power density. When the residence time becomes longer than the surface reaction half time, the silane concentration reaches its partial chemical equilibrium value, which is due to the balance between the rates of formation and decomposition of silane, and which is the basis for the deposition of a microcrystalline solide phase [9]. In our interpretation, this means that the condition:

$$\tau_{dwell} > \tau_{1/2} \qquad (2)$$

occurs when the experimental value of P increases from 69 to 156 mW/cm^2. Consequently, an equilibrium between etching and deposition takes place, with a rearrangement of the film structure.

The mechanism just illustrated fully applies to the deposition of the CTD films. The gas residence time for CTD was again 15 s, and the condition (2) is expected to be met. In fact, chemical transport was reported to produce crystalline silicon free of any noticeable amorphous tissue [10]. This deposition is in agreement with the results reported in ref. [3], and in our opinion, can also explain other results reported in the literature. It is possible that some research groups have produced films under uncontrolled CTD conditions. In particular, when deposition in extremely high H_2 dilution is performed [11], one should consider, in the interpretation of the experiment, that the silicon source could come from chemical transport.

The optical spectra measured on CTD samples are compatible with a film in which the μc-Si fraction extends down to the substrate. More experiments are needed to check this result with structural techniques, yet, this is a quite interesting result. The deposition rate obtained by CTD under our condition is very low, namely, 0.01 nm/s. This is of course too low for any production application, although, in some devices such as heterojunction solar cells, very thin µc-Si films are needed. However, the possibility of employing the CTD µc-Si film as seed layer for enhanced nucleation in the following deposition step using a different technique, should be considered. In fact, a thin seed layer is reported to improve the crystallinity of a subsequently deposited much thicker µc-Si:H film [11, 12].

CONCLUSIONS

Hydrogen etching of a-Si:H, in a VHF PECVD system, was studied by optical measurements associated with simulation, which have proven to be a very practical way for an initial sample structural characterization. Depending on the surface reaction half time compared to the gas residence time (controlled by power density, gas flow and pressure), we obtained etching or deposition, and, in between, a balance of the two mechanisms. This balance results in a rearrangement of the deposited material with consequent bulk modification and a transition from the amorphous to the microcrystalline phase.

Chemical transport deposition (CTD) was demonstrated to be able to produce film deposition without the introduction of silane in a Si-contaminated chamber. The films deposited under this very high dilution condition are microcrystalline, even at a very small thickness (7 nm). This is a new result that indicates that there is no intrinsic preclusion in depositing thin µc-Si layers. Care should be taken when evaluating results obtained in extremely high H_2 dilution of the gas mixture, to be sure that the deposition does not come from an unwanted CTD process.

ACKNOWLEDGEMENTS

This work is partially supported by the Progetto Finalizzato Materiali Speciali per Tecnologie Avanzate of CNR, Italy, and by the Ministero dell'Università e della Ricerca Scientifica e Tecnologica.

REFERENCES

1. P. Roca i Cabarrocas, S. Hamma, S.N. Sharma, G. Viera, E. Bertran, J. Costa, J. of Non-Cryst. Solids 227-230, (1998) 871.
2. N. Layadi, P. Roca i Cabarrocas, B. Drévillon, and I. Solomon, Phys. Rev. B 52, (1995) 5136.
3. K. Saitoh, M. Kondo, M. Fukawa, T. Nishimiya, and A. Matsuda, Appl. Phys. Lett 71, (1997) 3403.
4. I. Kaiser, N.H. Nickel, W. Fuhs, Phys. Rev. B 58, (1998) R1718.
5. "Properties of Amorphous Silicon", EMIS data review, INSPEC 1989.
6. G.E. Jellison, Jr., Appl. Phys. Lett., in press.
7. J.Perrin: in 'Plasma Deposition of Amorphous Silicon-Based Materials', edited by G. Bruno, P. Capezzuto, A. Madan (Academic Press, San Diego, 1995) p.177.
8. S. Veprek, M. Heintze, F.A. Sarott, M. Jurcik-Rajman and P. Willmott, Mat. Res. Soc. Symp. Proc. 118, (1988) 3.
9. K. Ensslen and S. Veprek, Plasma Chem. Plasma Proc. 7, (1987) 139.
10. S. Veprek, , F.A. Sarott, and Z. Igbal, Phys. Rev.B 36, (1987) 3344.
11. J.-H. Zhou, K. Ikuta, T. Yasuda, T. Umeda, S. Yamasaki, K. Tanaka, Appl. Phys. Lett. 71, (1997) 1535.
12. H.V. Nguyen, I. An, R.W. Collins, Y. Lu, M. Wakagi, and C.R. Wronski, Appl. Phys. Lett. 65, (1994) 3335.

ROBUST EXCITON POLARITON IN A QUANTUM-WELL WAVEGUIDE

M. Shirai, K. Hosomi, T. Mishima and   T. Katsuyama
Central Research Laboratory, Hitachi Ltd., Kokubunji Tokyo 185-8601, Japan,
shirai@crl.hitachi.co.jp

ABSTRACT

The refractive index change of polariton propagation in a GaAs quantum-well waveguide
was measured as a function of the electric field. Its temperature dependence was also
measured. The experimental results showed that the refractive index change of the polariton
propagation was at least three times as large as that of the conventional light propagation.
This effect remains up to 40 K, and coincides with the temperature dependence of the rate of
polariton scattering by phonons. We also fabricated directional-coupler-type switching
devices to apply this large refractive index change, and were able to demonstrate the operation
in a single quantum-well waveguide. Our results indicate that extremely small and low driving
voltage switching devices may be feasible.

INTRODUCTION

In nanometer-scale semiconductor structures, an exciton interacts strongly with a photon
due to the quantum confinement effect. This strong interaction leads to the stabilization of a
quasi-particle called an exciton polariton. The polariton is a composite particle comprised of
an exciton and a photon. Therefore, polaritons have various attractive properties, such as
excellent coherence due to their light-wave nature, high sensitivity under an electric field due
to the existence of the charged electron-hole pairs, and an ultra-short wavelength of
propagation due to the resonance state of the polariton. Thus, the application of polaritons in
nanometer-scale semiconductor structures could enable devices that are highly sensitive and
extremely small[1].
Basically, these devices would operate in the same manner as conventional optical devices,
such as directional coupler switches and Mach-Zehnder modulators, that are based on the
refractive index change of light under an electric field. Therefore, polariton propagation in a
device could reduce the driving voltage needed and the device size because of polariton's
high sensitivity to an electric field. Such a low-voltage modulator would be very effective in
high-bit-rate optical communication systems because it is difficult to realize a high-speed
signal driver with a large output-voltage swing.
In this report, we discuss the refractive index change of exciton polariton propagation, which
corresponds to the phase change of the polariton. In particular, we focus on its temperature
dependence. We also describe the application of polariton propagation to an experimental
directional-coupler-type switching device.

EXPERIMENTAL RESULTS

Refractive Index Change of Polariton Propagation

To apply polaritons in optical devices, we must precisely determine the magnitude of the
refractive index change of the polariton propagation under an electric field and its temperature
dependence. We measured the cross-over temperature at which polaritons disappear and the
refractive index change begins to decrease. Figure 1 shows our (a) sample structure, (b)
measurement setup, and (c) obtained output interference pattern when the phase change was
measured. The sample was basically a waveguide with 7.5-nm-thick GaAs single quantum-
well layer. The experimental setup consisted of a Mach-Zehnder-type interferometer that
enabled us to measure the refractive index change from the fringe pattern change.

Figure 2 shows the relationship between the refractive index change and the detuning wavelength. The refractive index change was derived from the phase change. The detuning shown here is that for the resonant energy of the polariton. The temperature was used as a parameter. The behavior shown in Fig. 2 is complicated because of the peak shift due to the quantum confined Stark effect. Therefore, we used the total index change, which is a summation of the index change over the resonant wavelength region. Figure 3 shows the relationship between the total index change and the temperature. The total index change remained about three times as larger as that at room temperature (300 K) up to 40 K. In other word, the polariton effect remains up to 40 K.

We also studied the index change theoretically. Figure 4 shows the relationship between the refractive index change and the detuning based on calculated results. In this case, the parameter was a damping factor of the polariton. The index change decreases drastically with an increasing damping factor. The relationship between the total index change and the damping factor is shown in Fig. 5. Comparing this result with Fig. 3, we can estimate that the damping factor of the polariton is still small at 40 K. Furthermore, we can compare our result with the theoretical work by Pau et al. [2] who determined relationship between the polariton scattering probability due to phonons and the temperature (Fig. 6). As shown, the temperature where the scattering probability rapidly increases coincides with the temperature where the polariton effect vanished in Fig. 3. Thus, scattering by phonons plays an important role in the temperature dependence of the polariton effect.

Thus, to increase the operating temperature of polariton devices, we need to reduce the scattering by phonons.

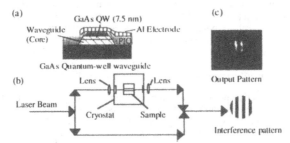

Fig. 1 System for measureing phase change under an electric field
(a) Sample structure, (b) Measurement setup, (c) Output pattern.

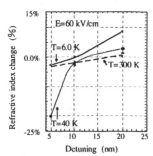

Fig. 2 Refractive index change
vs detuning (experiment).

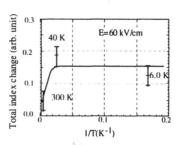

Fig. 3 Total index change vs
temperature (experiment).

524

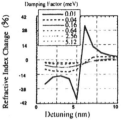 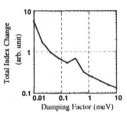 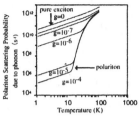

| Fig. 4 Refractive index change vs detuning (Theory). | Fig. 5 Total index change vs damping factor (Theory). | Fig. 6 Polariton-phonon scattering probability depending on the temperature (Pau et al.[2]) |

Application to Switching Devices

Here, we describe one example of the device applications of the polariton, i.e. a directional-coupler-type switching device. Figure 7 is a schematic of the device which consists of parallel adjacent waveguides. The layer structure is again a quantum-well waveguide with a 7.5 nm thick quantum-well. Photons propagating in the core layer interact with excitons, leading to polariton propagation.

The switching-operation principle of this device is the same as that of conventional optical directional coupler switches. The distance between the two waveguides was 0-2 μm to reduce the coupling length. The waveguide were composed of p-i-n structures with an i-layer width was 0.15 μm. This enabled efficient application of the electric field.

We simulated the switching characteristics by using a beam-propagation method [3]. Figure 8 shows the refractive index profile used for the simulation. In this simulation, the polariton propagation was represented by electromagnetic wave propagation. The characteristics of the polariton were represented by the macroscopic constant, i.e., the refractive index. Figure 9 shows the simulated results. The polariton beam was periodically transferred to another waveguide, and the periodic length was changed by applying the electric field to represent the directional-coupler-type operation of a switch. The output intensity is shown in Fig. 10 as a function of the applied electric field and the corresponding applied voltage. An on-off ratio of over 10 dB was obtained under an applied voltage as low as 0.2 V. Therefore, low-voltage operation of the switch appears feasible.

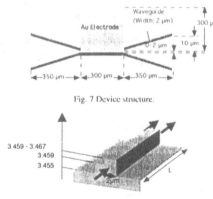 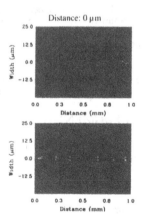

Fig. 7 Device structure.

Fig. 8 Refractive index profile.

Fig. 9 Simulated switching characteristics.

We also fabricated the device outlined in Fig. 7, mainly by electron beam lithography. We then measured switching performance of the fabricated device using a Ti-sapphire laser system whose light was injected into one end of the waveguide. Figure 11 shows the output intensity as a function of the applied voltage, at wavelengths of 800 nm and 810 nm. The measurement temperature was the liquid He temperature (LHeT). As shown in Fig. 11(a), the intensity of output A gradually decreased, while, the intensity of output B increased as voltage rose. This behavior was periodical, so the directional-coupler-type operation of the device could be observed. The on-off ratio of the signal from output A was about 6 dB. Figure 11(b) shows results similar to those of Fig. 11(a), but the overall shape of the results moved in the lower voltage direction due to the phase change corresponding to the wavelength change.

However, the applied voltage needed to obtain a change of phase was still a relatively large value, i.e., about 6 V. This value is one order of magnitude larger than the simulation result. We believe that this discrepancy was due to inefficient application of the electric field so the fabrication process needs to be improved. The discrepancy may also be due to the large amount of detuning we used to avoid a simultaneous absorption-coefficient change under the electric field.

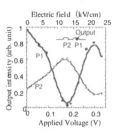

Fig. 10 Output intensity vs applied voltage (Simulation)

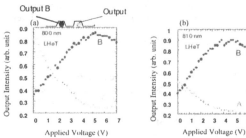

Fig. 11 Switching characteristics (Experiment)
(a) wavelength: 800 nm, (b) wavelength: 810 nm.

CONCLUSIONS

We measured the refractive index change of polariton propagation in a GaAs quantum well waveguide as a function of the electric field. Its temperature dependence was also measured. The refractive index change due to the polariton propagation up to 40 K was three times that of pure light propagation. This behavior can be explained by the temperature dependence of the rate of polariton scattering due to phonons. We applied this large refractive index change in directional-coupler-type switching devices where switching operation was successfully demonsrated in a single quantum-well waveguide. The on-off ratio was about 6 dB. These results indicate that extremely small and high-frequency switching devices may be attained.

ACKNOWLEDGMENT

This work was performed under the management of FED as apart of the MITI R&D Program Quantum Functional Devices Project supported by NEDO.

REFERENCES

[1] T. Katsuyama and K. Ogawa, J. Appl. Phys. **75**, 7607 (1994).
[2] S. Pau, G. Bjork, J. Jacobson, H. Cao, Y. Yamamoto, Phys. Rev. **B51**, 7090 (1995).
[3] J. A. Feck, J. R. Morris, and M. D. Feit, Appl. Phys. **10**, ¡29 (1976).

CHARACTERIZATION OF LASER ABLATED GERMANIUM NANOCLUSTERS

S. Vijayalakshmi[*], F. Shen, Y. Zhang , M. A. George[#], and H. Grebel
Nonlinear Nanostructures Laboratory, Electrical and Computer Engineering
New Jersey Institute Of Technology, Newark, NJ-07102
[#] University Of Alabama at Huntsville, Huntsville, AL-35899
[*] vijaya@megahertz.njit.edu

ABSTRACT

Morphological characterization and the nonlinear optical properties of laser ablated germanium nanoclusters are discussed in this paper. Laser ablated films contain micron sized droplets that are composed of nanoclusters. The clusters have cubic and tetragonal symmetry. Results from TEM, AFM, and XRD are presented for the surface characterization. Nonlinear absorption is observed at very low light intensities and nonlinear absorption and refraction are seen at peak light intensities of 18 KW/cm^2 at λ = 532 nm. The estimated change in refractive index is 0.05 at these intensity levels.

INTRODUCTION

Pulsed laser deposition (PLD) is a well established technique for growing thin films of semiconductors. Large laser intensities are required for high deposition rates. The interaction of the intense laser beam with the target material leads to two distinct properties of the deposited films: (i) micron sized droplets are produced and (ii) the film may have a different crystal structure than the original target due to the nonequilibrium state of the plasma. The uneven surface created by the presence of large droplets is typically considered as a disadvantage and much effort has been aimed at reducing the droplet size the use of high density targets [1], change in deposition geometry [2], use of size filters [3], and introduction of gases into the evaporation chamber [4]. It has been reported that the crystal structure of the target and the films were not the same in the case of laser ablated Si films [5]. Ongoing investigation of the Si films indicate that they may be composed of other crystal structures such as hexagonal.

We present the properties of Ge films grown by PLD and demonstrate that a typical film is composed of tetragonal and cubic crystallites and is highly optically nonlinear. The films were characterized using tunneling electron and atomic force microscopy (TEM and AFM) while the crystallinity was studied using x-ray diffraction (XRD) and Raman spectroscopy. We report in this paper that the presence of these droplets, and hence the cluster-cluster interaction in the deposited films might play a key role in the enhancement of the nonlinear optical properties of laser ablated films. Si films grown by laser ablation show nonlinear index of refraction as high as 0.17 at light intensities of 30 KW/cm^2 [6]. For Ge nanoparticles, the nonlinear refractive index is found to be approximately 0.05.

EXPERIMENTAL

Films containing Ge nanoclusters were prepared by laser ablation technique. A KrF excimer laser beam (λ = 248 nm, average power, <I>= 3 W, pulse duration = 8 ns, repetition rate

Mat. Res. Soc. Symp. Proc. Vol. 536 © 1999 Materials Research Society

= 50 Hz, deposition time=10 min.) was focused to a 100 μm spot-size on a crystalline Ge (100) target. The ablated material was collected onto a glass substrate or a TEM copper grid placed 3 cm from the target. The ambient pressure in the chamber was maintained at 8 x 10^{-6} Torr. The substrates were maintained at room temperature. The size of each substrate was about 1.5cm x 1.5cm. Due to the directional nature of the plume, the deposited films had varying thickness [7]. The experiments reported here were conducted over a small region where the film was gray in color and the estimated effective thickness was 400 nm as measured by laser scanning microscopy (LSM).

Optical absorption of the sample was measured using a tungsten lamp-spectrometer arrangement. Nonlinear transmission of the sample at low incident light intensities were measured by focusing the selected wavelengths from a white light source on the sample and by varying the intensity using neutral density filters at various wavelengths. Nonlinear refraction and absorption at higher intensities were measured using a frequency doubled Nd:YAG laser (pulse width = 8 ns; repetition rate = 10 Hz) by Z-scan technique. In this, a sample is scanned through the focal point of a lens and the transmission changes are measured as a function of sample position. For nonlinear absorption measurements, transmitted and reflected light are collected by two detectors (open aperture z-scan) while for refraction, transmitted light is measured through a small aperture (closed aperture z-scan).

RESULTS

Crystalline Nanostructures In Films

Films were grown on a TEM grid and the typical electron diffraction pattern is shown in figure 1. No diffraction pattern was observed in the region between the droplets indicating amorphous nature of the film. This amorphous region was studied using AFM and the result for a 1.5 μm x 1.5 μm region is shown in figure 2. The AFM measurements were performed in contact mode using a Digital Instruments Nanoscope III system with a piezoelectric tube scanner of effective scan range from 0.5 to 15 μm and calibrated against standard gold colloids. As is evident from the figure, this region contains several nanometer sized structures. Sectional analysis of the region indicates nanostructures ranging from 3 nm to 10 nm. It should be pointed that the cluster size estimate was obtained by measuring the height of the clusters instead of the width to avoid tip convolution.

Film crystallinity was also measured by x-ray diffraction and a typical diffraction pattern is shown in figure 3. A Rigaku D/MAX-B system with CuKα source and thin film attachment was used for XRD measurements. The x-ray beam is approximately 1cm^2 and so the entire film is sampled in these experiments. As can be seen, the film shows long range order and is highly crystalline. Lines marked by A, E, F, H, I and J correspond to the (111), (220), (311), (400), (331), and (422) planes respectively for normal diamond structure of Ge. Other lines do not correspond to the cubic structure and may be a result of different crystallographic symmetry of the nanostructures.

Figure 1: Electron diffraction pattern at the edge of a droplet. This pattern indicates cubic symmetry.

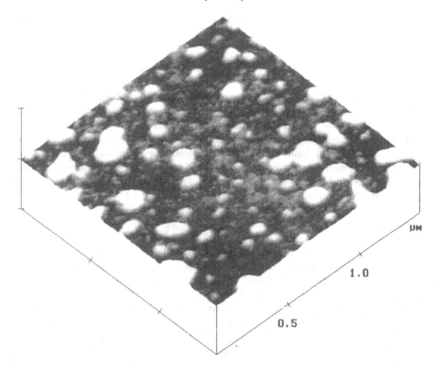

Figure 2: Nanostructures in between the droplets as seen by atomic force microscopy. The cluster heights were calculated to be 3-10 nm.

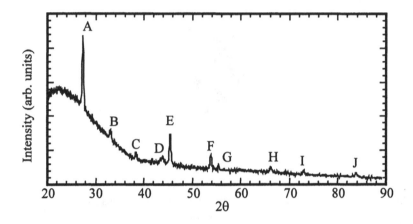

Figure 3: X-ray diffraction pattern of laser ablated Ge film. Refer to text for identification of the peaks

Optical Properties

Intensity dependent Transmission at wavelengths 500, 600, 700, and 800 nm are shown in figure 4. Transmission and reflection from the films were measured as a function of wavelength by attenuating the light using neutral density filters. Reflection of the sample at these wavelengths changed less than 5% as a function of intensity while transmission changes were observable as shown in figure 4. Figure 5 shows nonlinear absorption at higher intensities measured at $\lambda = 532$ nm by z-scan technique. The solid curve in figure 5 for transmission is the curve fitting obtained using linear approximation [6]. Under this approximation, the nonlinear absorption coefficient was calculated to be 1.3×10^4 cm^{-1}.

DISCUSSION

Crystallinity within the droplets is evident from the electron and x-ray diffraction patterns shown in figure 1 and 3. X-ray diffraction data shown in figure 3 indicates polycrystalline nature of the deposited films. Sato et. al [8] reported the formation of tetragonal phase of germanium in cluster beam evaporation method. Similar structure has also been verified in films grown by plasma enhanced chemical vapor deposition [9]. Since tetragonal structure is the high pressure form of Ge, most likely to occur in a nonequilibrium plasma created by laser ablation, we assign some of the unknown peaks to a tetragonal structure of Ge, namely peaks marked at B, C, and G corresponding to (201) or (112), (110), and (220) planes respectively. The d-spacing for peak marked as D does not correspond to any known structure of Ge. Additional confirmation of crystallinity and nanostructure in these films was obtained by Raman scattering experiments. Crystalline Ge shows Raman shift of 300 cm^{-1} with a line width

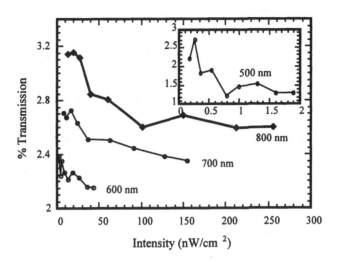

Figure 4: Intensity dependent transmission changes of the film at different wavelengths. Inset shows the expanded scale for 500 nm.

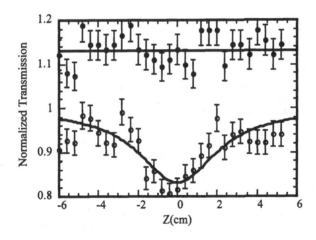

Figure 5: Nonlinear transmission (open circles) and reflection (solid circles) as a function of sample position from the focusing lens. Z = 0 indicates the focal point. Solid line shown through transmission data points is the curve fitting as in reference 5 while the line through reflection points are only guide for the eye.

of 3 cm^{-1} [10] while the films show a Raman shift of 298 cm^{-1} with corresponding line width of 6.6 cm^{-1}.

The electron diffraction pattern suggests cubic structure of the droplet. Figure 3 indicates a cubic structure coexisting with tetragonal structures. It must be noted that electron diffraction probes symmetry of a localized region (edge of a droplet) whereas the x-ray diffraction averages the entire film. As a result, XRD may exhibit additional structures within the film as compared to TEM data.

Laser ablated Ge films are highly nonlinear as can be seen from figures 4 and 5. At a peak intensity of 18 KW/cm^2, the nonlinear refractive index was calculated as $\Delta n = 0.05$. At $\lambda = 532$ nm (2.33 eV), Si exhibited negative refractive index (-0.17 at 30 KW/cm^2) while Ge has positive nonlinear change. Ge nanoclusters show large absorption near 2.5 and 4.5 eV [11]. Optical absorption of laser ablated Ge films also show an enhanced absorption around 2.2 eV. Based on these energy considerations, we believe that the large nonlinearity may be attributed to a resonant single photon absorption process. The nonlinearity may be further enhanced due to cluster-cluster interaction as in the case of Si.

CONCLUSION

In summary, based on the XRD and TEM measurements, we report a mixture of cubic and tetragonal phase for Ge films prepared by laser ablation. These films exhibit a large nonlinear refractive index, $\Delta n = 0.05$ at relatively low laser intensities. We propose that the enhancement of nonlinearity is due to resonance enhanced single photon absorption.

ACKNOWLEDGMENTS

We thank Prof. M. Libera, Stevens Institute Of Technology, Hoboken, NJ, for valuble discussions and analysis of TEM results.

REFERENCES

[1] R. K. Singh, D. Bhattacharya, and J. Narayan, Appl. Phys. Lett. **61**, 483 (1992)

[2] R. J. Kennedy, Thin Solid Films **214**, 223 (1994)

[3] R. Camata, H. A. Atwater, K. J. Vahala, and R. C. Flagan, Appl. Phys. Lett. **68**, 3162 (1996)

[4] T. Yoshida, S. Takeyama, Y. Yamada, and K. Mutoh Appl. Phys. Lett. **68**, 1772 (1996)

[5] S. Vijayalakshmi, Z. Iqbal, M. A. George, J. Federici, and H. Grebel, To appear in Thin Solid Films

[6] S. Vijayalakshmi, H. Grebel, Z. Iqbal, and C. W. White, J. Appl. Phys. to appear in Dec. 15, 1998 issue.

[7] F. Antoni, C. Fuchs, and E. Fogarassy, Appl. Surf. Sci. **96-98** 50 (1996)

[8] S. Sato, S. Nozaki, and H. Morisaki, Appl. Phys. Lett. **72** 2460 (1998)

[9] J. Jiang, K. Chen, X. Huang, Z. Li, and D. Feng Appl. Phys. Lett. **65** 1799 (1994)

[10] J. Gonzalez-Hernandez, G. H. Azerbayejani, R. Tsu, and F. H. Pollak, Appl. Phys. Lett. **47**, 1350 (1985)

[11] P. Tognini, L. C. Andreani, M. Geddo, A. Stella, P. Cheyssac, R. Kofman, and A. Migliori, Phys. Rev. B 53, 6992 (1996)

THEORETICAL INVESTIGATION OF EFFECTIVE QUANTUM DOTS INDUCED BY STRAIN IN SEMICONDUCTOR WIRES

KENJI SHIRAISHI, MASAO NAGASE, SEIJI HORIGUCHI, HIROYUKI KAGESHIMA
NTT Basic Research Laboratories, Atsugi-shi, Kanagawa 243-0198, Japan.
siraisi@will.brl.ntt.co.jp

ABSTRACT

A new fabrication technique for quantum dots is proposed on the basis of theoretical studies. It is known that the self-limiting phenomena of silicon oxidation are governed by the large stress induced by the volume expansion. We estimated the change in the band parameter of the silicon under the stress induced by the oxidation. The results show that the band-gap of the silicon considerably decreases under two-dimensional compressive strain. Accordingly, introducing strain to a small part of a Si wire results in the formation of a "strain-induced effective quantum dot". Moreover, we also propose theoretically a new procedure for fabricating a silicon single electron transistor by the combination of silicon oxidation and nitridation techniques.

INTRODUCTION

Quantum dot structures have attracted much attention in the field of mesoscopic physics as well as in general technological fields, since these structures involve a wide variety of interesting phenomena. A lot of studies were investigated for the fabrication techniques of quantum dot structures [1-3]. Recently, Tarucha et $al.$ succeeded in fabricating artificial atoms and molecules based on the combination of the GaAs epitaxial growth and lithography techniques and reported the Hund rule in the artificial atoms [1, 2]. Another technique is based on the application of self-organization phenomena. Leonard et $al.$ [3] reported the fabrication of self-organized InAs quantum dots in GaAs using the 3-dimensional growth mode of InAs/GaAs(100) heteroepitaxy. These techniques are based on the actual formation of dot-shaped structures in the host semiconductor, which has the wider band-gap than the dot. In this paper, we investigate the possibility of a new technique for fabricating effective quantum dots instead of actually synthesizing dot-shaped structures.

It has been reported that the critical thickness of heteroepitaxy is around 10-nm in the lattice-mismatched system of $Si_{0.5}Ge_{0.5}/Si(100)$ [4]. Therefore, large strain can be applied to the system by using the lattice-mismatched interface, provided the strained parts of the system is sufficiently small. Our technique is based on the strain effect induced by silicon oxidation. Oxidation occurs at the largely lattice-mismatched Si/SiO_2 interface, and the volume expansion of newly grown SiO_2 also generates large stress. This oxidation-induced stress results in the interesting phenomena of self-limiting oxidation, and recent first-principles calculations support the crucial role of the stress in oxidation processes [5]. Liu et $al.$ [6] studied the self-limiting phenomena of silicon oxidation in detail. They reported that two-dimensional compressive stress at the Si/SiO_2 interface is expected to be larger than 20000 atm, when the 10-nm-diameter Si wires are formed from 100-nm-diameter Si pillars by self-limiting oxidation [6].

In this paper, we estimate the change in band parameter, such as band-gaps and electron effective masses, caused by the strain using the first-principles band structure calculations, and investigate the possibility of the formation of a strain-induced quantum dot in a Si quantum wire. Moreover, we also propose a new procedure for fabricating a silicon single electron transistor by adopting our proposed strain-induced quantum dots with the technique of the Si oxidation and nitridation.

THEORY

The band structure calculation was performed by the local density functional formalism [7, 8]. The first-principles pseudopotential method was used [9, 10] and the electron exchange correlation potentials were approximated by the Ceperley-Alder potentials parametrized by Perdew and Zunger [11, 12]. The pseudo-wavefunctions were expanded by the planewave basis set (cut-off energy: 9.61 Ry), and the Brillouin zone integral is performed by the 512 k points.

Using the band parameters obtained from the first-principles calculations, we estimated the electron effective potentials in the Si quantum wires under the strain. In the electron effective potential calculations, the effects of six conduction band valleys are systematically involved within the effective mass approximation [13].

RESULTS

First, we estimated the strain induced by the silicon oxidation. When the two-dimensional compressive stress of 20000 atm is applied perpendicular to the [010] direction, which is the reported value of the oxidation induced stress in the self-limiting phenomena [6], Si crystal is two-dimensionally compressed by about 1% in the (010) plane. This is because the Si Young modulus is $\sim 1.30 \times 10^{12}$ dyne/cm^2. Due to the atomic position change caused by the strain, the fundamental band parameters are modulated.

We investigated the calculated electron effective masses and band-gaps as a function of the two-dimensional compressive stress perpendicular to the [010] direction. In this paper, we call the six Si conduction band valleys in the $\pm k_x$, $\pm k_y$ and $\pm k_z$ directions the [$\pm$a,0,0], [0,$\pm$a,0], and [0,0,$\pm$a] valleys, respectively. Although the changes in the calculated longitudinal and transverse masses are both small under the two-dimensional compressive strain less than 1%, the band-gap values are considerably modulated by the strain as seen in Fig. 1 (a) and (b). This can be explained as follows. The six-folded conduction valleys are split into four-folded valleys and two-folded valleys by the symmetry breaking due to the two-dimensional compressive strain perpendicular to the [010] direction. The four-folded valleys originate from the [$\pm$a,0,0] and [0,0,$\pm$a] valleys, and the two-folded valleys originate from the [0,$\pm$a,0] valleys. When the two-dimensional compressive strain is 1%, the band-gap values of the [$\pm$a,0,0] and [0,0,$\pm$a] valleys decrease by about 0.13 eV. On the other hand, the band-gap values of the [0,$\pm$a,0] valleys increase by about 0.06 eV. This symmetry breaking is the primary origin of the considerable band-gap-shrinkage. However, the change in the relative band dispersion just around the conduction band bottom is much smaller than that in the absolute values of band gaps. This is the reason why the change in the effective mass is smaller than the absolute band gap values. It is also noticeable that the above band-gap change can be approximated by the linear functions of the strain, when the magnitude of the strain is within 1%. In summary, the main band-gap values considerably decrease when the oxidation-induced two-dimensional stress is applied.

According to strain-induced-band-gap-shrinkage, we propose a new method by which quantum dots can be fabricated by using the oxidation-induced strain without actually synthesizing dot-shaped structures. First, we prepare a Si quantum wire along the [010] direction. Next, we partially oxidize this quantum wire. As a result, oxidation-induced compressive strain is applied to only the oxidized part. Accordingly, the strain-induced effective quantum dot formation is expected, although the actual dot-shaped structure is not fabricated.

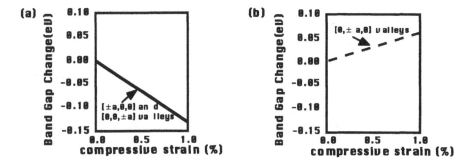

Fig. 1: Obtained band gaps as a function of the compressive strain perpendicular to the (010) direction. (a) Band-gap change of the [±a,0,0] and [0,0,±a] valleys. (b) Band-gap changes of the [0,±a,0] valleys.

The calculated electron effective potentials of this strain-induced effective quantum dot are shown in Fig. 2. Figure 2(a) indicates the electron effective potential of the strain-induced effective quantum dot obtained from a 10-nm-diameter quantum wire, and Fig. 2(b) describes the effective potential obtained from an elliptic quantum wire with 10-nm and 5-nm diameters. The zero value corresponds to the conduction band bottoms of bulk Si. In these calculations, all conduction valley effects are included within the effective mass approximation [13]. For simplicity, it is assumed that the oxidation induced strain is distributed with Gaussian shapes with the standard deviation of 7-nm around the maximum compressive strain of 1% along the [010] wire direction. As the conduction band offset ratio, we used the value of 75%, which was obtained for the conduction band offset value of the hypothetical superlattice of (hexagonal-diamond Si)/(cubic-diamond Si) by the first-principles calculations [14]. The calculated resultsclearly show that the partially oxidized parts of Si quantum wires act as a quantum dot because of the strain-induced-band-gap-shrinkage. It is noticeable that the bottoms of effective potentails are located lower than the bulk Si conduction band bottoms and that the unoxidized parts of the Si wire act as potential barriers. In the 10-nm-diameter circular Si wire, the potential bottom is located about 71 meV lower than the bulk Si conduction band bottom. In the elliptic wire with 10-nm and 5-nm diameters, the quantum dot level is 57 meV lower than the bulk value. The calculated potential barriers are both 98 meV, which corresponds to the electron band offset value. Moreover, these results also indicate that the strain-induced part can act as a quantum dot, even when it becomes considerably thinner by oxidation.

Finally, using the above concept of the strain-induced effective quantum dots, we propose a fabrication procedure for a Si single electron transistor based on the combination of Si oxidation and nitridation techniques. The procedure is schematically illustrated in Figs. 3(a)-(f). First, a SOI substrate is prepared [Fig. 3(a)]. Next, a Si wire is fabricated from the SOI substrate by

(a)　　　　　　　　　　　　　　　**(b)**

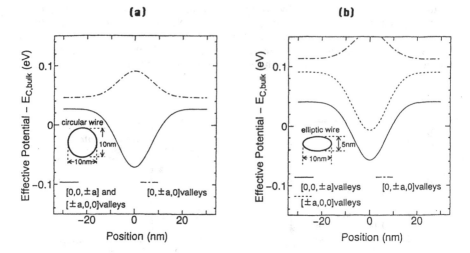

Fig. 2: Calculated effective potentials of electrons. (a) The electron effective potential obtained from a 10-nm-diameter quantum wire. (b) The electron effective potential obtained from a elliptic-shaped quantum wire with 10-nm and 5-nm diameters. The zero value corresponds to the conduction band bottom of bulk Si.

lithography [Fig. 3(b)]. Then, we deposit a nitride layer with a window, and the window is located at the center of the Si wire [Fig. 3(c)]. After the oxidation, only the small part of the Si wire under the nitride layer window, can be oxidized. Accordingly, compressive strain is applied only to the oxidized part [Fig. 3(d)]. After the removal of the nitride layer [Fig.3(e)], we obtained the band profile, which is described in Fig. 4. As seen in Fig. 4, unoxidized parts of the Si-wire act as a barrier for electrons. On the other hand, the oxidized part acts as a well. Finally, we attach a source and a drain electrode to the wide Si parts, and a poly-Si gate to a strain-induced quantum dot [Fig. 3(f)]. The obtained structure is the well-known Si single electron transistor (SET) that contains a strain-induced effective quantum dot. Since the procedure is based on the well-established silicon large scale integration (LSI) technologies and because the effect of stress on oxidation is actually used in pattern dependent oxidation (PADOX) [15], we believe this strain-induced SET fabrication procedure is not unrealistic. Moreover, the fabrication of the single electron devices with multiple dots (single electron pumps, single electron turnstiles) can also be realized by only increasing the number of nitride windows.

CONCLUSION

A new fabrication technique for quantum dots has been proposed on the basis of theoretical studies. We estimated the change in the band parameters of the silicon under large stress induced by the self-limiting phenomena of oxidation. The calculated results show that the band-gap of the silicon considerably decreases under the two-dimensional compressive strain. Accordingly, introduction of strain to a small part of a Si wire results in the formation of a "strain-induced effective quantum dot". This is also confirmed by the effective mass approximation that included the multiple conduction band valleys. Moreover, we also propose

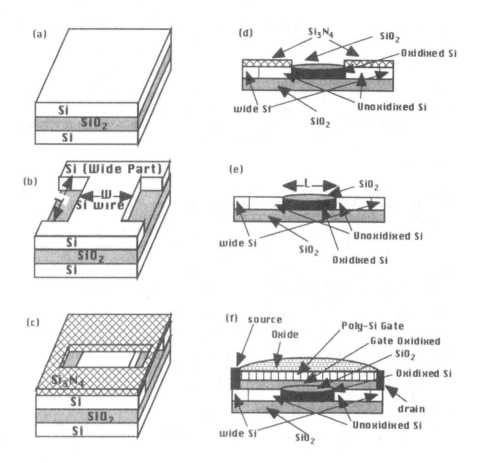

Fig. 3: The fabrication procedure for a Si single electron transistor. (d)~(f) are cross sections.

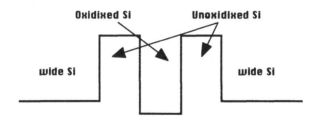

Fig. 4: The obtained band profiles of a partially oxidized Si quantum wire in Fig.3.

theoretically a new procedure for fabricating silicon single electron transistors by the combination of silicon oxidation and nitridation techniques.

ACKNOWLEDGEMENT

We would like to thank Dr. Katsumi Murase, Dr. Yasuo Takahashi, Dr. Yukinori Ono, Dr. Akira Fujiwara, Dr. Masashi Uematsu, and Dr. Hiroki Hibino for their stimulating discussions. We also acknowledge Dr. Takaaki Mukai and Dr. Takahiro Makino for their continues encouragement and fruitful comments throughout this work.

REFERENCES

[1] S. Tarucha et al. Phys. Rev. Lett., 77, 3613 (1996).

[2] T. Fujisawa, T. H. Oostrkamp, W.G. van der Wiel, B. W. Broer, R. Aguado, S. Tarucha, and L. P. Kouwenhoven, Science, 282, 932 (1998).

[3] D. Leonard, K. Pond, and P. M. Peroff, Phys. Rev. B, 50 11687 (1994).

[4] J. C. Bean, L. C. Feldman, A. T. Fiory, S. Nakahara, and I. K. Robinson, J. Vac. Sci. Technol. A 2, 436 (1984).

[5] H. Kageshima and K. Shiraishi, Phys. Rev. Lett., 81, 5936 (1998).

[6] H. I. Liu, D. K. Biegelsen, N. M. Johnson, F. A. Ponce, and R. F. W. Pease, J. Vac. Sci. Technol. B 11, 2532 (1993).

[7] P. Hohenberg and W. Kohn, Phys. Rev., B 136, 864 (1964).

[8] W. Kohn and L. J. Sham; Phys. Rev., A 140, 1133 (1965).

[9] J. Ihm, A. Zunger and M. L. Cohen , J. Phys. C 12, 4409 (1979).

[10] D. R. Hamann, M. Schluter and C. Chiang, Phys. Rev. Lett., 43, 1494 (1979).

[11] D. M. Ceperley and B. J. Alder; Phys. Rev. Lett., 45, 566 (1980).

[12] J. P. Perdew and A. Zunger, Phys. Rev., B 23, 5048 (1981).

[13] S. Horiguchi, unpublished.

[14] M. Murayama and T. Nakayama, J. Phys. Soc. Jpn., 61, 2419 (1992).

[15] Y. Takahashi, H. Namatsu, K. Kurihara, K. Iwadate, M. Nagase, K. Murase, and M. Tabe, IEEE Trans. Electron. Devices, 43, 1213 (1996).

ELECTRON BEAM EXCITED PLASMA CVD FOR SILICON GROWTH

K. Okitsu[1], M. Imaizumi[1], T. Ito[2], K. Yamaguchi[1], M. Yamaguchi[1], T. Hara[1], M. Ban[3], M. Tokai[3], and K. Kawamura[4]

1) Toyota Technological Institute, 2-12-1 Hisakata, Tempaku-ku, Nagoya 468, Japan
2) Toyota Central R&D Labs., Inc., Nagakute, Aichi 480-11, Japan
3) Kawasaki Heavy Industries Inc., 118 Futatsuzuka, Noda, Chiba 278, Japan
4) Chubu Electric Power Co., Inc., 20-1 Kitasekiyama, Ohdaka, Midori-ku, Nagoya 459, Japan
e-mail: oki2@toyota-ti.ac.jp

ABSTRACT

Electron beam excited plasma (EBEP) CVD is a novel fabrication route for poly Si. Deposition was carried out on Si and SiO_2 layer from pure SiH_4 without hydrogen dilution. Crystalline silicon (μc-Si:H) films were made with electron acceleration voltage, discharge current, source gas flow rate, chamber presser, substrate temperature varied systematically. Average grain size was about 10 nm. Crystalline ratio was up to 0.7 at the maximum. The films contain about 19 at% hydrogen in spite of no dilution. It is considered that EBEP supplies much about atomic hydrogen due to the high decomposability of the source gas.

INTRODUCTION

Recently, high density plasma methods have begun to be utilized for making polycrystalline silicon[1]. Electron beam excited plasma (EBEP) CVD is one of the novel methods of CVD[2-4]. The reason why we utilize EBEP is as follows; 1) Since the gas molecules have high ionization efficiency in high energy region (near 100 eV), they could be decomposed more efficiently compared with conventional plasma. 2) According to the principle of generation, we can control the plasma directly and well. We can set the electron accelerating voltage, the electron beam current in addition to the depositional conditions such as flow rate of SiH_4 and chamber pressure. Furthermore, the distance between substrate and plasma can be changed.

Another attractive feature is that the plasma is formed independent of electric or magnetic field. We can avoid acceleration to the substrate by electric and magnetic power, then EBEP-CVD is expected to grow good quality layer with much lower level of damage.

EXPERIMENTAL

Figure 1 is a schematic drawing of the EBEP-CVD system used in this study. The electron beam accelerated by about 100V was focused and introduced to the reactive ion chamber from the bottom. This generates the plasma by the collision with gas molecules. In our system, light emission was observed throughout the chamber via the window. The chamber height is about 60 cm.

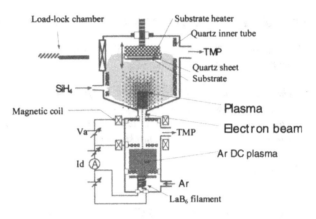

Figure 1. The schematic drawing of the EBEP-CVD system used in this study. The electron beam is introduced to the reactive chamber from the bottom.

The substrates used were 4-inch Si (100) and SiO_2 layer, which was thermally oxidized with 1μm thickness. They were cleaned with $H_2SO_4 + H_2O_2$ (= 4:1) at ~ 90 °C for 10 min, and dipped in diluted HF solution for 1 min. They were set near the ceiling of chamber, and faced downward to prevent falling of the dust. They were heated over the quartz tube plate. Their temperature (T_s) was varied from RT to 420 °C, which was measured, with thermocouple at the surface in the absence of plasma.

Experimental conditions were varied as shown in table I. SiH_4 was only used as source gas. As a plasma condition, electron acceleration voltage (V_a), the discharge current (I_d) which affect the electron beam current, the flow rate of SiH_4 (F) and the chamber pressure (P) were changed. We regarded the case of F=5.0sccm, P=3.5Pa, Ts=420, Va=100V, Id=15A as "standard condition". The deposition time was always 60 min.

Table I. Experimental condition investigated in this study. The marked are standard condition. ("sccm" is standard cc/min)

Parameter	Value
I_d : Discharge Current (A)	10, 15*, 20
V_a : Acceleration Voltage (V)	80, 90, 100*
P : Deposition Pressure (Pa)	2.0, 3.5*, 5.0, 6.5
F : SiH_4 Flow Rate (sccm)	2.5, 5.0*, 7.5, 10.0
T_s : Substrate Temperature (°C)	RT, 240, 300, 340, 380, 420*
T : Deposition Time (min)	60*

The grown films were analyzed by Raman scattering spectroscopy using an Ar⁺ ion laser with a wavelength of 488.0 nm. The crystalline properties, crystalline fraction and amorphous fraction and the grain size were characterized by Raman spectroscopy. The hydrogen content in the film was measured by the elastic recoil detection analysis (ERDA) with 2 MeV He⁺ ion beam [5].

RESULTS

The thickness of the silicon layer deposited under the standard condition was about 0.35 μm. Figure 2 shows the typical spectrum of the Si film deposited on a Si wafer under the standard condition. The peak position is located at the wavenumber of 520 cm^{-1}, which indicates that the film is crystalline. The component of the spectra caused by amorphous silicon was detected at 480 cm^{-1}, relatively weak. The crystalline ratio estimated by the ratio of the crystal peak intensity to the total is 0.7. The grain size of films were 10 nm evaluated from the spread of the Raman scattering spectra according to the equation by Sui *et al.* [6].

As mentioned in Table I, deposition conditions were varied systematically, while other parameters were kept fixed in each case. As results, SiH₄ flow rate and substrate temperature affected the deposition rate and characteristics of the films.

Figures 3 show the effects of the flow rate. As SiH₄ flow rate increased, deposition rate was increased accordingly. The highest deposition rate was about 0.3 nm/s when SiH₄ flow was 10 sccm. Judging from the deposition rate curve, the deposition is likely to be saturated above 10 sccm. As shown in figure 3-(2), the Raman scattering peaks were appeared around 520cm^{-1} on Si and SiO₂ layer in all samples, so all films were crystalline, grain size decreases from 10 to 3.5 as the flow rate increased, figure 3-(3).

If the substrate temperature was varied, the quality of film was changed from crystalline to amorphous shown in figures 4. When the temperature was from 300 to 420°C, the film

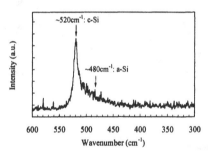

Figure 2. Typical Raman scattering spectrum from silicon film grown by EBEP-CVD under the standard condition.

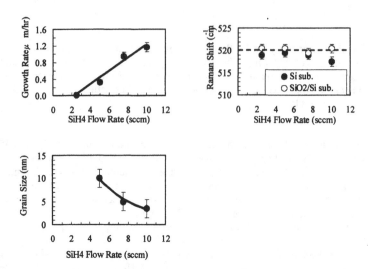

Figure 3. Effects of the SiH₄ flow rate on deposition rate and characteristics of the film.

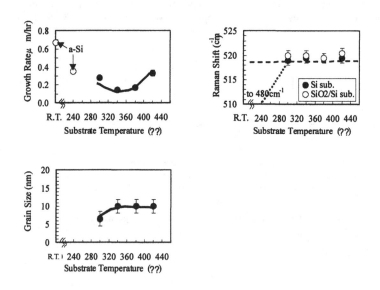

Figure 4. Effects of the substrate temperature on deposition rate and characteristics of the film.

was crystalline because Raman peak appeared in 520 cm⁻¹. Deposition rate was faster as the temperature was from 340 to 420 °C. On the other hand, in the cases of that the temperature was 240 °C or R.T., Raman peaks were at 480cm⁻¹. Then they were amorphous. These two results were the only cases, where amorphous layers were formed in all experiments. Critical temperature for microcrystalline Si deposition is 300 °C. In addition, the deposition rate in this region was increased by lowering the temperature. It is considered that etching of the amorphous components was done in high temperature.

Concerning the principle of the EBEP-CVD, electron acceleration voltage was decreased to 80V. The results were shown in figures 5. Making the voltage lower, deposition rate was increased from 0.35 to 0.52 (um/h). Although, it was not so much compared with the change of SiH₄ flowing rate. Layers were all highly crystalline. For the SiH₄ gas, electron's collision cross-section is maximum about 70 eV, and begin to decrease over there. Therefore, this result would be caused by the nature of the source gas.

While, when discharge current was varied from 10 to 20A, deposition rate was decreased little as increasing the current. Deposition films were all crystalline. It is concluded that the deposition is enough at 10A.

The content of hydrogen in the grown films were measured by ERDA. It was found that films contain as much as 19 at% hydrogen in spite of no dilution with H₂. From this result, the films grown by EBEP-CVD would have much hydrogen, because of high efficiency for decomposition. This must be the reason why crystalline silicon films were obtained in wide ranges of plasma condition.

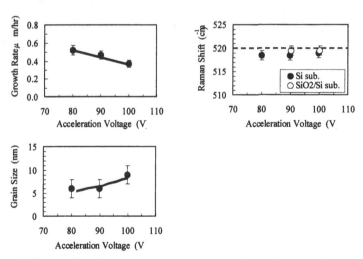

Figure 5. Effects of the electron acceleration voltage on deposition rate and characteristics of the film.

CONCLUSIONS

Electron beam excited plasma (EBEP) CVD was carried out for silicon deposition from 100% SiH_4 without hydrogen dilution on Si and SiO_2/Si substrate. Nano-crystalline silicon was made under various conditions except for cases of too low temperature (<300 °C). The grain size of the obtained films was about 10nm. The films contained about 19 at% hydrogen. Decomposition of the SiH_4 was sufficient enough in EBEP, to grow polycrystalline silicon.

ACKNOWLEDGEMENTS

We would like to thank Mr. Konomi of TCRL for ERDA measurement, to estimate the hydrogen content in the film. This work was performed under the Private University High-Tech. Research Center Program by the Ministry of Education of Japan for TTI.

REFERENCES

[1] M. Rostalsky, T. Kunze, N. Linke and J. Muller, IEEE P.V. Specialists Conference 1997, 743 (1997)
[2] T. Hara, M. Hamagaki, A. Sanda, Y. Aoyagi, S. Namba, J. Vac. Sci. Technol. **B5**, 366 (1987).
[3] T. Ohgo, T. Hara, M. Hamagaki, K. Ishii, M. Otsuka, J. Appl. Phys. **70**, 4050 (1991).
[4] T. Ito, I. Konomi, K. Okitsu, M. Imaizumi, K. Yamaguchi, M. Yamaguchi, T. Hara, M. Ban, M. Tokai, K. Kawamura, 2ed World Conf. and Exhibi. on PV Solar Energy Conversion (1998).
[5] M. Hamagaki, T.Hara, Y.Sadamoto, and T.Ohgo, Rev. Sci. Instrum. **66**, 3469 (1995)
[6] Z. Sui, P. P. Leong, I. P. Herman, G. S. Higashi, H. Temkin, Appl. Phys. Lett. **60**, 2086 (1992).

PRESSURE INDUCED STRUCTURAL TRANSFORMATIONS IN NANOCLUSTER ASSEMBLED GALLIUM ARSENIDE

S. KODIYALAM[1], A. CHATTERJEE[1], I. EBBSJÖ[2], R. K. KALIA[1], H. KIKUCHI[1], A. NAKANO[1], J. P. RINO[3], P. VASHISHTA[1]
[1]Concurrent Computing Laboratory for Materials Simulations, Dept. of Physics and Astronomy and Dept. of Computer Science, Louisiana State University. [2]Studsvik Neutron Research Laboratory, SWEDEN. [3]Universidade Federal de São Carlos, BRAZIL.

ABSTRACT

Pressure induced structural phase transformation in nanocluster assembled GaAs is studied using parallel molecular dynamics simulations in the isothermal-isobaric ensemble. In this system the spatial stress distribution is found to be inhomogeneous. As a result structural transformation initiates in the high stress regions at the interface between clusters. Structural and dynamical correlations in the nanophase system are characterized by calculating the spatially resolved bond angle and pair distribution functions and phonon density of states and comparing them with those for a single cluster and bulk crystalline and amorphous systems.

INTRODUCTION

Semiconductor nanocrystals show a variety of size dependent electronic and physical properties [1]. For example, 4-nm CdSe nanocrystals show a transformation from wurtzite to rock salt structures at a pressure of 6.7 GPa - much higher than the bulk transformation pressure of 2.7 GPa [1]. In this paper we investigate the initial stages of a high-pressure transformation in nanocluster assembled GaAs. Structural and dynamic correlations in the nanophase are computed and compared with those in an isolated nanocluster and bulk crystalline and amorphous GaAs systems.

SIMULATION PROCEDURE

Classical molecular dynamics simulations are carried out using an empirical potential. The potential has two and three body terms, with two body terms representing steric repulsion, screened Coulomb, charge-dipole and induced dipole-dipole interactions and three body terms representing bond bending and stretching forces. The potential is fitted to the experimental lattice constant and elastic moduli of crystalline (zinc blende) GaAs. With this potential the calculated pressure induced structural phase transformations in the bulk and the static structure factor of amorphous GaAs agree well with experiment [2]. Simulations are performed in the microcanonical (NVE - constant number of particles N, constant volume V and constant energy E) ensemble or in the isothermal (constant temperature T)-isobaric (constant pressure P) ensemble (NPT) [3]. We use recently proposed reversible integration schemes to propagate the equations of motion [4,5]. Parallel computation is based on a spatial decomposition scheme [6].

The initial configuration of the nanophase system, consisting of 1.6 million atoms, is constructed by randomly assembling 32 randomly oriented copies of a stoichiometric cluster of radius 65 Å that is thermalized at 1200 K. The assembling procedure avoids overlapping clusters - as a result the initial density of the system is only half the zero temperature crystalline density. The cluster is constructed by initially cutting a sphere out of the crystal (zinc blende) and then eliminating atoms with a coordination smaller than two. Overall stoichiometry is also maintained. It is then thermalized at 5 K by initializing and scaling velocities for 2000 integration time steps (one time step =2.18 fs) and then allowed to evolve at constant energy (NVE) for another 2000 steps. Conjugate gradient routine is then used to obtain a minimum energy configuration. Subsequently the temperature is reset to 5 K and increased to 300 K in 1000 steps using velocity scaling. It is then thermalized at 300 K by letting it evolve in NVE for 5000 steps. The same procedure of raising the temperature followed by thermalization is repeated in steps of 300 K until a temperature of 1200 K is reached.

The nanophase system is subjected to external hydrostatic pressure through NPT dynamics. The pressure is set to 0.4 GPa and the system run for 1000 steps using a modified NPT dynamics. Subsequently standard NPT simulation is carried out for another 5000 steps. The pressure is then increased to 2.5 GPa and a similar procedure is followed. Subsequent simulations are carried out using only 5000 steps of standard NPT dynamics at each value of external pressure: 5 GPa, 7.5 GPa and 10 GPa.

The nanophase system is also studied after reducing the external pressure to zero for comparing it with bulk systems and an isolated nanocluster. Such a "consolidated" system is constructed using the configuration that has experienced a maximum external pressure of 2.5 GPa. Before reducing the pressure, the temperature of the nanophase system is reduced to 300 K in steps of 300 K. The pressure is then reduced to 1 GPa after which it is reduced to 0.16 GPa. At each new temperature and pressure the simulation is carried out using NPT dynamics for 2000 steps. The modified NPT dynamics is used only during the reduction of pressure to 0.16 GPa. Finally, NVE simulation is carried out for 2500 steps.

For comparison with the consolidated nanophase system, a bulk crystalline system (of 10^5 atoms) at 300 K and zero pressure is constructed and run for 5000 steps in the NPT ensemble. Similarly, an isolated cluster is studied after thermalizing it at 300 K. Its construction follows the previously outlined procedure for cluster preparation - except that the increase in temperature is carried out over a much longer time period of 20000 steps. Finally, an amorphous bulk system (of 8000 atoms) at 300 K is prepared by quenching liquid GaAs at 3200K. The quenching schedule spanned a total of ~112000 steps (one time step =5 fs) and involved reducing the temperature (using velocity scaling) by a factor ~0.6-0.7 every 16000 steps - 6000 steps of which corresponded to thermalization (NVE simulation) at the reduced temperature.

RESULTS

Atomic coordination in a slice of the nanophase system under a pressure of 5 GPa is shown in Fig. 1. It is defined using a cutoff of 3.5 Å (~ average distance between the 1st and 2nd neighbors in the zero temperature zinc blende lattice). Structural transformation initiates at the interface between clusters. In these regions the coordination number of atoms is higher than that in the zinc blende lattice (4) and in some cases is equal to 6 corresponding to the coordination in the rock salt structure.

Figure 2 shows the spatial distribution of internal virial pressure for the system shown in Fig. 1. The pressure is inhomogeneous and shows large deviations from the external (average) pressure of 5 GPa. Larger stresses occur at the cluster-cluster interfaces which have higher coordinated atoms. This correlation between coordination and pressure suggests that the structural transformation initiates at the cluster-cluster interfaces due to the higher stresses in these regions.

We have calculated bond-angle distributions and pair-distribution functions in nanophase GaAs. As shown in Fig. 3(a) a small shoulder appears in the bond-angle distribution away from the 109° angle of the zinc blende lattice. To separate the effects of pressure on different regions in the nanoclusters, spatially resolved distributions are constructed by dividing the clusters into 10 Å wide shells around their respective centers of mass, with the 7th shell including all the atoms beyond 60 Å. The bond-angle distributions constructed in this manner are shown in Fig. 3(b). The presence of a shoulder in the distribution corresponding to the shell inside the clusters shows the onset of structural transformation even in the interior of the nanoclusters. The structure of the intercluster regions (defined as corresponding to shell 7) is also investigated by computing the spatially resolved Ga-As partial pair distribution function for the consolidated nanophase system. As seen in Fig. 4 there is a broadening of peaks relative to the crystalline system - similar to that in the isolated cluster and amorphous GaAs.

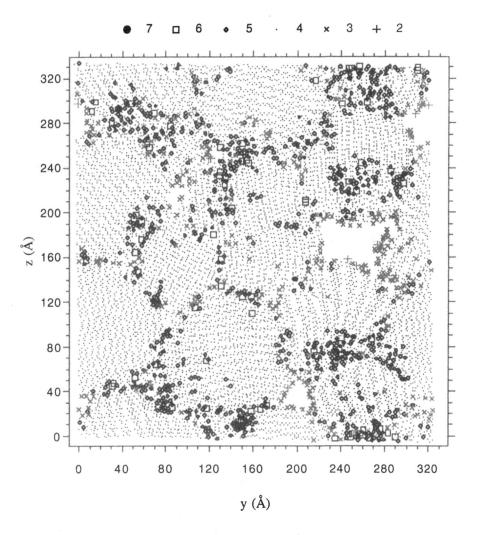

Fig. 1. Coordination number of atoms in a slice of the nanophase system at a pressure of 5 GPa. Individual clusters may be identified from the unique orientation of the 4-fold coordinated atoms. Cluster-cluster interfaces show evidence of structural transformation as they have higher coordinated atoms.

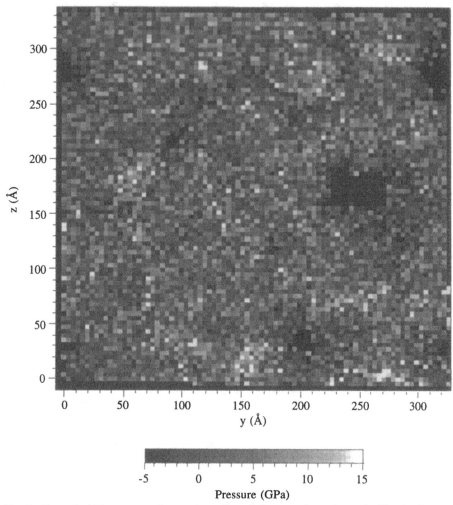

Fig. 2. Internal virial pressure of atoms in a slice of the nanophase shown in Fig. 1. Larger stresses at the cluster-cluster interfaces are correlated with the presence higher coordinated atoms in the same regions.

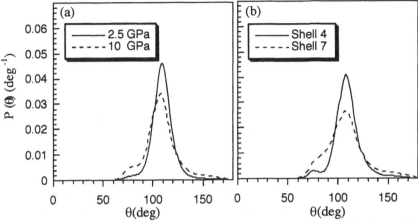

Fig. 3. (a) Normalized bond angle distribution in the nanophase system at 2.5 GPa and 10 GPa. The shoulder in the distributions indicates the initial stages of a structural transformation of the zinc blend lattice (with bond angle 109°). (b) Spatially resolved distributions (shell 4 extends between 30 Å to 40 Å radius within the clusters while shell 7 includes all atoms beyond 60 Å) for the nanophase system at 10 GPa. The transformation begins to occur even in the interior of the nanoclusters.

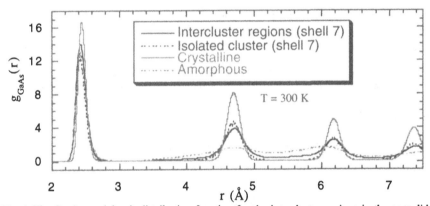

Fig. 4. The Ga-As partial pair-distribution function for the intercluster regions in the consolidated nanophase system is broadened relative to that in the crystalline bulk. Similar broadening is observed in the amorphous system and the outermost shell of the isolated cluster.

We have also determined the phonon density of states for nanophase GaAs. It is computed as the Fourier transform of the velocity autocorrelation function. As can be seen in Fig. 5, compared to the crystalline bulk, the intercluster regions of the consolidated nanophase system show broadened peaks similar to the isolated cluster and amorphous system. This is due to the breaking of symmetries that lead to degeneracies in the ideal crystal. Enhancement at low energies (2 - 7 meV) is due to the presence of floppy modes in the amorphous system and interfacial modes in the nanophase system.

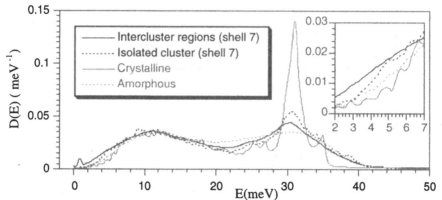

Fig. 5. Normalized phonon density of states (DOS) per degree of freedom for the consolidated nanophase, isolated cluster, crystalline GaAs and, amorphous GaAs. The inset shows the DOS at low phonon energies.

CONCLUSIONS

Initial stage of a pressure induced structural phase transformation has been studied in nanocluster assembled GaAs. Stress distributions are found to be inhomogeneous - greater stresses at the cluster-cluster interface are correlated with the presence of higher coordinated atoms in these regions. Structural and dynamic correlations of the intercluster regions in the consolidated nanophase system are similar to those in the outermost shell of an isolated cluster and those in amorphous GaAs.

ACKNOWLEDGMENTS

This work has been supported by NSF, DOE, AFOSR, USC-LSU MURI, ARO, PRF, NSF USA-Japan International Grant, and LEQSF. The authors thank Dr. Timothy Campbell for useful discussions.

REFERENCES

1. A.P. Alivisatos, MRS Bulletin, February 1998, pp. 18-23, and references therein.

2. J.W. Besson, J.P. Itie, A. Polian, G. Weil, J.L. Mansot, and J. Gonzalez, Phys. Rev. B **44**, p. 4214 (1991).

3. M. Parrinello and A. Rahman, J.Appl.Phys. **52**, p. 7182 (1981).

4. M. Tuckerman, B.J. Berne, and G.J. Martyna, J.Chem.Phys. **97**, p. 1990 (1992).

5. G.J. Martyna, M.E. Tuckerman, D.J. Tobias, and M.L.Klein, Mol.Phys. **87**, p. 1117 (1996).

6. D.C.Rapaport, Comput.Phys.Commun. **62**, p. 217 (1991).

STUDY OF ELECTRICAL PROPERTIES OF Ge-NANOCRYSTALLINE FILMS DEPOSITED BY CLUSTER-BEAM EVAPORATION TECHNIQUE

Souri Banerjee[+], H. Ono, S. Nozaki and H. Morisaki
Department of Communications and Systems, University of Electro-Communications, 1-5-1 Chofugaoka, Chofu-shi, Tokyo-182
[+] souri@cas.uec.ac.jp

ABSTRACT

Room temperature current-voltage (I-V) characteristics were studied across the thickness of the Ge nanocrystalline films, prepared by the cluster beam evaporation technique. The films thus prepared are deposited either at room temperature (Ge-RT) or at liquid nitrogen temperature (Ge-LNT). Ge-LNT nanofilm is subjected to oxidation while Ge-RT did not get oxidized. Steps were observed in the I-V characteristics of the thin Ge-LNT samples suggesting the Coulomb Blockade effect.

INTRODUCTION

The recent development in fabrication of nanostructured materials has made it possible to realize a Coulomb Blockade device [1] or a single electron transistor [SET] [2]. This has lead to a great interest in the electrical properties of nanocrystalline films.

In the present paper, we report electrical properties of Ge-nanocrystalline films (thereafter we call it as nanofilms in short) fabricated by cluster beam evaporation technique [3]. The Ge films have been deposited at room temperature and at liquid nitrogen temperatures, referred as Ge-RT and Ge-LNT, respectively. These nanofilms have different morphologies owing to a difference in migration at the surface. It was found [4-7] that, if a nanofilm is sufficiently thin enough to have only a few "islands" across the film, the Coulomb Blockade (CB) or Coulomb Staircase (CS) resulting from a large charging energy of the islands might occur. The purpose of this work is to investigate if Ge nanocrystals could be used as a linear chain of quantum dots to give rise to the CB effect so that they could be considered for potential application to SET's. To the best of our knowledge, very few efforts [7] have been made to study CB in Ge nanofilms. Here, we will show that the Ge nanofilms could exhibit the Coulomb Blockade, even at room temperature, by carefully selecting measurement conditions. Furthermore, it has been reported [8] that prolonged photo-oxidation of the Ge-LNT films would result in a smaller size distribution of Ge nanocrystals. By taking advantage of photo-oxidation, a comparative study of I-V characteristics of Ge-nanofilms with various sizes of Ge nanocrystals has also been carried out.

Mat. Res. Soc. Symp. Proc. Vol. 536 © 1999 Materials Research Society

EXPERIMENTAL

The cluster beam evaporation technique, which was described in detail elsewhere [3], was employed to deposit films consisting of Ge-nanocrystals on Si(100) or glass substrate. During deposition, the temperature of the crucible (containing small pieces of undoped Ge) was kept at 1700° C and the pressure of the chamber was fixed at 10^{-5} torr. The substrate temperature was kept at room temperature and liquid nitrogen temperatures to obtain Ge-RT and Ge-LNT nanocrystalline films respectively. The film thicknesses have been measured accurately by a Dektak[3] surface profiler.

The I-V characteristics were studied using a semiconductor parameter analyser (Model: HP4155) with Cascade Microtech probe station. A 313nm line of a Hg lamp selected with an optical band pass filter was used as an excitation source for photo-oxidation of the Ge-LNT nanocrystalline films. The power density of the excitation light was kept at 25μW/mm^2.

To measure the current through the vertical direction of the Ge-nanofilms, it is necessary to sandwich the film between two metal electrodes. To achieve this the following steps have been adopted: Prior to the film deposition, a thin gold film of about 100 Å was deposited on the substrate and then photolithography was used to make small circular holes (dia~500μm) in 300nm of resist on the gold film. Then the Ge-films of required thicknesses were deposited, followed by another deposition of a thin gold film. Finally each parallel plate capacitor was seperated from the other by scratching the upper gold film around the capacitor dot. The resist pattern was used to suppress leakage current. Thus each active device area uncovered with resist was $\pi[250\mu m]^2$. Figure1 shows the final sample structure.

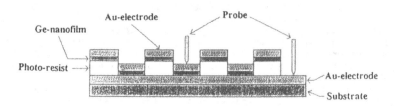

Fig.1 Structure of the samples prepared.

RESULTS AND DISCUSSIONS

The earlier TEM study revealed [3] that morphologies of Ge-RT and Ge-LNT nanocrystals are different. Ge-RT nanofilm was smooth and uniform containing nanoparticles having a narrow size distribution (an average size of 4.3nm) whereas Ge-LNT was rather rough and appeared to consist of nanostructures or an agglomeration of several Ge nanocrystals. The nanocrystals in Ge-LNT were surrounded by an oxide barrier.

In order to observe the CB effect in nanostructured semiconductor devices, it is necessary to have a series of very small islands, isolated by a potential barrier in the current path of the charge carrier. Since, a nanocrystal in Ge-LNT film is isolated from the adjacent nanocrystals by a surface oxide barrier, we felt that a Ge-LNT nanofilm would be a better material for the observation of CB effect than Ge-RT nanofilm, where the adjacent nanostructures are physically in contact with each other, lowering the chance for a carrier to be confined three-dimensionally in an island.

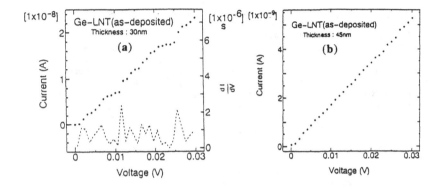

Fig.2 Room-temperature I-V characteristics of Ge-LNT sample of thickness (a) 30nm and (b) 45nm. Clear steps appear in (a). The dotted curve gives the differential conductance calculated numerically from the I-V plot. The appearence of sharp peaks in the conductance curve points towards the Coulomb Blockade effect.

Figures 2(a) and (b) are the I-V plots of Ge-LNT films with thicknesses 30nm and 45nm, respectively. Clearly, the Ge-LNT film with 30nm thickness has exhibited very prominent steps with a small Coulomb gap at zero current. The dotted line in fig 2(a) gives the differential conductance curve obtained numerically from the I-V curve. The curve exhibits sharp peaks. To confirm that the steps are not an artifact but a characteristic of the Ge nanofilms itself, I-V measurements were repeated and ramped up and down several times. Although small changes in the steps appear, no significant changes in the position of the steps were observed. The steps were also observed in the I-V characteristic for negative bias. Similar I-V characteristics with steps have already been reported for semiconductor nanocrystals [6,7]. The observed steps in the I-V characteristics of the Ge-nanofilms could be ascribed to the Coulomb Blockade where the electrons tunnels through the GeO_x barrier which surrounds the surface of a Ge nanocrystals. However such steps in a thin nanofilm would be detectable only if, 1) there is an well defined current path among the nanocrystals or, 2) nanocrystals are well ordered. The experimental results that the current in the Ge-LNT nanofilm does not scale with the active area or reciprocal of thickness eliminates the second possibility. Moreover, it is supported by the fact that I-V characteristics are strongly sample dependent. Thus we suggest that the carrier transport in the present Ge-LNT nanofilms is a highly selective process by which the charge moves along a special path which has the largest conductance. The highly selective conduction path for current results from a large fluctuation in the conductances of all the possible current paths because of randomness of position and sizes of Ge nanocrystals. If the film thickness is small enough for only a few clusters to exist in the current-path, the macroscopic conductance of the film under a very low bias (in the range of several mV) is likely to be determined by a selective path of highest conductance and CB effect could be observed even for a capacitor having an active area as large as ours. However, as seen, steps are not always present with sample-to-sample variation. If the oxide thickness is not large and the size of a Ge-nanocrystal core is large, there may exist several competing current paths of comparable conductances reducing the chance of observing steps in the I-V characteristics. Therefore, next we photo-oxidize the Ge-LNT nanofilm for reproducibility of pronounced steps as photo-oxidation could make size of a Ge nanocrystal smaller and more uniform in Ge-LNT nanofilms [8]

Figures 3(a) and (b) compare I-V characteristics of an as-deposited Ge-LNT film (40nm) and the same film after exposure to the UV light for 1hr, respectively. The as-deposited Ge-LNT sample which does not show any pronounced step in the I-V plot, exhibits noticeable steps in the I-V characteristics after photo-oxidation. Thus, one may conclude that photo-oxidation has increased the chance of having the well-defined path of

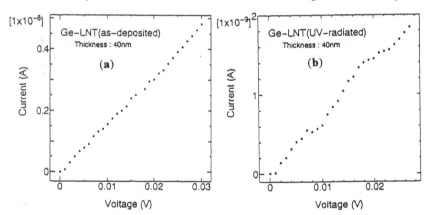

Fig.3 Room-temperature I-V characteristics of an as-deposited Ge-LNT (a) and the same sample after UV radiation for 1hr (b).

largest conductance than any other path in the Ge nanofilms so that the CB can be observed even for a comparatively thick Ge-LNT nanofilm. A slight increase in step width in the I-V characteristics may be attributed to size reduction of Ge nanocrystals resulting from the photo-oxidation.

Finally, to understand the nature of carrier transport in Ge-RT films (which do not seem to show CB effects at room temperature) we measured the temperature dependent conductivity. Fig.4 shows the plot of σ (in log scale) vs. $T^{-1/4}$. A good linear fit in the high temperature regime indicates a variable range hopping of carriers among the nanocrystals. In the low temperature region, where conductivity is independent of temperature suggests tunneling between the adjacent nanocrystals. The result encourages us to study whether the CB can be observed in thin Ge-RT films at sufficiently low temperature to considerably suppress hopping of carriers. Such investigations are already underway.

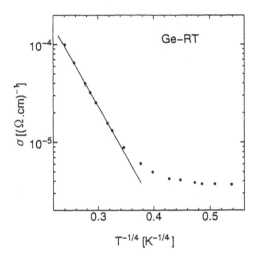

Fig.4 Plot of σ (in log scale) vs. $T^{-1/4}$ for Ge-RT films. A good linear fit in the high temperature regime indicates variable range hopping of carriers among the nanocrystals.

CONCLUSIONS

Room temperature I-V characteristics (across the film thickness) were studied for Ge-LNT films, prepared by cluster beam evaporation technique. Despite the relatively large device area, we are successful in observing pronounced Coulomb Blockade effect, even at room temperature, for the thin films. The 30nm Ge-LNT film is expected to have 3-4 nanoparticles (8-10nm diameter with the oxide layer) in the current path. More pronounced steps observed in the I-V characteristics after photo-oxidation suggests that presence of a well defined current path with larger conductance than any other path is required in order to observe the CB effect.

ACKNOWLEDGMENTS

This study was partly supported by research grant from the Mazda Foundation and Special Coordination Funds for promoting Science and Technology entitled "Research on New Materials Development by Nanospace Lab". One of the authors (SB) extends thanks to Dr. M.Moriya and Dr.Y.Show for useful suggestions and S Michimata for his continuous help while performing experiments.

REFERENCES

1. R.A.Smith and H.Ahmed, Appl.Phys.Lett. **71**, 3838 (1997).
2. K.Yano, T, Ishii, T Hashimoto et al. IEEE Trans. Elec.Device **41**, No.9 1997.
3. S.Sato, S.Nozaki and H.Morisaki, J Appl.Phys. **81**, 1518 (1997).
4. A.P.Alivisatos, MRS Bulletin, pp. 18, Feb 1998
5. R.P.Andres, T.Bein, M.Dorogi et al., Science **272**, 1323 (1996).
6. H.Ahmed and K.Nakazoto, Micro-Electr.Engn. **32**, 297 (1996).
7. Y.Inoue, M.Fujii, S.Hayashi and K.Yamamoto (Private Communications).
8. S.Sato, S.Nozaki, M.Iwase and H.Morisaki, Appl.Phys.Lett. **66**, 3176 (1995).

Development of a Porous Silicon Based Biosensor

Keiki-Pua S. Dancil, Douglas P. Greiner, and Michael J. Sailor[*]
Department of Chemistry and Biochemistry, The University of California San Diego, La Jolla, CA 92093-0358

ABSTRACT

In this paper we demonstrate that porous silicon (PS) can be used as an immobilization matrix and a transducer for biosensor applications. Thin layers of PS were fabricated showing fine structure in their reflection spectra, characteristic of longitudinal optical cavity modes, or Fabry-Perot interference fringes. The PS surface was modified by covalently bonding streptavidin to a heterobifunctional linker immobilized to the surface using common silane chemistry. The mode spacing and wavelength in the interference spectrum was modified, by displacing buffer and introducing proteins into the PS layer. Protein-protein interactions between immobilized Streptavidin and biotinylated Protein A followed by Protein A and IgG were detected. The surface was regenerated during the course of the experiment showing reversibility of the sensor at the third layer.

INTRODUCTION

Inexpensive biosensors designed to provide rapid multi-analyte detection are highly sought after for use in drug design and disease diagnosis, as well as detection of chemical and biological warfare agents. The basic design of most biosensors involves the incorporation of two components: a specific analyte recognition element (typically an immobilized biomolecule, i.e. oligonucleotide or protein) and a transducer which transforms a molecular recognition event into a quantifiable signal. Recently, our laboratories reported the discovery of a system which utilizes PS as an immobilization matrix for biomolecules and as an optical interferometric transducer of molecular binding events [1]. In addition to its unique optical and chemical properties, PS was chosen as the matrix material due to its high surface area. In comparison to flat surfaces, PS offers an immense increase in surface area, which allows for a higher amount of immobilized receptor molecules and hence increased sensitivity.

The sensor is based upon changes in the refractive index of the thin PS film. When incident white light is reflected from a PS sample, Fabry-Perot fringes in its reflectometric spectrum are observed. The fringe pattern is due to the constructive and destructive interference of light being reflected off the top interface (water/(PS + water)) and the bottom interface ((PS + water)/bulk silicon) of the thin film. The peaks in the interference spectrum (mλ) are related to the effective optical thickness (EOT) of the film by the following equation: [2]

$$m\lambda = 2nL \qquad (1)$$

where m is the spectral order, λ is the wavelength, n is the refractive index, and L is the thickness of the film. Since the thickness (L) is a constant parameter, changes in the fringe pattern are directly related to changes in the refractive index of the PS matrix (n_{PS}). For example, when a large biological molecule, such as a protein with $n_{protein} \sim 1.42$ is immobilized within the matrix, a corresponding volume of water with $n_{water} = 1.33$ is displaced out of the matrix. The slight increase in the overall n_{PS} causes an increase in the EOT, which is directly observed in the

557

interference pattern as a shift to longer wavelengths (red shift). Since this system is solely dependent on refractive index changes in the bulk film, response of the PS biosensor is independent of the analyte to surface distance, unlike sensors employing surface plasmon resonance (SPR).

EXPERIMENTAL

Etching procedure. Porous silicon samples were prepared by anodically etching p^{++}-type silicon (0.6-1.0mΩ-cm resistivity, (100) orientation, B-doped) in an ethanolic HF solution (HF:ethanol 3:1, v/v) at current densities ranging from 400-600 mA/cm^2 in the dark. Etching times varied to obtain constant coulombs of 4.5 C/cm^2. A Pt mesh counter electrode was used to ensure a homogenous electric field. After each etch, the sample was rinsed thoroughly with ethanol and methylene chloride and then dried under a stream of nitrogen.

SEM characterization of PS samples. A Cambridge 360 electron microscope, using an accelerating voltage of 20 keV was used to investigate pore dimensions and geometries. Each sample was sputtered with 10 nm of gold to reduce charging effects.

Derivatization of PS chip. Freshly etched chips were exposed to ozone for 10 minutes using an ozone generator (Fischer) to provide an oxidized surface for further functionalization. SPDP, N-succinimidyl 3-(2-pyridyldithio)propionate (PIERCE) (50 mg) was dissolved in 10 mL of methylene chloride and 4-aminobutyldimethylmethoxysilane (FLUKA) (19 μL) was added while stirring. The reaction was allowed to proceed overnight at room temperature. The product (2-pyridyldithiopropionamido)butyldimethylmonomethoxysilane) was purified via silica gel column using ethyl acetate as the eluent. The purified bifunctional silane was dissolved in 20 mL of toluene and was covalently attached to the oxidized PS surface by overnight reflux. The functionalized PS samples were characterized by FTIR. Total surface coverage was determined by exposure to 10 mM dithiothreitol, DTT, (ALDRICH) and detection of pyridine-2-thione by UV/VIS spectroscopy (343 nm, $\varepsilon = 8.08 \times 10^3$ M^{-1}cm^{-1})[3].

Streptavidin immobilization. 2 mg of streptavidin (PIERCE) was dissolved in 0.5 mL of 0.5 M sodium phosphate, pH 8.0. After cooling the protein solution on ice, 22 μL of 100 mM 2-iminothiolane (2-IT) (dissolved in water) was added and the reaction was allowed to proceed for 30 minutes at 0° C. To remove excess 2-IT, the modified protein was purified by centrifuge spin column (Sephadex G-50 equilibrated in phosphate buffer saline (PBS) pH 7.4[4]). The purified protein solution was combined with an equivalent volume of 20% (v/v) ethanol in PBS. The diluted streptavidin solution in ethanol was introduced to a PS chip, mounted in a flowcell, pre-equilibrated with PBS containing 10% (v/v) ethanol at a flow rate of 0.5 mL/min. Interferometric spectra were obtained and analyzed as described by Janshoff et al.[5].

Protein binding. Biotinylated Protein A (PIERCE) was dissolved in PBS pH 7.4 to a concentration of 2.5 mg/mL. Human IgG (CALBIOCHEM) was solubilized in PBS pH 7.4 at a concentration of 1.0 mg/mL. Each protein solution was introduced to the modified PS at a flowrate of 0.5 mL/min.

DISCUSSION

For biosensor applications it is essential that the pore dimensions be large enough to allow for diffusion of proteins in and out of the pores, but small enough to preserve reflective optical properties. By adjusting the current density, HF concentration, and dopant density of the

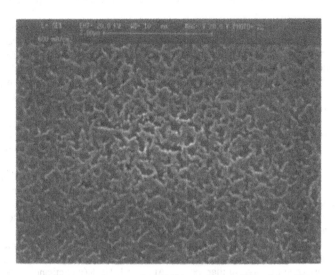

Figure 1. SEM top view image of a PS chip etched at 600 mA/cm^2.

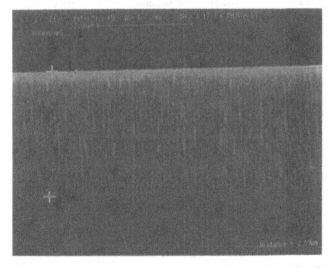

Figure 2. SEM cross sectional image of a PS chip etched at 600 mA/cm^2.

silicon, PS samples were fabricated having applicable dimensions. Figure 1 is a top view image of a PS sample etched at 600 mA/cm². Pores are visible with radii between 100 – 300 nm. Cross-sectional SEM images were obtained to investigate the pore geometry and depth of the porous layer. Figure 2 is a cross sectional image of the same sample. The pores appear to be 2.5 μm in depth and cylindrical in shape.

Before each binding curve was obtained, a baseline was determined by exposing the surface to PBS (pH = 7.4) containing 10% (v/v) ethanol. Ethanol was used to ensure wetting of the hydrophobic porous layer. Streptavidin was covalently attached to the PS matrix via a disulfide bond. Since native streptavidin lacks free cysteines it was necessary to introduce free thiols by reacting free primary amines on streptavidin with Traut's reagent. As shown in part A of Figure 3, an increase of approximately 75 nm in EOT was observed. This clearly indicated covalent immobilization of streptavidin to the surface. A buffer rinse of PBS, pH 7.4, containing 10% (v/v) ethanol ensured removal of non-covalently attached protein. After immobilization ethanol was removed by passing PBS over the sample. The removal of ethanol, at point B, is evident by a decrease in EOT due to the reduced refractive index of the PBS buffer. At point C, biotinylated Protein A was introduced. The EOT increased by 25 nm, indicative of an interaction between streptavidin and biotinylated Protein A. Control reactions (data not shown) with non-biotinylated Protein A or BSA show no increase in EOT. Human IgG was introduced at point D, showing an observed EOT change of approximately 11 nm. Acetic acid (1.0 M) was used at points E to selectively remove the bound IgG. The increase in EOT was due to higher refractive index of acetic acid buffer verses that of PBS. However, a PBS rinse at point F restored the curve to a baseline expected for a biotinylated Protein A surface. This decrease in EOT is characteristic of the released IgG. The procedure of adding IgG (point G), acetic acid (point H), and PBS rinse (point I) gave similar results. At point J, 10 % (v/v) ethanol in PBS was added followed by the addition of 10 mM DTT in 10% (v/v) ethanol in PBS (point K). DTT was added to reduce the disulfide bonds, removing all proteins from the surface. Figure 4 is an expanded view of Figure 3 showing the reversible interactions of the Protein A and Human IgG complex on the sensor surface.

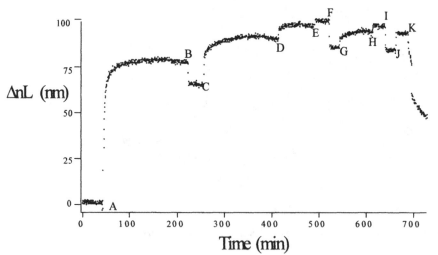

Figure 3: Binding curve showing changes in EOT over the course of an experiment. Between each step a buffer rinse step was conducted using PBS pH 7.4

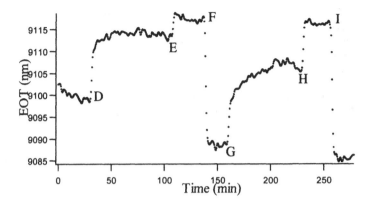

Figure 4: An expanded version of Figure **3** showing the reversibility of the sensor surface.

The sensor is based on changes in refractive index. Because most proteins have a refractive index of approximately 1.45, the PS sensor can be used to monitor changes in molecular weight. In the first layer an increase of 75 nm was detected by immobilization of streptavidin (MW = 60,000 g/mol). The next layer only showed an EOT change of 25 nm for biotinylated Protein A (MW = 42,000 g/mol). By introducing Human IgG (MW = 155,000 g/mol) to the surface, an EOT increase of 11 nm was detected. Each layer of Human IgG introduced gave similar changes in EOT. The mass of streptavidin and biotinylated Protein A are similar and should show equal changes in EOT. But, biotinylated Protein A gave a signal of one third that of streptavidin. We attribute the increased signal of streptavidin to wetting of the surface. After the first protein layer is bonded to the surface, it becomes more hydrophilic in character, allowing further layers of protein to interact. The molecular weight of Human IgG is approximately three times that of Protein A. Therefore, Human IgG should have shown an EOT increase three times that of Protein A. The signal from Human IgG was only half the signal of Protein A. We attribute the lower than expected shifts in signal to size exclusion effects; the large IgG protein is not able to access all of the available Protein A sites in the confined pores of the PS film.

CONCLUSIONS

We have demonstrated the use of PS as a transducing and immobilization matrix for biosensing using Protein A as a reversible binding substrate. Furthermore, we have detected reversible binding at the third protein layer of the sensor surface.

ACKNOWLEDGEMENTS

K-P.S.D. would like to thank NIH for a predoctoral fellowship.

REFERENCES

(1)   Lin, V. S.-Y.; Motesharei, K.; Dancil, K.-P. S.; Sailor, M. J.; Ghadiri, M. R. *Science* **1997**, *278*, 840-843.

(2)   Rossi, B. *Optics*; Addison-Wesley: Reading, MA, 1957.

(3)   Hermanson, G. T. *Bioconjugate Techniques*; Academic Press, Inc.: San Diego, 1996.

(4)   Penefsky, H. S. *Methods Enzymol.* **1979**, *56*, 527-530.

(5)   Janshoff, A.; Dancil, K.-P. S.; Steinem, C.; Greiner, D. P.; Lin, V. S.-Y.; Gurtner, C.; Motesharei, K.; Sailor, M. J.; Ghadiri, M. R. *Journal of the American Chemical Society* **1998**, *120*, 12108-12116.

AUTHOR INDEX

Aikens, J., 377
Akimov, I.A., 287
Alaql, A., 191
Albrecht, M., 81
Aldabergenova, S.B., 81
Allan, G., 39, 185
Alleman, J.L., 237
Allen, M.J., 173
Anderson, I.M., 317
Anderson, W.A., 129
Andreev, A.A., 81
Antonova, I.V., 499
Assanto, G., 469
Atwater, H.A., 223
Avalos, J., 123

Babinski, A., 269
Ban, M., 539
Banerjee, S., 551
Baranowski, J.M., 269
Barathieu, P., 481
Baraton, M-I., 341
Bawendi, M.G., 211
Bennett, J.C., 413
Boatner, L.A., 317
Bondarenko, V.P., 69, 135
Bonevich, J.E., 347
Bowditch, A.P., 149
Brehme, S., 457
Brotzman, Jr., R.W., 377
Bruno, G., 493
Budai, J.D., 251, 317
Buratto, S.K., 27
Burda, C., 419
Buriak, J.M., 173
Burt, M.G., 217, 365

Canham, L.T., 149
Cao, H., 477
Capezzuto, P., 493
Caussat, B., 481
Centurioni, E., 517
Chamard, V., 293
Chan, S., 21, 117, 135, 511
Chandrasekharan, N., 395
Chang, R.P.H., 477
Chatterjee, A., 545
Chazalviel, J-N., 155
Chikyow, T., 445
Cicala, G., 493
Colace, L., 469
Collins, R.W., 451
Colter, P.C., 245
Cox, T.I., 149
Craciun, G., 63
Credo, G.M., 27
Cruz, M., 123
Curtis, C.J., 237, 407

Dai, J.Y., 477
Dancil, K-P.S., 557
Datskos, P.G., 251
Delerue, C., 39, 185
Depaulis, L., 511
Desalvo, A., 517
Dinsmore, A.D., 205
Dolgyi, L., 69
Dolino, G., 293
Duscher, G., 359

Ebbsjö, I., 545
El-Sayed, M.A., 419
Elstner, L., 457
Erickson, L.E., 3
Estrada, A., 123
Etoh, K., 45
Eychmüller, A., 217, 257, 365
Eymery, J., 293

Fauchet, P.M., 9, 21, 69, 117, 135, 141, 185, 505, 511
Feldman, L.C., 251, 413
Feoktistov, N.A., 275
Finger, F., 299
Fonseca, L.F., 123, 197, 229
Fox, J.R., 287
Fuhs, W., 457
Fujiwara, H., 451

Gale, H., 149
Gall, S., 457
Galluzzi, F., 469
Gao, M.Y., 365
Gaskov, A.M., 389
Gelloz, B., 15
Geng, Y., 323
Geohegan, D.B., 359
George, M.A., 527
Ginley, D.S., 237, 407
Glozman, A., 257
Goldstein, Y., 229
Golubev, V.G., 275
Goossens, A., 337
Gorer, S., 395
Grebel, H., 527
Green, T.C., 419
Greiner, D.P., 557
Grob, J-J., 99
Grom, G.F., 69, 141
Guadalupe, A., 197
Gurdal, O., 191
Gutakovsky, A.K., 499

Halverson, W., 245
Hapke, P., 299
Hara, T., 539
Harrison, M.T., 217, 365

Hartnagel, H.L., 99
Hassan, K.M., 329
Hassan, Z., 245
Heinig, K-H., 251
Henderson, D.O., 317
Hinds, B., 111
Hirata, M., 305
Hirose, M., 45
Hirschman, K.D., 21, 141, 511
Hodes, G., 395
Horiguchi, S., 533
Hosomi, K., 523
Huang, M., 413
Huang, W., 401
Hwang, D.S., 439

Ichikawa, M., 487
Igarashi, T., 383
Ikuta, K., 439
Imaizumi, M., 539
Inglefield, C., 81
Irmer, G., 99
Isobe, T., 383
Isoya, J-I., 463
Ito, T., 539
Iwase, M., 57
Izumi, T., 57

Jadwisienzak, W.M., 245
Jagadish, D., 269
Ji, W., 401
Jones, K.M., 237
Jorne, J., 185

Kageshima, H., 533
Kalia, R.K., 545
Kalinin, S.V., 389
Kalkhoran, N.M., 511
Kanemitsu, Y., 223
Katsuyama, T., 523
Kawamura, K., 539
Kazuchits, N., 69
Keblinski, P., 311
Keller, P., 341
Kempton, B.G., 323
Kershaw, S.V., 217, 365
Khaselev, O., 237
Kheifets, L.I., 389
Khuravlev, K.S., 499
Kikuchi, H., 545
Kim, N.Y., 167
King, D.E., 407
Kippeny, T., 413
Klimov, V.I., 211
Kodiyalam, S., 545
Koguchi, N., 445
Koh, J., 451
Kohli, S., 263
Kohno, A., 45
Konagai, M., 487
Konishi, M., 383
Kordesch, M.E., 245

Kornowski, A., 217, 365
Koshida, N., 15, 105, 179
Koyama, H., 9, 135, 511
Kudryavtseva, S.M., 389
Kushida, T., 223

Labbé, H.J., 141
Laibinis, P.E., 167
Lannoo, M., 39
Leatherdale, C.A., 211
Lee, C., 75
Lee, Y., 451
Leon, R., 269
Lifshitz, E., 257
Ligonzo, T., 493
Link, S., 419
Lips, K., 457
Liu, X., 477
Lobo, C., 269
Lockwood, D.J., 3, 141
Loni, A., 149
Lopez, H.A., 69, 135, 511
Losurdo, M., 493
Lozykowski, H.J., 245
Lucovsky, G., 111

Maâref, H., 33
Makimura, T., 51
Malik, M.A., 353, 371
Malten, C., 463
Many, A., 229
Masini, G., 469
Mason, A., 237
Mason, M.D., 27
Matson, R.J., 237
McBranch, D.W., 211
McCaffrey, J.P., 141
McIlroy, D.N., 323
Meiling, H., 281
Meldrum, A., 251, 317, 359
Mendez, B., 63
Merhari, L., 341
Meshkov, L.L., 389
Meulenkamp, E.A., 337
Meyer, J-U., 341
M'ghaïeth, R., 33
Mihalcescu, I., 33
Mimura, S., 223
Min, K.S., 223
Minarini, C., 493
Mishima, T., 523
Miyazaki, S., 45
Miyoshi, N., 45
Mizuno, H., 179
Mizuta, T., 51
Monecke, J., 99
Montès, L., 505
Morante, J.R., 481
Morisaki, H., 57, 551
Morosov, I.I., 499
Mu, R., 317
Müller, J., 299

Murakami, K., 51
Muth, J.F., 329

Nagase, M., 533
Nakagawa, T., 15, 105
Nakajima, T., 105
Nakano, A., 545
Narayan, J., 329
Nastase, N., 63
Navarro, C., 123
Nayfeh, M.H., 191
Nesterenko, S.N., 389
Norton, M.G., 323
Nozaki, S., 57, 551

O'Brien, P., 353, 371
Ohno, Y., 305
Okamoto, S., 223
Okitsu, K., 539
Ong, H.C., 477
Ono, H., 551
Ozaki, N., 305
Ozanam, F., 155

Pangal, K., 505
Pennycook, S.J., 359
Peraza, L., 123
Petrovich, V., 69
Pevtsov, A.B., 275
Pinghini, R., 517
Piqueras, J., 63
Platen, J., 457
Plugaru, R., 63
Pokhil, G.P., 499
Popov, V.P., 499
Porteanu, H., 257
Puretzky, A.A., 359

Qadri, S.B., 205

Rastogi, A.C., 263
Ratna, B.R., 205
Reeves, C.L., 149
Rembeza, E.S., 389
Resto, O., 123, 197, 229
Revaprasadu, N., 353, 371
Ribelin, R., 237, 407
Rice, P., 149
Rino, J.P., 545
Rizzoli, R., 517
Rockenberger, J., 365
Rogach, A.L., 217, 365
Rosenthal, S.J., 413
Rossi, M.C., 493

Sailor, M.J., 557
Sameshima, T., 427
Sarkas, H.W., 377
Sato, K., 57
Scheid, E., 481
Schiavulli, L., 493
Schmuki, P., 3

Schoonman, J., 337
Schropp, R.E.I., 281
Schulz, D.L., 237, 407
Schwab, C., 99
Schwarz, Ch., 211
Scott, E.A.M., 149
Searson, P.C., 347
Seelig, E.W., 477
Senna, M., 383
Seo, S-Y., 75
Sharma, A.K., 329
Sharma, S.N., 263
Shen, F., 527
Shiba, K., 45
Shin, J.H., 75
Shinoda, H., 105
Shirai, M., 523
Shiraishi, K., 533
Show, Y., 57
Sieber, I., 457
Sirenko, A.A., 287
Song, Y.J., 129
Soni, R.K., 197
Spesivtsev, E.V., 499
Stewart, M.P., 173
Striemer, C.C., 21, 511
Strobel, M., 251
Strunk, H.P., 81
Sturm, J.C., 505
Sugiyama, Y., 57
Summonte, C., 517

Takeda, S., 305
Takeshita, J., 487
Tanaka, H., 223
Tanaka, K., 439, 463
Tang, S-H., 401
Taylor, J., 413
Taylor, P.C., 81
Teng, C.W., 329
Thomas, K.A., 251
Thönissen, M., 89
Tian, Y., 205
Tiginyanu, I.M., 99
Tokai, M., 539
Tomaszewicz, T., 269
Tröger, L., 365
Tsybeskov, L., 21, 69, 117, 135, 141,
    505, 511

Ueda, A., 317
Ueda, T., 51
Ueno, K., 105
Umeda, T., 463

Van de Krol, R., 337
Van Landuyt, J., 389
Vashishta, P., 545
Vertegel, A.A., 389
Vial, J.C., 33
Vijayalakshmi, S., 527
Vilà, A., 481

Viner, J., 81
Vogt, A., 99
Volchek, S., 69
Vorozov, N., 69

Wagner, H., 299
Wagner, S., 505
Wakefield, G., 371
Warmack, R.J., 251
Waters, K., 149
Weisz, S.Z., 123, 197, 229
Weller, H., 217, 257, 365
Weston, K.D., 27
White, C.W., 251, 317
Withrow, S.P., 317
Wolfe, D., 111
Wolkin, M.V., 185
Wong, E.M., 347
Wronski, C.R., 451
Wu, M., 317
Wysmolek, A., 269

Xi, X.X., 287
Xu, J., 401

Yakovtseva, V., 69
Yamada, A., 487
Yamaguchi, K., 539
Yamaguchi, M., 539
Yamani, Z., 191
Yamasaki, S., 439, 463
Yang, X., 211
Yasuda, T., 439
Yoshiyama, M., 105

Zhang, D., 323
Zhang, Y., 527
Zhao, Y.G., 477
Zhu, Q., 511
Zignani, F., 517
Zuhr, R.A., 251, 317
Zweiacker, K., 341

SUBJECT INDEX

AlN, 329
amorphous GaN, 245
anodization, 99
array, nanometer, 251
atomic force microscopy (AFM), 141
Auger effect, 33

BaTiO$_3$-CuO, 341
biocompatibility, 149
biosensor, 557

calcification, 149
carboxyl groups, 383
CdSe, 365
CdTe, 263, 365, 407
channealing, 499
chemical transport, 517
colloidal, 217
conductivity, 275
contact, 237
Coulomb blockade effect, 551
coupling, exciton-phonon, 223
cryosol method, 389
crystalline volume fraction, 299
crystallization, 505
    laser, 427
CuInSe$_2$, 371
CuSe, 371, 419
Cu-Te, 407
CVD, 439, 539
    ERC-, 457

decomposition method, 51
defects, 39, 57
depletion, 105
deposition
    chemical solution, 395
    electrophoretic, 347
    hot-wire, 281
    low-temperature, 487
    plasma-enhanced chemical vapor, 451
    thin-film, 487
derivatization, 155
detector, UV, 123
diffraction, in situ x-ray, 337
dimer model, 191
doped, 353
doping, 395

ECR plasma assisted MOCVD, 245
EELS, 229, 287
effect, field, 281
efficiency
    power, 15
    quantum, 15
electroluminescence, 15, 135, 179, 197, 511
electron beam, 539

electronic
    devices, 245
    structure, 185
electroreflectance, 269
emission, visible light, 57
energy relaxation, 211
erbium, 69, 135
ESR, 463
etching, 517
excitation, 69

femtosecond, 211
    dynamics, 419
film, 275
    amorphous silicon, 451
    deposited, 377
    microcrystalline silicon, 451
    focused ion beam, 251
formation
    mechanisms, 293
    process, 51
free carrier absorption, 427

GaAs, 251, 545
Ge-dots, 329
Ge-nanocrystalline films, 551
germanium, 469
grain boundary, 427
growth, 323, 463

Hall effect, 427
heat treatment w/high pressure water, 427
HgTe, 217, 407
homoepitaxy, 457
hot wire, 487
hydrogen, 499
    dilution ratio, 129
    ·hydrosilylation, 173

illumination, 293
implantation, 317, 499
indium tin oxide, 377
induced absorption, 9
infrared, 217
in situ, 463
integrated, 21
interfaces
    SiC-SiO$_2$, 111
    Si-SiO$_2$, 111
interfacial electron transfer, 419
interferometric sensing, 557
in-vivo, 149
ion implantation, 63, 223

kinetic(s), 311
    growth, 347

laser
  ablation, 51, 527
  synthesis, 359
  ultraviolet, 477
LED, 117, 135
  arrays, 21
light emitting, 3
lithium intercalation, 337
luminescence, 123, 229
  cathodo-, 63
  fast, 33

Meyer, Neidel rule, 281
microcavities, 117
microcrystallinity, 493
micropatterns, 3
microsphere, 205
microstructure, 317
mirrorless, 477
mirrors, 117
molecular dynamics simulation, 311
multilayers, 89, 117, 135

nanoclusters, 527
nanocrystal, GaAs, 63, 223, 317, 481
nanocrystalline, 21, 141, 275
  GaN, 81, 377
nanocrystallites, 33
nanocrystals, 27, 39, 57, 111, 129, 185, 211,
  217, 251, 257, 263, 323, 383, 499
nanofiber, 205
nanomaterials, 389
nanoparticle(s), 51, 205, 237, 353, 371, 419
  gas-suspended, 359
nanophase, 545
nanosized powder, 341
nanostructures, 191, 197, 305
  Ge, 527
nanowires, 323
near-infrared, 469
n-type, 293
nucleation, 439

optical
  integrated waveguide, 69
  nonlinearity, 527
  properties, 269
organic film, 167
oxidation, 9, 69, 533
  electrochemical, 15

parallel molecular dynamics, 545
passivation, 155
  oxide, 21
  surface, 45, 179, 185
PECVD, 323, 493
  VHF-, 517
phonon
  assisted transition, 39
  engineering, 99
photochemical etching, 179
photodetector, 469

photoelectrochemistry, 237
photoluminescence, 27, 45, 57, 69, 173, 197,
  217, 245, 269, 329, 365, 505, 511
  Er^{3+}, 81
  in situ, 359
  oxide-free blue, 179
  spectra, 69
  time resolved, 33
photovoltaic, 237
plasmon, 99, 229
polariton, 523
polycrystalline, 469
polymerization, 205
polysilicon, porous, 511
porosity, 99
porous
  gallium arsenide (GaAs), 3
  silicon (Si), 3, 9, 15, 21, 27, 69, 89, 105,
    117, 135, 149, 155, 167, 173, 179,
    191, 197, 229, 275, 293, 511, 557
powder, 477
protein-protein interactions, 557
prototype, 123

quantum(-)
  confinement, 329
  dot(s), 257, 263, 269, 533
  semiconductor, 395
  silicon, 45, 287
  particle, 347
  well, 257, 523

radiative recombination, 39, 45
RAMAN scattering, 505
rate, deposition, 299
Rayleigh-scattering imaging, 359
recombination mechanism, 185
refractive index, 523
resonance
  electron
    paramagnetic, 383
    spin, 57, 299
  optically detected magnetic, 257
restructured islands, 191
room-temperature current-voltage
  characteristics, 551

S^{2-} vacancies, 383
Sb-Te, 407
scanned probe microscopy, 27
Schottky barrier diodes, 243
seed layer, 129
segregation, 481
self-assembled Si, 45
semiconductor, 317
sensor
  CO_2, 341
  H_2S, 389
separation, phase, 481
SiH_4, 539

silicon(/), 123, 149, 197, 251, 287, 305, 311, 439, 493, 499, 533
   amorphous, 63, 487, 505
   cluster, 311
   crystalline, 539
   heterogeneous, 281
   hydrogenated, 463
   microcrystalline, 135, 299, 457, 517
   nanocrystalline hydrogenated, 129
   oxide, 439
   polycrystalline, 487
   silicon oxide superlattices, 141
single molecule detection, 27
SIPOS, 481
size-dependence, 287
solar cell, 129, 237, 451
   heterostructure, 457
sol-gel synthesis, 205
spectroscopy
   ellipsometry, 451, 493
   infrared, 167, 341
   photothermal deflection, 81
   RAMAN, 141, 287, 305, 493
   selective excitation, 223
sputtering, radio frequency, 263
stability, 3, 155, 173
strain, 533
stress distribution, 545
structural transformation, 545
sulfide, 317
superlattices, Si/SiO$_2$, 141
superlinear light emission, 9
surface
   chemistry, 155, 557
   defects, 3
   electron donor, 419
   modification, 167
   reaction, 463
   states, 395

switching device, 523
synthesis, 365

thermal
   annealing, 111
   conductivity, 105
   effect, 9
thermo-acoustic effect, 105
thin, 275
film
   amorphous GaN, 81
   TiO$_2$, 263
thiols, 365
III-V compounds, 99
tin dioxide, 389
titanium dioxide, 263, 337
TOPO, 353, 371
transistor
   single electron, 533
   thin film, 281
transition, intra-band, 211
transmission electron microscopy (TEM), 141, 305
   high resolution, 81
trapping, 211

ultrasound, 105

waveguide, 523

x-ray reflectivity, 293
XTEM, 191

zinc oxide, 347
Zn-O, 359, 477
ZnS, 353, 383

Printed in the United States
By Bookmasters